ISOTOPE HYDROLOGY 1978

VOL. II

PROCEEDINGS SERIES

ISOTOPE HYDROLOGY 1978

PROCEEDINGS OF AN INTERNATIONAL SYMPOSIUM
ON ISOTOPE HYDROLOGY
JOINTLY ORGANIZED BY THE
INTERNATIONAL ATOMIC ENERGY AGENCY
AND THE
UNITED NATIONS EDUCATIONAL, SCIENTIFIC AND
CULTURAL ORGANIZATION
AND HELD IN NEUHERBERG,
19–23 JUNE 1978

In two volumes

VOL.II

INTERNATIONAL ATOMIC ENERGY AGENCY
VIENNA, 1979

ISOTOPE HYDROLOGY 1978
IAEA, VIENNA, 1979
STI/PUB/493
ISBN 92—0—040179—1

Printed by the IAEA in Austria
January 1979

FOREWORD

The International Symposium on Isotope Hydrology, jointly organized with UNESCO, in co-operation with the National Committee of the Federal Republic of Germany for the International Hydrological Programme and the Gesellschaft für Strahlen- und Umweltforschung mbH, was held at Neuherberg near Munich on 19–23 June 1978. It was the fifth on the subject to have been held by the Agency, its predecessors having taken place in 1963, 1967, 1970 and 1974.

Among the 46 papers presented, the majority reported on field studies encountered in all stages of the hydrological cycle. A growing use is noted in developing areas and this is evidence of the economic value of isotope hydrology techniques when applied to water systems where conventional hydrological data are scarce.

The results gathered prove isotope techniques to be reliable tools for the following fields: origin, mechanism and amount of recharge; groundwater movement and storage; snow and ice hydrology; surface run-off and hydrograph separation; dispersion in surface waters; dynamics of lakes; and fluvial and coastal sedimentology.

The conclusion which can be drawn from the symposium and its proceedings is that many isotope techniques have reached a final stage of development as analytical procedures in the field and the laboratory, and interpretation of the data gained has reached an advanced stage of refinement. A drawback for the wider use of isotope techniques is that certain field equipment, hitherto only constructed by a few institutes for their own use, is not yet commercially available. Stress is laid on the need to rectify this situation by encouraging the production of the equipment for wider use.

The various groundwater dating methods based on the decay of different radioactive isotopes, require further research since different existing dating techniques do not always give the same ages of the same groundwater body. When this happens, a correct interpretation of the raw data is only possible with the support of additional geochemical data indispensable for studying mixing problems and for gaining a better knowledge of the interaction of the environmental isotopes with the aquifer matrix.

The proceedings are intended as a comprehensive introduction to the use of such techniques in hydrology for all those concerned with the design and management of water projects in which isotope techniques are appropriate as a supplementary tool.

EDITORIAL NOTE

CONTENTS OF VOLUME II

GROUNDWATER RECHARGE STUDIES — UNSATURATED ZONE
(Session VIII)

Session VI

GROUNDWATER IN ARID ZONES AND PALAEOWATERS

Chairman

J.Ch. FONTES
France

AN EVALUATION OF
ISOTOPE CONCENTRATIONS
IN THE GROUNDWATER OF
SAUDI ARABIA

W.J. SHAMPINE
US Geological Survey,
Ministry of Agriculture and Water,
Riyadh

T. DINÇER
FAO Trust Fund Project,
Riyadh

M. NOORY
Water Resources Development Department,
Ministry of Agriculture and Water,
Riyadh,
Saudi Arabia

Abstract

AN EVALUATION OF ISOTOPE CONCENTRATIONS IN THE GROUNDWATER OF
SAUDI ARABIA.

Varying amounts of data have been collected on concentrations of oxygen-18,
deuterium, tritium, sulphur-34, carbon-13, and carbon-14 at 400 sites in an area of about
800 000 km² in Saudi Arabia. Carbon-13 concentrations cluster into groups related to the
aquifer matrix, with mean values of -9.89‰ for sandstone aquifers and -6.08‰ for limestone
aquifers. Most of the recharge to the aquifers in Saudi Arabia took place about 20 000 years
ago, and relatively little water is being recharged currently. The oxygen-18 and deuterium
concentrations also cluster into groups related to the aquifer matrix, with heavier concentrations being found in the sandstone aquifers and lighter concentrations in the limestone aquifers.
There are significant areal variations in the oxygen-18 and deuterium concentrations although
the isopleths for these two isotopes tend to be essentially parallel. The oxygen-18 concentration varies areally from about -2.5‰ in the outcrop areas to about -6.5‰ downdip. Stable
isotope data show that the Umm-Er-Radhuma aquifer is leaking upwards into the Khobar and
Neogene aquifers in the Hofuf area. The Wasia aquifer, however, is hydraulically isolated from
the overlying aquifers.

INTRODUCTION

Data on concentrations of oxygen-18 deuterium, tritium, sulfur-34, carbon-13, and carbon-14 have been collected in varying amounts at more than 400 sites in Saudi Arabia. These data were collected during several hydrologic studies conducted in Saudi Arabia since 1967 [1-11], most of which studied selected areas of the Arabian Shelf area of the Kingdom.

No new data were collected for this study, but all of the hydrologically oriented isotope data available in the Kingdom have been compiled and summarized. A rigorous interpretation of these data would be very difficult because, in general, the sampling techniques and analytical methods used by the many different reporting groups are unknown. A general assessment of the data can be made, however, which is useful to evaluate the quantity and quality of the data available as well as to indicate potential areas for future work. The compiled data also have permitted a general evaluation of the isotopic characteristics of the water in large, regional aquifer systems (about 800,000 km^2).

The ^{34}S data are limited in number and very scattered areally, consequently, they were not very useful in the overall aquifer evaluations. Nevertheless, the summarized data were included in this paper to illustrate the quantity of data available and the range of concentrations measured.

Physical Setting

The Arabian Peninsula can be divided into two main geologic provinces: The Arabian Shield and the Arabian Shelf. The Shield area is in the western part of the peninsula. It consists of crystalline rocks of the Precambrian basement complex and small local aquifers from which few isotope data have been collected.

The Shelf area is in the eastern part of the peninsula, and consists of a series of sedimentary formations dipping gently toward the Arabian Gulf. Some of the more important aquifers include the Tabuk and Saq Formations (Paleozoic sandstones), the Minjur, Biyadh and Wasia Formations (Mesozoic sandstones), the Umm-Er-Radhuma Formation (Paleocene and lower Eocene limestone and dolomite), the Alat and Khobar members (Eocene limestone and dolomite) of the Dammam Formation, the Neogene complex (Pliocene sandy limestone, sands, and sandstones), and wadi alluvium. Most of the available isotope data were collected in the Shelf area.

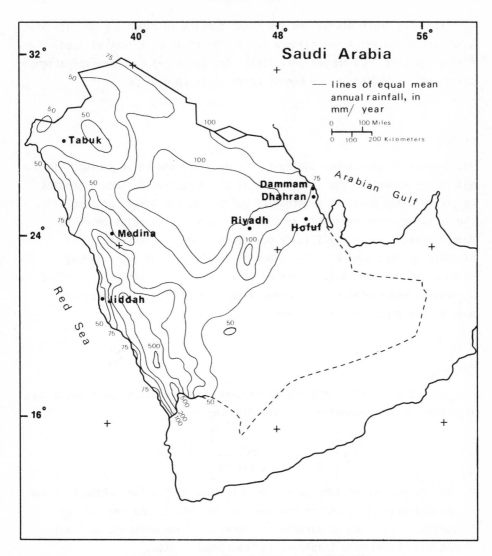

FIG.1.   Mean annual precipitation in Saudi Arabia.

Climatic Setting

Saudi Arabia has a warm, arid climate. Mean daily air temperatures range from about 0° to $35^{\circ}C$, with extremes ranging from -10° to more than $47^{\circ}C$. Precipitation generally occurs as convective thunderstorms and is areally discontinuous. Mean annual precipitation ranges from 50 mm to greater than 600 mm per year (Fig. 1).

The high temperatures and dry winds combine to cause a high rate of evaporation. Class A pan evaporation at Hofuf, in East-central Saudi Arabia, exceeds 3500 mm per year [11]. The potential evapotranspiration, calculated according to the Penman Formula, is 2800 mm per year.

CARBON

Measurement of the carbon isotopes, ^{14}C and ^{13}C, can be very useful in hydrologic studies. The ^{14}C content of the dissolved carbonate species in ground water provides a method of estimating the length of time since the water was recharged into an aquifer. Accurate calculation of the "age" however, is very difficult due to many complex factors. Many of these factors have been described by Vogel [12] and evaluated by Pearson and Hanshaw [13]. A method utilizing $^{13}C/^{12}C$ ratios, described by Pearson and Hanshaw [13], commonly is used to adjust the apparent age, as measured, to the real age.

Carbon-13

Ratios of $^{13}C/^{12}C$ reported in this paper are calculated as deviations from the PDB (Peedee belemnite) standard where:

$$\delta^{13}C = \left[\frac{^{13}C \ / \ ^{12}C \ \text{sample}}{^{13}C \ / \ ^{13}C \ \text{standard}} \right] - 1 \times 1000$$

On this scale, marine limestones commonly have $\delta^{13}C$ values close to 0 per mille, and land plant values are near -25 per mille. As reported by MacDonald [7], values comparable to these have been measured in Saudi Arabia and listed here (Table 1) as "background" data.
In humid regions, -25 per mille commonly is used to represent the $\delta^{13}C$ concentration occurring in soil air. Pearson and Hanshaw [13], however, describe data showing that the $\delta^{13}C$ concentration in soil air of arid zones commonly has been reported as -16 to -18 per mille. This higher range of values is supported by the data collected in Saudi Arabia and described below.

In addition to the "background" data shown in Table 1, $\delta^{13}C$ values have been collected at 162 other sites in Saudi Arabia, ranging in concentration from -0.4 per mille to -16.2 per mille. Histograms of the

TABLE I.   SELECTED CARBON-13 "BACKGROUND" VALUES IN
SAUDI ARABIA [7]

Source	$(\delta^{13}C$ per mille)
Tuwayq Mountain Limestone	0.8
Upper Dhruma Limestone	0.9
Upper Wasia Sandstone	-1.3
Carbonaceous shale, from 241 m	-25.7
Lignite, from 1450 m	-23.4
Thorn plant, living	-24.5

number of occurrences of specific concentrations (Fig. 2) show the distribution of the data for selected aquifers and for the total of the compiled data. Of the 162 samples, only three had concentrations that were
considered exceptional. Although possibly the result of some unknown
mechanism, the three concentrations reported as less than -22 per mille
were not used in this report. These three values, each from a different
aquifer, were more than five standard deviations away from the mean of the
data for the aquifers in which they were measured. Because the values
were such excessive outliers and because they would unduly bias the
limited data available for individual aquifers, they were disregarded by
the authors.

     In addition to the well-established influence of humic material on
^{13}C values, the data show an apparent influence by the aquifer matrix.
When the mean $\delta^{13}C$ values for each aquifer are ranked in order of increasing concentration, the data cluster into distinct groups. The Saq,
Biyadh, and Minjur aquifers, which are dominantly sandstones, have similar
values with a mean $\delta^{13}C$ concentration of -9.89 per mille. The limestone
aquifers, Alat, Umm-Er-Radhuma and Khobar, also have similar values and
have a mean $\delta^{13}C$ value of -6.08 per mille. The Wasia and Neogene aquifers
contain both limestone and sandstone and have an intermediate mean concentration of -7.08 per mille. The reason for the $\delta^{13}C$ differences found
in the various aquifers is not readily apparent. It may be a function of
the carbonate geochemistry, the result of some unknown mechanism associated
with the aquifer matrix, or possibly a function of differential vegetation

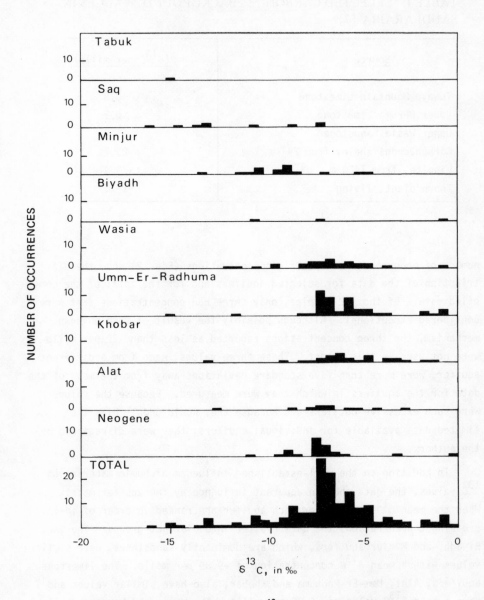

FIG.2. Frequency distribution of the ^{13}C values for selected aquifers.

patterns. Sufficient data currently are not available to evaluate properly these possibilities, although the authors believe the difference is
related to the carbonate geochemistry.

Carbon-14

Carbon-14 analyses have been made on 160 water samples collected in
Saudi Arabia. The "ages" calculated from these data in various reports
are not relatable, however, because different correction factors generally
were used in each report. To overcome this problem, all the "ages" were
recalculated using a uniform system of corrections.

The correction factor used in this paper was based on the ^{13}C concentration, where -17 per mille was arbitrarily selected as a value for the
concentration of ^{13}C in soil air. The correction was made as follows:

$$t = 8267 \ln \left[\frac{100}{^{14}C} \times \frac{^{13}C_m}{^{13}C_b} \right]$$

where t = calculated age, in years

^{14}C = measured ^{14}C concentration, as percent of modern,

$^{13}C_m$ = measured ^{13}C concentration,

and $^{13}C_b$ = concentration of ^{13}C in soil air.

Variations in ^{14}C "ages" measured in an aquifer can result from either
the way the samples were collected or as a function of the hydrologic
characteristics of the aquifer. For example, the distribution of the
sample locations can have a significant effect on the data. Accepting
the assumption that water remote from the area of recharge should be
older than water found near outcrop areas, then the data would be unduly biased if the samples were not relatively evenly distributed areally.
To check this possibility for the data used in this report, a histogram
was prepared showing the number of occurrences of sample location and the
outcrop area of the sampled aquifer (Fig.3). It seems that distribution
probably is not a significant problem for the data used in this report
because the sample locations are fairly evenly distributed, at least up
to about 150 km from the outcrop area.

A histogram of 143 of the "uniformly" corrected "ages" has been prepared (Fig. 4) to help assess the water "ages" distribution within Saudi

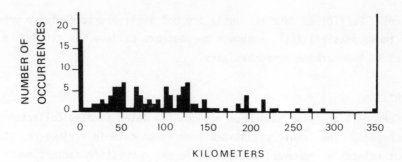

FIG.3. *Frequency distribution of the distance of the ^{14}C sample locations from the outcrop areas.*

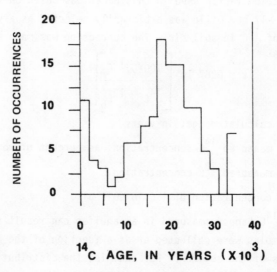

FIG.4. *Frequency distribution of the calculated ^{14}C ages.*

Arabia. Although the frequency distributions shown in Figure 4 are a function of the sample distribution, the authors feel that the samples are distributed well enough (Fig. 3) to yield useful information. Of all the available data, only a few samples were calculated to be modern water, and most of those were from large-diameter, hand-dug wells. Water in these wells is very susceptible to direct recharge by rainfall and may not represent true aquifer conditions.

Only 8 percent of the samples have a calculated age of less than 2000 years. Many of the data tended to center around a calculated age of about 20,000 years. Almost half (47 percent) of the samples have a calculated age ranging from 18,000 to 28,000 years.

STABLE ISOTOPES

Mean values have been calculated from the data collected in Saudi Arabia for the stable isotopes ^{18}O, D, and ^{34}S and are summarized in table II in order of decreasing geologic age.

Areal variations occur in the stable isotope concentrations in water from all the aquifers but only data from the Biyadh aquifer were subdivided and considered as more than one group. Although only five samples are available from the Biyadh aquifer, the range of the measure ^{18}O and D values is more extreme than for any of the other aquifers. Nevertheless, the data seem to cluster into two distinct groups (table II). The samples obtained from near the outcrop area contain oxygen and deuterium values in agreement with the other sandstone aquifers, whereas the samples located far from the outcrop area contain oxygen and deuterium values considerably lighter than any other group of samples.

The oxygen and deuterium data collected for the Saq are biased in that 21 of the 29 samples came from wells located in the outcrop area. Considering only the eight samples collected away from the outcrop area, the mean $\delta^{18}O$ does not change much, only decreasing from -2.4 per mille to -3.06 per mille $\pm$ 1.76 per mille. The δD, however, decreases from -12.6 per mille to -20.49 per mille $\pm$ 8.35 per mille.

If the mean values of ^{18}O and D were ranked, by aquifer, in order of ^{18}O depletion, they would form groups similar to the ^{13}C groups previously described. The concentrations found in the sandstone aquifers cluster into a group distinct from the concentrations found in the group of limestone aquifers. The sandstone units -- the Saq, Minjur, and Tabuk aquifers and the alluvial sand and gravel -- contain water heavier with respect to ^{18}O and D (and lighter in ^{13}C) than water from the limestone units -- the Alat, Umm-Er-Radhuma and Khobar aquifers. The aquifers classified as containing both limestone and sandstone, -- the Wasia and Neogene aquifers, -- contain water with middle values. As with the ^{13}C data, the reason for this grouping is not readily apparent.

TABLE II. MEAN CONCENTRATIONS OF STABLE ISOTOPES, BY AQUIFER

Aquifer	Number of Samples	$\delta^{18}O$ Mean	St. Dev.	Number of Samples	δD Mean	St. Dev.	Number of Samples	$\delta^{34}S$ Mean	St. Dev.
Tabuk	4	-4.1	1.18	4	-26.6	5.56	-		
Saq	29	-2.4	1.10	29	-12.6	7.95	-		
Minjur	13	-2.9	1.92	13	-21.5	14.85	-		
Biyadh (0)[a]	3	-3.0	0.47	3	-10.4	4.03	-		
Biyadh (R)[a]	2	-7.3	0.57	2	-57.8	5.52	2	22.4	1.48
Wasia	28	-4.9	1.07	22	-29.1	13.24	22	15.5	5.04
Umm-Er-Rad-huma	56	-5.2	0.85	52	-38.5	7.77	45	13.3	2.28
Khobar	34	-5.4	0.85	27	-39.5	4.35	30	13.4	2.70
Alat	15	-5.1	0.85	9	-35.8	5.95	9	11.3	2.78
Neogene	140	-4.6	1.70	26	-35.0	8.88	28	12.3	1.24
Alluvial	36	-2.7	0.73						

[a] Biyadh — 0 denotes outcrop wells, R denotes wells far from the outcrop area.

Sulfur-34 data are less conclusive than the ^{18}O and D data because
there are fewer data available and fewer aquifers are represented. Without data from the sandstone aquifers, it is not possible to determine if
^{34}S values cluster into groups as do the other isotopes.

Although the mean concentrations of ^{18}O and D cluster into groups
related to the aquifer matrix, the means for the various aquifers tend
to be distinct from each other. A graphical statistical test for the
difference of the means has been prepared (Fig. 5).

In Figure 5 the mean concentrations of ^{18}O and D were plotted for
each aquifer along with the associated error envelope. The error envelopes
were determined by calculating the standard deviation for ^{18}O and D values
for each aquifer and dividing these deviations by the square root of the
number of samples for each aquifer. Thus, there is a 67 percent chance of
the true mean occurring within the error envelope shown for each aquifer.

Data from the Saq aquifer, the three outcrop wells in the Biyadh
aquifer, and the alluvial material all show enriched values of ^{18}O and D
similar to waters of recent precipitation origin, about -2.5 per mille
and -10 per mille respectively [5]. Although their means are different
and they are quite distinct from the other aquifers, statistically the
data from these three groups could be from the same population; that is,
water recharged under similar climatic conditions.

The error envelopes for the Tabuk and Minjur aquifers are relatively
large. The error for the Tabuk data is large probably because there are
only four samples available. The large error in the Minjur is a reflection
of data variability.

The data from the Wasia aquifer samples have a mean value which seems
to be different from the closely related group formed by the Umm-Er-Radhuma,
Khobar and Alat aquifers. The values from the Alat and the Wasia aquifers,
however, are not statistically distinct and the hypothesis of these waters
having been recharged under similar climatic conditions cannot be rejected
on the basis of available data. Similarly, data from the Neogene aquifer
also has a mean value different from the other aquifers studied but is not
statistically distinct from the Wasia and Alat aquifers. However, the
hypothesis that the Neogene, Wasia, and Alat aquifers are from the same
population is hydrologically unreasonable.

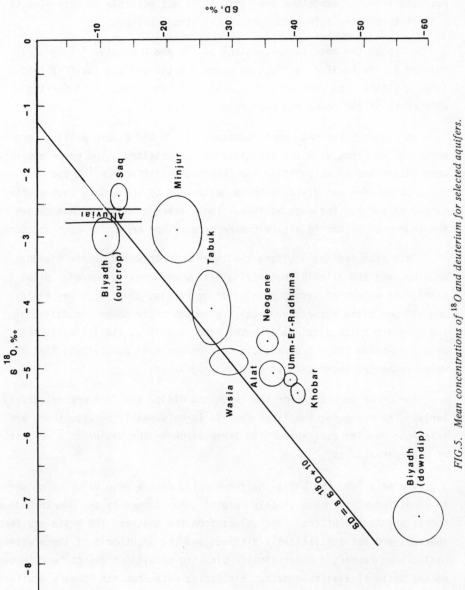

FIG.5.   Mean concentrations of ^{18}O and deuterium for selected aquifers.

The Neogene data are believed to be slightly biased in that some of the data are thought to represent a mixture with Umm-Er-Radhuma waters. If these samples were disregarded, the mean values probably would be slightly heavier and, thus, more distinct.

The mean values of the data collected from the Biyadh aquifer remote from the outcrop area are very light. This group of data (although only two samples) is very distinct from the other data available and represents water recharged during a period much cooler than the present.

All the mean values, except that for the Biyadh outcrop samples, and most of the error envelopes fall below the Craig precipitation line. Values below the Craig line, where

$$\delta D = 8 \, \delta^{18}O + 10,$$

seem to be a common feature of old ground water in many other regions of the world [14]. Some work has been done in Saudi Arabia in an attempt to "localize" the slope and intercept included in Craig's equation. To date, the work is inconclusive.

Areal Variations

Areal distribution plots of ^{18}O, D, and ^{34}S concentrations have been prepared for selected aquifers. Isopleths of equal concentrations were drawn for the ^{18}O and D values, but the data were too scarce and too variable to draw meaningful isopleths of the ^{34}S values. The D and ^{18}O isopleths are essentially parallel. Because of the similarity only the ^{18}O isopleths have been shown in this paper. (Figs. 6 - 9).

Figures 6 to 9 show that significant areal variations of the ^{18}O concentration occur within all the aquifers. In general, concentrations of about -2.5 per mille occur in the outcrop areas and decrease downdip to concentrations of about -6.5 per mille. The higher values in the outcrop areas probably represent an average of the values in modern precipitation, and indicate modern recharge.

Differences in the altitude of outcrop areas and paleoclimatic temperature changes are possible explanations for the observed variations in the concentrations of the stable isotopes. As the altitude increases, the ^{18}O concentration will vary about 0.2 per mille per 100 m, Mazor [15].

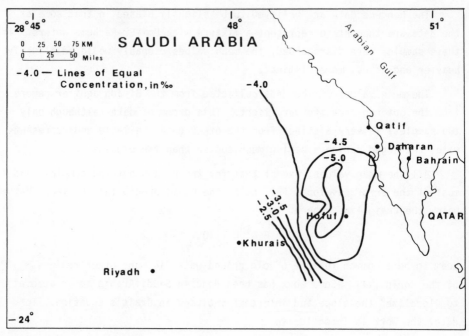

FIG.6. ^{18}O *variations in the Neogene aquifer.*

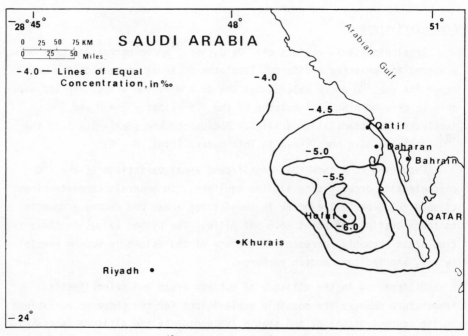

FIG.7. ^{18}O *variations in the Khobar aquifer.*

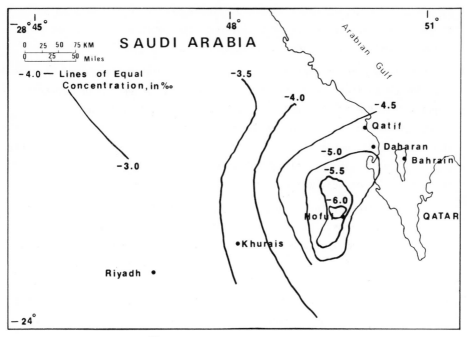

FIG.8. ^{18}O variations in the Umm-Er-Radhuma aquifer.

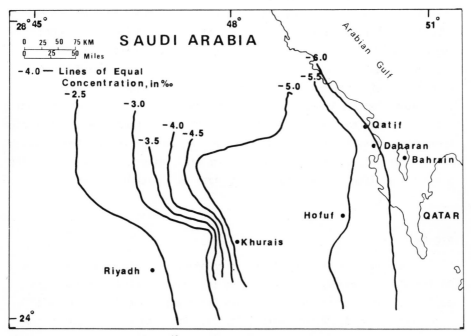

FIG.9. ^{18}O variations in the Wasia aquifer.

Altitude, however, is not a very significant factor in the Arabian Shelf
of Saudi Arabia. The range of altitude of the outcrop areas is about
250 m, which represents an ^{18}O variation of only 0.5 per mille.

It would seem, therefore, that the large variations measured throughout a given aquifer are mostly due to paleoclimatic temperature changes.
Dansgaard [16] has shown that ^{18}O concentrations will vary about 0.7 per
mille per ^{o}C increase in the mean annual temperature. Thus, to account for
the 4 per mille change in the ^{18}O concentration found throughout the Wasia
aquifer (Fig. 9), a paleoclimatic change in the mean annual temperature of
about 5.7^{o}C is needed, which is not an unreasonable value.

The paleotemperature change probably is not the only factor causing
areal ^{18}O variations. The poor correlation between water ages and ^{18}O
concentrations implies some other factor also is involved. For example,
in the Wasia aquifer, the ^{18}O isopleths increase continuously from west
to east, (Fig. 9), although the oldest water measured in the aquifer was
west of the middle of the aquifer. This "pocket" of old water, surrounded
by younger water, is not reflected in the ^{18}O isopleths.

The stable isotope data can be used to help describe complex groundwater flow patterns, such as occur in the area around Hofuf, Saudi Arabia.
The Hofuf region is a ground-water discharge area that is known for the
occurrence of many springs. The effect of the discharge can be noted by
the closed ^{18}O isopleths for the Neogene, Khobar, and Umm-Er-Radhuma
aquifers (Figs. 6 - 8). The data also indicate that near Hofuf there is
some degree of intercommunication among these three aquifers. It appears
that water from the Umm-Er-Radhuma is moving upward in this area, spreading
into the Khobar and Neogene aquifers. On the other hand, the Wasia aquifer
seems to be completely isolated from the overlying aquifers. The isopleth
pattern for this aquifer (Fig. 9) is completely dissimilar to those of the
overlying aquifers. The dissimilarity implies a lack of communication.

RECHARGE

Water-age data indicate there is relatively little modern recharge
taking place. Seventy five percent of the water sampled from the major
aquifer systems throughout the Arabian Shelf area ranged in corrected
"age" from 8,000 to 32,000 years (Fig. 3). Presumably, this time period
was a relatively humid period in Saudi Arabia, resulting in a considerable

amount of recharge. Water "ages" of samples from outcrop areas, assumed to
be recharge areas, range from 0 to greater than 13,000 years, and one well
located only 6 km from the outcrop area yields water with a corrected age
of 26,000 years. The relatively old water found in some outcrop areas as
well as the general scarcity of water with an age between very modern and
about 8,000 years suggests that recharge to the major aquifers may have
been significantly reduced in the past 8,000 years from what it was 20,000
years ago. These data are in agreement with the known climatic history of
the Arabian Peninsula.

Other isotope data, however, provide some evidence that at least
limited recharge is taking place. Detailed tritium data collected by
Dincer and others [11] on sand dunes in central Saudi Arabia indicate
that about 20 mm of water annually is infiltrating the dunes, at least to
a depth of 7 m. It is not known, however, whether all, or any, of this
water actually reaches the deeper aquifer formations underlying the sand
dunes.

Tritium data, which are useful to show very modern recharge, have
been collected by many consultants in Saudi Arabia. In general, these
data are negative. Of 137 samples, only those from 6 wells, all in outcrop areas, contained water with 15 T.U. (tritium units) or more, some
of which may be the result of atmospheric contamination. The lack of
tritium in samples collected in the outcrop areas suggests that modern
recharge may be limited.

Streamflow records show that the wadi alluvial aquifers in Saudi
Arabia are being recharged by intermittent floods, as would be expected.
Although not yet documented, it is thought that some of these alluvial
aquifers then recharge some of the more extensive deeper aquifers.

The mean ^{18}O concentration in the alluvium of Wadi Bishah, which
certainly contains modern precipitation, is -2.7 per mille. Similarily,
an ^{18}O concentration of about -2.5 per mille was found in water from the
outcrop areas of all the other aquifers studied. The similarity of these
values implies that some recharge probably is taking place.

SUMMARY AND CONCLUSIONS

The soil air in arid zones has been reported as having a $\delta^{13}C$ concentration of -16 to -18 per mille rather than the value of -25 per mille

commonly used as a standard. Data collected in Saudi Arabia are in agreement with the higher values, although there appears to be a difference
based on the aquifer matrix. The mean $\delta^{13}C$ of sandstone aquifers is
-9.89 per mille, whereas the mean $\delta^{13}C$ of limestone aquifers is -6.08 per
mille. It is thought that the difference is a function of the carbonate
geochemistry, but adequate field data are not available to evaluate this
possibility.

In a manner similar to the ^{13}C data, the mean values of ^{18}O and D
cluster into groups seemingly related to the aquifer matrix. With the
exception of the Biyadh downdip samples, the sandstone aquifers yield
water heavier with respect to ^{18}O and D (and lighter in ^{13}C) than water
from the limestone aquifers.

A graphical statistical test has been prepared for the ^{18}O and D data
to test for the significance of the difference of the means. This plot
shows that the mean values for most of the aquifers are statistically
distinct and that the data were not derived from the same population.
However, there are some aquifers, such as the Saq aquifer, the alluvium,
and the group represented by the three Biyadh outcrop wells, that have
different means but which are not distinct statistically.

Although significant variations occur in the areal distribution of
^{18}O and D for all the aquifers studied, the D and ^{18}O isopleths were
virtually parallel. In general, the ^{18}O concentration is about -2.5 per
mille in outcrop areas and decreases downdip to about -6.5 per mille. The
variations in concentration in each aquifer are thought to be due largely
to paleoclimatic temperature changes.

Stable isotope data can be used to help describe the complex groundwater flow patterns in the Hofuf area of Eastern Saudi Arabia. The ^{18}O
data indicate that water from the Umm-Er-Radhuma aquifer is moving upward
into overlying aquifers and eventually being discharged by the many springs
in the area. The data also indicate that the Wasia aquifer (underlying
the Umm-Er-Radhuma) is not in communication with any of the overlying
aquifers.

Water-age data suggest there is relatively little modern recharge
taking place. Calculations based on ^{14}C-dating techniques suggest that
most of the ground water in eastern Saudi Arabia was recharged about
20,000 to 25,000 years ago. Very few samples were calculated as being
modern water. Other isotope data, however, indicate that some recharge is

occurring. Detailed tritium data collected at one site on a sand dune show a downward movement of about 20 mm of water per year, although the total depth of penetration of this water is not known. Tritium data collected over the entire basin are generally negative. Only six samples contained over 15 TU and the potential for atmospheric contamination of these six samples was high.

The ^{18}O concentration of about -2.5 per mille measured in outcrop areas of the large aquifers is very similar to that found in the alluvial aquifers. The similarity of these values suggests that some recharge into the large aquifers may also be taking place in the outcrop areas.

REFERENCES

1. ITALCONSULT, Water and Agricultural Development Studies for Area IV, Report to the Ministry of Agriculture and Water, Saudi Arabia, (1969).

2. ITALCONSULT, Final Report, Wadi Bishah Selected Area; Geohydrological Investigations, Report to the Ministry of Agriculture and Water, Saudi Arabia (1969).

3. PARSONS BASIL, Agriculture and Water Resources, The Great Nafud Sedimentary Basin, Kingdom of Saudi Arabia, Report to the Ministry of Agriculture and Water, Saudi Arabia, (1968).

4. BUREAU DE RECHERCHES GEOLOGIQUES ET MINIERES, Hydrogeological Investigation of the Al Wasia Aquifer in the Eastern Province of Saudi Araiba, Regional Study, Report to the Ministry of Agriculture and Water, Saudi Arabia, (1976).

5. BUREAU DE RECHERCHES GEOLOGIQUES ET MINIERES, Al Hasa Groundwater Resources and Management Programme, Report to the Ministry of Agriculture and Water, Saudi Arabia, (1978).

6. SOGREAH, Water and Agriculture Development Studies, Area V, Report to the Ministry of Agriculture and Water, Saudi Arabia, (1968).

7. SIR M. MACDONALD & PARTNERS, Riyadh Additional Water Resources Study, Report to the Ministry of Agriculture and Water, Saudi Arabia, (1975).

8. OTKUN, GALIP, Outlines of Hydrogeology of Al Hassa Springs and the Results of Some Recent Observations, Report to the Ministry of Agriculture and Water, Saudi Arabia, (1974).

9.   OTKUN, GALIP, Personal Communication.

10. RAMIREZ, L.F., and BOWERS, S., Personal Communication.

11. DINCER, T., AL-MUGRIN, A., AND ZIMMERMAN, U., Study of the infiltra-
     tion and recharge through the sand dunes in arid zones with special
     reference to the stable isotopes and thermonuclear tritium, J. Hydrol.
     23 (1974), 79-109.

12. VOGEL, J. C., Carbon-14 dating of groundwater, (Proc. Isotope
     Hydrology 1970), IAEA, Vienna (1970).

13. PEARSON, F. J., and HANSHAW, B. B., Sources of dissolved carbonate
     species in groundwater and their effects on carbon-14 dating (Proc.
     Isotope Hydrology 1970), IAEA, Vienna (1970), 271-286.

14. CRAIG, H., Isotopic Variations in Meteoric Waters. Science 133
     (1961), 1702-1703.

15. MAZOR, E., Multitracing and multisampling in hydrological studies,
     Interpretation of environmental isotope and hydrochemical data in
     groundwater hydrology, (Proc. Techn. Comm., Vienna, 1975), IAEA,
     Vienna (1976).

16. DANSGAARD, W., Stable isotopes in precipitation, Tellus 16 4 (1964),
     436-468.

DISCUSSION

B.Th. VERHAGEN: Did you consider the influence of physiography on the stable isotope data? In semi-arid and arid environments rainfall variability and the resulting selection of recharge might play an important rôle.

W.J. SHAMPINE: We considered this only to the extent of showing that the recharge area had an elevation range of 250 m, which would affect $\delta^{18}O$ concentration by no more than 0.5‰.

B.Th. VERHAGEN: Was there any correlation between the ^{14}C and stable isotope data?

W.J. SHAMPINE: Unfortunately, there appeared to be no correlation. A mathematical correlation was not attempted.

P.L. AIREY: Although no correlation was observed between ^{14}C ages and stable isotope ratios, did you endeavour to find systematic variations in the stable isotope ratios with ^{14}C ages using, say, a moving average procedure? Any such variations might correlate with other palaeoclimatic parameters.

W.J. SHAMPINE: No, we did not try to use a moving average procedure.

P.L. AIREY: Did you observe an inverse correlation between groundwater age and the standard deviations associated with δD and $\delta^{18}O$, as is sometimes found in Australia?

W.J. SHAMPINE: No. We have never thought of looking for such a correlation.

B.Th. VERHAGEN: With regard to Mr. Airey's second question, I might mention that a reduction in the standard deviation of stable isotope values with radiocarbon age is an effect we have seen in groundwater in the Kalahari.

H. HÜBNER: In your first slide you showed $\delta^{13}C$ values and their distribution frequencies. There were in fact some very low ^{13}C values, between about -20 and $-22‰$. May I ask how you explain such negative values in groundwater? I would say that only biological processes can explain them.

W.J. SHAMPINE: I cannot explain these values but have simply repeated what has been published previously. The values are five standard deviations away from the mean and, I think, are probably incorrect. In fact, I did not use those three values for the data analyses in my paper.

A.R. MARCE: Do the very low ^{13}C values for the bicarbonates of deep aquifer waters relate perhaps to very saline waters in the Wasia or the Biyadh, which are to a greater or lesser extent contaminated by waters coming from petroleum structures?

W.J. SHAMPINE: As I recall, one of the samples was from the Biyadh aquifer, but the other two were from the Umm-Er-Radhuma. Although I agree that the very low values are probably erroneous, there are no data available to prove this. There certainly are no data to show movement of waters from the petroleum structures.

A.R. MARCE: In the aquifers of Saudi Arabia there is an enormous difference in $^{18}O/D$ ratio between recent waters, whether tritiated or not, and non-tritiated old waters. Recent waters always have ^{14}C contents higher than 15% modern and fit a line $\delta D‰ = 6\,\delta^{18}O‰ + 10$, as calculated from isotope data for precipitation on Bahrain Island and confirmed by data for rain recorded at Riyadh and Hofuf (enriched isotope values). Non-tritiated old waters, on the other hand, always have ^{14}C contents lower than 15% modern and their stable isotopic ratio fits the line $\delta D‰ = 8\,\delta^{18}O‰ + 3$ (depleted isotope values, $< -3.5‰$ for $\delta^{18}O$ and $< -20‰$ for δD).

APPLICATION OF ENVIRONMENTAL ISOTOPES TO GROUNDWATER INVESTIGATIONS IN QATAR

Y. YURTSEVER, B.R. PAYNE
International Atomic Energy Agency,
Vienna

Abstract

APPLICATION OF ENVIRONMENTAL ISOTOPES TO GROUNDWATER INVESTIGATIONS
IN QATAR.

Environmental isotope techniques have been used to study the dynamic behaviour of the
groundwater systems in Qatar. The aim was to identify the mixing processes between the
shallow phreatic aquifer, the deep confined aquifers and sea-water. Environmental tritium was
used to estimate the average recharge of the shallow fresh-water aquifer. The results of the
environmental isotope study, supplemented by hydrochemical data, have clearly indicated that
in the southwestern part of the peninsula the shallow aquifer is in direct hydraulic connection
with the deeper artesian aquifers. Furthermore, the poor quality of water in that area is mainly
due to substantial upward leakage of water from the deeper aquifers. In some parts of the
peninsula sea-water was identified as an additional source of salinity. A model was developed
for estimating the turnover rate and average recharge of the phreatic aquifer from the tritium
data. The results of this model provide an independent estimate of the average annual recharge.

INTRODUCTION

The Qatar peninsula has a surface area of about 11 600 km^2 with its NS
axis being approximately 190 km in length and its EW width being about 85 km
at its widest central part.

The northern part of the peninsula is characterized by a gently rolling terrain
with shallow depressions, while in the south the terrain is moderately rough with
deep depressions and steep mesa-type hills.

The peninsula lies within the desert belt and the average annual rainfall is
about 80 mm in the northern part and about 50 mm in the south. The temporal
and spatial distribution of rainfall is extremely variable and usually occurs as
intense rainfall of short duration mostly during the winter months. Such significant individual storms are probably responsible for replenishment of groundwater.
The frequency of such storms is quite random and may vary from a few days up
to several years. The surface drainage pattern is mainly internal towards many
of the depressions through poorly developed drainage channels.

The environmental isotope study, which was carried out within the scope
of a UNDP/FAO project, was aimed at providing information on various dynamic
characteristics of the groundwater system which is the only potential source of
water in Qatar.

1. HYDROGEOLOGY

The basic geological setting of the Qatar peninsula is part of the regional
stratigraphic sequence observed in Saudi Arabia. The regional geological structures
of Saudi Arabia and Qatar are shown in Fig.1 [1]. The igneous and metamorphic
basement complex, which outcrops in the highlands of the interior of Saudi
Arabia, slopes eastwards in the direction of Qatar where it is overlain by sedimentary deposits. The stratigraphic sequence in the peninsula includes the sand-
dunes of Quaternary age, alluvial deposits in the surface depressions, the
Miopliocene Hofuf formation, the Miocene Dam formation, the Eocene Damman
and Rus formations and the Paleocene Umm-Er-Rhaduma formation [1].
The uniformity of the eastern slope of the sedimentary formations is later
subjected to structural disturbances to form parallel anticlinal and domed
structures in the whole region including Qatar. The NNW-SSE trending anticlines
and synclines intersect the broad anticlinal arch of the peninsula in southern
Qatar (Fig.2). These structural features and the inferred faults are highly significant as regards the occurrence, movement and quality of groundwater and produce
different hydrogeological conditions in the southern and northern parts of Qatar.
In particular the two parallel faults trending WNW-ESE across central Qatar
divide the peninsula into two distinct groundwater systems. In the south the
movement of groundwater from the outcrop in Saudi Arabia is significantly
affected by the underlying structural features. However, in the northern part
the groundwater is known to occur as a fresh-water lens floating on saline water
with the interface being at a depth of about 100 m according to geophysical
measurements.
The major water-bearing formations are the Eocene and particularly the
Rus formations which provide potable water at depths down to about 100 m.
The younger Hofuf and Dam formations and the surface deposits generally lie
above the water table. The deeper aquifers of the Umm-Er-Rhaduma formation
produce brackish water. While this is a generally valid condition in the northern
part of the peninsula, the wells in the southern part which reach to a depth of
about 100 m generally yield brackish water. In the Abu-Samra area near the
Salwa Gulf in the south, even wells drilled to a depth of 30 m in the Damman
formation yield brackish water under artesian pressure. It is postulated that the
brackish water of the deeper formations flowing eastward from Saudi Arabia
under artesian pressure moves across the Salwa syncline into southern Qatar where

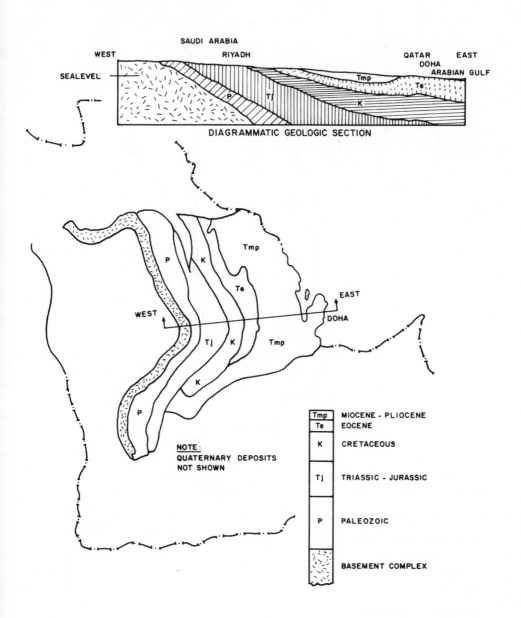

FIG.1.   Regional geology.

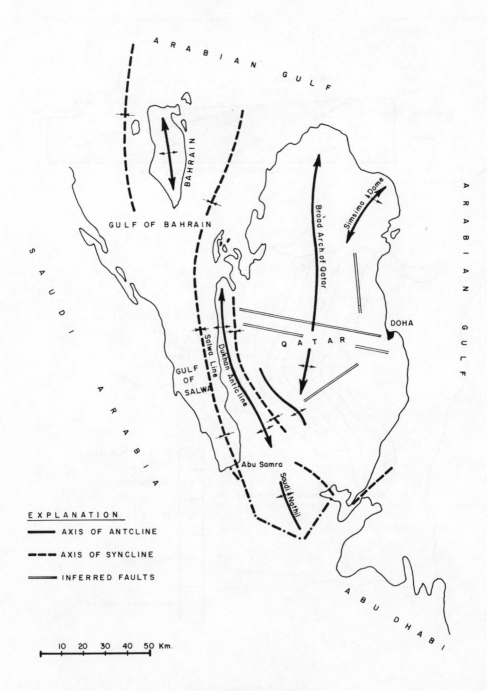

FIG. 2.   Regional structural features of study area.

it leaks upwards into the overlying formations giving rise to the poor quality water. In the northern part of the peninsula it is believed that when significant rainfall occurs the groundwater is recharged forming a fresh-water lens lying on top of saline water.

2. THE PROBLEM

A knowledge of the hydraulic interconnections between the deeper aquifers containing brackish water and the upper fresh-water aquifers and also the identification of possible sea-water intrusion are of utmost importance for engineering decisions in the development and use of available potable ground-water in the peninsula. In addition, an estimate of the recharge of the upper aquifers is an important factor in the development of the groundwater resources of the peninsula. This information is essential not only in estimating the available safe yield of the fresh-water resources, but also has a direct bearing on the preservation of the dynamic balance of the groundwater system in order to avoid excessive sea-water intrusion and/or leakage of brackish water into the overlying fresh-water aquifers.

3. ENVIRONMENTAL ISOTOPE DATA

The investigation was based on the sampling in 1975 and 1976 of 43 sources, mostly wells and boreholes, from different parts of the peninsula. Selected wells were sampled more than once. All the 71 samples were analysed for their ^{18}O, deuterium and tritium content and ^{14}C and ^{13}C analyses were made on a few selected samples. Details of the analyses are given in Table I and the locations of sampling points are shown in Fig.3.

The $^{18}O/^{16}O$ and D/H ratios are expressed as delta values versus SMOW and the $^{13}C/^{12}C$ ratios versus PDB. The analytical errors are 0.1, 1 and 0.3‰ respectively. The tritium concentrations are expressed in tritium units (1 TU = 1 ^3H atom per 10^{18} H atoms) and the ^{14}C data are expressed in percentage modern ^{14}C.

A few isotopic data prior to this investigation and hydrochemical data were also taken into consideration. Data on the ^{18}O, deuterium and tritium content of monthly precipitation since 1962 were available for Bahrain Island which is the closest station to Qatar in the IAEA/WMO global precipitation monitoring network [2].

Text continued on p.477.

TABLE I. RESULTS OF ALL THE ISOTOPIC ANALYSES AND Cl⁻ DATA FROM THE PROJECT AREA

Location	Name	IAEA Sample No.	Sampling Date	Cl^- (ppm)	$\delta^{18}O$ (‰)	δD (‰)	Tritium (TU)	^{14}C % Mod.	^{13}C (‰)
Al Sufairiah	Well A-19	QAT-1ᵃ (+15.0)	26.10.72	58	-2.66	-6.3	0.6 ± 0.2		
		QAT-2ᵃ (+10.0)	26.10.72	950	-2.76	-12.1	0.3 ± 0.3		
Simsima	Well A-5	QAT-23	09.03.75	1300ᵇ	-2.95	-19.3	1.1 ± 0.2		
		QAT-71	27.04.76		-3.49	-20.7	0.7 ± 0.2		
North of Doha	Well A-1	QAT-3ᵃ	26.10.72	3400	-2.33	-12.3	3.2 ± 0.2		
		QAT-74	27.04.76		+2.38	+13.2	15.8 ± 0.5		
Near Doha	seawater	QAT-4ᵃ	26.10.72	25000	+2.16	+15.5			
		QAT-5ᵃ	26.10.72	1600	-4.10	-34.6	0.1 ± 0.3		
		QAT-35	12.07.75	2500	-2.66	-28.9	0.4 ± 0.2		
		QAT-57	31.03.76					8.9 ± 1.3	-7.1
		QAT-63	12.04.76		-2.73	-30.1	0.8 ± 0.2		
		QAT-69	12.05.76		-1.37	- 8.0	2.0 ± 0.3		

Location	Name	IAEA Sample No.	Sampling Date	Cl⁻ (ppm)	$\delta^{18}O$ (‰)	δD (‰)	Tritium (TU)	^{14}C % Mod.	^{13}C (‰)
Salwa Area Abu Samra	Well No. 1	QAT-6[a]	26.10.72		-5.17	-38.9	1.2 ± 0.3		
		QAT-16	27.02.75	780	-5.13	-41.0			-7.6
		QAT-54	? 02.76					2.7 ± 1.5	-7.6
Salwa Area	Salwa Well	QAT-6a[a]	28.01.73	801 830	-5.10	-38.9	0.7 ± 0.4		
Hashim Depression	Observation Well No. 1	QAT-7	25.02.75	2940[b]	-1.45	- 0.5	23.9 ± 1.0		
Government Farm	Observation Well	QAT-8	25.02.75	380[b]	-2.32	-10.0	5.8 ± 0.3		
Government Farm	Well A-10	QAT-9	25.02.75	375[b]	-2.18	- 8.6	5.6 ± 0.3		
Farm	Well B-24	QAT-10	26.02.75	590[b]	-2.03	-11.6	11.6 ± 0.4		
Farm	Well B-26	QAT-11	26.02.75	420[b]	-1.77	- 8.6	4.2 ± 0.3		
Farm	Well B-34a	QAT-12	26.02.75	980[b]	-1.76	- 9.1	6.3 ± 0.3		
Rudait Rashid Well field	Storage Tank of the Wells	QAT-13	26.02.75		-1.95	- 6.6	6.7 ± 0.3		
El Oturiyan Well Field	Well B-31	QAT-14	26.02.75	29[b]	-1.96	-4.7	17.5 ± 0.7		
El Oturiyan Well Field	Storage Tank of the Wells	QAT-15	26.02.75		-1.89	- 2.9	2.3 ± 0.2		
Karaannah Depression	Well C-6	QAT-18	27.02.75	100[b]	-1.66	- 1.3	59.4 ± 1.4		
		QAT-28	18.03.75		-1.96	- 3.3	57.5 ± 1.9		
		QAT-60	12.04.76		-1.51	- 1.5	46.3 ± 1.1		

TABLE I (cont.)

Location	Name	IAEA Sample No.	Sampling Date	Cl⁻ (ppm)	$\delta^{18}O$ (‰)	δD (‰)	Tritium (TU)	^{14}C % Mod.	^{13}C (‰)
Karaannah Depression	surface pond	QAT-19	27.02.75		-1.25	+ 1.5	21.4 ± 0.6		
		QAT-21	01.03.75		-0.60	+ 2.4			
Southern Qatar	Saudi Nathil Deep Well (Artesian)	QAT-20	01.03.75	42900ᵇ		-39.4			
		QAT-20a	14.04.75					5.8 ± 1.1	-0.4
Karaannah Depression	Precipitation	QAT-22	28.01.75		-0.72	+14.4	33.5 ± 0.9		
	Well A-25	QAT-24	09.03.75	230ᵇ	+1.40	+16.4	16.1 ± 0.5		
		QAT-25	09.03.75	2320ᵇ	-1.66	- 5.3	9.6 ± 0.3		
	Well A-26	QAT-75	25.04.76		-2.09	- 6.4	11.7 ± 0.4		
	Well A-32	QAT-26	09.03.75	590ᵇ	+0.10	+ 4.7	38.4 ± 1.5		
	Well B-5	QAT-27	09.03.75	920ᵇ	-3.62	-23.8	5.2 ± 0.3		
		QAT-73	05.05.76		-3.85	-22.7	3.2 ± 0.3		
Karaannah Depression	Temporary Surface Pond	QAT-29	04.03.75		+0.31	+ 8.7			
		QAT-30	10.03.75		+2.04	+15.3			
		QAT-31	18.03.75		+5.57	+33.9			
		QAT-32	25.03.75		+11.51	+59.3			

Location	Name	IAEA Sample No.	Sampling Date	Cl⁻ (ppm)	$\delta^{18}O$ (‰)	δD (‰)	Tritium (TU)	^{14}C % Mod.	^{13}C (‰)
South Qatar	Well C-10	QAT-33	12.07.75	2200	-4.59	-39.0	0.1 ± 0.2		
		QAT-55	10.03.76	2060				4.8 ± 1.9	-7.2
		QAT-61	12.04.76		-4.79	-38.8	0.8 ± 0.2		
South Qatar	Well C-11	QAT-34	12.07.75	1640	-4.81	-37.2	0.4 ± 0.2		
		QAT-56	18.04.76					-0.7 ± 3.3	-7.2
		QAT-62	12.04.76		-4.74	-38.9	0.2 ± 0.2		
South Qatar	Well C-13	QAT-36	13.07.75	1680	-4.58	-36.9	1.3 ± 0.2		
		QAT-58	27.04.76					3.0 ± 1.8	-7.7
		QAT-64	27.04.76		-5.20	-38.5	0.4 ± 0.2		
South Qatar	Well C-14	QAT-37	13.07.75	1300	-3.79	-34.4	2.4 ± 0.2		
		QAT-65	13.04.76		-3.79	-31.6	0.9 ± 0.3		
South Qatar	Well C-15	QAT-38	13.07.75	2150	-4.07	-34.5	1.1 ± 0.2		
		QAT-66	13.04.76		-4.07	-32.3	0.4 ± 0.2		
South Qatar	Well C-16	QAT-39	13.07.75	1460	-0.82	- 6.0	5.4 ± 0.3		
		QAT-67	13.04.76		-1.00	-4.6	4.6 ± 0.2		

TABLE I (cont.)

Location	Name	IAEA Sample No.	Sampling Date	Cl⁻ (ppm)	$\delta^{18}O$ (‰)	δD (‰)	Tritium (TU)	^{14}C % Mod.	^{13}C (‰)
South Qatar	Well C-18	QAT-40	12.07.75	7850	-3.57	-42.3	-1.0 ± 3.2 / 0.3 ± 0.2		
		QAT-70	27.04.76		-3.76	-41.3	0.1 ± 0.2		
	Well A-6	QAT-41	23.12.75	1620	-3.05	-21.9	0.5 ± 0.2		
		QAT-72	27.04.76	1600	-2.92	-19.5	0.3 ± 0.3		
	Well A-18	QAT-42	23.12.75	130	-2.66	-16.2	0.1 ± 0.2		
	Well B-2	QAT-43	24.12.75	2800	-1.97	-9.6	5.1 ± 0.3		
	Well B-22	QAT-44	17.12.75	52	-2.29	-10.3	5.0 ± 0.3		
	Well B-29	QAT-45	24.12.75	470	-1.90	-8.4	0.3 ± 0.2		
		QAT-59	10.04.76		-1.95	-8.0	0.3 ± 0.3		
	Well B-37	QAT-46	23.12.75	122	-1.51	-4.0	0.1 ± 0.2		
	Well B-39	QAT-47	23.12.75	150	-2.20	-4.2	12.4 ± 0.4		
	Well C-1a	QAT-48	23.12.75	3700	-1.44	-12.1	1.6 ± 0.3		
		QAT-49	23.12.75	880	-2.20	-8.8	0.5 ± 0.2		
	Well C-2	QAT-68	12.05.76		-2.29	-9.2	1.1 ± 0.2		

Location	Name	IAEA Sample No.	Sampling Date	Cl$^-$ (ppm)	$\delta^{18}O$ (‰)	δD (‰)	Tritium (TU)	^{14}C % Mod.	^{13}C (‰)
Traina Farm	Well C-3	QAT-50	23.12.75	1260	-2.52	-12.5	1.4 ± 0.2		
Al Muaizir	Well	QAT-51	27.12.75		-1.65	- 9.8	2.0 ± 0.2		
Al Aameriya	Well C-8	QAT-52	27.12.75	266	-1.56	- 7.0	1.3 + 0.3		
Afforstation	Well No. 1	QAT-53	27.12.75		-2.47	- 6.3	0.7 ± 0.2		

a Results available from an earlier study.
b Sampling date is the same as for isotope analyses.

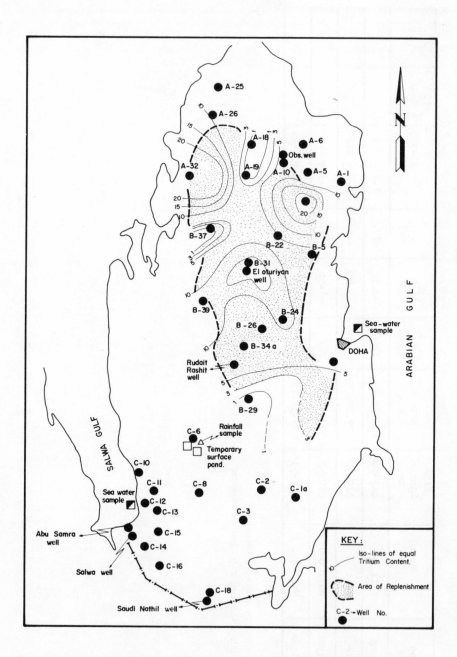

FIG.3.   *Location of sampled sources and isolines of equal tritium content for shallow groundwater.*

4.   INTERPRETATION

4.1. Stable isotope data

From the available stable isotope data for precipitation in Bahrain Island the mean values, weighted for the amount of precipitation, are:

$\overline{\delta^{18}O} = -1.86\%o$

$\overline{\delta D} = -3.4\%o$

The data relating to months in which the precipitation was less than 10 mm were excluded since their delta values were indicative of a significant evaporation effect. Even some of the months with amounts of precipitation up to 20 mm also exhibited an evaporation effect. If these values are also excluded the weighted mean stable isotopic composition of rainfall for Bahrain Island is:

$\overline{\delta^{18}O} = -2.27\%o$

$\overline{\delta D} = -7.2\%o$

The $\delta^{18}O-\delta D$ relation, with an a priori assumed slope of 8, defined by the least-squares fit of precipitation data, excluding those exhibiting a significant evaporation effect, was found to be as follows and was used in the overall evaluation of the data:

$$\delta D = 8 \; \delta^{18}O + 10.9 \tag{1}$$

The $\delta^{18}O$ values for all the sampled wells and for sea-water are shown in Fig.4 as a histogram in order to provide an overall picture of the ^{18}O data. The majority of the wells located in the northern part of the peninsula have $\delta^{18}O$ values ranging from -1.5 to $-3.0\%o$, while most of the wells located in the southern area have more depleted values in the range of -3.8 to $-5.2\%o$. Clearly the large spread in $\delta^{18}O$ values of groundwater must be the result of various possible processes such as evaporation prior to recharge, sea-water intrusion and mixing between water bodies having different $\delta^{18}O$ values.

The stable isotopic composition of all samples is plotted in Fig.5 from which it will be seen that the wells located in northern Qatar cluster around the precipitation line defined by Eq. (1).

The two samples of sea-water have an enriched stable isotope content which enables tracing of any sea-water intrusion to the aquifers. The sample from the Salwa Gulf ($\delta^{18}O = 4.53\%o$, $\delta D = 28.9\%o$) is more enriched than that from near Doha ($\delta^{18}O = 2.16\%o$, $\delta D = 15.5\%o$) as a result of the greater effect of evaporation at the former location.

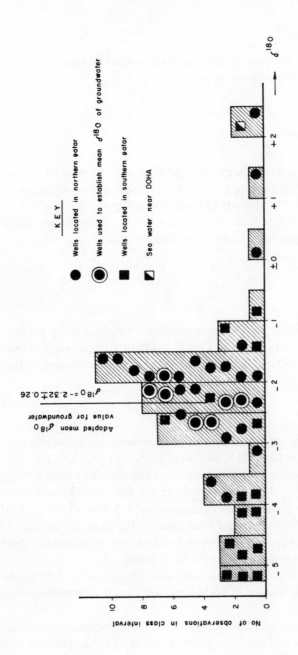

FIG.4.   Histogram of δ¹⁸O values of the sampled wells.

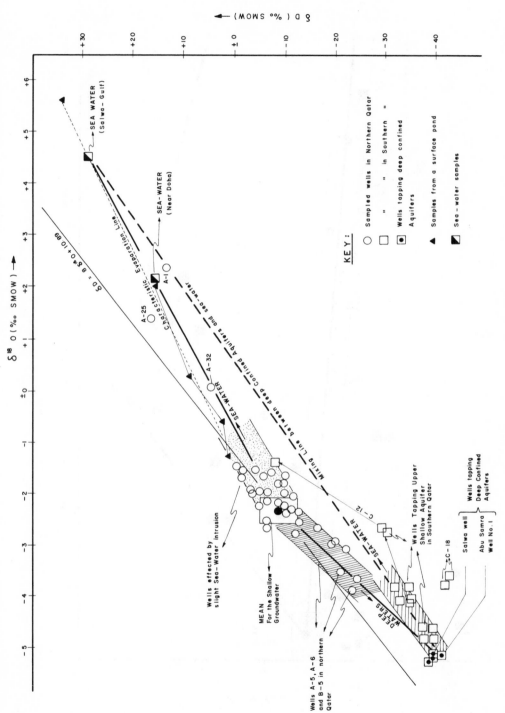

FIG.5. δ¹⁸O-δD plot of the samples collected in the study area.

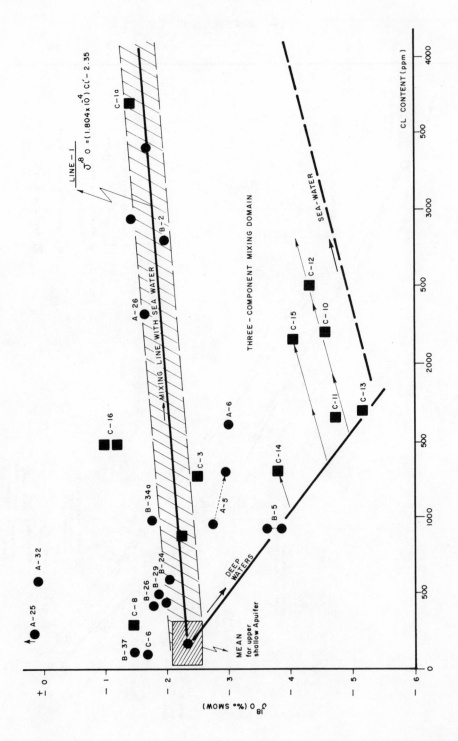

FIG. 6. $\delta^{18}O$-Cl^{-} plot for the sampled wells.

Four successive samples collected from a temporary surface pond in the southern part of the peninsula, resulting from the heavy rain of March 1975, define a typical evaporation line for the area with a slope of 5. Both the evaporation effect and sea-water intrusion shift the points on the $\delta^{18}O-\delta D$ diagram in the same direction. Therefore, hydrochemical data, particularly chloride data, were used to differentiate between these two processes.

The plots of $\delta^{18}O$ versus Cl (Fig.6) and δD versus $\delta^{18}O$ (Fig.5) were used to derive the following interpretations of the dynamics of the groundwater systems in the peninsula.

4.1.1. Direct recharge by rainfall and the mean isotopic composition of the upper fresh-water aquifer

The wells which plot close to the precipitation line defined by Eq.(1) do not show any appreciable evaporation effect and also have quite low chloride concentrations. The mean composition of these wells provides the following mean values:

$$\overline{\delta^{18}O} = -2.32 \pm 0.26\%o$$
$$\overline{\delta D} = -8.61 \pm 4.1\%o$$
$$\overline{Cl^-} = 168 \pm 150 \text{ ppm}$$

If evaporation from the temporary surface ponds prior to recharge, mixing with water from the deeper aquifers and/or mixing with sea-water do not occur, the fresh-water aquifer could be expected to have values similar to the above mean values. The fact that the above mean delta values do not differ significantly from the weighted mean stable isotopic composition of rain, excluding months with less than 20 mm of rainfall, provides confirmation that the source of recharge of the upper aquifer is direct infiltration of rainfall. Furthermore, it is most unlikely that there is any significant recharge during months having a total rainfall of less than 20 mm.

4.1.2. Sea-water intrusion

In Fig.6 sea-water is characterized by a $\delta^{18}O$ of 2.16%o and a chloride concentration of 25 000 ppm, while groundwater in the upper shallow aquifer is defined by the mean values derived in the previous paragraph, i.e. $\delta^{18}O = -2.32\%o$ and $Cl^- = 168$ ppm. Line 1 defines the mixing between these two sources of water. Therefore, those wells which are scattered along this line are partially contaminated with sea-water. This mixing process between fresh-water and sea-water is also supported by the stable isotope data for these wells where, from Fig.5, it will be seen that they deviate slightly from the precipitation line in the direction of the indices for sea-water. Some of these wells (e.g. A-1, A-26 and C-1) are located near the coast where sea-water intrusion could be expected. However, for the wells located comparatively far away from the coast, such as B-24, B-29 and B-34a, the contribution of saline water may be derived from upward diffusion from the fresh-water/sea-water interface, possibly as a result of local fluctuations of the water table due to pumping. The relative positions of the points along Line 1 of Fig.6 places an upper limit of 10–15% on the contribution of sea-water to the aquifer at the locations of these wells.

The significantly enriched stable isotopic composition of water from the wells A-25 and A-32 and their departure from the precipitation line in Fig.5 are indicative of evaporation. Although sea-water intrusion to these wells might have been suspected in view of their proximity to the coast, their relatively low chloride content (Fig.6) excludes the existence of any significant sea-water intrusion. The quite high tritium values of these wells is evidence for the occurrence of recent recharge. Therefore, the enriched stable isotopic composition may have resulted from partial evaporation during the possible ponding of water prior to recharge. It is also possible that evaporation occurs directly from the water table since both of these wells are very shallow and the depth to water is very small.

4.1.3. Leakage from the deeper confined aquifers to the upper phreatic aquifer in southern Qatar

On the basis of the Abu-Samra well No.1 and the Salwa well, the deep confined aquifer has a stable isotopic composition of −5.1 and −40‰ for $\delta^{18}O$ and δD respectively. The only result available for water in the Umm-Er-Rhaduma formation is a δD value of −39.4‰ for the very deep Saudi-Nathil well, which is similar to the data for the previous two wells and also to earlier data reported for this formation [3, 4]. This depleted stable isotopic composition of the deeper aquifers is interpreted as being due to recharge at higher elevations in Saudi Arabia and possibly also to palaeoclimatic effects since the water in the Umm-Er-Rhaduma formation has a low ^{14}C content of 5.8 ± 1.1% modern indicating an uncorrected age of about 23 000 a. The difference in stable isotopic composition between the deep aquifers and the shallow aquifer enables a study of the leakage from the former to the latter.

All the wells tapping the upper shallow aquifer in southwestern Qatar (C-10 to C-18) have a depleted stable isotopic composition not too different from that of the deeper aquifer (Fig.5). Therefore, it is concluded that the upward leakage from the deep aquifer does occur and in fact is the dominant source of recharge to these relatively shallow wells. It is also seen from Fig.5 that the isotopic composition of some of these wells may be considered as falling on a mixing line (the dashed line in Fig.5) between the stable isotope indices for sea-water in the Salwa Gulf and for the deep confined aquifers. This therefore suggests a small intrusion of sea-water to the water pumped from these wells.

These conclusions are also supported by the chloride data. From Fig.6 it is seen that the points representing the wells in southwestern Qatar are shifted from the mixing line between the upper and deep aquifers in the direction of sea-water. Thus, the data suggest a three-component mixing model with sea-water, the upper aquifer, and the deeper aquifer as the components. The first two components are rather well defined by their $\delta^{18}O$ and chloride data, but the chloride content of the deeper aquifer is not well known. Nevertheless, the mixing lines in Fig.6 define a triangular mixing diagram. Points falling on mixing lines are indicative of two-component mixtures, while the position of points falling within the mixing triangle is a function of the mixing between three components. The water from wells C-10, C-11, C-14 and C-15 appears to be the result of such a three-component mixing model. However, from Fig.6 it is clear that the contribution of sea-water is the minor component.

Turning to the northern part of Qatar, the isotopic data suggest that upward leakage from the deep aquifers may occur in a region parallel to the eastern coast. The wells A-5, A-6 and B-5 have a more depleted stable isotopic composition than other wells in northern Qatar and from Fig.5 it is seen that they fall on a mixing line between the deep aquifers and the overlying fresh-water aquifer. Since the only source of water in the peninsula with a depleted stable isotopic composition are the deep aquifers, it is concluded that upward leakage from the deeper aquifers does occur in the area represented by these wells. A possible alternative source of depleted water might be the occurrence of very recent recharge by rainfall having a rather depleted stable isotopic composition. This possibility was rejected on account of the very low tritium values observed for these wells.

The chloride data support the above conclusion. The position of B-5 in Fig.6 is on the mixing line between the upper and the deeper aquifers. The positions of A-5 and A-6 in Fig.6 suggest a three-component mixing model in which sea-water is the minor component.

These wells are located along an inferred fault (Fig.2), so the conclusion based on the isotope data suggests that this fault may provide the motivation for leakage from the deeper confined aquifers to the upper phreatic aquifer.

4.2. ^{14}C data

A few selected wells in southwestern Qatar were sampled for ^{14}C analysis to provide an additional tracing parameter to confirm the conclusions based on the stable isotope and chloride data.

The ^{14}C contents of the Saudi Nathil and Abu-Samra No.1 wells are 5.8 ± 1.1 and $2.7 \pm 1.5\%$ modern, respectively. The uncorrected ages of 23 000 and 30 000 a, respectively, are in agreement with the quite large distance from southwestern Qatar to the outcrop areas in the interior of Saudi Arabia. Similar ^{14}C values were measured for the wells (C-10, C-11, C-12 and C-13) tapping the upper aquifer, therefore providing support for the conclusion that most of the water from these wells originates from the deep confined aquifers.

4.3. Tritium data

Data on the tritium concentration of monthly rainfall at Bahrain Island since 1962 are available [2]. The maximum concentration (1420 TU) was in April 1963. In common with other stations in the northern hemisphere the concentration decreased thereafter. During 1974 and 1975 the concentration was in the range of 15–50 TU. Prior to 1952 the tritium concentration in precipitation in this region was probably in the range of 5–10 TU.

The tritium concentrations of the groundwater samples (see Table I) range from zero to 60 TU. Looking at the spatial variations of tritium in groundwater it is found that most of the wells containing more than 3 TU, thus indicating recharge since 1952, are located in the central and northern parts of the peninsula. The iso-curves of equal tritium concentrations for the phreatic aquifer are shown in Fig.3. Regions having more than 3 TU are shown as shaded areas in order to indicate where recharge has occurred. A comparison of the pattern of the iso-curves of tritium with the general patterns of change of water levels in the peninsula indicates that regions where a significant rise of the water table has occurred are associated with relatively higher tritium concentrations, thus providing evidence for recharge in the areas indicated on Fig.3.

4.4. Estimates of the turnover rate of the groundwater and average annual recharge

The tritium data for the groundwater samples were used to estimate the turnover rate of the phreatic aquifer in order to arrive at a quantitative estimate of the average annual recharge and consequently of the potential average safe yield.

The basis of the approach is the fact that the observed tritium concentration of the groundwater is the result of the response of the system to the known tritium input by precipitation. If the response of the system can be described by certain dynamic parameters, then the tritium input can be related to the observed output to estimate the dynamic parameters.

A mathematical expression for such a linear lumped parameter model is the following convolution integral:

$$C_0 = \int_0^t f_i(t-T) \cdot f(T) \cdot e^{-\lambda t} \cdot dT \tag{2}$$

TABLE II. ANNUAL MINIMUM AND MAXIMUM TRITIUM CONTENT
OF PRECIPITATION FOR BAHRAIN ISLAND

Year	Minimum Observed Tritium Content (TU)	Maximum Observed Tritium Content (TU)
1956	8[a]	30[a]
1957	25[a]	60[a]
1958	40[a]	400[a]
1959	30[a]	400[a]
1960	20[a]	100[a]
1961	70[a]	205
1962	300	810
1963	470	1420
1964	126	720
1965	105	315
1966	85[b]	249[b]
1967	64[b]	184[b]
1968	43	119
1969	29	79
1970	73	122
1971	52	124
1972	28	74
1973	16	48
1974	15[a]	50

[a] Estimated from OTTOWA DATA.
[b] Linear interpolation between consecutive values.

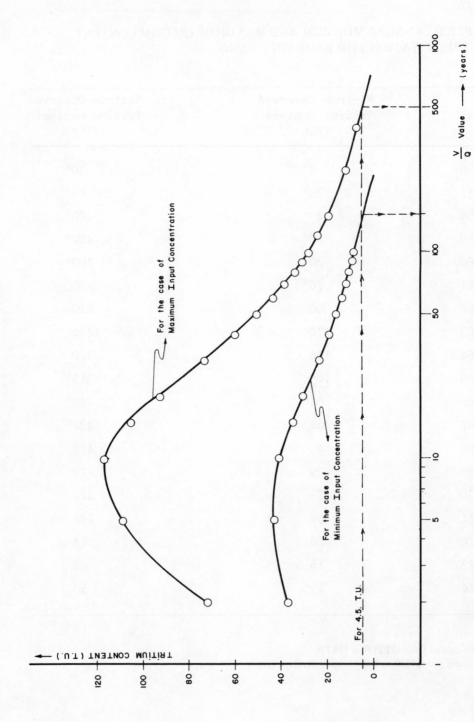

FIG. 7. Computed tritium-output-concentration curves for the year 1975 as a function of V/Q parameter.

where C_0 = tritium output concentration
     $f_i(t - T)$ = tritium input concentration
     $f(T)$ = system response function (transit time distribution)
     $e^{-\lambda t}$ = radioactive decay correction factor
     t   = time
     T   = transit time

An "exponential" type of system response function was adopted which is described by the following equation:

$$f(T) = \frac{Q}{V} \cdot e^{-\frac{Q \cdot t}{VQ \ V}} \tag{3}$$

In this case the system response function is described by a single parameter, V/Q, which has the dimensions of time and signifies the turnover rate of the system.

It is assumed that each sampled well has its own sphere of influence defining a compartmental volume within which the transit time distribution of the water is described by Eq.(3). Thus, the overall groundwater system is visualized as being made up of these individual compartments and it is assumed that the sampled wells are adequate to represent the phreatic aquifer. The adopted response function describes the case of a "well-mixed" single-storage element. Thus, each compartmental volume defined by a well can be treated as "well-mixed" for which the mixing takes place in the well itself. Consequently, the parameter V/Q for each well may be estimated by Eq.(2). Since the overall phreatic aquifer is assumed to be made up of these individual compartments, an average value of V/Q can be obtained for the whole system.

The tritium input concentration for the area was obtained using the available tritium data for precipitation at Bahrain Island. The computation was made for dT equal to one year. Since the recharge of the aquifer is mainly by individual rainfall events of significant amount but irregular occurrence, it is difficult to estimate the tritium input concentration of the actual recharging water for each year. Therefore, two extreme cases of minimum and maximum input concentration values were used in the computations in order to set lower and upper limits to the tritium concentration of the recharging water. These were defined by taking the minimum and maximum observed tritium concentrations of precipitation for each year and are given in Table II. The two extreme cases of the tritium output concentrations in 1975 for various values of V/Q are illustrated in Fig.7. The measured tritium concentration of each well was then used to estimate the maximum and minimum values of the parameter V/Q. The results are listed in Table III. The wells in southwestern Qatar (C-10 to C-18), which have a tritium concentration close to zero, were not included since their low tritium concentrations are no doubt due to the substantial leakage of tritium-free water from the deeper aquifer.

TABLE III.   COMPUTATION OF AVERAGE TURNOVER TIME FOR
THE WELLS AND THE AQUIFER

Well No.	Tritium Content (TU)	Estimated $(\frac{V}{Q})$ Values (years)	
		For Minimum Input	For Maximum Input
B-39	12.4	55	235
A-26	11.7	70	250
B-24	11.6	70	250
Rudait Rashid	6.7	120	420
B-34a	6.3	125	425
Obs. well No. 1	5.8	135	430
A-10	5.6	135	435
B-2	5.1	140	470
B-22	5.0	140	470
B-26	4.2	150	520
B-5	4.2	150	520
A-1	3.2	170	560
El-oturiyan	2.3	180	600
A-5	1	210	700
A-19	1	210	700
A-6	1	210	700
B-29	1	210	700
A-18	1	210	700
B-37	1	210	700
		Mean 152	515

The average turnover rate for the phreatic aquifer is defined by the mean of the V/Q values given in Table III and is 152 a and 515 a for the limiting cases. During the later stages of the project, geophysical results indicated a total storage volume of 3.6×10^9 m^3 in the northern fresh-water lens where the potential recharge area is 2180 km^2 [5]. The limiting values of the average annual recharge are 7×10^6 m^3 and 24×10^6 m^3 for the maximum and minimum input conditions, respectively. The equivalent depths would therefore be 3.2 and 11.0 mm, respectively. The total annual storm rainfall over the northern area has varied from about 35 to 200 mm [5]. The average recharge is thus about 8% of the annual precipitation.

In view of the assumptions made in the above quantitative evaluation, the estimates of the average recharge are not intended as a precise answer for engineering decisions, but rather as an independent estimate of the approximate recharge to be expected in the area.

5. CONCLUSION

The environmental isotope investigations carried out in relation to the various hydrological and hydrogeological problems in Qatar have provided very useful qualitative and quantitative information on the dynamics of the groundwater systems. The information obtained on the mixing processes in the complex ground-water systems of the peninsula are of practical importance for the development of groundwater which is the only source of water in Qatar. The study illustrates the value of the conjunctive use of environmental isotopes and hydrochemical data.

ACKNOWLEDGEMENTS

The authors express their appreciation to the staff of the UNDP project in Qatar, particularly the project manager, Mr. J.G. Pike, who posed the problems and carried out most of the sampling. The environmental isotope analyses were carried out in the Isotope Hydrology Laboratory of the Agency.

REFERENCES

[1] Hydro-agricultural Resources Survey — Qatar, Report prepared for UNDP/FAO Project (1974).
[2] INTERNATIONAL ATOMIC ENERGY AGENCY, World Survey of Isotope Concentration in Precipitation, Environmental Isotope Data Nos 1, 2, 3, 4, 5, TRS 96, 117, 129, 147, 165, IAEA, Vienna (1953–1971).

[3] DINÇER, T. et al., "Study of groundwater recharge and movement in shallow and deep
    aquifers in Saudi Arabia with stable isotopes and salinity data", Isotope Techniques in
    Groundwater Hydrology 1974 (Proc. Symp. Vienna, 1974) 1, IAEA, Vienna (1974) 363.

[4] OTKUN, G. et al., "Recognition and dating of fossil waters in Saudi Arabia", Hydro-
    geological Problems in Arid Regions, Section 11-Hydrogeology, 25th Int. Geol. Congr.,
    Sydney, Australia, 1976.

[5] PIKE, J.G., Personal communication (1977).

DISCUSSION

W.J. SHAMPINE: Can you give us details about the elevation of the water table or piezometric surface in Qatar? This should be an important factor with regard to fresh-water lenses and salt-water encroachment.

Y. YURTSEVER: The fresh water in the upper aquifer is in fact a lense of groundwater, and the elevation of the water-table is about three to four metres above sea-level in the central part of its lense.

W.J. SHAMPINE: Experience in Saudi Arabia, supported by the interpretation of aerial photographs from Qatar, suggests that the inferred faults and minor folds referred to in the section of your paper entitled "Hydrogeology" and in Fig.2 are in fact solution collapse features. Have these features been considered and would not solution collapse features affect the isotope and ionic chemistry of the groundwater in a different way than a fault would?

Y. YURTSEVER: It is true that in the peninsula of Qatar, too, there are numerous collapsed structural features which might possibly affect the chemical composition of the water. However, the project geologists and hydrogeologists working in the area believed that there were two fault lines across central Qatar. In any case, the findings of the environmental isotope data have certainly provided enough evidence that the groundwaters in the north and south have their origin in two distinct systems with different dynamic characteristics, regardless of the existence of faults.

K.U. WEYER: Did you mention in your oral presentation that the ^{14}C data in the southern part of the Qatar peninsula are the same as in the northern part?

Y. YURTSEVER: I think there is a misunderstanding here. We compared the ^{14}C contents of the wells in southern Qatar with those of the deeper confined aquifers, merely to confirm the upward leakage from deep artesian aquifers into the overlying shallow aquifer in southern Qatar.

R.G. THOMAS: An estimate of recharge made by using standard hydrological methods for seven years is about 20 to 30 million m^3/a. The isotope method has therefore for all practical purposes matched the hydrological method. Work is continuing on the geology and hydrology as well as on isotopes and chemistry, and new information should be available by 1979.

CONTRIBUTION DES ISOTOPES DU MILIEU A L'ETUDE DE GRANDS AQUIFERES DU MAROC

A. KABBAJ, I. ZERYOUHI, Ph. CARLIER
Division des ressources en eau,
Direction de l'hydraulique,
Rabat, Maroc

A. MARCE
Bureau de recherches géologiques et minières,
Orléans, France

Abstract–Résumé

THE CONTRIBUTION OF ENVIRONMENTAL ISOTOPES TO STUDIES OF LARGE AQUIFERS IN MOROCCO.

The geochemistry of environmental isotopes has been used for the study of various aquifers in Morocco, some of which are large, such as the Charf el Akab in the Tangiers area, the Oum er Rbia basin and the Turonian aquifer of the Tadla, the free groundwater of the Quaternary lacustrine limestones of the Saïs Plain and the Lias limestone aquifer. These isotope studies take hydrogeochemical data into account and have made it possible to determine the conditions of recharge of the aquifers, to distinguish waters of different origin from the Atlas Mountains or from the Phosphate Plateau in the Tadla Basin and the Saïs plain, to estimate the recharge of one aquifer by another — for example groundwater of the Lias limestones passing via the folds of the Saïs Plain into the lacustrine limestone aquifer — and to test the homogeneity or heterogeneity of these aquifers and their tightness (e.g. the Turonian aquifer of the Tadla and the special case of the Charf el Akab in relation to the marine environment). Altogether, these results made it possible to test the value of the techniques used and to specify the general conditions in which they can profitably be used.

CONTRIBUTION DES ISOTOPES DU MILIEU A L'ETUDE DE GRANDS AQUIFERES DU MAROC.

La géochimie des isotopes du milieu a été appliquée sur différents aquifères du Maroc, dont certains de grandes dimensions: le Charf el Akab de la région de Tanger, le bassin de l'Oum er Rbia et l'aquifère turonien du Tadla, la nappe libre des calcaires lacustres quaternaires de la plaine du Saïs et l'aquifère des calcaires du Lias. Les études isotopiques tenant compte des données hydrogéochimiques ont permis de préciser les conditions d'alimentation de ces aquifères, de distinguer les types d'eau d'origine différente en provenance de l'Atlas ou du plateau des phosphates dans le bassin du Tadla ou de la plaine du Saïs, d'évaluer les apports d'un aquifère à un autre — nappe des calcaires du Lias alimentant par les flexures de la plaine du Saïs la nappe des calcaires lacustres —, de contrôler l'homogénéité ou l'hétérogénéité de ces aquifères ainsi que leur étanchéité — aquifère turonien du Tadla et cas particulier du Charf el Akab vis-à-vis du domaine marin. L'ensemble des résultats a permis d'éprouver l'utilité de ces techniques et de préciser les conditions générales de leur emploi.

491

INTRODUCTION

La Division des Ressources en Eau de la Direction de l'Hydraulique du Royaume du MAROC a fait appel aux techniques isotopiques pour apporter des éléments souvent déterminants à la solution de problèmes hydrogéologiques de natures très différentes (figure 1).

- Le bassin miopliocène du CHARF EL AKAB de la région de TANGER qui renferme une nappe d'eau souterraine utilisée pour l'alimentation en eau potable. Le renouvellement des eaux de cet aquifère est complété par une recharge artificielle, au moyen de bassins d'infiltration.

- Les nappes phréatiques du bassin sédimentaire de l'OUM ER RBIA et l'aquifère turonien du synclinal profond du TADLA (BENI MELLAL, KASBA TADLA)

- La nappe libre des calcaires lacustres villafranchiens et l'aquifère des calcaires du Lias du synclinal profond du SAIS (FES).

LE BASSIN DU CHARF EL AKAB - TANGER

Le bassin du CHARF EL AKAB est situé à 20 km environ au sud-ouest de la ville de TANGER à proximité du littoral de l'Océan Atlantique.

S'étendant sur une superficie d'environ 22 km^2, le bassin miopliocène du CHARF EL AKAB renferme une nappe d'eau souterraine utilisée depuis longtemps pour l'alimentation en eau potable de TANGER. A l'origine, cette alimentation était effectuée par le débit naturel des sources en bordure est du bassin, dont le débit moyen était de 2700 m^3/j, mais très rapidement il s'est avéré nécessaire de réaliser des forages d'exploitation par pompage.

Le bassin du CHARF EL AKAB est un bassin fermé, sans réseau hydrographique, ni débouché superficiel vers l'océan. Déprimé dans sa partie centrale, les eaux de ruissellement se rassemblent en période humide dans les lacs temporaires que sont les DAYA SEGUERA et DAYDAT (figure 2).

Ce bassin, entouré par une ligne de crêtes constituées de bancs gréseux de l'Unité numidienne (Eocène-Oligocène), occupe une dépression creusée dans les assises marneuses gris-bleues de l'unité de TANGER (Paléocène-Eocène).

Les formations détritiques constituant le bassin du CHARF EL AKAB se sont déposées entre le Tortonien supérieur et le Pliocène, au cours d'un cycle sédimentaire complet. Cependant, de fréquents accidents apparaissent dans la sédimentation créant des discontinuités dans les formations aquifères. L'étude lithostratigraphique des affleurements et des coupes de sondages a permis de distinguer trois aquifères (figure 3).

Les besoins en eau potable de la ville de TANGER (évalués à 12 millions de m^3/an en 1977) sont assurés en période humide (novembre à avril) par les eaux de surface après traitement (Oueds MHARHAR et BOUGADOU), en période sèche (mai à octobre) par l'exploitation des eaux souterraines.

Dans les conditions naturelles, avant 1948, le niveau piézométrique moyen était de + 15 m NGM.[1] Après la mise en exploitation de la nappe par pompage, le niveau piézométrique s'était abaissé à - 2 m NGM en 1954, soit un rabattement de 17 m en 7 ans, date à laquelle la recharge artificielle de l'aquifère avec l'excédent des eaux de surface par bassins d'infiltration a été décidée.

Dans un premier temps, la recharge artificielle de la nappe fut réalisée par des bassins d'infiltration à l'aide de l'eau des dayas de 1955 à 1957, puis à partir des eaux de l'Oued MHARHAR, sur lequel fut construit un barrage en 1958 permettant des débits de recharge plus importants, de l'ordre de 2500 m^3/jour (novembre à avril). Cette intervention permit de rétablir le niveau piézométrique naturel à + 15 m NGM en 1961.

L'accroissement de la consommation en eau potable de la ville de TANGER a entraîné l'utilisation de la totalité des eaux de la station de traitement (capacité 20 à 22 000 m^3/jour), supprimant toutes possibilités de recharge depuis 1972. Simultanément, l'augmentation des prélèvements par pompage (5 millions de m^3/an) est devenue nécessaire.

Ces conditions d'exploitation de l'aquifère ont provoqué un abaissement important du niveau piézométrique qui atteignait la cote - 2 m NGM en 1971, - 10 m NGM en 1973 et - 29 m NGM en 1977, faisant craindre une invasion des eaux salées marines.

Une étude géologique réalisée en 1974 a montré l'étanchéité de ce réservoir par rapport à l'océan, en raison du biseau existant entre les formations aquifères et le substratum, ainsi que la présence des "marnes bleues" le long du littoral. Cependant, un doute subsiste en raison de la présence d'eaux salées dans la partie nord du CHARF EL AKAB.

1. Précipitations

Les données climatologiques de la région de TANGER ont fait l'objet de relevés sur une période de 30 ans de 1933 à 1963.

Du point de vue des caractéristiques isotopiques, il n'existe pas de données pour la zone de TANGER. Cependant, l'Agence Internationale de l'Energie Atomique de Vienne a effectué des mesures sur les précipitations relevées à la station météorologique de GIBRALTAR. Ces données ne sont complètes que sur une période de 4 ans de 1962 à 1965 (1) : $\delta D \%_o$ $_{smow}$ = - 25,0 $\delta {}^{18}O \%_o$ $_{smow}$ = - 4,2.

2. Forages étudiés

Le choix des forages a été réalisé en fonction de la profondeur d'exploitation de l'aquifère et de leur distance à l'affleurement et plus particulièrement aux bassins d'infiltration. Leur situation est indiquée sur la carte géologique (figure 2) et leur position est reportée sur la coupe schématique nord-sud passant par le centre du bassin (figure 3).

---

[1] NGM: niveau général marin.

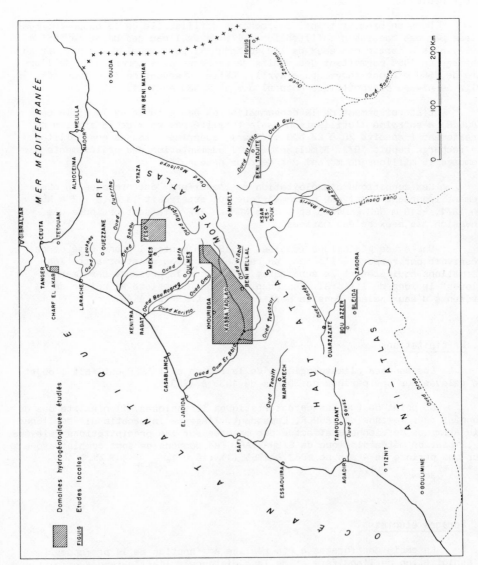

FIG.1. Plan de situation des études.

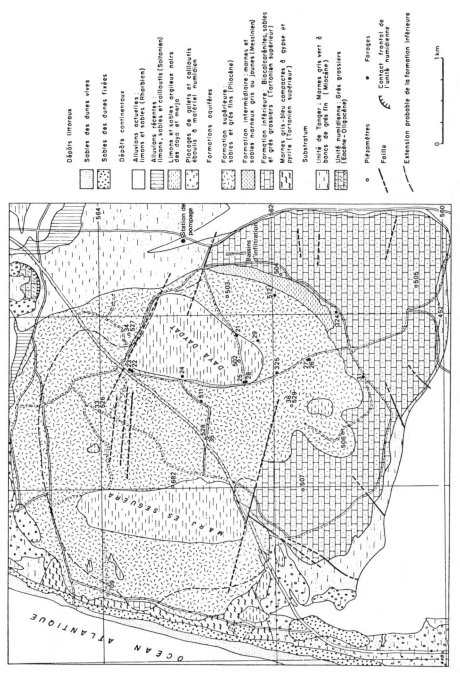

FIG.2.   Carte géologique du bassin aquifère du Charf el Akab (Tanger) (d'après R. Médioni, 1974).

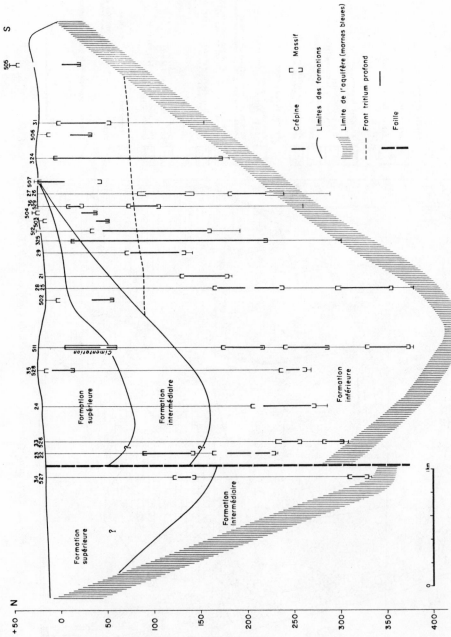

FIG.3. Charf el Akab: coupe schématique nord-sud.

Sur cette coupe, certains puits sont en projection, leur situation en latitude étant respectée approximativement. L'indication des cotes des crépines
ainsi que des massifs filtrants permet de déterminer le ou les aquifères
intéressés, suivant les limites des formations.

Les eaux sont caractérisées par deux faciès chimiques principaux
(tableau I)[2] :

- un faciès carbonaté calcique : général
- un faciès chloruré sodique   : forage 23/1.

L'étude isotopique montre une certaine uniformité de l'ensemble du
bassin (tableau II). La nappe principale profonde présente une composition
isotopique constante en hydrogène (D/H) et en oxygène ($^{18}O/^{16}O$), parfaitement
bien représentée par le puits 24/1, identique à celle des eaux de pluie
relevées à la station météorologique de GIBRALTAR, indiquant une homogénéité de l'alimentation de l'aquifère.

L'eau des puits et piézomètres profonds présente des teneurs en
tritium nulles, montrant l'absence de contribution des eaux récentes au
renouvellement de cet aquifère. Celle-ci est révélée par contre sur l'eau
des puits crépinés sur toute l'épaisseur de l'aquifère tels que 324/1,
325/1, 512/1 et 511/1 avant la cimentation de la partie supérieure de ce
forage en 1973.

L'observation de ces teneurs en tritium sur les différents puits
permet ainsi de tracer la limite du front des eaux récentes, postérieures
à 1950, ayant affecté l'aquifère dans la zone de recharge. Cette limite
n'atteint pas la cote - 100 m NGM (figure 3).

Les teneurs en tritium décroissent en fonction de l'éloignement
des bassins d'infiltration, puisqu'elles n'excèdent pas 10 UT sur les
piézomètres supérieurs 529/1, 505/1, 506/1, 507/1.

Les mesures de radiocarbone confirment l'influence des eaux
récentes sur certains forages. Elles permettent de calculer un "âge"
moyen d'environ 5000 ans pour les eaux anciennes (Ao = 85 % $\delta^{13}Co$ = - 17‰)
et d'estimer le volume total des réserves de l'ordre de 500 millions de m^3.

L'étude plus particulière du puits 23/1 dont l'eau est caractérisée par une salinité élevée a permis d'éliminer l'éventualité d'une invasion marine récente, sur la base de la composition isotopique $^{18}O/D$ et de
celle du soufre des sulfates dissous. Cette salinité pourrait être attri-
buée au lessivage d'une lentille évaporite à localisation indéterminée,
présente dans les formations supérieure ou intermédiaire.

L'étude du système aquifère du CHARF EL AKAB a permis de montrer
son homogénéité du point de vue chimique et isotopique. Cependant, une dif-
férence existe entre les trois formations et seule la formation inférieure
renferme une eau de bonne qualité. Les mesures de radioactivité naturelle
tritium et radiocarbone ont mis en évidence l'existence d'eaux récentes
contribuant à la recharge du système.

---

[2] Les tableaux sont groupés dans l'appendice.

LE BASSIN DE L'OUM ER RBIA - L'AQUIFERE TURONIEN DU TADLA

Le bassin sédimentaire de l'OUM ER RBIA, d'une superficie de 34 335 km^2, s'étend entre le Moyen-Atlas, au sud, le plateau des phosphates au nord et l'Océan Atlantique à l'ouest. Le fleuve, d'une longueur de 555 km, l'un des plus réguliers du Maroc, naît dans le Causse moyen atlasique au nord-est de KHENIFRA. Ses sources alimentent le débit de base du Haut OUM ER RBIA, soit environ 11 à 15 m^3/s.

Au débouché de l'ATLAS, l'OUM ER RBIA pénètre dans le bassin du TADLA où il est utilisé pour l'alimentation en eau potable, ainsi que pour l'irrigation (figure 4).

La plaine du TADLA correspond à une puissante série sédimentaire allant du Trias au Plio-Quaternaire, où l'on note la présence des calcaires du Turonien, formation très perméable et aquifère de grandes dimensions.

1. Les sources (tableaux III et V)

1.1. Les sources de l'OUM ER RBIA

La zone d'alimentation des sources de l'OUM ER RBIA est constituée par les calcaires karstiques du Lias du Causse Moyen Atlasique. Elle est représentée par le plateau du GUIGOU dont l'altitude moyenne est de l'ordre de 1800 à 2000 m, celle des sources étant de 1300 mètres environ.

La zone d'émergence des sources karstiques est caractérisée par un accident géologique, provoquant le chevauchement du Trias sur le Lias. Au droit de cet accident naissent des sources d'eau douce, alimentées directement par le Lias inférieur, présentant un débit relativement constant toujours supérieur à 11 m^3/s, avec des salinités n'excédant pas 700 mg/l. Mais un apport de 0,5 m^3/s d'eaux salées d'origine triasique modifie notablement la composition chimique des eaux des sources de l'OUM ER RBIA, qui acquièrent après mélange une salinité moyenne de l'ordre de 1200 à 1400 mg/l.

L'influence de ces sources salées est nette sur tout le cours du fleuve, particulièrement en période d'étiage, limitant l'utilisation de ces eaux pour l'irrigation et l'alimentation en eau potable.

Les compositions isotopiques ^{18}O et D sont en bonne concordance avec la droite des précipitations définie par CRAIG (2) et DANSGAARD (3) (figure 5).

La composition isotopique en ^{18}O de la source principale douce 147/30, située à 1300 m d'altitude, confirme que l'altitude moyenne de l'aire d'alimentation est de l'ordre de 1800 mètres (figure 6). Cette valeur est minimale, puisqu'il faut tenir compte de l'apport local des précipitations.

Les teneurs isotopiques des sources salées, plus enrichies, indiquent que l'aire d'alimentation est moins élevée, comprise entre 1300 et 1500 mètres d'altitude (figure 6).

Les teneurs en tritium pour ces trois sources sont relativement basses, compte tenu des valeurs mesurées, en 1972, dans les eaux de surface de cette région. La teneur de 23 U.T. de la source principale 147/30

indique la présence d'un mélange d'eaux récentes superficielles (60 %) et
d'eaux anciennes à circulation lente (teneur en tritium nulle).

Etant donné que l'on se trouve en présence d'un mélange, il est
impossible de calculer un "âge moyen" pour l'eau des sources. Cependant,
la teneur en ^{14}C, nettement inférieure aux teneurs actuelles (120 %) montre l'importance des réserves, nécessaires pour alimenter des débits de
l'ordre de 350 millions de m^3/an dont 210 millions proviennent, en consi-
dérant les teneurs en tritium, d'apports récents.

1.2. L'AIN ASSERDOUN (210/37)

Cette source karstique émerge à une altitude de 580 m au-dessus
de BENI MELLAL. L'eau de cette source est peu minéralisée, avec un faciès
bicarbonaté calcique. Sa composition isotopique en oxygène (δ ^{18}O ‰ =
- 6,8)indique que l'altitude de son aire d'alimentation, de l'ordre de
1400 m (figure 6), correspond au DJEBEL TASSEMIT. La teneur en tritium de
20 U.T. permet une évaluation approximative de la proportion d'eaux récentes de l'ordre de 50 %.

1.3. Les sources de la Bahira orientale

Le groupe des sources AÏN IGLI (18/36), AÏN BISSI BISSA (9/36) et
AÏN MESSALA (379/36) est localisé en aval de DAR OULD ZIDOUH, en rive gau-
che de l'OUM ER RBIA entre les Oueds EL ABID (522/45) et TESSAOUT (831/45)
d'origine atlasique. La composition chimique de ces eaux, au faciès mixte
chloruré sodique et bicarbonaté calcique, est assez proche de celles de
ces oueds(très variables dans le temps). Cependant, l'augmentation relative des chlorures et des sulfates est attribuable au lessivage des terrains
triasiques constituant le piedmont de l'Atlas.

Ces sources ont des teneurs en tritium élevées, AÏN IGLI (18/36)
avec 32 U.T. pour la plus importante. Ces mesures montrent que ces sources
correspondent à un exutoire de la nappe phréatique alimentée par épandages
d'irrigation à partir des Oueds EL ABID et TESSAOUT. Les compositions isotopiques en oxygène confirment cette hypothèse, puisqu'elles sont analogues à celles du domaine atlasique, avec une altitude de l'aire d'alimentation de l'ordre de 1400 à 1700 mètres (figure 6).

L'ensemble de ces eaux fait partie du domaine atlasique. Leur
composition isotopique oxygène - deutérium, caractérisant une altitude
de l'aire d'alimentation supérieure à 1000 mètres, permettra une diffé-
renciation dans la plaine du TADLA, par rapport aux eaux provenant du
plateau des phosphates (800 m).

2. L'aquifère des calcaires turoniens (tableaux IV et V)

Le substratum du synclinal du TADLA est constitué de formations
paléozoïques sur lesquelles repose, en discordance, un ensemble épais
de couches sédimentaires allant du Trias au Plio - quaternaire.

De nombreux forages réalisés dans la plaine du TADLA ont atteint
l'aquifère des calcaires blancs du Turonien.

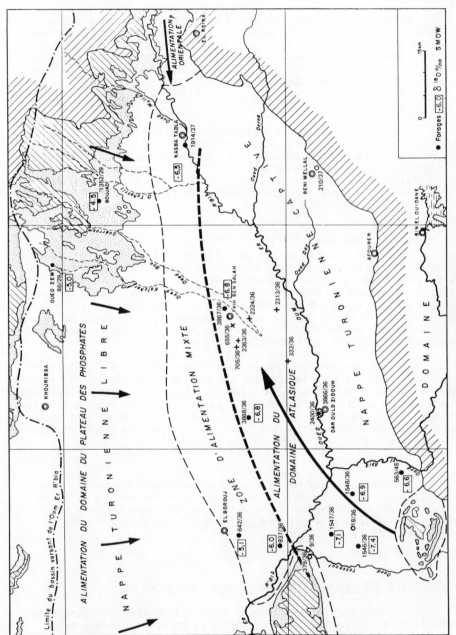

FIG.4. Bassin du Tadla: schéma d'alimentation de la nappe turonienne.

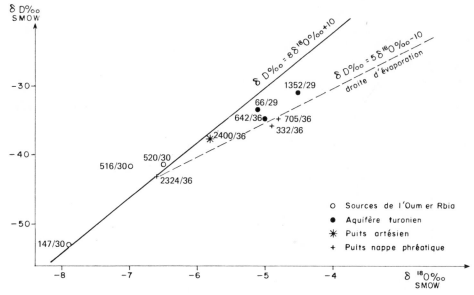

FIG.5.   *Relation oxygène-deutérium (bassin du Tadla).*

        Les affleurements au nord, plateau des phosphates, région de
KHOURIBGA, OUED ZEM et BOUJAD, et à l'ouest d'EL BOROUJ de part de d'autre
de l'OUM ER RBIA couvrent une superficie d'environ 1100 km², permettant
une recharge naturelle importante de cette nappe.

        Sous la plaine du TADLA, le Turonien s'enfonce très rapidement
formant un synclinal profond, chevauché, d'est en ouest par l'Atlas dans
sa partie orientale.

        En bordure du plateau des phosphates, le Turonien est recouvert
par des séries peu épaisses : la nappe turonienne est libre du fait de la
faible épaisseur des terrains de couverture. Mais elle devient captive au
sud d'une ligne est-ouest passant au nord de KASBA - TADLA (profondeur du
Turonien 189 m à 221 m), de FKIH BEN SALAH (profondeur du Turonien 358 m à
400 m), d'EL BOROUJ (profondeur du Turonien 76 m à 110 m). Plus au sud, vers
le coeur du synclinal, le Turonien a été reconnu à l'aplomb de DAR OULD
ZIDOUH à une profondeur de 823 mètres.

        La puissance des calcaires turoniens, de l'ordre de 25 à 30 mètres
au nord et à l'ouest, augmente jusqu'à 60 mètres au coeur du synclinal.

        Les eaux de la nappe turonienne libre sont caractérisées par un
faciès bicarbonaté calcique et magnésien. Vers le sud, la nappe devient
captive et son hydrochimie apparaît plus complexe.

        Deux puits ont fait l'objet de prélèvements dans la région de OUED
ZEM (86/29) et ·BOUJAD (1 352/29). Les eaux de ce puits ont des teneurs en
tritium élevées, montrant l'importance de la recharge annuelle dans cette
zone d'affleurement.

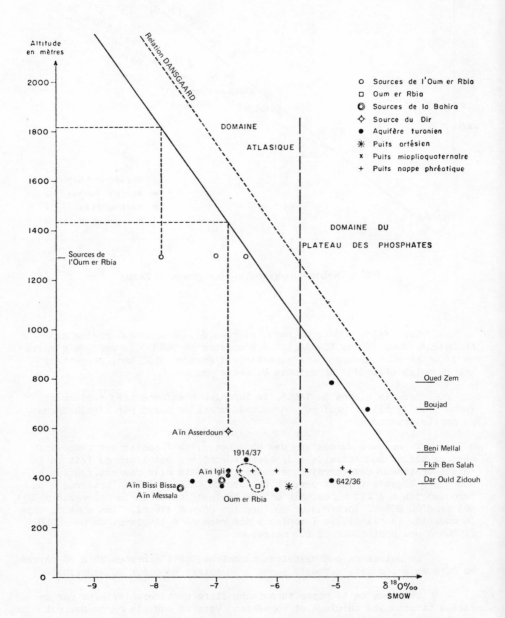

FIG.6. Relation composition isotopique en oxygène et altitude (bassin du Tadla).

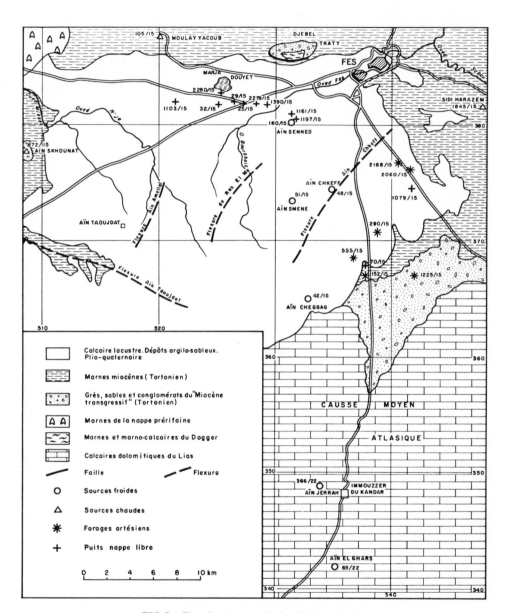

FIG.7. Plan de situation de la plaine du Saïs.

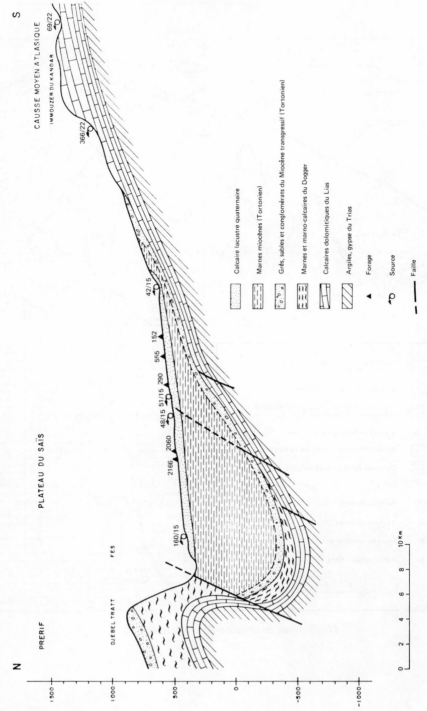

FIG. 8. Coupe schématique du synclinal du Saïs.

Les caractéristiques du puits de OUED ZEM 86/29 à l'altitude de
780 m confirment que dans la zone d'affleurement les eaux réalimentant
l'aquifère turonien ont une composition isotopique en oxygène voisine de
- 5 ‰ (figures 5 et 6).

On retrouve une certaine analogie entre les compositions isotopiques des eaux du forage d'EL BOROUJ et celles des puits creusés dans les
affleurements sur le plateau des phosphates. La teneur en ^{14}C (45,5 %),
compte tenu de la faible teneur en tritium permet d'estimer un "âge moyen"
de l'ordre de 4000 ans (A_O = 85 % $\delta^{13}C_O$ = - 17 ‰).

Cette valeur, qui n'est qu'indicative, montre un temps de transfert
relativement long depuis la zone d'affleurement jusqu'au puits exploité
soit une distance de 55 km environ. Sur cette base, la vitesse de transfert
dans la nappe serait de l'ordre de 14 m/an.

L'ensemble de cette étude a permis de distinguer dans le bassin
versant du TADLA, deux types d'eau d'origine différente (figure 6) :

    - les eaux du domaine atlasique dont les compositions isoto-
piques sont appauvries en isotopes lourds, en fonction de l'altitude de
l'impluvium d'alimentation

    - les eaux du domaine du plateau des phosphates et de la plaine du
TADLA dont les compositions isotopiques sont enrichies en isotopes lourds
par comparaison aux précédentes.

Ces caractéristiques isotopiques permettent de différencier les
eaux dans la plaine du TADLA, et d'affirmer l'absence de drainance du Turonien vers la nappe phréatique dans les zones étudiées.

L'aquifère turonien présente des faciès chimiques des eaux très
différents en relation avec des variations lithologiques des formations
passant progressivement du nord ouest au sud est d'un domaine marin ouvert
à un domaine évaporitique confiné : les méthodes isotopiques montrent qu'à
l'alimentation de la nappe turonienne au nord sur le plateau des phosphates
se superposent d'autres possibilités importantes de recharge, à partir du
domaine atlasique, dont l'influence semble plus étendue que prévue.

LA PLAINE DU SAIS - LA NAPPE LIBRE DES CALCAIRES LACUSTRES QUATERNAIRES
L'AQUIFERE DES CALCAIRES DU LIAS

Le bassin de MEKNES-FES est bien individualisé entre le Causse
moyen atlasique calcaire au sud et à l'est, et le domaine des rides prérifaines au nord. Cette région se subdivise en deux unités : le plateau de
MEKNES à l'ouest d'une superficie de 1 200 km^2 et la plaine du SAIS à
l'est d'une superficie de 800 km^2. Cette dernière est seule l'objet de
cette étude (figure 7).

Le bassin hydrogéologique de la plaine du SAIS renferme deux formations aquifères importantes, séparées par une formation semi-perméable
de marnes miocènes, s'épaississant du sud vers le nord (figure 8) :

    - les calcaires dolomitiques du Lias, en profondeur, prolongement
du Causse vers le nord et contenant une importante nappe captive artésienne

- les calcaires lacustres plioquaternaires, au sommet, avec une nappe libre; cette formation est affectée de flexures superposées à des failles de chevauchement dans les calcaires du Lias.

A l'emplacement des flexures, des sources importantes jaillissant dans les calcaires lacustres sont considérées comme des exutoires de la nappe des calcaires du Lias.

Par ailleurs, deux sources chaudes ont été aménagées en stations thermales :
        - la source de SIDI HARAZEM, à 12 km à l'est de FES, dont les eaux sont peu minéralisées et sulfureuses
        - la source de MOULAY YACOUB à une vingtaine de km à l'ouest de FES, dont les eaux sont fortement minéralisées et sulfureuses.

La plaine du SAIS présente une altitude décroissante du Sud aux environs de 900 mètres depuis le Causse moyen atlasique, au nord dans la région de FES, aux environs de 400 mètres.

Le bassin est constitué par un synclinal dissymétrique s'enfonçant progressivement à partir du Causse du sud vers le nord et se redressant brutalement par failles pour former les anticlinaux des rides prérifaines, DJEBELS TRATT et ZALAGH, à l'altitude de 900 mètres environ.

La plaine du SAIS est drainée par des oueds dont le sens d'écoulement est sud-nord, ayant pour origine les sources du Causse et les émergences de la nappe libre, par l'oued FES, affluent du SEBOU à écoulement ouest-est et par l'oued N'JA dont le sens d'écoulement est est-ouest.

A 10 km à l'ouest de FES, la daya de DOUYET atteint une superficie de 1 km^2 en hiver. Cet étang correspond à une cuvette synclinale subsidente comblée par des dépôts limoneux imperméables quaternaires.

Le bilan hydraulique de la plaine du SAIS montre que la somme des émissions d'eau de la nappe libre dépasse largement la recharge due à la pluie et à l'irrigation. Le complément de l'alimentation de cette nappe est réalisé par les eaux des calcaires du Lias, à la faveur des flexures affectant les formations.

Les caractéristiques physico-chimiques des eaux souterraines et superficielles circulant dans la plaine du SAIS ont été déterminées à partir d'analyses complètes (tableau VI) sur différents prélèvements intéressant :
        - la nappe libre des calcaires lacustres
        - l'aquifère profond des calcaires du Lias
        - les sources thermales dans la partie nord de la plaine.

Toutes ces eaux, exceptées les eaux thermales de la zone nord, sont faiblement minéralisées, le résidu sec étant compris entre 300 et 700 mg/l. Leur faciès est généralement du type bicarbonaté magnésien et calcique ; le rapport magnésium/calcium égal ou supérieur à l'unité indique que les assises dolomitiques du Lias jouent un rôle important dans l'acquisition de la minéralisation des eaux de ce bassin hydrogéologique.

Les sources thermales sont situées dans la partie aval de la plaine du SAIS en bordure de la zone de charriage du RIF où des extrusions de Trias salifères sont connues. Les sources de MOULAY YACOUB (105/15) et AIN SKHOUNAT (872/15) acquièrent une minéralisation élevée dont l'origine triasique l'emporte sur celle du Lias, avec un faciès chloruré sodique

bicarbonaté magnésien et calcique. Par contre, la source de SIDI HARAZEM
(1 845/15) conserve le faciès bicarbonaté magnésien et calcique caracté-
ristique des calcaires dolomitiques du Lias.

1. Les forages profonds des calcaires dolomitiques du Lias (tableau VII)

Les compositions isotopiques de l'oxygène et du deutérium montrent
un groupement de valeurs pour les eaux des forages 290/15, 555/15, 2 060/15,
2 168/15 (figure 9).

La valeur de la composition isotopique en oxygène indique que l'altitude moyenne de l'aire d'alimentation des calcaires du Lias est de l'ordre de 1400 mètres (figure 10). Etant donné que la limite inférieure d'affleurement de ces calcaires est de l'ordre de 900 mètres (figure 8), la
limite supérieure peut être estimée entre 1900 et 2000 mètres, ce qui est
concordant avec l'altitude maximale du Causse.

Les teneurs en radiocarbone indiquent que l'on se trouve en présence d'eaux récentes, mais non affectées par les explosions thermonuclé-
aires, avec des teneurs inférieures à 85 %.

2. Les sources de la plaine du SAIS (tableau VII)

Les compositions isotopiques oxygène et deutérium montrent une
analogie entre les eaux des sources et celles des forages profonds des
calcaires du Lias (figure 9). Les teneurs en tritium indiquent que l'on
se trouve en présence d'eaux récentes.

Les teneurs en radiocarbone des eaux de ces sources confirment
bien le caractère actuel de ces eaux.

3. Les sources chaudes

La source de SIDI HARAZEM située à une douzaine de kilomètres à
l'est de FES à une altitude de 262 mètres est en fait un captage artésien,
dont les eaux tièdes (30 à 35°C) ont une minéralisation légèrement supérieure (800 mg/l) à celle des eaux des calcaires dolomitiques du Lias.

Les compositions isotopiques de l'oxygène et du deutérium montrent
l'analogie existant entre cette eau et celle des forages profonds étudiés
dans la plaine du SAIS   (figures 9 et 10).

L'âge moyen de cette eau peut être estimé de l'ordre de 4000 ans
(A_O = 85 % $\delta^{13} C_O$ = - 17 %o).

La source de MOULAY YACOUB située à une vingtaine de kilomètres à
l'ouest de FES, à une altitude de 283 mètres, jaillit dans les formations
marneuses tortoniennes et les marnes gypsifères des nappes prérifaines.
Elle est caractérisée par des eaux sulfureuses chaudes (54°C) très minéralisées.

Les compositions isotopiques de l'oxygène et du deutérium diffé-
rencient cette eau de celles des forages profonds des calcaires dolomitiques du Lias, et de la source de SIDI HARAZEM (figure 9).

La composition isotopique en oxygène indique que l'altitude moyenne de l'aire d'alimentation de cette source est de l'ordre de 800 à 900

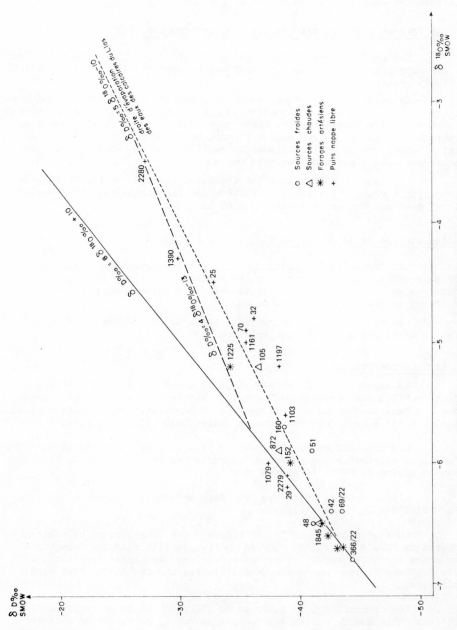

FIG. 9. *Relation oxygène-deutérium (bassin de la plaine du Saïs).*

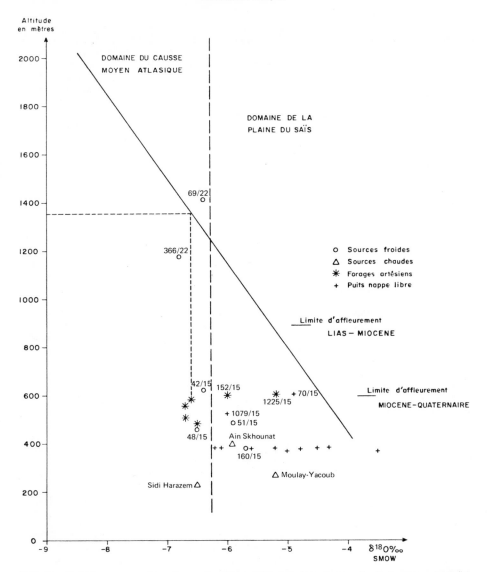

FIG.10. Relation composition isotopique en oxygène et altitude (bassin de la plaine du Saïs).

mètres (figure 10). Les données actuelles ne permettent pas de déterminer l'origine exacte des eaux de cette source, soit en bordure du domaine atlasique, soit dans le domaine rifain. L'âge moyen de cette eau serait de l'ordre de 19 000 ans (A_0 = 85 %)

Les analyses isotopiques ont permis de caractériser les sources de flexure de la plaine du SAIS, dont l'alimentation est réalisée en quasi-totalité par les eaux profondes des calcaires dolomitiques du Lias.

510   KABBAJ et al.

L'ensemble de l'étude de la plaine du SAIS a confirmé l'importance de la recharge de la nappe libre des calcaires lacustres plioquaternaires par les eaux profondes (70 %) soit directement par les failles, soit indirectement par l'irrigation, ce que l'on supposait par des critères hydrogéologiques, mais que l'on démontre par les isotopes. A l'aide des techniques isotopiques, l'altitude moyenne de l'aire d'alimentation des calcaires du Lias a.été calculée aux environs de 1400 mètres, valeur médiane de l'altitude des bassins versants du Causse Moyen atlasique.

CONCLUSIONS

Les études effectuées au Maroc illustrent bien les types de contribution apportée par les techniques isotopiques à la connaissance des systèmes aquifères.

La géochimie isotopique permet le contrôle de l'homogénéité ou de l'hétérogénéité des réservoirs. Ainsi, la nappe profonde des calcaires du Lias dans la plaine du Saïs est homogène, tandis que celle des calcaires du Turonien dans le bassin du Tadla montre une certaine hétérogénéité, déjà mise en évidence par l'hydrogéochimie.

Ces méthodes ont déterminé avec précision les conditions d'alimentation des aquifères, notamment leur degré d'actualité et leur province d'origine, données primordiales pour définir correctement les conditions aux limites. Ces informations sont d'autant plus utiles que les mesures piézométriques sont souvent insuffisantes dans le cas de certains grands aquifères sédimentaires. Ainsi il a été clairement démontré qu'à l'alimentation de la nappe turonienne au nord sur le plateau des phosphates se superposent d'autres possibilités importantes de recharge à partir du domaine atlasique, dont l'influence semble plus étendue que prévu.

Des indications ont pu être obtenues sur l'existence de drainance dans les systèmes aquifères. Ainsi dans le bassin du Tadla a été prouvée l'absence de transferts d'eau par drainance des calcaires turoniens avec les nappes phréatiques des Beni Amir et des Beni Moussa. Dans la plaine du Saïs les isotopes du milieu ont permis de quantifier les apports par les failles de la nappe profonde des calcaires du Lias à la nappe libre des calcaires lacustres quaternaires.

Enfin, dans le cas du Charf El Akab, une étude géologique minutieuse et un choix judicieux des mesures isotopiques ont montré l'étanchéité du système aquifère par rapport aux influences marines et ont permis d'évaluer le volume total des réserves.

REFERENCES

[1]   INTERNATIONAL ATOMIC ENERGY AGENCY, Environmental Isotope Data Nos 1, 2, 3, 4, 5, Technical Reports Series Nos 96, 117, 129, 147, 165, IAEA, Vienna (1969, 1970, 1971, 1973, 1975).
[2]   CRAIG, H., Isotopic variations in meteoric waters, Science 133 (1961) 1702.
[3]   DANSGAARD, W., Stable isotopes in precipitation, Tellus 16 4 (1964) 436—68.

Appendice
TABLEAUX I A VII

TABLEAU I. ANALYSES CHIMIQUES DU CHARF EL AKAB

REFERENCE N° IRE		Ca++	Mg++	Na+	K+	Cl-	SO4--	HCO3-	R.S. mg/l	C20°C µmhos	T°C	pH	CO2 libre mg/l	dH d°F	TAC d°F
		CATIONS				ANIONS									
24/1 29.08.73	mg/l	79,2	1,2	14,0	0,6	26,6	3,5	220	292	416	20°4	7,7	15,0	20,3	18,1
	meq	3,96	0,10	0,61	0,01	0,75	0,07	3,62							
31/1 30.08.73	mg/l	65,6	0,97	12,0	0,2	21,3	3,5	174	269	336	21°1	7,7	8,0	16,8	14,3
	meq	3,28	0,08	0,52	0,005	0,60	0,07	2,86							
324/1 30.08.73	mg/l	76,3	5,3	16,0	0,5	24,8	4,5	228	299	420	21°2	7,6	12,0	20,6	18,7
	meq	3,68	0,44	0,69	0,01	0,69	0,09	3,75							
325/1 30.08.73	mg/l	88,0	2,4	29,0	0,5	44,4	5,5	248	356	509	20°5	7,4	22,0	23,0	20,3
	meq	4,40	0,20	1,26	0,01	1,25	0,11	4,07							
511/1 29.08.73	mg/l	100,0	21,2	39,0	0,95	60,3	11,5	359	477	688	20°6	7,6	40,0	32,2	29,4
	meq	5,04	1,75	1,69	0,02	1,69	0,24	5,89							
512/1 30.08.73	mg/l	75,4	1,4	15,0	0,79	35,5	5,5	187	342	407	21°2	7,7	8,5	19,4	15,8
	meq	3,76	0,12	0,65	0,02	1,00	0,11	3,07							
22/1 29.08.73	mg/l	84,0	9,7	35,0	1,67	49,7	24,5	302	404	591	20°8	7,5	20,0	25,2	24,8
	meq	4,20	0,80	1,52	0,04	1,40	0,51	4,95							
23/1 30.08.73	mg/l	111,0	102,0	1840	11,1	2545	260,0	844	5456	8003	20°2	7,3	75,0	70,2	69,2
	meq	5,56	8,40	80,00	0,28	71,69	5,41	13,84							
682/1 20.03.74	mg/l	60,0	37	34,5	1,56	64,0	48,0	320	/	629	/	8,2	/	30,0	/
	meq	3,0	3,0	1,50	0,04	1,8	1,0	5,2							

KABBAJ et al.

TABLEAU II. ANALYSES ISOTOPIQUES

REFERENCE N° IRE	T°C	pH	$C_{20°C}$ µmhos	Cl^- mg/l	CO_3H^- mg/l	CO_2 libre mg/l	dH d°F
24/1 (11.05.72)	21°	/	/	41,2	244	25,5	23
(29.08.73)	20°4	7,7	420	26,6	244	15,0	20
31/1 (11.05.72)	21°	/	/	31,3	298	21,7	27
(30.08.73)	21°1	7,7	336	21,3	189	8,0	17
324/1 (13.05.72)	/	/	/	24,7	302	21,7	24
(30.08.73)	21°2	7,6	396	24,8	238	12,0	21
325/1 (16.05.72)	/	/	/	60,0	372	43,1	29
(30.08.73)	20°5	7,4	500	44,4	262	22,0	23
511/1 (12.05.72)	/	/	/	82,0	404	78,3	15
(29.08.73)	20°6	7,6	678	60,3	378	40,0	32
512/1 (17.05.72)	/	/	/	40,0	207	12,3	20
(30.08.73)	21°2	7,7	393	35,5	187	8,5	19
22/1 (16.05.72)	/	/	/	62,5	353	25,5	12
(30.08.73)	20°8	7,5	603	49,7	372	20,0	25
(05.06.74)	/	/	/	/	/	/	/
23/1 (13.05.72)	20°	/	/	2500	893	69,5	/
(30.08.73)	20°2	7,3	8309	2545	878	75,0	70
(08.10.74)	/	/	7680	/	/	/	/
EAU D'INJECTION (15.05.72) OUED MAHRHAR	/	/	/	60	210	7,9	24
502/1 (30.08.73) (19.06.74)	22°	/	4422	1331	198	/	/

DU CHARF EL AKAB

TAC d°F	³H U.T	ANALYSES ISOTOPIQUES				
		D δ‰ SMOW	¹⁸O δ‰ SMOW	¹⁴C % NBS	¹³C δ‰ PDB	³⁴S δ‰ C.D.
20	≤ 1	-24,1±0,5	-4,2±0,1	46,1±0,9	-17,1±0,5	
20	≤ 1	/	-4,2±0,1			
25	3 ± 1	-23,1±0,5	-4,0±0,1	17,2±0,6	-11,4±0,5	
15	29 ± 3	/	-4,1±0,1			
25	2 ± 1	-23,5±0,5	-4,1±0,1	10,2±0,5	-11,2±0,5	
19	9 ± 2	/	-4,4±0,1			
33	15 ± 3	-23,3±0,5	-3,9±0,1	60,3±1,0	-18,6±0,5	
21	≤ 1	/	-4,3±0,1			
33	7 ± 2	-23,9±0,5	-4,3±0,1	83,7±1,2	-21,2±0,5	
31	≤ 1	/	-4,3±0,1			
18	9 ± 2	-22,9±0,5	-4,2±0,1	65,8±1,2	-15,8±0,5	
16	16 ± 3	/	-4,1±0,1			
30	3 ± 1	-23,5±0,5	-4,2±0,1	19,1±0,6	/	
30	3 ± 1	/	-4,3±0,1			+ 8,8±0,2
/	3 ± 1	/	-4,3±0,1			
85	7 ± 2	-13,3±0,5	-2,1±0,1	54,3±0,9	-16,7±0,5	
72	1 ± 1	-12,5±0,5	-2,3±0,1			+15,3±0,2
/	≤ 1	-14,9	-2,7±0,1			+14,3±0,2
16	41 ± 4	-17,6	-2,9±0,1	122,9±1,6	-13,3±0,5	
16	12 ± 2	-16,7	-2,8±0,1			+ 8,1±0,2
	16 ± 2		-2,8±0,1			

KABBAJ et al.

TABLEAU III. ANALYSES CHIMIQUES DES EAUX

REFERENCE N° IRE			Ca++	Mg++	Na+	K+
SOURCES	147/30 04.12.70	mg/l meq	84,0 4,2	32,8 2,7	97,0 4,2	1,6 0,04
DE	516/30 05.08.70	mg/l meq	238,0 11,9	48,6 4,0	6831,0 297,0	7,8 0,2
L'OUM ER RBIA	520/30	mg/l meq	170,0 8,5	66,9 5,5	3669,0 159,5	2,4 0,03
OUM ER RBIA DAR OULD ZIDOUH	3866/36 10.09.73	mg/l meq	73,6 3,7	47,7 3,92	160,0 6,95	3,15 0,08
AIN ASSERDOUN	210/37 03.07.73	mg/l meq	68,0 3,4	17,9 1,5	18,0 0,74	0,7 0,02
DAR OULD ZIDOUH	2400/36 10.09.73	mg/l meq	76,0 3,8	45,7 3,76	26,0 1,13	1,12 0,03
	705/36 25.02.69	mg/l meq	140,0 7,0	36,5 3,0	33,0 1,43	0,95 0,03
FKIH BEN SALAH	2363/36 18.08.71	mg/l meq	146,0 7,3	26,7 2,2	32,5 1,41	1,40 0,04
EL KHIAS	2324/36 17.08.71	mg/l meq	120,0 6,0	109,4 9,0	352,5 15,32	3,0 0,08
PUITS A 43	2313/36 23.08.73	mg/l meq	690,0 34,5	1283,0 105,5	3224,0 140,2	62,4 1,6
BIR GHEJJAT	332/36 20.08.71	mg/l meq	260,0 13,0	248,1 20,4	165 7,2	2,65 0,07
FKIH BEN SALAH FORAGE OCP	655/36 28.04.69	mg/l meq	138,0 6,9	79,0 6,5	66,5 2,9	1,45 0,04
AIN IGLI	18/36 11.05.64	mg/l meq	122,0 6,1	62,0 5,1	115,0 5,0	3,0 0,06
AIN BISSI BISSA	9/36 14.09.72	mg/l meq	60,0 3,0	60,8 5,0	142,6 6,2	/ /
AIN MESSALA	379/36 18.02.53	mg/l meq	65,0 3,2	67,0 5,5	131,5 5,7	/ /

SUPERFICIELLES DU BASSIN DU TADLA

Cl^-	SO_4^{--}	HCO_3^-	Résidu sec mg/l (calculé)	$C_{20^\circ C}$ µmhos (calculée)	pH
177,5	48,0	329	690	1025	7,6
5,0	1,0	5,4			
10614,5	259,2	475	18380	24730	8,0
299,0	5,4	7,8			
5813,1	139,2	448	10210	/	/
163,7	2,9	7,3		(14590)	
298,0	48,0	268	702	1334	/
8,4	1,0	4,4			
24,8	9,0	298	320	/	/
0,7	0,19	4,9		(460)	
46,1	30,0	415	/	700	/
1,3	0,62	6,8	(490)		
219,4	4,8	256	650	/	/
6,2	0,1	4,2		(930)	
150,9	67,2	262	660	/	/
4,2	1,4	4,3		(940)	
798,8	62,4	366	1740	/	7,3
22,5	1,3	6,0		(2490)	
9649,0	67,0	622	15600	25450	6,6
271,8	1,4	10,2			
1304,6	19,2	219	2160	3786	7,2
36,8	0,4	3,6			
379,1	24,0	293	920	/	7,1
10,7	0,5	4,8		(1315)	
178,0	319,0	329	⌐1000	/	7,6
5,0	6,5	5,4		(1430)	
266,0	216,0	/	/	/	
7,5	4,5	/			
216,5	190,0	/	865	/	
6,1	4,0	/		(1240)	

KABBAJ et al.

TABLEAU IV. ANALYSES CHIMIQUES DE L'AQUIFERE

REFERENCE N° IRE		Ca++	Mg++	Na+	K+
OUED ZEM	86/29 mg/l	108,0	40,0	17,0	4,0
	30.09.59 meq	5,4	3	0,74	0,10
BOUJAD	1352/29 mg/l	50,0	57,2	62,0	2,4
	15.07.72 meq	2,5	4,7	2,7	0,06
KASBA TADLA	1914/37 mg/l	46,0	41,3	53,0	4,2
	30.09.71 meq	2,3	3,4	2,3	0,11
EL BOROUJ	642/36 mg/l	62,0	59,0	53,0	/
	19.08.59 meq	3,1	4,9	2,3	/
FKIH BEN SALAH	3867/36 mg/l	202,0	58,4	145,0	5,8
	09.11.73 meq	10,1	4,8	6,3	0,15
	3868/36 mg/l	260,0	77,8	133,0	8,8
	.73 meq	13,0	6,4	5,8	0,22
	837/36 mg/l	152,0	34,0	/	/
	21.06.74 meq	7,6	2,8	/	/
	1547/36 mg/l	92,0	34,0	/	/
	30.12.73 meq	4,6	2,8	/	/
	1546/36 mg/l	96,0	36,0	/	/
	15.11.73 meq	4,8	3,0	/	/
	1548/36 mg/l	976,0	340,0	/	/
	25.05.74 meq	48,8	28,0	/	/
	563/45 mg/l	1060,0	389,1	3565,0	12,7
	14.03.74 meq	53,0	32,0	155,0	0,32

TURONIEN DU BASSIN DU TADLA

Cl^-	SO_4^{--}	HCO_3^-	Résidu sec mg/l (calculé)	$C_{20°C}$ µmhos (calculée)	pH
48,0 1,4	170,0 3,5	281 4,6	642	/ (920)	/
113,6 3,2	48,0 1,0	354 5,8	550	983	/
97,6 2,7	24,0 0,5	299 4,9	500	/ (715)	7,6
104,0 2,9	77,0 1,6	349 5,7	538	/ (770)	/
275,1 7,8	432,0 9,0	287 4,7	1330	/ (1900)	7,8
310,6 8,8	614,4 12,8	262 4,3	1590	/ (2270)	/
112,0 3,4	278,0 5,8	287 4,7	1050	1236	/
114,0 3,2	163,0 3,4	262 4,3	700	839	/
121,0 3,4	163,0 3,4	274 4,5	760	965	/
5897,0 166,1	1142,0 23,8	183 3,0	11200	17105	/
7047,0 198,5	1944,0 40,5	177 2,9	14120	20081	8,1

KABBAJ et al.

TABLEAU V. ANALYSES ISOTOPIQUES

REFERENCE			T °C	Cl⁻ mg/l	HCO₃⁻ mg/l	CO₂ libre mg/l	dH d°F
	N° IRE						
SOURCES	147/30	06.07.72	14°2	163	311	18,5	29,5
DE	516/30	07.07.72	13°5	10153	451	20,8	71
L'OUM ER RBIA	520/30	07.07.72	15°0	7242	427	20,8	64
OUM ER RBIA	3866/36	10.09.73					
AÏN ASSERDOUN	210/37	03.09.73					
DAR OULD ZIDOUH	2400/36	11.07.72 10.09.73	/	53	433	39,6	75
FKIH BEN SALAH	705/36 2363/36	14.07.72 19.04.74	22°3	284	262	25,5	115
EL KHIAS	2324/36	10.07.72	21°5	799	360	34,8	71
PUITS A 43	2313/36	19.04.74					
BIR GHEJJAT	332/36	11.07.72	22°0	1367	207	23,1	333
FKIH BEN SALAH FORAGE OCP	655/36	19.04.74					
AÏN IGLI	18/36	30.05.74					
AÏN BISSI BISSA	9/36	30.05.74					
AÏN MESSALA	379/36	30.05.74					
OUED ZEM	86.29	17.07.72	19°5	53	384	25,5	95
BOUJAD	1352/29	17.07.72	17°0	110	372	20,8	89
KASBA TADLA	1914/37	15.04.74					
EL BOROUJ	642/36	12.07.72 10.09.73 03.07.74	28°0	135	366	46,2	80
FKIH BEN SALAH	3867/36 3868/36 837/36 1547/36 1546/36 1548/36 563/45	15.04.74 15.04.74 21.06.74 30.05.74 30.05.74 30.05.74 14.03.74	37°2 38°8	/	/	/	/

U BASSIN DU TADLA

^{3}H U.T.	D δ‰ SMOW	^{18}O δ‰ SMOW	^{14}C % N.B.S.	^{13}C δ‰ P.D.B.	^{34}S δ‰ C.D.
23±3	-52,7±0,5	-7,9±0,1	65,2±1,9	-15,9±0,5	/
9±2	-41,6±0,5	-7,0±0,1	71,2±1,1	-22,0±0,5	/
25±3	-41,3±0,5	-6,5±0,1	/	/	/
36±3		-6,3±0,1			
19±2		-6,8±0,1			
/	-37,1±0,5	-5,8±0,1	54,7±1,0	-18,0±0,5	/
≤1		-6,3±0,1			
10±2	-34,4±0,5	-4,8±0,1	70,9±1,1	-15,0±0,5	/
21±2		-6,0±0,1			+11,7±0,2
33±4	-42,8±0,5	-6,6±0,1	76,1±1,7	-14,9±0,5	/
28±3		-6,4±0,1			
24±3	-35,8±0,5	-4,9±0,1	61,9±1,0	-18,2±0,5	/
21±2		-5,5±0,1			
32±3		-6,9±0,1			
11±1		-7,6±0,1			
15±2		-7,6±0,1			
34±4	-33,1±0,5	-5,1±0,1	/	/	/
56±5	-31,1±0,5	-4,5±0,1	/	/	/
≤1		-6,5±0,1			+7,3±0,2
/	-34,5±0,5	-5,0±0,1	45,5±0,9	-15,6±0,5	/
3±1		-5,1±0,1			+4,3±0,2
2±1		-5,2±0,1			
≤1		-6,8±0,1			+16,1±0,2
≤1		-6,8±0,1			
4±1		-6,0±0,1			
12±1		-7,1±0,1			
30±3		-7,4±0,1			
10±1		-6,9±0,1			
2±1		-6,6±0,1			

KABBAJ et al.

TABLEAU VI. ANALYSES CHIMIQUES

REFERENCE N° IRE		Ca⁺⁺	Mg⁺⁺	Na⁺	K⁺	Cl⁻
	290/15 mg/l 1951 meq	55,0 2,75	43,0 3,54	8,3 0,36	/ /	21,0 0,59
	555/15 mg/l 08.11.51 meq	78,5 3,92	43,5 3,58	6,8 0,30	/ /	28,5 0,80
	2060/15 mg/l 23.01.59 meq	59,0 2,95	39,0 3,21	13,0 0,57	/ /	28,0 0,79
	2168/15 mg/l 03.07.64 meq	60,0 3,00	39,5 3,25	30,0 1,30	0,90 0,02	53,2 1,50
	152/15 mg/l meq	48,0 2,40	54,5 4,48	10,5 0,46	/ /	22,5 0,64
	1225/15 mg/l 10.04.54 meq	28,0 1,40	58,5 4,81	29,0 1,26	/ /	32,0 0,90
Aïn Cheggag	42/15 mg/l 10.06.69 meq	30,0 1,50	39,3 3,23	49,5 2,15	0,78 0,02	106,5 3,00
Aïn Chkeff	48/15 mg/l 13.06.69 meq	45,0 2,25	45,6 3,75	36,8 1,60	0,78 0,02	62,1 1,75
Aïn Sméne	51/15 mg/l 13.06.69 meq	50,0 2,50	39,3 3,23	59,8 2,60	0,78 0,02	88,8 2,50
Aïn Senned	160/15 mg/l 10.11.53 meq	66,5 3,33	52,0 4,27	101,0 4,39	/ /	116,0 3,27
Aïn El Ghars	69/22 mg/l 07.09.69 meq	75,0 3,75	45,6 3,75	2,76 0,12	1,17 0,03	8,8 0,25
Aïn Jerrah	366/22 mg/l 23.04.69 meq	65,0 3,25	42,6 3,50	5,52 0,24	0,78 0,02	8,9 0,25
Sidi Harazem	1845/15 mg/l 14.06.70 meq	74,0 3,7	46,0 3,78	125 5,43	3,00 0,08	252,0 7,10
Aïn Skhounat	872/15 mg/l 10.02.62 meq	60,0 3,0	54,7 4,50	245 10,65	/ /	372,8 10,50
-Moulay Yacoub	105/15 mg/l 21.02.62 meq	1080 54,0	310 25,5	9300 404	/ /	17129 482,5
	1079/15 mg/l 13.06.72 meq	50 2,5	48,6 4,0	90,8 3,95	/ /	133,1 3,75

DE LA PLAINE DU SAÏS

SO₄⁻⁻	CO₃H⁻	Résidu sec mg/l (calculé)	C₂₀°C µmhos (calculée)	T°C	dH d°F
6,0 0,13	300 4,92	290	/ (415)	/	27,5
47,5 0,99	300 4,92	375	/ (535)	/	35,0
4,0 0,08	176 2,88	325	/ (465)	/	31,0
24,0 0,50	366 6,00	390	/ (560)	/	31,3
7,3 0,15		480	/ (685)	/	/
77,0 1,60		394	/ (560)	/	31,0
9,6 0,20	195 3,20	/ (405)	578	18°9	23,8
33,6 0,70	256 4,20	/ (455)	648	19°	30,0 -
33,6 0,70	232 3,80	/ (490)	700	19°6	28,8
134,5 2,80	180 2,95	670	/ (960)	/	38,0
38,4 0,80	402 6,59	/	/	/	37,5
38,4 0,80	360 5,90	/ (400)	571	/	33,8
29,0 0,60	372 6,10	720	1116	33°8	
28,8 0,60		1000	/ (1430)	/	35,5
9,8 0,20		33400	/ (47700)	/	/
24,0 0,50	305,0 5,0	/	/	/	/

KABBAJ et al.

TABLEAU VII. ANALYSES ISOTOPIQUES

REFERENCE N° IRE	Profondeur m	Niveau piézo m	T°C	Cl⁻ mg/l	CO_3H^- mg/l
290/15 6/6/72	414,3	Artésien	19°	18	378
555/15 6/6/72	243,7	Artésien	19°	25	378
2060/15 5/6/72	675,0	Artésien	22°	28	354
2168/15 31/5/72	752,0	Artésien	26°	53	335
152/15 15/6/72	229,4	Artésien	18°	18	378
1225/15 21/6/72	311,5	Artésien	18°5	32	305
Aïn Cheggag 42/15 8/6/72	Source	(630)	18°	110	378
Aïn Chkeff 48/15 9/6/72	Source	(460)	19°5	50	354
Aïn Sméne 51/15 23/6/72	Source	(490)	17°5	106	360
Aïn Senned 160/15 21/6/72	Source	(381)	18°	99	378
Aïn El Ghars 69/22 7/6/72	Source	(1417)	11°5	14	354
Aïn Jerrah 366/22 7/6/72	Source	(1178)	14°	32	409
Sidi Harazem1845/15 26/6/72	84,1	Artésien	33°	217	348
Aïn Skhounat 872/15 24/6/72	Source	(400)	38°	390	354
Moulay Yacoub 105/15 22/6/72	Source	(283)	54°	17857	287
1079/15 16/6/72	4,3	2,3	16°	124	457
70/15 15/6/72	6,3	0,6	16°	43	396
25/15 9/6/72	9,0	8,4	18°	231	275
29/15 12/6/72	4,7	3,9	19°	28	354
32/15 9/6/72	9,9	7,6	19°	46	336
1103/15 14/6/72	6,4	3,7	18°5	28	348
1161/15 13/6/72	5,2	1,8	16°	188	683
1197/15 10/6/72	6,3	3,3	17°5	114	476
1390/15 13/6/72	8,2	4,8	18°	220	348
2279/15 14/6/72	4,9	1,9	16°	21	396
2280/15 13/6/72	6,5	3,0	17°	330	323

DE LA PLAINE DU SAÏS

CO_2 libre mg/l	dH d°Γ	^3H UT	^2D δ‰ SMOW	^{18}O δ‰ SMOW	^{14}C % N.B.S.	^{13}C δ‰ P.D.B.
20,9	32,5	/	-43,4±0,5	-6,7±0,1	79,1±1,4	-17,5±0,5
20,9	31,5	/	-42,2±0,5	-6,6±0,1	81,1±1,3	-15,3±0,5
16,3	36,5	/	-43,0±0,5	-6,7±0,1	72,3±1,4	-16,3±0,5
20,9	33,5	/	-41,9±0,5	-6,5±0,1	57,5±1,3	-14,0±0,5
25,5	32,5	/	-39,1±0,5	-6,0±0,1	51,1±1,1	-14,0±0,5
16,3	29,0	/	-34,1±0,5	-5,2±0,1	5,2±0,9	-15,5±0,5
20,9	32,5	6 ±2	-42,6±0,5	-6,4±0,1	85,6±1,5	-15,1±0,5
20,9	31,0	3 ±1	-41,1±0,5	-6,5±0,1	83,2±2,1	-16,0±0,5
20,9	30,0	9 ±2	-41,0±0,5	-5,9±0,1	/	/
30,1	33,5	12±2	-38,7±0,5	-5,7±0,1	/	/
16,3	30,0	42±4	-43,6±0,5	-6,4±0,1	/	/
32,3	35,0	33±4	-44,2±0,5	-6,8±0,1	/	/
37,0	33,0	/	-41,6±0,5	-6,5±0,1	50,3±1,2	-17,2±0,5
30,1	30,0	4±2	-38,2±0,5	-5,9±0,1	/	/
/	/	/	-36,7±0,5	-5,2±0,1	8,7±0,9	/
32,3	36,0	16±3	-37,2±0,5	-6,0±0,1	/	/
23,1	40,0	11±2	-35,5±0,5	-4,9±0,1	/	/
20,9	51,0	19±3	-32,7±0,5	-4,5±0,1	/	/
25,5	30,0	8±2	-38,7±0,5	-6,2±0,1	/	/
18,5	34,0	18±3	-36,1±0,5	-4,8±0,1	/	/
27,7	30,0	10±2	-38,7±0,5	-5,6±0,1	/	/
55,7	59,0	20±3	-35,4±0,5	-5,0±0,1	/	/
35,2	44,5	20±3	-38,1±0,5	-5,2±0,1	/	/
20,9	43,1	28±4	-29,9±0,5	-4,3±0,1	/	/
30,1	33,0	13±2	-38,9±0,5	-6,1±0,1	/	/
20,9	57,0	42±4	-27,1±0,5	-3,5±0,1	/	/

DISCUSSION

L. KOSTOV: I believe that your research is of great importance, especially in view of the potential use of phreatic water for irrigation. The chemistry and the ^{18}O content (due to evaporation) of groundwater in the irrigated part of the Tadla area are strongly influenced by return flow to the aquifer. Did you take this influence into account in your study, especially when using data from wells that are within or affected by the irrigated area, in establishing the ^{18}O-altitude relation, for example?

A. MARCE: Measurements of the groundwater of the Tadla have shown both the influence of return flow of the waters of the Oum er Rbia and the importance of evaporation. Certain waters are rendered saline by the solubilizing of salt deposits and have not in fact undergone evaporation processes. The ^{18}O-altitude correlation line was calculated without taking the isotopic data for the latter waters into account.

U. SIEGENTHALER: You established the $\delta^{18}O$-altitude relation from Dansgaard's relation between temperature and $\delta^{18}O$ which he found for coastal stations at medium and high altitudes. Continental stations are, however, often off Dansgaard's line, as can be seen for example from the data and Fig.4 of the paper presented by Mr. Hübner[1], and it seems dangerous to me simply to take Dansgaard's relation for a regional problem. I wonder, therefore, whether for your $\delta^{18}O$-altitude relation you also used data from your study region.

A. MARCE: Some isotopic measurements were made on precipitation in the Tadla and Saïs areas, and these agree quite well with the calculated lines. It should be realized that these lines were calculated from temperatures as a function of rainfall, and for altitudes between 300 and 2000 m. Finally, the lines were adjusted to take into account isotopic measurements on the waters of very localized boreholes.

J.Ch. FONTES (Chairman): Why did you use an A_0 of 85% for your age calculations when ^{13}C data were available?

A. MARCE: This work was done about five years ago, and the measured value has been corrected for $\delta^{13}C$, a value of $-17‰$ having been chosen for the initial $\delta^{13}C_0$. In fact, calculations done with our new techniques designed to determine the initial ^{14}C content have produced identical results to within the measurement uncertainty.

---

[1] HÜBNER, H., et al., IAEA-SM-228/16, these Proceedings.

ISOTOPE HYDROLOGY IN NORTHERN CHILE

P. FRITZ
Department of Earth Sciences,
University of Waterloo, Waterloo,
Canada

C. SILVA HENNINGS, O. SUZUKI
Comisión Chilena de Energía Nuclear,
Santiago, Chile

E. SALATI
CENA, Piracicaba, S.P., Brazil

Abstract

ISOTOPE HYDROLOGY IN NORTHERN CHILE.
    Environmental isotope analyses were done on samples from aquifers in the Pampa del
Tamarugal and the Salar de Atacama drainage basin in northern Chile. In the Pampa it is
possible to delineate individual groundwater bodies on the basis of their ^{18}O and deuterium
contents and, in some cases, to relate these to specific recharge areas. A marked displacement
from the meteoric water line indicates that river recharge is an important mechanism for
groundwater renewal. Groundwater ages appear high at distance from the Andes and much
of the water found in the Pampa may have to be treated as a non-renewable resource. The
groundwaters, springs and rivers of the Salar de Atacama drainage basin vary between -6.9
and $-8.6‰$. No difference between the different waters can be recognized and an evaporative
isotope enrichment indicates that also here river recharge is an important process. Some
groundwaters adjacent to the Salar are very salty but ^{18}O and deuterium data show that these
waters are not refluxed brines but simply salty freshwater. The ^{14}C contents in groundwaters
and springs are very low but their δ^{13}C values are high. It is concluded that this is probably due to
the uptake of volcanic CO_2. ^{14}C age dating is thus not possible unless the δ^{13}C values of all
possible carbon sources can be defined and the geochemical evolution of the groundwaters
is better understood.

1. INTRODUCTION

    The expansion and often the survival of agriculture and industry
in Northern Chile depend to a large degree on the limited availability of
fresh water resources. Precipitation is absent or negligable throughout
most of the area with the exception of the higher Andes. There, above
3000 m some vegetation is found and in higher parts the average annual
precipitation may amount to 400–600 mm. These are the only regions where
at present significant amounts of surface and ground water can be generated.

    The need for further development of these water resources has
long been recognized and during the past two decades both national (CORFO,
CCHEN) and international organizations (OAS, UNDP, IAEA) have undertaken

or assisted in extensive groundwater exploration programmes. This search
for groundwater resources, and their evaluation and possible exploitation
are generally persued with the tools of classical hydrology, i.e. drilling,
pumptesting and some geophysical and geochemical studies. With these,
however, only limited information about origin and age of the waters
encountered can be obtained. Environmental isotope techniques can here
play a major role. This report presents the results of such investigations
which have been carried out in the Pampa del Tamarugal and the drainage
basin of the Salar de Atacama (figure 1).

1.1 Pampa del Tamarugal

    · The Pampa del Tamarugal is a gently sloped plain which stretches
from the front ranges (Cordilleras de la Costa) to the foot of the Andes
(Cordillera de Los Andes). Its medium elevation is close to 1200 m.a.s.l.[1]
No precipitation falls in the Pampa but up to 300 mm/a are recorded above
4000 m.a.s.l. (figure 2). Virtually all of this arrives from the Atlantic
via the Amazon Basin and therefore has a rather complex history. This is
reflected in the isotope data discussed below.

    For all practical purposes one can consider the Pampa del
Tamarugal to be a closed basin recharged (?) by water arriving from the
higher Andes but with no significant discharge other than evaporation.
Ground water found in these aquifers is of good quality and is used for the
municipal water supplies of Iquique, the old nitrate capital of Northern
Chile, and other, less important towns. The success of attempts to re-
establish tamarugo forests and to develop some oasis in the Pampa (e.g.
Esmeralda) also depends exclusively on the future availability of these
resources.

1.2 Salar de Atacama

    The Salar de Atacama, a tectonic depression [2], also belongs
to a closed basin in which the evaporation from the Salars is the ultimate
point of discharge for the waters arriving from the northern and eastern
mountains. The average elevation of the Salar de Atacama is 2350 m.a.s.l.,
a typical desert climate prevails. Figure 3 shows the isohyetas of the
area emphasizing again that significant rainfalls occur only above 3000 m.
a.s.l. Almost all precipitation originates in the Atlantic regions and
arrives via the Bolivian highlands in the drainage basin. Most rainfall
occurs during the Bolivian winter (December to March) and only seldom do
Pacific disturbances enter the area.

    At least two active volcanoes exist within or adjacent to the
eastern part of the drainage basin: Tumiza (5287 m.a.s.l.) and Legia
(5790 m.a.s.l.). Such volcanic activity could have some influence on the
aqueous geochemistry of the local groundwaters especially through the
contribution of volcanic-magmatic carbon dioxide.

    Only this eastern part of the drainage basin was investigated
(figure 1) since only here a certain groundwater potential does exist.
Most aquifers are found in alluvial fans stretching towards the Salar.
Cemented ash layers (ignimbrites) within the gravels and sands act as
confining layers between aquifers. A schematic geologic cross-section is
shown in figure 4 and also emphasizes that in many parts the freshwater in
the sands and gravels are underlain by salt water.

------

[1] m.a.s.l. = metres above sea level throughout this paper.

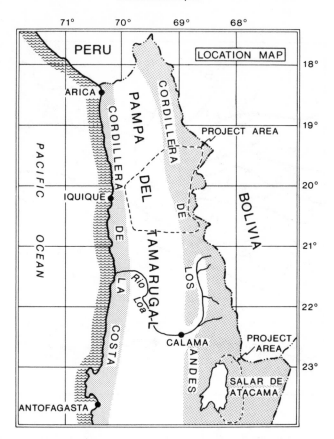

FIG.1.   Location of project areas in northern Chile.

2. PRECIPITATIONS

A key element in many environmental isotope studies is a knowledge of ^{18}O and deuterium contents in precipitations. For this study it was especially important to establish the δD vs $\delta^{18}O$ relationships since evaporation is an extremely important process affecting virtually all systems in these regions. Furthermore, in a mountain terrain, altitude affects should be recognized in order to be able to characterize potential recharge areas.

The collection of precipitation in these remote areas is not an easy task and continuous records can only be expected from the few permanent stations. Further efforts will be undertaken to accumulate the required information and the results presented here should be regarded as preliminary.

Data collected since 1970 (1) for the study area in and adjacent to the Pampa del Tamarugal result in a well defined meteoric waterline, were

$$\delta D = (7.8\ \delta^{18}O + 9.5)\ ^o/_{oo}{}^2$$

---

[2] The ^{18}O and deuterium analyses are expressed in the conventional $\delta^o/_{oo}$ notation and refer to SMOW.

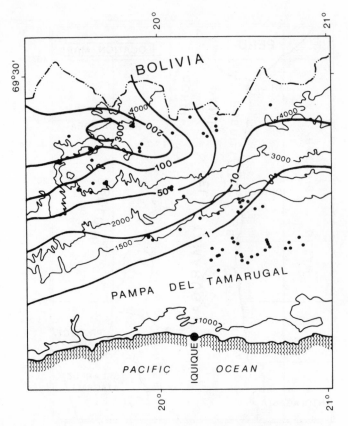

*FIG. 2. Average annual rainfall in the project area of the Pampa del Tamarugal. Isohyeta
lines show mm/a precipitation. The filled circles are well- and spring-sampling stations.*

        Much fewer data are available for the drainage basin of the Salar
de Atacama, although the first samples were already taken in 1970 too. The
data available define a meteoric waterline were

$$\delta D = (7.3 \; \delta^{18}O + 8.0) \; {}^{o}/_{oo}$$

        Attempts were also made to resolve altitude effects as they may
exist in this region. As it turned out this is not a simple task since the
complex history of the vapormasses arriving from Bolivia does not permit
the immediate recognition of characteristic slopes and, before any detailed
discussions on the altitude of recharge of a given system is possible it
will be necessary to define in detail these altitude effects.

3.   DEUTERIUM AND ^{18}O IN GROUNDWATERS AND SPRINGS

        Since 1970 several hundred stable isotope analyses were done on
springs and well samples throughout northern Chile. A summary of the data
relevant to the project of the Pampa del Tamarugal and the Salar de
Atacama is presented in Figures 5, 6 and 7.

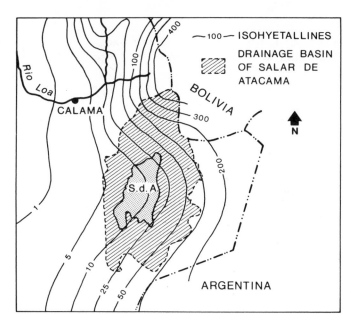

FIG.3. Average annual rainfall in the region of the drainage basin of the Salar de Atacama.

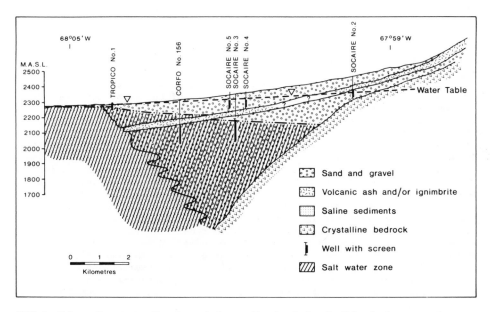

FIG.4. Schematic cross-section through the aquifers bordering the Salar de Atacama. An unconfined aquifer with good-quality water overlies the ignimbrites whereas the confined aquifer beneath it carries an increasing salt load as the salar is approached.

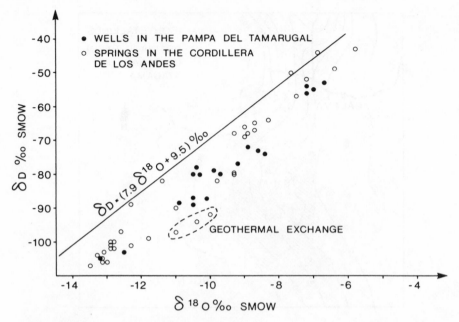

FIG.5. ^{18}O and deuterium contents in the groundwaters and springs of the Pampa del
Tamarugal. Note the displacement to the right of the meteoric waterline which is mostly due
to surface evaporation effects. Geothermal exchange indicated for samples from the
Puchuldiza geothermal fields is a local phenomenon.

3.1 Pampa del Tamarugal

A comparison of the "meteoric waterline" with data obtained on
groundwater from wells in the alluvial deposits of the Pampa and numerous
springs which discharge in the higher parts of the project area shows a
marked displacement of these samples to the right of this line (figure 5).
This can have two causes: evaporation and/or ^{18}O-exchange between rock
and water at elevated temperatures. The latter is known from the geothermal
areas of Chile (3). Three samples from the Puchuldiza geothermal fields
in the project area are identified in figure 5. Their salt contents is
well above the one of "normal" groundwaters in the area exceeding 20,000
mg/l and it is thus unlikely that many of the samples analysed for this
study are affected by such a geothermal ^{18}O shift. Instead it is believed
that groundwater recharge in these areas occurs primarily through infiltration into alluvial deposits associated with running surface water. Most
creeks descending from the high Andes show indeed an evaporative isotope
enrichment and displacement from the meteoric waterlines similar to the one
noted for these springs and well waters (1). Field observations of
"disappearing rivers" as well as the fact that the potential evapotranspiration in this area amounts to 2500 to 3000 mm/a support this assumption.

If the displacement were only observed in the well waters of the
Pampa one could argue that periodic floods which cover vast areas of the
Pampas could also contribute to the renewal of groundwater resources.

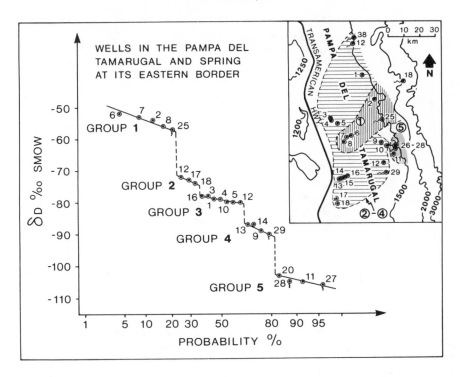

FIG.6.   This figure shows a probability vs δD plot for the well-waters from the Pampa del Tamarugal and shows their regional distribution in the area.

However, the vertical permeability of these stratified sand and gravel aquifers is very low because of interbedded silty and argillaceous deposits. Thus any vertical infiltration must be small, a fact supported by the observation that the water ponded after flooding evaporates from isolated bodies of water. We do not have any analyses of such evaporated water but the isotopic enrichment would be probably much greater than the one observed on the river waters and the displacement from the meteoric water-line would be more significant.

An important characteristic of these groundwaters is their very large spread in ^{18}O and deuterium contents. If only a single, confined aquifer would exist and if recharge had a unique source than much smaller variations would be expected. On closer inspection individual groundwater bodies with rather well defined compositions can be recognized, i.e. the isotope contents do not vary at random. This is emphasized on figure 6 which shows a probability vs δD plot and in the insert the regional distribution of the individual groups of values.

Because of the lack of wells with screens set at specific horizons it is at present not possible to fully delineate each groundwater body and it is probable that not only horizontal but also vertical differences do exist. Despite these shortcomings the origin of some

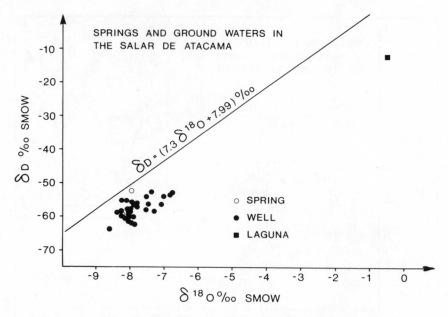

FIG.7. ^{18}O and deuterium contents in wells and springs within the drainage basin of the
Salar de Atacama. Again a displacement from the meteoric water line is noted and suggests
that evaporation from open water surfaces is important. Note also the heavily enriched sample
from the one lagoon analysed so far for ^{18}O and deuterium.

groundwaters can be assigned to recharge close to deeply incised river
valleys (quebradas) which descend from the Andes. For example, group 1
waters probably originate in the quebrada de Juan de Morales, and the
wells in group 5 tap the same system as the springs discharging at the
oasis of Pica at the foot of the Andes. There groundwaters are prevented
from entering the Pampa proper by major tectonic features and thus form
a very well defined groundwater body.

Less well defined is at present the origin of the waters in
groups 2-4. Mixing of water from different aquifers may well take place
within many wells and much more hydrogeologic work will be required before
any conclusive statements are possible.

3.2 Salar de Atacama (Fig.8)

At least two aquifers are recognized in the alluvial and
volcanic deposits bordering the Salar de Atacama: an unconfined one overlying the very impermeable ignimbrites and a confined one below these
strata. Furthermore it is common that freshwater in alluvial deposit
bordering the salares overlies salt water (figure 4). The origin of this
salt water is not very clear since it could be refluxed and/or old brine
or just "salty" freshwater.

The salars are the final discharge point of any hydrologic
system within the drainage basin and waterloss through evaporation will

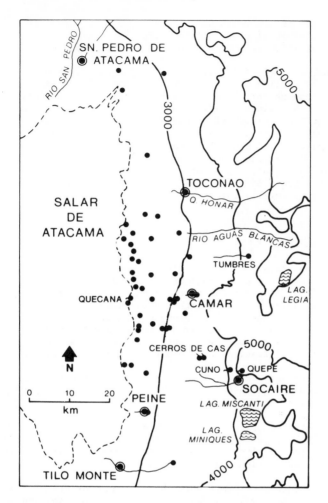

*FIG.8.   Location map of wells and springs sampled in the basin of the Salar de Atacama.
Note that all rivers disappear in alluvial deposits before reaching the salar.*

lead to significant isotope enrichments.   However, no correlation exists
between isotopic composition and salinity which signifies that the salty
groundwater is not refluxed brine but "freshwater" which picked up a
substantial dissolved load from the evaporitic minerals present in the
sedimentary rocks adjacent to the salar.

        An inspection of $\delta^{18}O$ and δD values of the well samples (figure
7) reveals the same pattern as observed fro the groundwaters in the Pampa,
that is a slight shift to the right of the meteoric waterline.   (However,
it has to be emphasized that only few precipitation samples were available
to define this line).   This again indicates that the recharge mechanism of
these groundwaters is the same as in the other project area.   Also here all

creeks disappear before reaching the salar and thus certainly contribute
to the renewal of local groundwaters. The ^{18}O contents of the creeks cannot be distinguished from the ones of the well waters.

The $\delta^{18}O$ values for rivers vary between -7.6 and -8.4 $^o/oo$
(n = 4), in the shallow unconfined aquifer from -7.0 to -8.6 $^o/oo$ (n = 28)
which is the same as for the deeper groundwater (salty or not, n=16) and
the range for the springs is -7.3 to -8.6 $^o/oo$ (n = 11). This similarity
indicates that the altitude of recharge for all of them is very similar--
unless altitude effects are very small in this area. Again, as in the
Pampa del Tamarugal, it probably also signifies that the amounts of
vertical recharge through the alluvial and volcanic deposits bordering
the salar are very small. The opportunities for such recharge are anyway
limited since between 1958 and 1968 there were only two rainstorms with
more than 10 mm rainfall, three with 5 to 10 mm and six with less than
5 mm.

4. GROUNDWATER AGES

The comparison of the stable isotope contents of groundwaters
drawn from wells with modern precipitation is only justified if one can
show that no major climatic shifts occurred. Historic records from
Northern Chile indicate, however, that such changes took place. Spanish
maps show forests and lakes in the Pampa del Tamarugal, in an area where
the last rainfall supposedly occurred somewhat over a hundred years ago.
Precipitation records also show a continuous decline of the amounts of
average annual precipitations throughout Northern Chile and one must
conclude that a comparison of modern precipitation with the groundwaters
is only valid for very young waters--although the magnitude of any
possible isotope shifts caused by these climatic changes can at present
not be assessed.

Age determinations have thus been attempted using tritium and
carbon-14 in both project areas. Unfortunately there is some doubt about
the quality of the results obtained for Pampa samples and new samples will
be taken, expected tritium contents are below the detection limit.

Below detection limit were the tritium contents of five samples
from the Salar di Atacama: S. Pedro 5, Tropico 1, Quelana and Socaire 3
and 5. Tropico 1 is a very shallow well (figure 4), whereas the other
represented deeper groundwaters and a spring (Quelana). The data indicate
that all groundwater in this region is more than 20 years old and that
vertical recharge through the alluvial fill is indeed not very important.

4.1 Carbon-14 dating

The groundwaters in the project areas can have a multiple origin:
either they originate from "disappearing" creeks (as is certainly the case
for some of them) or they arrive as fracture flow etc. from recharge
areas with sufficient precipitation. The initial geochemical conditions
may thereby be quite different as would be their geochemical evolution in
the subsurface. Unfortunately we have no chemical analyses for the
samples collected for age determinations and therefore the theoretical
discussion is somewhat limited.

Carbon-14 and carbon-13 analyses were done on surface waters in
order to have some knowledge of the initial conditions of the newly formed

TABLE 1. ^{13}C ADJUSTED WATER AGES FOR THE PAMPA DEL TAMARUGAL

| | ^{14}CpmC/δ^{13}C-TIC^o/oo | δ^{13}C-soil = C_1 = -18 ^{o}/oo | |
		$\delta^{13}C_2$ = -7 ^{o}/oo	$\delta^{13}C_2$ = +3 ^{o}/oo
Canchones CORFO	23.1/-10.1	1060 y	7900 y
Esmeralda	98.1/-13.0	future	future
Pintados	6.1/-10.5	13070 y	19160 y
Duplijza	62.3/-8.4	future	future

ground water. Also investigated were potential recharge areas in the
higher parts of the Andes with special emphasis on the soil-CO_2 composition.
Both types of information are needed for the evaluation of ^{14}C contents
in groundwaters in terms of water ages.

The results from wells in the Pampa del Tamarugal are shown in
table 1, those from the Salar de Atacama drainage basin in table 2,
figure 8 is a location map for wells adjacent to the Salar.

4.1.1 Rivers in the Salar de Atacama drainage basin

An inspection of table 3 shows that the ^{14}C contents of all
creeks is much higher than the one in the groundwaters: the latter have
^{14}C contents below 15 pmC whereas modern values (100 pmC) are approached
in at least one creek. This difference is almost certainly due to isotope
exchange with the atmosphere which occurs during surface runoff. Such
exchange also affects the ^{13}C contents of the dissolved inorganic carbon
which, however, is also dependent on biologic activities in the creeks:
Aquatic plant respiration-CO_2 would be ^{13}C depleted but contain the same
^{14}C contents as the atmosphere.

Interesting are thus the data obtained for Rio Socaire which has
a ^{14}C content of only 32.6 pmC. Its δ^{13}C is very close to the one
expected for a system open to the atmosphere (as shown by the calculated
δ^{13}C-CO_2) and also its calculated partial pressure of CO_2 (pCO_2) turned
out to be close to atmospheric values. The "low" ^{14}C content; documents
that the creek carries a significant portion of groundwater low in ^{14}C.
The fact that it has not reached an atmospheric equilibrium for the ^{14}C
probably is explained by different rates of approach towards equilibrium
for ^{13}C and ^{14}C since the concentration gradients are very different.

If the groundwaters formed from infiltrating rivers than the
carbon isotopic composition of the river reflects the "initial" conditions.
If no "dilution" of the ^{14}C contents takes place during the following
subsurface history then the differences in ^{14}C contents between groundwaters and surface waters reflect ^{14}C decay. The minimum age of any of
these waters would then between 8000 and 9000 years assuming A_o = 32.6 pmC
(the lowest creek value) and A_{final} = 12.2 pmC (the highest groundwater
value).

However, the chemical load of the infiltrating rivers is lower and
their δ^{13}C's are higher than in many groundwaters. It is impossible to
change the ^{13}C contents of the dissolved inorganic carbon of a groundwater

TABLE 2. CARBON ISOTOPIC COMPOSITION OF SPRING, WELL AND RIVER WATERS IN THE SALAR DE ATACAMA (COLLECTED JUNE–NOV 1977)

Sample No.	Locality	^{14}C-pmC	$\delta^{13}C$-DIC	Calculated $\delta^{13}C$-CO$_2$	t°C	field-pH	Calculated pCO$_2$
Spring samples							
151	Tumbres	7.1, 7.5	$-8.4, -8.4$	-11.9	15	6.3	$10^{-1.1}$
167	Peine	6.8	$+1.5$	-5.5	24.5	7.4	$10^{-2.2}$
169	Quelana	n.d.	-5.2	-12.3	14	7.15	$10^{-2.1}$
171	Laguna Meniques	45	$+1.0$	-6.3	20	7.4	$10^{-2.2}$
172	Tilomonte	n.d.	$+1.0$	-5.5	26	7.2	n.d. $10^{-1.4}$
173	Cerros de Cas sup.	3.7	-1.1	-7.7	20	7.1	$10^{-1.4}$
182	Cerros de Cas inf.	n.d.	-2.1	-8.5	16.5	7.0	n.d.
174	Quepe-Socaire	6.9	-1.9	-7.7	27	6.95	n.d.
176	Cuno-Socaire sup.	n.d.	-3.9	-12.2	12	7.65	n.d.
177	Cuno-Socaire inj.	n.d.	-2.5	-10.5	15	7.6	n.d. $10^{-1.5}$
181	Camar	13.5	$+1.8$	-3.6	21	6.8	$10^{-1.5}$
Well samples							
147	Tropico	1.0	-1.1	? -8.5[a]	n.d.	7.6[a]	n.d.
150	Tambillo	3.2	-5.5	n.d.	n.d.	n.d.	n.d.
152	Socaire 3	1.4	-0.2	? -7.8[a]	n.d.	7.8[a]	n.d.
160	Zerzo 1	12.2	-3.6	? -11.2[a]	n.d.	7.7[a]	n.d.
165	Socaire 5	1.1	$-0.2, -0.2$	-6.7	27	7.2	n.d. $10^{-1.2}$
178	San Pedro 5	11.1	-5.5	-10.4	30	6.75	$10^{-1.2}$
180	San Pedro 3	12.1	-5.5	-11.2	29	6.95	n.d.
Creeks							
159	Tilomonte	83.6	$+0.0, +0.6$	n.d.	n.d.	n.d.	n.d.
163	Socaire	32.6	-2.6	-6.3	12	8.7	n.d. $10^{-3.4}$
166	Aguas Blancas	52.6	-1.8	-10.3	16	8.6	n.d.
--	San Pedro	n.d.	-2.0	n.d.	n.d.	n.d.	n.d.

[a] These values were calculated assuming calcite saturation.

TABLE 3. ^{13}C IN SOIL-CO$_2$ AND ASSOCIATED PLANTS

	δ^{13}C SOIL	PLANT	δ^{13}C PLANT
Station 1			
close to Salar de Atacama, not a recharge area. pCO$_2$ ⩽10$^{-3.5}$atm	-14.7°/$_{oo}$	Atriplex sp.	-14.0°/$_{oo}$
Station 2			
under Tamarugo tree close to Salar de Atacama. Not a recharge area pCO$_2$ >10$^{-3.5}$atm	-22.3°/$_{oo}$	Prosopis Tamarugo	-28.1°/$_{oo}$
Station 3		Lepidophyllum sp.	-21.3°/$_{oo}$
above "Tumbres". Potential recharge area with active vegetation. pCO$_2$ >10$^{-3.5}$atm	-18.3°/$_{oo}$	Baccharis sp.	-20.0°/$_{oo}$
		Unidentified species	-22.1°/$_{oo}$
		"Añagua"	-23.6°/$_{oo}$

without adding, subtracting or exchanging carbon of this reservoir with another carbon reservoir and, therefore, ^{14}C-corrections based on chemical and isotopic evidence have then to be introduced. Whether this can be done depends to a large degree on the complexity of the system. In the following an attempt is made to describe the possible carbon sources that may exist in the drainage basins of Northern Chile and then to make some statements about possible groundwater ages.

4.1.2 Soil-CO$_2$ and plants

Practically all subsurface and surface waters contain carbon that originated in biologically active soils. This is also true in Northern Chile and therefore we attempted to learn something about its carbon isotopic composition. The area is very remote and the soil-CO$_2$ had to be "concentrated" in the field. This was done by sucking soil-gas through a stainless steel probe (1 to 1.5 m long) and stripping with an alkaline BaCl$_2$ solution. The precipitate was separated in a small collection flask and carried back to the laboratory.

Unfortunately the sampling was done during winter (June, 1977) when plant activity was rather low (although not nonexistant) and therefore at many sites the pCO$_2$'s were so low that this sampling became impractical. In order to overcome this, all dominant plant species were collected and analyzed for their ^{13}C concentrations. The data are shown in summary form in figure 9 and table 3 lists the ^{13}C results of three soil-CO$_2$ samples and their associated plants.

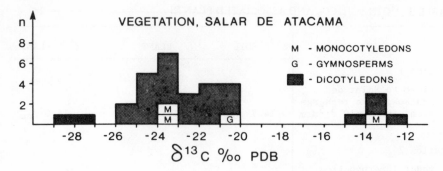

FIG. 9. $\delta^{13}C$ values of major plant species collected in the drainage basin of the Salar de Atacama.

The most interesting features of the data obtained can be summarized as follows: a) the soil-CO_2 has a higher ^{13}C content than the associated plant material. This has been noticed previoulsy a.o. by Rightmire & Hanshaw (4) and Fritz et al. (5) who observed enrichments in the soil-CO_2 of up to 9 $^{O}/_{OO}$ due to a mixture of respiration and decay-CO_2 in the soils. b) The recharge areas are dominated by plants which use the C_3-photo-synthetic cycle and produce organic matter with $\delta^{13}C$'s below -20 $^{O}/_{OO}$. This is also true for the potential recharge areas above 3500 m.a.s.l. where grasses of the Stipa-group dominate. Thus Stipa sp. from 3700 m.a.s. l. above Socaire gave a $\delta^{13}C$ = -23.4 $^{O}/_{OO}$ and the grass "Paja brava" from Socaire had a $\delta^{13}C$ = -23.7 $^{O}/_{OO}$. This contrasts with grasses growing adjacent to the Salar deAtacana where Grama salada had a $\delta^{13}C$ = -13.4 $^{O}/_{OO}$. Thus it is safe to assume that if infiltration occurs in the mountains the water will take up a soil-CO_2 with $\delta^{13}C$-values somewhere close to -18 $^{O}/_{OO}$ as measured at station 3 (table 3).

4.1.3 Rockcarbon

The volcanic-magmatic rocks forming the Andes are free of major carbonate sequences although locally minor amounts may occur. These are either volcanic-magmatic carbonate minerals or freshwater carbonates such as found at the orifice of many springs; they are rather extensive at Tilomonte, Peine and Camar [2].

The ^{13}C contents of the magmatic carbonates has not been measured but it can be assumed that the values are close to those of magmatic CO_2 ($\delta^{13}C$=-5 to -7 $^{O}/_{OO}$) or slightly more positive because of isotope effects during deposition. Such carbonates could be dissolved by migrating ground waters, but at present we cannot judge how important they might be for the carbonate budget of these waters.

The freshwater carbonates found as spring deposits will not participate in the geochemical evolution of these groundwaters and it is not known whether similar deposits exist in the subsurface, buried by volcanic ash etc. Nevertheless, some analyses have been made and it is interesting to note that the $\delta^{13}C$'s of three samples from Peine and Camar vary between +2.0 and +5.5 $^{O}/_{OO}$, indicating deposition from waters with ^{13}C contents similar to those discharged today at these localities (table 2).

One might also note that the extensiveness of these deposits indicates much greater discharges from these springs at the time of their formation than from today's. The hydrologic activity of these systems has clearly decreased since.

4.1.4 Volcanic CO_2

Within the drainage basin of the Salar de Atacama and adjacent to it one finds active fumaroles associated with young volcanoes. Therefore it is possible that magmatic CO_2 could play a role in the carbon budget of these groundwaters. The ^{13}C content of such a gas would be close to -7 $^o/_{oo}$. The dissolved inorganic carbon in a water with pH values between 7 and 8 with such a gas would be close to +2 $^o/_{oo}$; as some of our samples have values close to this, they do not contradict such reasoning.

During sample collection field pH values were measured at many springs and wells and used to calculate the isotopic composition of a gas phase that would be in equilibrium with these waters, combined with alkalinity measurement it yields the pCO_2 of this gas phase (table 2).

For the $\delta^{13}C$'s one recognizes two groups: one where the $\delta^{13}C$ of the calculated CO_2 varies between -10.5 $^o/_{oo}$ and -12.2 $^o/_{oo}$ and another where it ranges from -5.5 $^o/_{oo}$ to -8.5 $^o/_{oo}$ with one additional value at -3.6 $^o/_{oo}$. The pCO_2's are rather high and close to or above 10^{-2} atm. This does not prove that an actual gas phase exists—a pop-bottle can hold 10's of atmospheres without it—, and similar pCO_2's could be generated in a closed system such as a confined aquifer in which the pH drops significantly after carbonate uptake because of geochemical reactions. For each pH-unit drop the pCO_2 would increase by one order of magnitude. Although geochemically such a decrease in pH is not impossible it is judged not to be important here.

High alkalinities determined in the field and corresponding to bicarbonate values between 120 and 500 ppm cannot be generated in a soil-CO_2 environment with pCO_2 barely above $10^{-3.5}$ atm (as it exists in the potential recharge areas) without pH values higher than those measured. Thus a secondary carbon source is likely to exist. It is thus interesting to note that the calculated $\delta^{13}C$-CO_2 values show a regional distribution: the wells and springs in the northern half of the drainage basin have the low $\delta^{13}C$'s whereas in the southern half of the project area one observes high $\delta^{13}C$ values. This could suggest that different carbon sources exist within the drainage basin and the similarity of calculate $\delta^{13}C$-CO_2 values with expected magmatic CO_2 values might indicate that magmatic CO_2 plays a major role in the carbon budget of the waters at least in the southern half of this project area.

To generate positive $\delta^{13}C$ values a corresponding source for the carbon would have to be found. If one could show that a redissolution of freshwater carbonates takes place then they could be the source since their $\delta^{13}C$-values seem to be high enough (see above). However, these freshwater carbonates occur as spring-deposits and it appears doubtful that they were buried and available for dissolution only in one part of the drainage basin. Normal magmatic carbonates are unlikely to have sufficiently high ^{13}C contents.

In conclusion one can state that there is not yet any physical or clear chemical evidence to show that volcanic CO_2 plays an important part in the carbon budget of these groundwaters but isotope and chemical

data do not exclude such a carbon source. This is important in the
following discussion on ^{14}C-water ages.

4.1.5 Groundwater ages in the Salar de Atacama drainage basin

The participation of magmatic volcanic CO_2 in the geochemical evolution of the local groundwaters has serious implication with respect to
the usefulness of the ^{14}C-method in such environments. The following
situation appears possible:

During and after movement of surface water from rivers or other
recharge areas towards the saturated zone silicate reactions provide the
groundwater with much of its chemical load and also lead to relatively
well buffered pH-values above 7. Such water could absorb significant
amounts of carbon dioxide, and should a gas phase with more than 1% CO_2
($pCO_2 = 10^{-2}$) be present than the observed alkalinities in most groundwaters could be accounted for. With time a carbon isotope equilibrium
between this gas phase and the dissolved carbonates would be established,
essentially governed by the CO_2 - HCO_3^- isotope fractionation factor.
Depending on temperature the bicarbonate would be enriched by 8-10 $^o/oo$
and the observed δ^{13}C-DIC values would be explained if this CO_2-gas had a
δ^{13}C close to -7 $^o/oo$. Being magmatic in origin it has very low or no
[14]C activity and thus, regardless of water ages, also the aqueous carbon
would be virtually free of it. Whatever activity is measured could be the
residual of the very minor amounts of soil or atmospheric CO_2 left after
this major addition of or isotope exchange with magmatic CO_2.

Under the outlined conditions carbon-14 dating is not possible.
This may well be the case for most, if not all samples collected in the
Atacama basin. However, some other possibilities should be explored.

There can be little doubt that geochemistry of these waters is
dominated by silicate reactions and, therefore, the carbonate dissolution
models cannot be applied. However, if one assumes that only a mixing of
different carbons takes place (neglecting exchange or carbonate precipitation) then for a two source system one could write that

$$\delta^{13}C\text{-TIC} = x\delta^{13}C_1 + (1-x)\delta^{13}C_2$$

where TIC = total dissolved inorganic carbon, C_1 and C_2 the first and
second carbon source which could be soil carbondioxide, rockcarbon or
volcanic CO_2. The δ^{13}C would have to be specified and, for example, $\delta^{13}C_1$
could be the soil-CO_2 value and $\delta^{13}C_2$ the rockcarbon or volcanic CO_2 or a
mixture of both.

The ratio x/1 corresponds to the ratio initial/final carbonate
content and thus defines the ^{14}C dilution, provided C_2 is free of [14]C. It
is equivalent to the correction factor q used in many carbon-14 correction
models [10] where

$$A_m = qA_o \exp - \lambda t$$

since also

$$x = q = \frac{\delta^{13}C\text{-DIC} - \delta^{13}C_2}{\delta^{13}C_1 - \delta^{13}C_2}$$

an attempt can be made to correct the measured ^{14}C ages on the basis of the
δ^{13}C values of the various components.

TABLE 4. ^{14}C AGES OF GROUNDWATERS IN THE SALAR DE ATACAMA BASIN

	^{14}CpmC/δ^{13}C $^o/_{oo}$	$\delta^{13}C_1 = -18.5$, recharge pH = 5 t = 20°	
		$\delta^{13}C_2 = -2$ $^o/_{oo}$	$\delta^{13}C = +2$ $^o/_{oo}$
Tumbres	7.3/-8.4	13810 y B.P.	16030 y B.P.
Tambillo	3.2/-5.5	15635 y B.P.	20140 y B.P.
Zarzo 1	12.2/-3.6	future	6630 y B.P.
San Pedro 5	11.1/-5.5	5350 y B.P.	9860 y B.P.
San Pedro 6	12.1/-5.5	4640 y B.P.	9150 y B.P.

Commonly one assumes that $C_1 = \delta^{13}C$ of the soil-CO_2. This, however, implies that the CO_2 uptake occurs without fractionation effects. Yet at pH >7 bicarbonate becomes the dominant aqueous species and C_1 will be several $^o/_{oo}$ higher than if pH<6. This shift affects the q value and the calculated difference in ^{14}C age between a water initially at pH = 4 and one at pH = 7 is close to 3500 years, the latter yielding older ages.

Very critical is the determination of $\delta^{13}C_2$ which could be any-where between ~-7 $^o/_{oo}$ and +5 $^o/_{oo}$ (see above). However, maximum ^{14}C dilution is achieved if δ^{13}C-TIC = $\delta^{13}C_2$ and modern as future ages will be calculated. This may well be the case for all samples with δ^{13}C's close to zero or above it. Those waters with lower ^{13}C contents (Tumbres, Tambillo, San Pedro 3 and 5 and Zarzo 1) would then be the older ones for which one could calculate the following ages (table 4). The more negative the δ^{13}C of second carbon source the younger will be the corrected ages, and the values choses are somewhat arbitrary and do not necessary reflect the composition of this source.

The significance of these "ages" can at present not be fully assessed but they tend to indicate that the wells listed in table 4 yield the oldest water in the vicinity of the Salar, whereas most groundwaters in the alluvial fills east of the Salar (figure 8) would be rather young. Also, the high "age" for Tumbres calculated with these input data is probably unrealistic since this water has a very low pH. Any volcanic-CO_2 uptake would occur with much smaller fractionation effects and the $\delta^{13}C_2$ might well be close to -7 $^o/_{oo}$, making this water much younger (a $\delta^{13}C_2 = -7.6$ $^o/_{oo}$ would give a modern age).

4.1.6   Groundwater ages in the Pampa del Tamarugal

The results of the isotope analyses recorded in table 5 show δ^{13}C values quite different, because much lower, from those of the Salar de Atacama. It thus appears that the geochemical evolution follows more normal paths and that we look essentially at a carbonate system dominated by soil-CO_2 (assumed δ^{13}C ≃ -18 $^o/_{oo}$) and rock carbonate. Again the choice of the δ^{13}C value for the rock carbonate becomes important if one wishes to use the ^{13}C correction method for the adjustment of the raw ^{14}C dates. From the above discussions one might assume that the extreme possible values for these carbonates lie between -7 and +3 $^o/_{oo}$ the first being a repre-sentative for magmatic calcites, the second for freshwater precipitates. The following table 1 presents a summary of the results using a δ^{13}C correction as outlined above. The data show that these model calculations

are not fully applicable since two of the samples gave for both types of
calculation future ages. For Esmeralda this can be explained by return
flow from irrigation, whereby exchange with modern soil-CO_2 generated in
the orchards of this artificial oasis in the Pampas might have occurred.

Duplijza is very close to modern if high ^{13}C-carbonates are
available for dissolution and a very young age is not unrealistic, as this
well is located at the foot of the Andes and receives modern recharge.

Interesting are the water ages of the wells at Pintados and
Canchones where also from a strictly hydrologic point of view one would
have expected older water--provided no significant vertical recharge occurs:
The aquifers in the Pampa have an average transmissivity between 500 m^2/day
and 1000 m^2/day which would lead to water ages between 16,500 and 33,000
years for Pintados and between 5,000 and 11,000 Y for Canchones assuming
both originate at the Quebrada Juan de Morales. The ^{14}C-ages range from
13,070 to 19,160 Y for Pintados and 1060 Y to 7900 Y for Canchones. This
agreement is encouraging and indicates that in this project area ^{14}C-age
determinations can fulfill an important task. It also shows that much
of the groundwater in the Pampa del Tamarugal has to be treated as a non-
renewable resource since minor active recharge occurs apparently only in
the vicinity of quebradas. More detailed studies are warranted.

ACKNOWLEDGEMENTS

        The field and isotope work have been sponsored and supported by
the Comision Chilena de Energia Nuclear and received assistance to various
degrees from UNDP and IAEA. Specifically we would like to thank Dr. B. R.
Payne (IAEA, Vienna) and his collaborators for their assistance during
some phases of the analytical work and the discussion of the results
obtained. Also acknowledged is the assistance provided by the University
of Waterloo, Canada, where some of the isotope analyses were carried out
with the help of Mr. R. J. Drimmie and Ms. J. Catterall. Drafting was done
by Ms. M. Maziarz. Financial assistance was also provided through the
National Research Council of Canada to P.F. (grant # A7954).

REFERENCES

[1] SILVA, G., SALATI, E., SUZUKI, O., FRITZ, P., Origen de los aguas subterráneas de
    la Pampa del Tamarugal., Rep. Tech. Sess., Confaqua, Mar Del Plata, Argentina,
    March 1977, 23 pp.
[2] STOERTZ, G.E., ERICKSEN, G.E., US Geol. Survey Prof. Pap. 811, Washington (1974)
    65 pp.
[3] Report to the Geothermal Coordinator, D.S.I.R., Wellington, New Zealand on: Hydrogen
    and oxygen isotope ratios in the waters of the El Tatio hydrothermal area, Chile
    (July 1969).
[4] RIGHTMIRE, C.I., HANSHAW, B.B., Wat. Resour. Res. 9 (1973) 958.
[5] FRITZ, P., REARDON, E.J., BARKER, J., BROWN, R.M., CHERRY, J.A., KILLEY,
    R.W.O., MCNAUGHTON, D., Wat. Resour. Res. (1978) in press.
[6] WINOGRAD, I.J., FARLEKAS, G.M., in Isotope Techniques in Groundwater Hydrology
    1974 (Proc. Symp. Vienna, 1974) 2, IAEA, Vienna (1974).

[7] REARDON, E.J., FRITZ, P., J. Hydrol. **36** (1978) 701.

[8] FONTES, J.C., GARNIER, M.J., C.R. Rénn. Ann. Sc. Terre, Geol. Soc. Fr., Paris (1976).

[9] MOOK, W.G., in Interpretation of Environmental Isotope and Hydrochemical Data in Groundwater Hydrology (Proc. Advisory Group Meeting, Vienna, 1975), IAEA, Vienna (1976) 213.

[10] PEARSON, J.F., WHITE, D.E., Wat. Resour. Res. **3** (1967) 251.

DISCUSSION

H. ZOJER: Since the sediments of the Pampa del Tamarugal can be characterized as horizontally well-permeable, I would like to know the depth at which aquifers can be reached by boreholes.

C. SILVA HENNINGS: Most of the wells sampled in the study area are between 80 and 200 m deep.

H. ZOJER: Can you tell us anything about borehole logging in this area, for example about results of resistivity or self-potential logging?

C. SILVA HENNINGS: The system of aquifers in the Pampa del Tamarugal is rather complex. We have no information on borehole logging, although geophysical studies have been carried out in the areas mentioned in the paper.

S.K. GUPTA: How did you collect the sample of CO_2 in equilibrium with the dissolved carbonate in water?

P. FRITZ: This CO_2 was not sampled but both pCO_2 and its $\delta^{13}C$ were calculated using field pH, alkalinities and ^{13}C analyses of the dissolved inorganic carbon.

S.K. GUPTA: Did you attempt any correlation between the elevation of a given well and the ^{18}O content of the water in the well?

P. FRITZ: All these wells are located in the low-lying areas of the drainage basins and so no altitude-$\delta^{18}O$ relationships can be observed. However, a regional distribution of ^{18}O or deuterium can be found.

T. TORGERSEN: Since evaluation of water-supply problems in arid areas depends on information about yearly variability, can you say what methods can be used to extend the time frame of your study? In particular, could the trees shown in the slides during your oral presentation be used for dendrochronological reconstruction of past water supply?

P. FRITZ: Most of the trees shown are very young, since the old forests were burned in the nitrate mines. However, the United States Geological Survey (USGS) has undertaken an extensive study of ancient water levels in lakes which today are salars. Both the study and historical records show that the hydrological activity of these systems has decreased with time and there is reason to believe that it is continuing to do so. Little, however, is known about yearly cycles, and dendrochronological studies would be difficult as trees are very rare in the areas of higher rainfall. Nevertheless, it is not impossible that suitable trees might be found.

Y. YURTSEVER: I was most interested in your use of cumulative probability distribution curves in Fig.6 for differentiating groups of waters, and I believe that such curves can be very useful for the interpretation of isotope data. However, one has to be very cautious when making the interpretations. For example, a hydrologist would perhaps draw a single line from the data you have shown, rather than drawing a broken line defining different groups of waters.

P. FRITZ: This distribution or grouping of deuterium values has been verified by careful interpretation of the ^{18}O curve. A further verification can be made by direct comparison between ^{18}O and deuterium as well as the regional distribution patterns.

PALAEOCLIMATIC TRENDS DEDUCED FROM THE HYDROCHEMISTRY OF A TRIASSIC SANDSTONE AQUIFER, UNITED KINGDOM

A.H. BATH, W.M. EDMUNDS
Institute of Geological Sciences,
Hydrogeology Unit,
Wallingford

J.N. ANDREWS
School of Chemistry,
University of Bath, Bath,
United Kingdom

Abstract

PALAEOCLIMATIC TRENDS DEDUCED FROM THE HYDROCHEMISTRY OF A TRIASSIC SANDSTONE AQUIFER, UNITED KINGDOM.

A detailed geochemical study (elemental, isotopic and dissolved inert gases) of unconfined and confined sections of the Triassic non-marine sandstone aquifer in eastern England has been undertaken. Aspects of the recharge history of this aquifer over the past 40 000 years are revealed by examination of the data. ^{14}C activity and δ^{13}C values show downgradient decrease and increase, respectively, the latter from -12 to $-13‰$ (PDB) at outcrop to $-8.5‰$ in deep confined groundwaters, indicating a continuing reaction between water and carbonate mineral phases. Although the bulk carbonate contained in sandstone samples gives δ^{13}C around $-7‰$, modelling of the carbon isotopic evolution and consideration of the resulting ^{14}C age corrections suggest that a carbonate with δ^{13}C closer to $0‰$ has played a major rôle in at least the earlier stages of hydrochemical evolution. The corrected radiocarbon age ranges, generated by the computer program WATEQF−ISOTOP, are used as a framework in which the palaeo-environmental information from oxygen and hydrogen isotope data, inert gas contents, and chloride levels are discussed. The measurement of dissolved helium levels demonstrates an excess of ^{4}He in many samples, which correlates with radiocarbon ages. The assumption of bulk chemical and physical properties for the aquifer rock allows independent 'excess ^{4}He' ages to be computed, which are mostly in excess of the corrected ^{14}C ages. The trend of δ^{18}O is from about $-8‰$ at outcrop to $-9.7‰$ (SMOW) downgradient, which is significantly more depleted than has been found in previous studies of UK basins. The δ^2H and δ^{18}O values are related by the regression line $\delta^2H = 6.6\, \delta^{18}O + 1$; data from other UK studies also lie close to this trend. The heavy isotope depletion down-dip indicates recharge in a colder climate. The dissolved contents of argon and krypton also indicate a similar change in palaeotemperatures, though the calculated magnitude of change $(5-6°C)$ is greater. Chloride levels in recharging water were considerably lower in the early Holocene then they have been in recent precipitation implying changes in atmospheric circulation. All this palaeo-environmental information is shown to be consistent with the present accepted model for climatic change over the British Isles during this period, which includes the most recent glacial episode of the late Pleistocene (Devensian).

1.   INTRODUCTION

This paper presents aspects of interpretation arising from the
chemical and isotopic study of one of the major British Triassic sandstone
aquifers. A parallel paper on this aquifer [1] aims to identify the
significant geochemical processes - solution-mineral equilibria, redox, and
exchange reactions - operating in the aquifer profile and to use this
framework to interpret the distribution and mobility of minor and trace
elements. The Triassic sandstones are quantitatively the second most
important aquifer system in the U.K., and similar hydrochemical reconnais-
sance studies have been made of other areas [2].

A major feature of the present study is the application of many of the
chemical and isotopic measurements available, including the analysis of
inert gas contents, and the opportunity to assess and compare their significance in determining the recharge and evolutionary history of the aquifer.
Further detailed discussion of inert gas measurement and interpretation
appears elsewhere [3, 4].

2.   HYDROGEOLOGY

The outline hydrogeology of the 2000 km^2 area has been described by
Land [5] and the geological setting of the boreholes mentioned in this
study are shown in Figure 1, where the degree of penetration of the aquifer
is indicated.

The Bunter Sandstone is generally red brown or yellow with a poorly
cemented quartzose lithology, varying in thickness from around 120 m in
the south to 300 m in the north. The formation is fairly homogeneous
although an upper and lower sandstone separated by a marl can be recognised
whilst the aquifer still acts effectively as a single hydraulic unit.
The aquifer dips gently to the east at about 1 in 50. The sandstone is
confined effectively in the east by a thick marl sequence (Keuper) although
fine grained interbedded sandstones near the base of the Keuper can locally
act as a transitional minor aquifer. The Bunter Sandstone is underlain by
a mixed sequence of Permian mudstones and marls which in this region form
an impermeable basal seal except perhaps where cut by faults.

The Bunter Sandstone is believed to have been deposited in a fluviatile
environment during semi-arid continental conditions [6]. The Bunter is
bounded above and below by transitional non-marine lithologies and therefore represents a unique example of an aquifer with little or no marine
influence on its lithology; this has an important effect not only on the
groundwater chemistry as a whole, but also on the isotope hydrogeochemistry.
A cored borehole was used in this study to investigate the controlling
lithochemistry in detail and the results are discussed elsewhere [1]. It
was found that dolomite and calcite cements and/or detritus were both
present in roughly equal amounts and that the carbonate content
increased from under 1% to around 4% towards the base of the borehole. The
carbonate of the rock is shown isotopically to be non-marine in origin and
the significance of this is discussed below.

The intergranular hydraulic conductivity of the sandstone averages
3.4 metres day^{-1} with a mean porosity of 30% [7]; the hydraulic conduc-
tivities determined from pumping tests on the other hand are typically 5
times higher than this reflecting the dominance of fissure flow over
intergranular flow in the system [8]. The natural hydraulic gradients are
very low - around 1 in 250 in the outcrop area - and in the confined

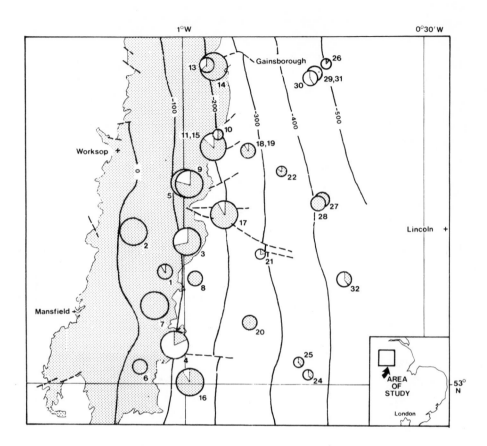

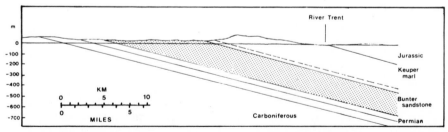

FIG.1.   Hydrogeological map of Triassic sandstone study area in the English East Midlands
showing sampled localities, numbered as in Table I.   Circle diameters indicate approximate daily
abstraction rate in $m^3 d^{-1}$.   Small diameter $< 10^3 m^3$, median $10^3 - 10^4 m^3$, large $> 10^4 m^3 d^{-1}$.
Segment size indicates percentage penetration of borehole into the Bunter Sandstone aquifer.
Contours are in 100-m intervals on the base of the Bunter aquifer.

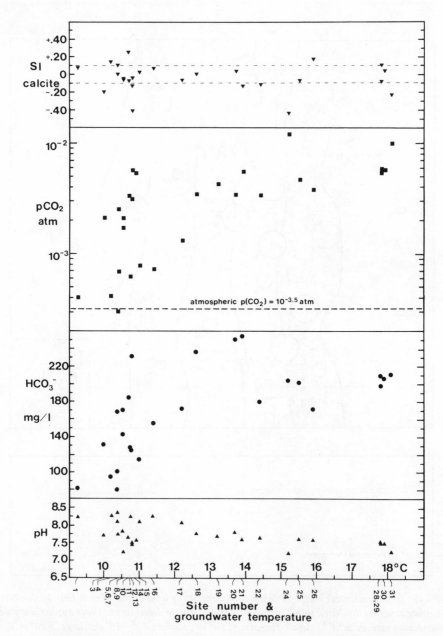

FIG.2. *Downgradient trends of pH, HCO$_3^-$, PCO$_2$ and saturation index of calcite (SI$_c$) in Bunter groundwaters. See text (Section 2) for explanation of use of temperature scale on abscissa, and Table I for identification of sites.*

section the piezometric surface is subhorizontal. The natural gradients,
especially in confined sections are significantly modified and extensive
cones of depression exist around major borehole sites so that to a large
extent the water samples from each pumped well represent a 'radial
scavenging' rather than interception of down-gradient flow.

Although there is evidence of considerable pollution in some of the
outcrop areas, much of the area is considered to represent an uncontaminated body of groundwater where natural processes predominate.

The geochemical data presented in the following discussion are derived
from pumped sources each of which differs considerably in depth and in
pumping régime, within an aquifer block which is essentially hydrogeologically uniform. In order to account for three dimensional differences
in aquifer characteristics across this area and to reduce the data to a
single line of section, the groundwater temperature is used as a 'distance'
scale in subsequent diagrams. This has the advantage of separating wells
of different depth at the same site and takes account of mixing.

3. CARBONATE HYDROCHEMISTRY

3.1 Salient features

The carbonate and sulphate hydrochemistry of this aquifer is described
in detail elsewhere [1]. The salient features of the carbonate system are
the growth of HCO_3^- contents from about 80 mg/l in the recharge zone to
over 200 mg/l down-dip in the confined aquifer, and relatively constant
levels of Ca^{2+} and Mg^{2+} throughout the aquifer. Field measurements of pH
show a slight but significant decrease down-gradient (about pH 8.0 to about
pH 7.5; Figure 2) and the computed state of equilibrium with respect to
calcite is found to be consistently close to saturation (Figure 2;
SI_c = saturation index = log $[a_{Ca^{2+}} \cdot a_{CO_3^{2-}}/K_{calcite}]$). Thus, equilibrium

with respect to calcite is attained at an early stage following infiltration; reaction coefficients for Ca^{2+} and HCO_3^- suggest that the dominant
reaction by which equilibrium is attained is:

$$CaCO_3 + H_2CO_3^0 \longrightarrow Ca^{2+} + 2HCO_3^-$$

Following the attainment of equilibrium, this stoichiometry may be
subsequently altered by external influences on pH and by incongruent
solution reactions of carbonate minerals. Additionally the influx of Ca^{2+}
associated with SO_4^{2-} from gypsum-bearing formations occurs at some points
in the confined aquifer, resulting in disturbance of the carbonate equilibrium and the possibility of carbonate precipitation to maintain this
equilibrium.

3.2 CO_2, $^{13}C/^{12}C$ and the carbonate reaction system

In a carbonate aquifer, it has been demonstrated that the availability
of a buffering reservoir of gaseous CO_2 and the evolution of the aqueous
$^{13}C/^{12}C$ ratio are closely inter-linked [9]. The concept of 'open' and
'closed' system behaviours, as discussed by Langmuir [10], may be applied
to such aquifers using as criteria the evolutionary trends of the chemical
and isotopic parameters (pH, pCO_2, $\delta^{13}C$). Figure 2 shows the variation in
pH, HCO_3^- and pCO_2 in groundwaters from the area under investigation. pCO_2
values are computed as the partial pressures of a gas phase in equilibrium
with (i.e. potentially buffering or being buffered by) the particular

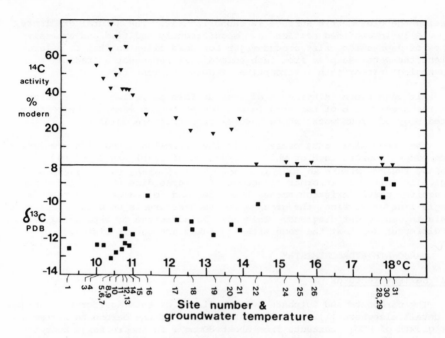

FIG.3. *Downgradient trends of $\delta^{13}C$ and ^{14}C measurements in Bunter groundwaters.*

groundwaters; equilibrium constant expressions relating pCO_2 with measured pH and $m(HCO_3^-)$ values are taken from Harned and Davis [11]. $\delta^{13}C$ values (Figure 3) represent analyses for <u>total</u> dissolved inorganic carbon species, and are measured on aliquot samples of the barium carbonate collected for ^{14}C determinations.

The pattern of pCO_2 values (Figure 2) within the aquifer is one of a general increase from close to atmospheric values ($10^{-3.5}$ atm) in parts of the unconfined aquifer up to 10^{-2} atm in the deeper parts of the confined aquifer. Within the unconfined aquifer, however, extensive variation in computed pCO_2 values is found, which suggests that a closed system model might describe the behaviour more closely. Although a spatially homogeneous buffer of CO_2 seems unlikely, it must be admitted that the possibility of reaction 'domains' each with its own externally buffered CO_2 reservoir is not ruled out by these observations.

The suggestion of closed system behaviour in the unconfined, as well as the confined aquifer, is borne out by computations of the $\delta^{13}C$ values for hypothetical gaseous CO_2 in equilibrium with each water sample. This calculation is possible by means of applying established fractionation factors for $^{13}C/^{12}C$ amongst the inorganic carbon species to the isotopic mass balance over all these species which takes the form:

$$mC_T \cdot \delta^{13}C_T = (mHCO_3^- \cdot \delta^{13}C_{HCO_3^-}) + (mCO_3^{2-} \cdot \delta^{13}C_{CO_3^{2-}}) + (mH_2CO_3^0 \cdot \delta^{13}C_{H_2CO_3^0})$$

The fractionation factors (from Deines et al., [9]) are used to relate the required parameter $\delta^{13}C_{CO_2(gas)}$ to each of the $\delta^{13}C$ values on the right-

hand side of the mass balance equation, and permit solution for $\delta^{13}C_{CO_2}$. Computed values for $\delta^{13}C_{CO_2}$ lie in the range -19.7 to -22.0 ^{o}/oo (vs PDB) in the unconfined aquifer, decreasing to a minimum of -15.8 ^{o}/oo in the confined section. Such values of $\delta^{13}C$ are somewhat higher (i.e. less negative) than values which might obtain for soil CO_2 (i.e. the potential 'open' system reservoir) in a temperate region. Thus the closed system model tends to be supported by isotopic as well as chemical measurements.

4. ^{14}C MEASUREMENTS AND GROUNDWATER MODELLING

    The computer program WATEQF-ISOTOP [12] has been applied to the present set of chemical and isotopic data to assist in the elucidation of carbonate system behaviour. One objective of modelling the groundwater evolution is to enable more reliable and rigorous corrections to be applied to age information derived from ^{14}C measurements. WATEQF-ISOTOP utilises the equilibrium specification of the carbonate system generated by the program WATEQ [13, 14] to generate models for carbon isotope behaviour corresponding to various assumed (Model 1) or imposed (Model 2) initial conditions. The identification of plausible pathways by which observed chemical and isotopic parameters might have evolved permits also the computation of specific corrected ^{14}C ages for the matrix of pathways. The inspection of generated ages, as well as the elimination of unlikely pathways as the number of sampling points with which to test the model in the same aquifer increases, should reduce the choice further. The assumptions and limitations of the ISOTOP subroutine should also be stressed, as they are by Reardon and Fritz [12]. In particular, it should be emphasised that this model considers only the transition, by closed-system reaction, between initial solution conditions (which may themselves represent open-system conditions) and the observed groundwater composition. The model is independent of the incremental pathway of reaction, within the assumptions implicit in its application (i.e. closed-system carbonate reaction only and continued isotopic disequilibrium between dissolved and solid carbonate). In the initial application of the WATEQF-ISOTOP model, the values of $\delta^{13}C$ for soil CO_2 and the reacting carbonate mineral phase were estimated to be -26.1 ^{o}/oo and 0 ^{o}/oo (vs PDB) respectively. The value assumed for $\delta^{13}C$ (soil CO_2) is based on a reported average value [15] for organic matter in U.K. soils. Table I lists the corrected ^{14}C data computed for each set of possible initial conditions for each groundwater sample, as suggested by the model. The listing demonstrates clearly how proportionately substantial are the computed corrections to ^{14}C data, particularly at an early stage of groundwater evolution in this case.

    Seven samples of drillcore from the Bunter Sandstone gave $\delta^{13}C$ values around -7 ^{o}/oo (PDB) for their carbonate content. The ISOTOP model was tested using this new value for the parameter representing new inorganic carbon entering the solution phase. As one would expect for computations assuming an isotopically lighter carbonate rock, the corrections computed for the ^{14}C data are greater than in the initial computations using $\delta^{13}C$ (rock) = 0 ^{o}/oo as Table II shows. However it is seen that the ranges of corrected ^{14}C 'model' ages given by many of the samples, including all those from the unconfined aquifer, using the lighter value for $\delta^{13}C$ (rock) are seriously negative or 'future' ages. It is clear, in spite of the evidence for the presence of carbonate minerals at -7 ^{o}/oo $\delta^{13}C$, that the

TABLE I. TRITIUM AND ¹⁴C DATA FOR EAST MIDLANDS TRIASSIC AQUIFER

Corrected ¹⁴C ages are tabulated for assumption of closed system evolution from initial pH and pCO₂ conditions stated. Assumed values for δ¹³C of rock carbonate and soil CO₂ are 0.0 and −26.1‰ versus PDB, respectively: modern ¹⁴C = 100%

No	Location	^{3}H (TU)	^{14}C Activity (% modern)	$\delta^{13}C_T$ (‰ PDB)	^{14}C Ages, Corrected by Closed System Model (WATEQ-ISOTOP)							
					$pH_i = 5$		$pH_i = 6$		$pH_i = 7$		$pH_i = 8$	
					Log pCO_{2i} (atm)	Age (yr)	Log pCO_{2i} (atm)	Age (yr)	Log pCO_{2i} (atm)	Age (yr)	Log pCO_{2i} (atm)	Age (yr)
1	Amen Corner		57.2	-12.6	-1.9	-1850	-2.0	-790	-2.4	1580		
3	Boughton	15±2	55.1	-12.4	-1.7	-1590	-1.8	-1110	-2.2	1210	-3.1	1380
4	Farnsfield		47.6	-12.4	-1.8	-610	-1.9	390	-2.4	2100		
5	Elkesley, No 6	8±2	77.6	-13.5	-1.5	-3730	-1.6	-2880	-2.1	-1000		
6	Far Baulker, No 3		63.7	-13.1	-1.8	-2640	-1.9	-1790	-2.4	120		
7	Rufford		42.2	-11.6	-2.0	190	-2.1	760	-2.5	2910		
8	Ompton, No 2		50.0	-12.8	-1.6	-510	-1.7	410	-2.1	2220		
9	Elkesley, No 5	25±2	70.8	-12.9	-1.7	-3180	-1.7	-2730	-2.1	-190	-3.1	50
10	Retford, Clarks No 2	4±2	52.6	-12.6	-1.5	-1100	-1.6	-300	-2.0	1690		
11	Ordsall, No 1	10±2	42.2	-11.9	-1.8	40	-1.9	720	-2.3	2810	-3.2	2980
12	Clarks, No 1	2±2	65.1	-12.3	-1.5	-2900	-1.5	-2300	-1.9	-10	-2.9	320
13	Everton, No 1	0±2	42.1	-11.5	-1.8	-190	-1.9	460	-2.3	2530	-3.2	2690
14	Everton, No 3		42.0	-12.4	-1.7	830	-1.8	1430	-1.9	3810	-2.9	4050
15	Retford, Whisker Hill		39.2	-11.8	-1.8	500	-1.9	1360	-2.3	3250		

Sample											
16 Halam	0±2	28.6	-9.7	-1.8	1560	-1.9	2220	-2.3	4260		
17 Markham Clinton, No 1		26.7	-11.0	-1.7	3530	-1.7	4020	-2.2	6290	-3.1	6840
18 Grove, No 3		19.8	-11.8	-0.5	6680	-1.5	7210	-2.0	9410	-2.9	9830
19 Grove, No 2		18.0	-11.8	-1.5	7350	-1.6	7960	-2.0	10100	-2.9	10600
20 Caunton		20.9	-11.2	-1.5	5760	-1.6	6300	-2.0	8490	-2.9	8910
21 Egmanton, B.P.	0.5±2	26.3	-11.5	-1.5	4000	-1.5	4700	-1.9	7060	-2.9	7270
22 Rampton	0.3±2	0.7	-10.1	-1.7	33000	-1.7	33400	-2.2	35700	-3.1	36100
24 Newark, British Gypsum		2.6	-8.5	-1.7	20600	-1.7	21300	-2.2	23400	-3.0	24200
25 Newark, Castle Brewery	0±2	1.6	-8.6	-1.7	24900	-1.8	25200	-2.2	27600	-3.1	27900
26 Gainsborough, B.P.		2.7	-9.4	-1.7	21100	-1.8	21500	-2.3	23900	-3.1	24100
28 Newton, No 2	0±2	1.5	-9.2	-1.7	25700	-1.7	26400	-2.2	27800	-3.0	28000
29 Gainsborough, Lea Road No 2	0±2	4.8	-9.7	-1.6	16600	-1.7	17400	-2.2	19200	-3.0	20700
30 Gainsborough, Humble Carr	2±2	1.6	-8.7	-1.7	24900	-1.8	25300	-2.2	27600	-3.1	28200
31 Gainsborough, Lea Road No 3		1.1	-9.0	-1.6	29800	-1.7	30600	-2.2	32500	-3.0	33700

TABLE II. TRITIUM AND CARBON-14 DATA FOR EAST MIDLANDS TRIASSIC AQUIFER

Corrected ^{14}C ages are tabulated for assumption of closed system evolution from initial pH and pCO_2 conditions stated. Assumed values for $\delta^{13}C$ of rock carbonate and soil CO_2 and -7.0 and $-26.1\permil$ versus PDB, respectively; modern $^{14}C = 100\%$

| No | Location | ^3H (TU) | ^{14}C Activity (% modern) | $\delta^{13}C_T$ (°/oo PDB) | ^{14}C Ages, Corrected by Closed System Model (WATEQ-ISOTOP) | | | | | | | | |
| | | | | | $pH_i = 5$ | | $pH_i = 6$ | | $pH_i = 7$ | | $pH_i = 8$ | |
					Log pCO$_2$i (atm)	Age (yr)	log pCO$_2$i (atm)	Age (yr)	log pCO$_2$i (atm)	Age (yr)	log pCO$_2$i (atm)	Age (yr)
1	Amen Corner		57.2	-12.6	-2.2	-5880	-2.2	-4950	-2.6	-1990		
3	Boughton	15±2	55.1	-12.4	-2.0	-5940	-2.0	-4870	-2.4	-2260	-3.3	-1740
4	Farnsfield		47.6	-12.4	-2.1	-4800	-2.2	-3640	-2.5	-850		
5	Elkesley, No 6	8±2	77.6	-13.5	-1.8	-7450	-1.9	-6270	-2.2	-3250	-3.1	-2110
6	Far Baulker, No 3		63.7	-13.1	-2.1	-6130	-2.1	-5150	-2.5	-?230		
7	Rufford		42.2	-11.6	-2.3	-5290	-2.4	-4470	-2.7	-1140		
8	Ompton, No 2		50.0	-12.8	-1.9	-4800	-1.9	-3770	-2.3	-690	-3.1	100
9	Elkesley, No 5	25±2	70.8	-12.9	-1.9	-7480	-2.0	-6250	-2.4	-3530	-3.3	-2940
10	Retford, Clarks No 2	4±2	52.6	-12.6	-1.8	-5490	-1.9	-4570	-2.2	-1330	-3.1	-350
11	Retford, Ordsall No 1	10±2	42.4	-11.9	-2.0	-4450	-2.1	-3570	-2.5	-550	-3.3	-240
12	Retford, Clarks No 1	2±2	65.1	-12.3	-1.7	-7410	-1.8	-6750	-2.1	-3500	-3.0	-2020
13	Everton, No 1	0±2	42.1	-11.5	-2.1	-5240	-2.2	-4310	-2.5	-1200	-3.4	-880
14	Everton, No 3		42.0	-12.4	-1.7	-3470	-1.8	-2840	-2.2	330	-3.0	1820
15	Retford, Whisker Hill		39.2	-11.8	-2.1	-4070	-2.2	-3020	-2.5	-130		

	Site												
16	Halam		28.6	-9.7	-2.2	-6240	-2.4	-5530	-2.7	-2150	-3.5	-1030	
17	Markham Clinton, No 1	0±2	26.7	-11.0	-2.0	-2280	-2.1	-1360	-2.4	1410	-3.3	2050	
18	Grove, No 3		19.8	-11.8	-1.8	1480	-1.8	2300	-2.2	5740	-3.0	7030	
19	Grove, No 2		18.0	-11.8	-1.8	2130	-1.9	3040	-2.2	6360	-3.1	7500	
20	Caunton		20.9	-11.2	-1.8	-370	-1.9	630	-2.2	3900	-3.1	4890	
21	Egmanton, B.P.	0.5±2	26.3	-11.5	-1.8	-1310	-1.8	-530	-2.2	2770	-3.0	4150	
22	Rampton	0.3±2	0.7	-10.1	-2.1	25400	-2.1	26400	-2.5	29700	-3.4	30000	
24	Newark, British Gypsum		2.6	-8.5	-2.3	8630	-2.4	9300	-2.7	12500			
25	Newark, Castle Brewery	0±2	1.6	-8.6	-2.3	13100	-2.4	14000	-2.7	17400	-3.5	17600	
26	Gainsborough, B.P.		2.7	-9.4	-2.2	12700	-2.3	13400	-2.6	16300			
28	Newton, No 2	0±2	1.5	-9.2	-2.2	14200	-2.3	15100	-2.6	18600	-3.5	19800	
29	Gainsborough, Lea Road No 2	0±2	4.8	-9.7	-2.1	8360	-2.1	9400	-2.5	12600	-3.4	12800	
30	Gainsborough, Humble Carr	2±2	1.6	-8.7	-2.3	13400	-2.4	14200	-2.7	17800			
31	Gainsborough, Lea Road No 3		1.1	-9.0	-2.2	20000	-2.2	20900	-2.6	23600	-3.5	24700	

ISOTOP model does not give a satisfactory representation of the hydrochemical evolution if that value is used for source material for dissolving
inorganic carbon. A more concordant picture of corrected ages is given by
computations using a heavier $\delta^{13}C$ (rock) value (e.g. 0 ^{o}/oo as in Table I).
This contention is further supported by the observation that in the
unconfined aquifer many of the samples show significant levels of tritium
combined with ^{14}C activities around 50 ± 10 pmc (percent modern carbon
activity) and $\delta^{13}C$ values around -12 to -13 ^{o}/oo. This suggests that (on a
simple model setting aside the possible effects of mixing of waters of
different ages) the equilibrium with respect to calcite which obtains for
many of these samples has been attained with relative rapidity and by the
stoichiometric carbonate dissolution of a carbonate with $\delta^{13}C$ around 0 ^{o}/oo
PDB. Indeed, the WATEQF-ISOTOP model allows more explicit definition of
the initial conditions of pH and pCO_2 which might lead by congruent solution
under closed-system conditions to the observed ^{14}C activity whilst also
showing concordance with the positive tritium levels (i.e. near-zero
corrected ^{14}C age).

     It is therefore necessary to postulate the existence of carbonate
material within the unsaturated zone or shallow saturated aquifer which is
isotopically distinct from the bulk carbonate content of Bunter material
which has been found to have $\delta^{13}C$ around -7 ^{o}/oo. Evidence suggesting the
possibility of two coexisting carbonate phases in the Bunter is given by
analysis of solutions resulting from acid-leaching of the carbonate
material. The molar Ca:Mg ratios in these solutions lie in the range
1.5-1.0 [1] which indicates that dolomite may be present in addition to
calcite. The activity ratio of Ca:Mg in groundwater samples also indicates
that equilibrium with respect to dolomite as well as calcite is achieved as
the groundwater enters the confined part of the aquifer [1]. Thus the
carbonate phase with $\delta^{13}C$ = 0 ^{o}/oo could be a dolomite or calcite occurring
as a minor component but more susceptible to solution than the bulk carbonate present as cementing material within the sandstone. An alternative
source for a carbonate of different isotope composition is the soil zone
itself in which it is possible that substantial amounts of detrital
carbonate material possibly transported by ice or fluvially may have
persisted.

     The possibility that the hydrochemical evolution of groundwaters in
the confined aquifer has taken place in two (or more) stages with reactions
with isotopically distinct carbonates must also be considered. The first
stage might be the early reaction with carbonate ($\delta^{13}C \sim 0$ ^{o}/oo) which is
strongly suggested by the above data, followed by a second stage in which
further reaction (possibly incongruent dissolution) with the bulk carbonate
material ($\delta^{13}C \sim -7$ ^{o}/oo) takes place. The second stage could continue to
shift the $\delta^{13}C$ in solution until a state of isotopic equilibrium between
solution and rock was reached (when $\delta^{13}C_{solution} - \delta^{13}C_{rock} \sim -2.5$ ^{o}/oo
based on fractionation data for 15°C, [9, 17]. It is possible to model the
second reaction stage and thereby obtain a corrected ^{14}C age for its
duration, by means of the Model 2 subroutine of ISOTOP [12]. For present
purposes initial solution conditions for Model 2 computations were
represented by the young groundwaters in the unconfined aquifer. Table III
lists examples of computed age differences between such a groundwater just within the confined aquifer and some of the more evolved groundwaters, assuming
a closed system reaction with a carbonate mineral phase at $\delta^{13}C \sim -7$ ^{o}/oo.
It may be noted that age differences computed using Model 2 fall within the
range of corrected Model 1 ages when $\delta^{13}C$ (rock) is assumed to be 0 ^{o}/oo.
Table III also shows that there is little difference in computed ages

TABLE III.   WATEQF–ISOTOP MODEL 2 AGES COMPUTED FOR GROUNDWATERS IN THE CONFINED
TRIASSIC AQUIFER.
Initial conditions for closed system model are set equivalent to those observed at Ompton in the confined aquifer.
$\delta^{18}C$ (rock) value set at $-7‰$

No	Location	Model 2 ages (years)		Moles of calcite ppt (Ca balance)
		Congruent solution	Incongruent solution	
22	Rampton	34700	33900	0.00052
24	Newark, British Gypsum	22000	22000	0.00002
25	Newark Castle Brewery	26900	26800	0.00002
28	Newton No 2	27400	26700	0.00050
30	Gainsborough, Humble Carr	26600	24500	0.00122
31	Gainsborough, Lea Road No 3	30700	29300	0.00090

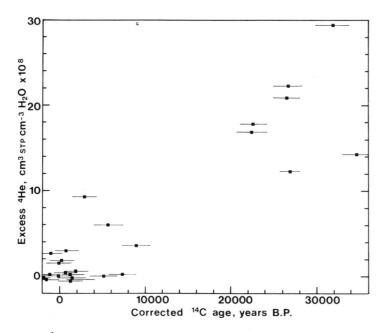

*FIG.4.   'Excess' ^4He in Bunter groundwaters plotted as a function of 'corrected' radiocarbon
age (see Section 5).*

TABLE IV. ^{4}He EXCESS IN GROUNDWATERS AS A RESULT OF ACCUMULATION DUE TO RADIOACTIVE
DECAY OF U, Th AND DAUGHTER PRODUCTS

No	Location	Excess ^{4}He cm^3 STP cm^{-3} H$_2$O x10^8	^{4}He age[a] (yr)	Corrected ^{14}C age range[b] (yr) min	max
1	Amen Corner	0.0	−	−1800	1600
2	Boughton	1.5	4400	−1600	1400
4	Farnsfield	0.4	1100	−600	2100
5	Elkesley, No 6	0.0	−	−3700	−1000
6	Far Baulker, No 3	0.0	−	−2600	100
7	Rufford	0.2	600	200	2900
8	Ompton, No 2	3.0	8800	−500	2200
9	Elkesley, No 5	0.0	−	−3200	50
10	Retford, Clarks No 2	1.8	5300	−1100	1700
11	Retford, Ordsall No 1	0.0	−	40	3000
12	Retford, Clarks No 1	2.7	8000	−2900	300
13	Everton, No 1	0.0	−	−200	2700
14	Everton, No 3	0.0	−	800	4100
15	Retford, Whisker Hill	0.0	−	500	3300
16	Halam, No 1	9.3	27000	1500	4300
17	Markham Clinton, No 1	0.0	−	3500	6800
19	Grove, No 2	3.6	10600	7300	10600
20	Caunton	0.0	−	5800	8900
21	Egmanton, B.P.	6.0	17600	4000	7300
22	Rampton	14.3	42100	33000	36100
24	Newark, British Gypsum	16.9	49800	20600	24200
25	Newark, Castle Brewery	20.9	61500	24900	27900
26	Gainsborough, B.P.	18.8	55300	21100	24100
28	Newton, No 2	12.3	36200	25700	28000
30	Gainsborough, Humble Carr	22.3	65600	24900	28200
31	Gainsborough, Lea Road No 3	29.4	86500	29800	33700
32	South Scarle	12.7	37400		

Notes: [a] ^{4}He age calculated from relationship t = 2.94 x 10^{11} [He] where constant is
derived from rock properties and U, Th contents and decay constants; see text.

[b] Corrected ^{14}C age range generated by WATEQF-ISOTOP assuming δ^{13}C (rock) = 0 ^{o}/oo;
see Table I.

between models which assume congruent and incongruent carbonate solution
processes. Consideration of mass balance of Ca between initial and final
states shows that generally small amounts of secondary calcite precipitation are suggested by this model (last column of Table III); in most cases,
consideration of inorganic C mass balance suggests that no such precipitation has occurred.

5.   GROUNDWATER AGE AND ^{4}He CONTENT

It has been shown in the preceding section how attempts may be made to
correct ^{14}C measurements for the effects of dilution during groundwater
evolution by means of chemical and isotopic modelling techniques. However,
it is also clear that the simple, though rigorous, models used are unable
to give precisely defined corrected ^{14}C ages for the complex hydrochemical
situation described here. At best, we are able only to infer corrected
'age-bands', (Figure 4), and the interpretation is further complicated by
the rôles of assumed δ^{13}C values for rock and open-system CO_2 in the
overall model. In the following discussion of palaeoclimatic indicators,
the corrected 'age-bands' computed on the basis of a δ^{13}C (rock) value of
∿ 0 ^{o}/oo (Table I) will be referred to. Though these may result from
over-simplified modelling, it is considered that the introduction of more
detailed models for corrected ages, as seen in the previous section, would
only confuse the discussion. To a very large extent, the relative age
distribution of the groundwaters is unaffected by the choice and detail of
models.

Levels of ^{4}He dissolved in these groundwaters have been measured [4],
and demonstrate a correlation with ^{14}C age (Figure 4). The secular increase
in ^{4}He is caused by the accumulation in the groundwater of radiogenic ^{4}He
resulting from the radioactive decay of U, Th and their daughter products.
Analyses for total ^{4}He in the water samples are corrected for the amount of
atmospheric ^{4}He which would have become dissolved in the water prior to
infiltration; values for 'excess' ^{4}He contents corrected in this manner are
listed in Table IV. If all ^{4}He generated by decay of radioelements in
aquifer material is assumed to dissolve in the groundwater it can be shown
that the content of ^{4}He in water is a function of density, porosity and U
and Th contents of the aquifer material through which the water has passed
and also of residence time in the aquifer [4]. The mode of water movement
in Bunter Sandstone is considered to provide suitable conditions for this
assumption to be justified, and therefore by using mean values for the
above parameters the age values in Table IV have been calculated. The
groundwater ages calculated from ^{4}He contents are considerably greater than
the corresponding corrected ^{14}C ages, though the correlation of ^{4}He content
with age is established (Figure 4). The ^{4}He age calculation is strongly
dependent on assumed values for bulk properties of the host-rock, and gives
rise to considerable scope for error. Values for these properties used in
the present computations (Table IV) are 30% intergranular porosity, bulk
density 2.15 g cm^{-3}, and U and Th contents of 1.9 and 8.7 µg/g respectively
in the sandstone (average values with about 70% scatter for drillcore
material). It is clear that refinements of both the procedure for correct-
ing ^{14}C age data and that for calculating ^{4}He ages might lead to better
agreement.

6.   OXYGEN AND HYDROGEN ISOTOPES AND INERT GAS CONTENTS

Stable oxygen and hydrogen isotope analyses of groundwater from many
of the sites are shown in Table V. In Figure 5, these data are plotted on

TABLE V.　$^{18}O/^{16}O$ AND $^{2}H/^{1}H$ ANALYSES OF TRIASSIC AQUIFER GROUNDWATERS.
Chloride levels are also shown

No	Location	$\delta^{18}O$ $^{o}/oo$ vs SMOW	$\delta^{2}H$ $^{o}/oo$ vs SMOW	Cl^{-} (mg/l)
1	Amen Corner	-8.6	-54	76
2	Budby, No 1			94
3	Boughton			35
4	Farnsfield			24
5	Elkesley, No 6	-8.3	-55	174
6	Far Baulker, No 3			18
7	Rufford			34
8	Ompton, No 2			70
9	Elkesley, No 5	-8.6	-56	99
10	Retford, Clarks No 2			34
12	Retford, Clarks No 1	-7.7	-51	63
13	Everton, No 1			18
14	Everton, No 3			19
15	Retford, Whisker Hill	-8.4	-55	
16	Halam, No 1			7.4
17	Markham Clinton, No 1	-8.3	-51	7.1
18	Grove, No 3	-8.3	-50	5.2
19	Grove, No 2	-8.4	-55	6.3
20	Caunton			5.6
21	Egmanton, B.P.	-8.1	-51	6.8
22	Rampton	-9.6	-62	6.9
24	Newark, British Gypsum	-9.8	-66	11.6
25	Newark, Castle Brewery	-9.8	-66	15
26	Gainsborough, B.P.	-9.3	-61	11.0
27	Newton, No 3	-9.1	-58	10.2
28	Newton, No 2	-9.2	-57	9.6
29	Gainsborough, Lea Road No 2	-9.3	-63	21
30	Gainsborough, Humble Carr	-9.8	-59	20
31	Gainsborough, Lea Road No 3			18
32	South Scarle	-9.3	-61	8.6

the conventional δ-diagram, also shown in this diagram are representative ranges for other U.K. groundwaters. The present set of data is considerably more depleted in isotopic composition than other data (which are from areas further south in the U.K. than the area under investigation). Additionally, the isotope compositions form two groupings, both falling above the world-wide precipitation line with slope 8. On closer inspection, it is found that the lower, more depleted, group of data comprises all the groundwaters with inferred ^{14}C ages older than at least 16600 years (it is stressed that this figure represents a <u>minimum</u> age only for this phenomenon).

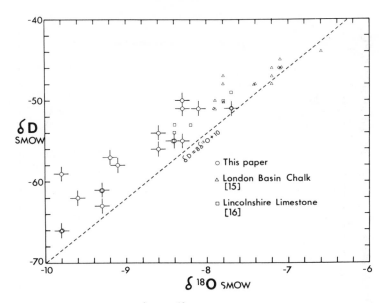

FIG.5.   Conventional δ-diagram of δ^2H vs $\delta^{18}O$ for Bunter groundwaters. Other data from UK aquifers are also shown.
Data from this study give a regression line $\delta^2H = 6.6\ \delta^{18}O + 1.3\ (r^2 = 0.83)$.

This relationship is more clearly depicted in Figure 6a, in which $\delta^{18}O$ values are plotted against ^{14}C age (corrected).

It is believed that the depleted isotopic composition of the pre-Holocene groundwaters is due to palaeoclimatic variation within the recharge history of this aquifer. Further support for this contention comes from studies of the inert gas contents of the groundwater [4]. The levels of the inert gases dissolved in a groundwater should reflect the temperature at which that water was last equilibrated with atmospheric gases during the recharge process, i.e. presumably within the soil zone. Contents of argon and krypton were measured by isotope dilution mass spectrometry [4] and palaeotemperature values were calculated from inert gas solubility data [18]. When the average of palaeotemperatures from argon and krypton contents (generally within 1-2°C of each other) are plotted against ^{14}C age data, a grouping very similar to that found for oxygen and hydrogen isotope analyses becomes apparent (Figure 6b), with the older group of groundwaters suggesting palaeotemperatures during recharge some 5-6°C lower than those for groundwaters recharged during the past 10000 years at least.

Two observed phenomena require discussion - (i) the gap between at least 10000 and 17000 years occurring in measured ^{14}C ages (though of course this may be in part due to coincidental lack of relevant sampling points), and (ii) the significance of the secular temperature changes between Upper Pleistocene and Holocene indicated by both the oxygen and hydrogen isotope and the inert gas data. It is important that the processes responsible for the O/H isotope systematics and the dissolved inert gas contents are distinguished. As mentioned above, inert gas contents attain

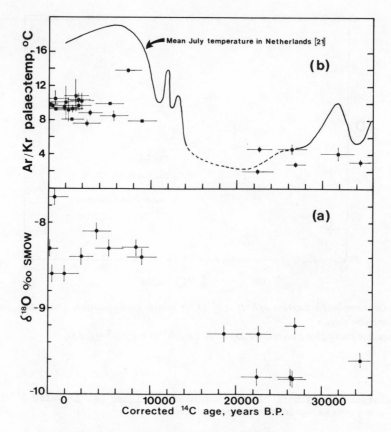

FIG.6. (a) *Plot of* $\delta^{18}O$ *values against corrected radiocarbon ages.*
        (b) *Plot of palaeotemperatures calculated from argon and krypton solubility data*
            against corrected radiocarbon ages.

equilibrium at the air-soil interface, and therefore would be expected to
reflect a mean soil temperature weighted for the annual distribution of recharge density. On the other hand the stable isotopic composition of rainfall is determined primarily by the temperature at which the condensation
process occurs, as well as by subsequent equilibrium or non-equilibrium
modifications.

A correlation exists between $\delta^{18}O$ variations and mean annual surface
air temperature changes:

$$\frac{\Delta\,\delta^{18}O}{\Delta T} \;\simeq\; 0.7\;^{o}/oo\;{}^{o}C^{-1}$$

which agrees well with theoretical predictions based on an isobaric cooling
model **if** it is assumed that condensation temperature varies in parallel
with surface temperature [19]. Therefore the change in $\delta^{18}O$ values shown
in Figure 6a may be considered to indicate a change in cloud-formation
temperature of at least 1-2°C over the time scale involved.

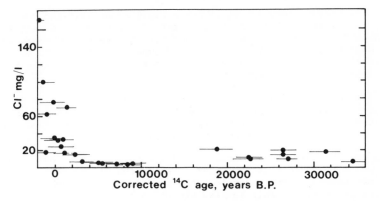

FIG.7.   Levels of chloride in Bunter groundwaters as a function of corrected radiocarbon age.

The interruption to recharge implied by the absence of corrected radio-carbon ages between 10000 and 17000 years must be considered in the context of the late Devensian (Weichselian) glaciation which affected northern Europe and North America. The glacial maximum is considered to have occurred around 20000 − 18000 years, although the advance of the ice sheets halted just to the north and east of this area [20]. It is therefore suggested that the recharge 'gap' may be caused by continuous permafrost conditions over the Bunter outcrop zone during this period. This suggestion will be considered later in the light of other palaeoenvironmental evidence.

The secular change of 5-6°C in surface temperature indicated by inert gas contents is similar to the rapid climatic amelioration documented by van der Hammen et al [21] for the Netherlands using pollen analyses (see curve on Figure 6b), and which is endorsed by analyses of English fossil beetle assemblages [22]. The smaller but contemporaneous change in temperature inferred from the oxygen and hydrogen isotopic compositions requires further discussion. Whilst the above relationship with tempera-ture was derived by Dansgaard [19] from total precipitation compositions and mean surface air temperature data, the present groundwater data represents an isotopic composition only for that fraction of total precipitation which infiltrated the aquifer. Thus the temperature coefficient of 0.7 °/oo °C^{-1} must be applied with caution in the present context. It is also possible that the magnitude of changes in cloud-formation temperature were significantly lower than concomitant changes in ground temperature; major changes in the patterns of air-mass movement are known to have occurred over this period.

7.   CHLORIDE

The groundwater chloride follows an unusuallbut consistent pattern along the profile, from high levels at outcrop to very low values down-gradient (Table V). With the exception of sites at Gainsborough, where some leakage from the Keuper series is suspected, the chloride values for confined waters lie between 5.2 and 15 mg/l. This is remarkable for two reasons - (i) the apparent negligible uptake of chloride within the aquifer system (South Scarle, the furthest east, has only 8.6 mg/l Cl$^-$), and (ii)

the very low values (around 6 mg/l), especially after allowing for enrichment due to evapotranspiration, in the west of the unconfined section.
Assuming that chloride has been conserved during movement down-gradient
then it must be concluded that the older groundwaters represent a
significantly less saline recharge than at present (present-day rainfall
in this area contains chloride at levels around 10 mg/l, [24]).

In Figure 7 the chloride data are plotted against the adjusted radiocarbon ages, showing that high chloride levels are restricted to recent
groundwaters, and that a consistently low Cl⁻ group is found for early
Holocene groundwaters (recharged after the last glacial maximum).

Two possible explanations for this distribution can be put forward.
Firstly it is possible that the water owes its low Cl⁻ to recharge from
glacial meltwaters either under ice sheets or from outwash channels behind
retreating ice sheets. This may explain the origin of some of the oldest
water in the profile but cannot explain the low Cl⁻ of the Holocene
groundwater since this conflicts with the enrichment of ^{18}O and depleted
inert gas levels in the Holocene waters indicating recharge from a warmer
precipitation. In addition, there is no evidence of ice cover over this
area during the latest glacial advance [20].

The most likely explanation for the low Cl⁻ groundwaters in particular
those of the early Holocene is that they were recharged in a period when
rainfall was derived from continental air masses, dominated by easterly
winds [23].

8.   CONCLUSION

The integrated geochemical and isotopic study of an aquifer system can
yield important information to both the hydrogeologist and the palaeo-
climatologist. The present study highlights some of the problems
encountered in the interpretation of carbon isotope data in an aquifer of
non-marine lithology; the relationship between carbonate cementing phases
within the Triassic sandstone and the evolving groundwater chemistry remains
unclear in view of the evidence in support of an isotopically heavier
source of carbon contributing at least to the early stage of hydrochemical
evolution. Clearly, the uncertainties in the parameters used in the
correction procedure remain an obstacle to obtaining corrected ^{14}C ages
with more confidence. The non-uniform nature of groundwater evolution,
including the possibility in various parts of the aquifer of incongruent
dolomite or calcite solution or gypsum reaction, may be more amenable to
realistic modelling by incremental mass transfer techniques (e.g. [25]).
Nevertheless, it is belived that the present procedures provide a framework of groundwater ages in which the significance of isotope and inert
gas data can be examined usefully.

The palaeoenvironment of the British Isles during the late Devensian
has been investigated by many different techniques - geomorphological,
palaeobotanical, palaeontological and so on (summarised by various authors,
[26]). The suggestion seems reasonable that the lobes of ice sheets advan-
cing southwards did so non-contemporaneously, bringing the apparent 'peak'
of glacial or periglacial conditions at different times in different places
[27]. The inference of periglacial permafrost in the East Midlands from
17000 years onwards may be compared with other evidence for permafrost
conditions in the West Midlands from 18000-13500 years BP following a
maximum ice advance there at 20000-18000 years BP [28]. Analogy of the
Devensian environment with the present-day arctic and sub-arctic North

American environment suggests that aquifer recharge would have occurred, if
at all, predominantly during the summer months. Therefore the 5-6°C
decrease in soil-air temperatures during the recharge of pre-glacial groundwaters, as suggested by inert gas solubilities, is typical of a very cold
tundra environment in which mean annual temperatures were considerably
lower overall [29].   Similarly, the magnitude of atmospheric temperature
changes reflected by the oxygen and hydrogen isotope data in recharge
waters is even smaller (1-2°C) than that of surface temperatures. Chloride
concentrations taken together with the isotopic data suggest that they may
relate to a change in atmospheric circulation patterns over the British
Isles during the Holocene.

        Having used the geochemical and isotopic data to derive a palaeo-
climatic model it is then possible to examine its impact on the hydrogeological interpretation and to use it as a framework for considering the
solute movement downgradient - a topic to be dealt with in a subsequent
paper. However, it is clear that the large proportion of groundwater
stored in the Bunter was recharged during the late Pleistocene and early
Holocene and that the chemically distinct current recharge has not moved
significantly into the confined aquifer. It is also apparent that water
containing radiocarbon has moved eastwards under natural gradients for
some 20 km, this implies that there is an outlet to the east within the
North Sea basin. The palaeoclimatic model can also be useful in the
interpretation of the overall water quality, in delineating zones of
ancient "non-polluted" water and the extent of mixing with recent waters.
In fact the multiple evidence of the palaeoclimatic model of the aquifer
provides a much more reliable basis for dating the groundwater over the
past 35000 years than does radiocarbon data alone.

ACKNOWLEDGEMENTS

        This investigation was supported by the Natural Environment Research
Council through the Institute of Geological Sciences and a grant to the
University of Bath. Radiocarbon measurements were carried out at the NERC
Radiocarbon Laboratory at East Kilbride, Scotland, and oxygen and hydrogen
isotope analyses were carried out in the Stable Isotope Laboratory at AERE
Harwell. Dr D Harkness (NERC) and Mr E W Clipsham (AERE) are thanked for
their assistance. Mr D Lee assisted with analyses and interpretation of
inert gas contents. Drs E J Reardon and P Fritz are thanked for providing
a copy of WATEQF-ISOTOP, and for discussions on the interpretation of
carbon isotope data. The staff of the Severn-Trent Water Authority
co-operated in the provision of access to pumped public supply boreholes,
without which the study would not have been possible. This paper is
published by permission of the Director, Institute of Geological Sciences.

REFERENCES

[1]   EDMUNDS, W. M., et al. Geochemical processes and controls on minor
            element distribution in a sandstone aquifer. In preparation.
[2]   EDMUNDS, W. M., MORGAN-JONES, M. Q. Jl. Engng. Geol., London, 9,
            (1976) 73.
[3]   ANDREWS, J. N. Radiogenic and inert gases in groundwaters. Proc.
            2nd International Symposium on Water-Rock Interaction,
            Strasbourg. (1977) I-334.

[4]   ANDREWS, J. N., LEE, D. J. Inert gases in Bunter Sandstone ground-
       waters as indicators of age and palaeoclimatic trends. In
       preparation.

[5]   LAND, D. H. Hydrogeology of the Bunter Sandstone in Nottinghamshire.
       Hydrogeological Rept. No 1, Water Supply Papers of Geol. Surv.
       G.B., London (1966).

[6]   SMITH, E. G. et al. Geology of the Country around East Retford,
       Worksop and Gainsborough, Mem. Geol. Surv. G.B., H.M.S.O.,
       London (1973).

[7]   LOVELOCK, P. E. R. Aquifer properties of Permo-Triassic sandstones
       in the U.K., Inst. of Geol. Sciences Bulletin No 56, H.M.S.O.,
       London (1977)

[8]   WILLIAMS, B. P. J., DOWNING, R. A., LOVELOCK, P. E. R. Proc. 24th
       Intl. Geol. Cong., Montreal, Section 11 (1972) 169.

[9]   DEINES, P., et al., Geochim. Cosmochim. Acta, 38 (1974), 1147.

[10]   LANGMUIR, D., Geochim. Cosmochim. Acta, 35 (1971), 1023.

[11]   HARNED, H. S., DAVIS, R., J. Amer. Chem. Soc., 65 (1943), 2030.

[12]   REARDON, E. J., FRITZ, P., J. Hydrol, 36, (1978), 201.

[13]   TRUESDELL, A. H., JONES, B. F., J. Res. U.S. Geol. Surv., 2, (1974),
       233.

[14]   PLUMMER, L. N., et al., U.S. Geol. Surv., Water Res. Invest., 76-13,
       (1976).

[15]   SMITH, D. B., et. al., Water Resour. Res., 12, 3 (1976), 392.

[16]   DOWNING, R. A., et al., J. Hydrol., 33, (1977), 201.

[17]   BOTTINGA, Y., J. Phys. Chem. 72, (1968), 800.

[18]   MORRISON, T. J., JOHNSTONE, N. B., J. Chem. Soc., (1954), 3441.

[19]   DANSGAARD, W., Tellus, 16, (1964), 436.

[20]   WEST, R. G., Phil. Trans. R. Soc. Lond. B., 280, (177), 229.

[21]   HAMMEN, T. van der, et al., Geol. En Mijnb., 46, 3, (1967), 79.

[22]   COOPE, G. R., Phil. Trans. R. Soc. Lond. B., 280, (1977), 313

[23]   LAMB, H. H. "The Changing Climate", Methuen, London (1966)

[24]   CAWSE, P. A., UKAEA Report, AERE-R 7669, H.M.S.O. London (1974).

[25]   PLUMMER, L. N., Water Resour. Res., 13, 5, (1977), 801.

[26]   MITCHELL, G. F., WEST, R. G., Eds., Discussion on "The changing
       environmental conditions in Great Britain and Ireland during
       the Devensian (Last) cold stage", Phil. Trans. R. Soc. Lond.
       B., 280, 972, (1977), 103-374.

[27]   SHOTTON, F. W., Phil. Trans. R. Soc. Lond. B., 280, (1977), 107.

[28]   MORGAN, A. V., Phil. Trans. R. Soc. Lond. B., 280, (1977), 197.

[29]   WATSON, E., Phil. Trans. R. Soc. Lond. B., 280, (1977), 183.

DISCUSSION

T. FLORKOWSKI: With regard to the question of excess ^{4}He, have you measured the total U/Th content of water or rock in different wells? I wonder whether the excess ^{4}He might not be related to the total uranium content. May be it would be even better to correlate excess ^{4}He with age. In any case, your correlation of excess ^{4}He with radiocarbon is very remarkable (as is to be expected) and this is the first time it has been seen so distinctly, I believe.

A.H. BATH: The total U/Th content of the aquifer material has so far been measured in only 14 samples from a single drill core, and the scope for error in calculating ^{4}He accumulation ages from such limited data is clearly rather large.

Uranium determinations in some of these groundwater samples indicated levels of less than 1 ppb, if I remember correctly. At this order of magnitude the contribution of dissolved U/Th nuclides to the total ^{4}He load is small compared with that of the solid phase nuclides (less than 10% according to my calculations using the parameters in the text).

K.U. WEYER: I quite agree that a realistic conceptual hydrological framework should always be used for interpreting the results of isotope studies. However, I would like to propose a second possibile interpretation of the results you have presented in your paper. As I understand it, you used the down-dip flow concept for the Bunter sandstone aquifer with recharge occurring in the outcrop area. It might, however, very well be the case that the chain of hills shown in your Fig. 1 acts as a major recharge area. From them, water would flow through the Bunter sandstone aquifer in a westerly direction to the outcrop area and in an easterly direction to the River Trent. This concept might also be supported by some of your diagrams, as it might be possible to divide the data into two groups, west and east, instead of regarding them as a manifestation of a trend. I would therefore like to ask you whether your model of down-dip flow is supported by piezometric head data?

A.H. BATH: The area of higher elevation to which you refer involves a gradient of only 50 metres or less, as you will see from the scale of Fig. 1. Your suggestion of recharge in this zone is intriguing, though judging from the lithology of the Keuper marl this formation in fact acts as an effective aquiclude. As far as I know, the information available from hydrogeological analysis of this area does not indicate any possibility of such recharge, though the piezometric surface is locally depressed in zones of high abstraction. This conclusion is supported by the small number of tritium and ^{14}C analyses that have been done on waters from this area.

S.K. GUPTA: The helium measurements your report are very exciting, as this is probably the first time that ^{4}He dating of groundwater has been seen to be successful. You mention in Section 5 of your paper that ^{4}He accumulation ages are consistently higher than the corresponding ^{14}C ages. Apart from the several factors outlined by you as an explanation for this, another possible mechanism might be small leaks of relatively young water to the main aquifer water. Whenever such a mechanism is operative, the age indicated by the isotope with the shorter half-life would be smaller than that indicated by the isotope with the longer half-life. This observation is supported by our measurements of ^{32}Si and ^{14}C ages of groundwater in India: we find that ^{32}Si ages become progressively smaller than the ^{14}C age.

A.H. BATH: Leakage of ^{4}He or of solutions enriched in U/Th from adjacent formations needs to be considered as a source of error. Anomalous U/Th should be detected in groundwater analyses, while ^{4}He leakage (which may account for some scatter in Fig. 4) must be a uniform process in order to maintain the correlation between radiocarbon age and ^{4}He content as the water moves down-dip.

I agree that the mixing even of small amounts of younger water can cause a significant reduction in the composite apparent radiocarbon age, the size of this effect varying with the ^{14}C activity of the old component as well as with the proportions of mixing.

P.L. AIREY: Do the δ^2H-^{14}C age curves reflect the palaeotemperature variation as clearly as the δ^{18}O data do?

A.H. BATH: Yes, Fig. 5 shows that the δ^2H versus δ^{18}O correlation has a correlation coefficient of 0.83, which gives an indication of how similar the δ^2H versus ^{14}C plot would be.

P.L. AIREY: Could the variability of the groundwater chloride concentration with ^{14}C age also reflect changes in catchment runoff coefficients? If so, there might be a correlation with palaeo lake levels which also reflect net catchment runoff.

A.H. BATH: The early Holocene chloride levels are still lower than the average found in present rainfall, and so it is not necessary, when using the most simple approach, to discuss variations in evapotranspiration or runoff. Of course, with more data it would be interesting to consider the effects of such processes on water chemistry.

PALAEOCLIMATIC INFORMATION FROM DEUTERIUM AND OXYGEN-18 IN CARBON-14-DATED NORTH SAHARIAN GROUNDWATERS

*Groundwater formation in the past**

C. SONNTAG, E. KLITZSCH, E.P. LÖHNERT,
E.M. EL-SHAZLY, K.O. MÜNNICH, Ch. JUNGHANS,
U. THORWEIHE, K. WEISTROFFER, F.M. SWAILEM

Abstract

PALAEOCLIMATIC INFORMATION FROM DEUTERIUM AND OXYGEN-18 IN
CARBON-14-DATED NORTH SAHARIAN GROUNDWATERS: GROUNDWATER
FORMATION IN THE PAST.

A statistical presentation of ^{14}C groundwater ages for various regions of the northern
Sahara reflects the alternating sequence of humid and arid periods in the late Pleistocene
and Holocene. Groundwaters older than 20 000 years BP are found all over the Sahara.
Isoline-presentation of the continental effect in deuterium and oxygen-18 of Saharian
groundwater is similar to the one in modern European groundwater. This similarity proves
the Western Drift influence when, in the past, winter rain was sufficient for groundwater
formation in the Sahara (great pluvial). The postpluvial humid phases of the Sahara during
the Holocene were probably of decreasing importance from west to east. The lower deuterium
excess d = δD − 8 $\times$ δ^{18}O observed in old Saharian groundwaters is interpreted to be due to a
lower moisture deficit of the air over the ocean during the last ice-age. Extremely high D and
^{18}O contents of modern groundwater in the Sahel zone south of the Sahara are probably due
to summer rain originating from tropical rain forest evapotranspiration.

INTRODUCTION

With increasing effort in development projects in the Sahara, estimates of
the Saharian groundwater reserves have increased from 15 $\times$ 10^{12} m^3 in 1966 [1]
to 60 $\times$ 10^{12} m^3 in 1976 [2]. If the amount of groundwater is assumed to be
homogeneously distributed over the whole Sahara area of 4.5 $\times$ 10^6 km^2, these

---

* This work was prepared by **C. Sonntag, K.O. Münnich** and **Ch. Junghans**, *Institut für
Umweltphysik, Universität Heidelberg, Federal Republic of Germany*; **E. Klitzsch, U. Thorweihe**
and **K. Weistroffer**, *Geologisches Institut Berlin;* **E.P. Löhnert**, *Geology Dept., University of Ife,
Nigeria;* **E.M. El-Shazly**, *Nuclear Materials Corporation, Cairo, Egypt; and* **F.M. Swailem**,
Middle Eastern Regional Radioisotope Centre, Cairo, Egypt.

figures correspond to a hypothetic water layer of 3.3 m and 13 m thickness respectively, to be compared with the world-wide average value of 55 m [3]. As far as the inner Sahara is concerned, there now is an increasing tendency to believe that the Saharian groundwater is not recharged under the present climatic conditions, which would mean that the water forms a fossil deposit, and its exploitation would be equivalent to mining. Twelve years ago, however, a recharge rate of 4×10^9 m^3/a was still estimated under the assumption of a long-scale groundwater movement from recharge areas in rainy mountainous terrain of Central Africa through the Sahara [1]. Isotope dating (^{14}C in combination with tritium, deuterium and ^{18}O) of groundwater in the Central and Eastern Sahara does not indicate any significant age increase with groundwater flow direction as was to be expected on the basis of the above recharge estimate [4,5]. Only on a regional scale has natural groundwater motion been found, for example south of the Sahara Atlas [6].

About two years ago a significant west-east decrease of the heavy stable isotope content in the fossil Saharian groundwaters was found [7], which is quite similar to the decrease found in European winter precipitation, known as "continental effect". The continental effect is caused by isotope fractionation during condensation of water vapour, where the heavy water molecules HDO and $H_2^{18}O$ are favoured, i.e. the condensate has a heavy isotope content higher than the remaining water vapour. If wet oceanic air masses are carried over the continent under successive water loss by precipitation, the heavy isotope content of its water vapour, and consequently also that of the rain derived from it, decreases with increasing distance from the coast.

Interpretation of the west-east decrease, in the case of the Sahara waters, as the "continental effect in groundwater", leads to the conclusion that North Africa must have been influenced by the western drift in the past. This is especially true in the case of the very old waters (age > 20 000 years BP). Thus it must be assumed that the westerly winds must have brought rain by carrying wet Atlantic air masses across the Sahara.

A model treatment of the continental effect, based on successive Rayleigh condensation steps in a closed air-mass system (no vertical water vapour exchange and no exchange between rain and water vapour assumed) has given an estimate of the mean palaeowinter precipitation as well as its west to east variation across the Sahara [8].

On the whole, the isotopes in groundwater contain considerable palaeo-climatic information and encouraged us to undertake a more detailed study of Saharian groundwaters, and also of modern European groundwater. In the course of this comprehensive study a data centre is being established, where relevant data for the reconstruction of the groundwater formation history will be stored, i.e. isotopic, geographical and hydrogeological data and hydrochemical data. In this paper a short extract from this data collection is presented.

¹⁴C GROUNDWATER AGES

Figure 1 shows frequency distributions of the ¹⁴C groundwater ages for different regions (based on an initial ¹⁴C content of 85% modern — no carbonate exchange or other age corrections). Following Geyh and Jäkel [9] the maxima and minima of this statistical presentation of the groundwater ages are interpreted as footprints of humid and arid climate phases in the Sahara. This interpretation agrees surprisingly well with current palaeoclimatic ideas about the Sahara for the late Pleistocene and Holocene [10].

Starting from high ages, a broad maximum between 50 000 and 20 000 years BP is observed. These ages are related to artesian and deep groundwaters which are obviously found in all parts of the Sahara. During the time-span preceding the ice-age maximum (about 18 000 years BP) the Sahara received sufficient winter rain from the western drift to produce groundwater. This great pluvial is characterized by variable temperatures. The age interval between 20 000 and 14 000 years BP is less densely populated. During this time the Sahara was cold and dry and groundwater formation under these semi-arid conditions was low. Strong climatic variations with winter rain from the western drift followed at between 14 000 and 10 000 years BP. This variable climate then changed to the warm postglacial period with rain carried by winds of convective origin. Figure 1 shows a sequence of relatively short humid and arid phases and is surprisingly similar to the ¹⁴C age distributions of Saharian archaeological material of Geyh and Jäkel [9]. As can be seen in the figure, however, the frequency maxima of postpluvial ¹⁴C ages decrease from west to east, being obviously absent in the Western Desert of Egypt. It might be concluded from this that the postpluvial humid phases with rain of convective origin have not reached the eastern Sahara. Another conclusion would be that the shallow groundwaters which these humid phases had formed have already been lost by natural discharge.

CONTINENTAL EFFECT IN DEUTERIUM AND ¹⁸O

For a better understanding of the mechanisms involved and for the design of an improved model estimate of the palaeowinter precipitation and its spatial variation in the Sahara, a more detailed study of the continental effect in the ¹⁴C-dated Saharian groundwaters is needed. In addition, for a comparative study of analogues, a comprehensive investigation of the continental effect in modern European groundwater is indispensable. In the case of modern groundwater, open-model parameters like winter precipitation rate and its spatial variation are known. This makes it possible to study their influence on the continental effect. This information is then transposed to the model of the palaeocontinental effect in the Sahara.

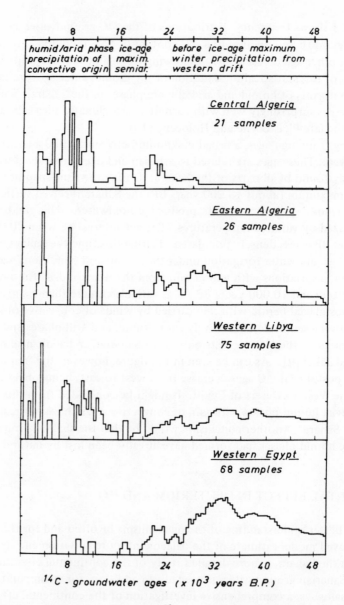

FIG.1. Abundance distribution of apparent ^{14}C ages of Saharian groundwaters. (Data from Algeria taken from Gonfiantini et al. [6].)

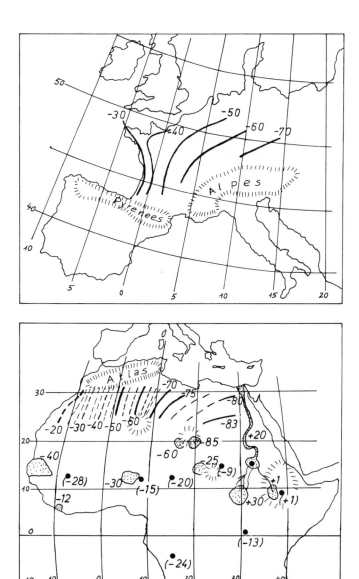

FIG.2.   Preliminary isoline- presentation of δD of modern European and fossil Saharian
groundwaters (data of Ref.[6] included).   Mean δD of modern (and fossil) groundwater
(in dotted areas) and of mean weighted annual precipitation (black points, numbers in brackets)
in Central Africa.   The European data points uniquely fall into distinct isozones, the resolution
across the isolines (excluding the influence of altitude effects) may turn out to be not much
more than σ ≈ ± 50 km or about ± 2‰ in δD.

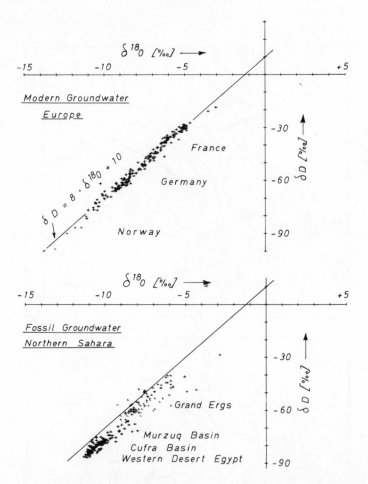

FIG.3. δD versus δ¹⁸O diagram of modern European and fossil Saharian groundwater (data
of Ref.[6] included; all European data points measured by Mrs. Christel Junghans; the
spread around the regression line is predominantly due to our present analytical precision
of ± 1‰ for deuterium).

Therefore, parallel to our work in the Sahara, an extended groundwater
sampling campaign was started in Europe. Experience shows that the tap-water
of villages or small towns far from large rivers and from industrial centres with
high water consumption, is suited for sampling local groundwater. This arises
because the use of local shallow groundwater or spring water is the most economic
and therefore the preferred water supply policy.

The preliminary results of our continuing investigations in the Sahara and
in Europe are also summarized in the isoline-presentation of the deuterium data

in Fig.2. The isolines are based on about 300 deuterium data (up to now only about half the European samples have been analysed for both heavy isotopes) in either case. While, in the case of Europe the sampling locations are homogeneously spread over the whole area, the Sahara data is concentrated in relatively small areas where wells exist. Thus, the broken isolines are more intuitive in character. As can be seen from Fig.2, in Europe the winter rain is carried by air masses, which come from WNW direction with increasing northerly direction towards eastern Europe, from where, unfortunately, no water samples were available. In the case of the western part of the Sahara the palaeowinter rain obviously came from west, the Atlas mountains suppressing the northerly component. In the Central Sahara a slight north/south continental effect is found, which increasingly comes into play in the eastern part. The significant north/south continental effect observed in the Western Desert oasis waters of Egypt may also apply to eastern Europe.

The existence of the continental effect in the Saharian groundwaters indicates that these water bodies were formed essentially by local rainfall. Long-range groundwater movement from the rainy mountainous regions of Central Africa to the various Sahara basins is therefore unlikely.

On a regional basis, however, groundwater movement may exist, for example along the isolines from Tibesti via Cufra and Farafra to Bahariya. But here a significant increase of the ^{14}C ages in this flow direction has not so far been found. Clarification of this problem would be, of course, of great importance for the Egyptian New Valley Project.

DEUTERIUM-^{18}O RELATIONSHIP

The stable isotope data of the Saharian and European groundwaters are presented in the δD versus δ^{18}O diagrams of Fig.3. As can be seen from these figures, the modern European groundwaters fall on the classical "meteoric water line" which, at δ^{18}O = 0, shows a deuterium excess of d = + 10‰. The variance of the data points from this regression line is very small. The Sahara waters, however, fall on a straight line parallel to this, cutting the δD-axis at d $\cong$ + 5‰ only. Moreover, the variance is considerably larger here; this may be due to slight kinetic evaporation effects. We do, however, not believe that the lower apparent deuterium excess is exclusively due to kinetic re-evaporation, since the eastern Sahara data (Cufra, Western Desert) show a variance as low as the European waters, but nevertheless a smaller deuterium excess. It is commonly assumed that the deuterium excess of meteoric water is due to kinetic isotope fractionation in the evaporation of sea-water. The data of a wind/water tunnel experiment in this laboratory is in agreement with the issue that the deuterium excess decreases with decreasing moisture deficit of the air over the ocean [11].

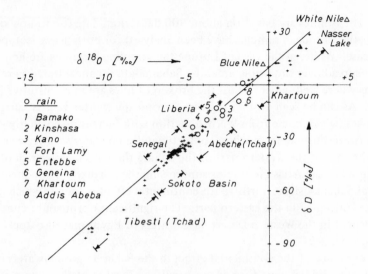

FIG.4. δD versus $\delta^{18}O$ diagram of modern Central African groundwater and rain, together with groundwater of high ^{14}C ages (Tibesti, Senegal).

The smaller d of the Sahara waters would correspond to a moisture deficit of only 15% in the marine atmosphere to be compared with approximately 22% today. Present palaeoclimatic ideas are not in contradiction to this conclusion [10].

STABLE ISOTOPES IN MODERN AND FOSSIL GROUNDWATER SOUTH OF THE SAHARA

Among other points, our future palaeoclimatic studies will be devoted to the question regarding to which extent the climate belts were shifted towards the equator during the last ice-age. In other words — down to which latitude can one find a west-east slope of the heavy isotope content in fossil groundwater? Up to now, isotope data of fossil and modern groundwater are only available from sporadic studies in Northern Senegal [12], Nigeria, Tchad and Sudan — areas belonging to the Sahel Zone (see dotted spots in Fig.2). Moreover, the mean annual deuterium content of precipitation from various IAEA/WMO stations in Central Africa are presented in Fig.2. Apart from the fossil waters of the Tibesti area of northern Tchad, which show very low deuterium content and are thus to be accounted to the North Saharian groundwaters of western drift influence, in all other data a trend of increasing heavy stable isotope content from Western to Eastern Africa is found. The rain data show a similar trend. The modern (bomb-tritium-containing) groundwater near Khartoum has a higher heavy isotope content

than SMOW (Standard Mean Ocean Water) — positive delta value. The same is true
for the Nile and thus for Nile valley groundwater. The δD versus $\delta^{18}O$ presentation
of this data (Fig.4) gives no indication for more than moderate heavy isotope
enrichment by kinetic evaporation before water infiltration. Thus, the only explanation we have for this west/east trend to high positive delta values in D and ^{18}O is
to assume an increasing influence of water vapour from the tropical rain forest to
groundwater. If so, positive delta values of rain may come about as follows. The
deuterium content of groundwater of the tropical rain forest is expected to be
around -50 to $-30\%_0$. Evapotranspiration, which is expected to cause no
isotope fractionation, leads to heavy water vapour with the same isotope composition as that of the groundwater. Condensation of this continental water vapour
under equilibrium isotope fractionation, with a fractionation factor of around
1.06, yields δD values of $+10$ to $+30\%_0$ for the condensate (rain). These isotopically heavy rains are to be expected during summer when the inner tropical
convergence (ITC) lies at about $12°$ northern latitude. Tropical water vapour
is carried northwards by the southern Passat, air masses are lifted and produce
rain in the ITC zone.

FUTURE PROGRAMME

The groundwater data collection is to be expanded with data of northern,
southern and particularly of eastern Europe. Groundwater sampling expeditions
to the western and southern Sahara (Sahel Zone included), areas from which
only very poor data are available, are in preparation. Thus an improved spatial
distribution record of stable isotopes in modern European and ^{14}C dated Saharian
groundwaters (fine structure of the continental effect with particular attention to
superimposed altitude effects) is to be expected. The isotope analyses will be
supplemented by noble gas measurements which it is hoped will give information
about the local temperature of the lower atmosphere and of the ground [13]
at time of groundwater formation, i.e. the spatial palaeotemperature pattern of
the Sahara. With this data (stable isotopes and noble gases) an improved model
reconstruction of the formation history of the fossil Saharian groundwater reserves
should be possible with the help of better hydrogeological knowledge of groundwater volume and infiltration areas of the various groundwater bodies. Particular
attention will be paid to the question regarding to which extent present-day
groundwater recharge might exist.

At present we are trying to simulate the continental effect by a general
water-vapour transport model which considers the local vertical distribution of
stable isotopes in atmospheric vapour as well as the molecular exchange between
the local rain and this vapour. This model is more realistic than the previous
Rayleigh condensation model mentioned earlier. Special attention will be paid

to the deuterium excess d, i.e. to the $D/^{18}O$ relationship. The deuterium excess
seems to be a tracer for the seawater evaporation pattern and thus for the water-
vapour origin. In this new model treatment the isotope data of the IAEA/WMO
Precipitation Network will play an essential role.

ACKNOWLEDGEMENTS

   Special thanks are due to Dr. M.A. Geyh, Niedersächsisches Landesamt
für Bodenforschung, Hanover, who has supplied all his unpublished ^{14}C and
tritium data of Saharian groundwaters and has provided stable isotope analyses
of these water samples in Heidelberg. Moreover, he arranged for modern European
groundwater sampling.
   Thanks are also due to the following persons who have supplied water samples
from North Africa and Europe for isotope analyses:

		Sampling Location
Dipl. Geol. D. Klussmann	Geologisches und Ingenieur-büro Dr. Pichel, Fuldatal	Senegal
Dr. K. Wirth	Institut für Bodenkunde Universität Göttingen	Nigeria
Dr. Puschmann	German Water Engineering Lingen	Libya/Liberia
Prof. G. Matthess	Geol. Institut Universität Kiel	Tunisia
Dipl. Chem. Ragaa El Shoheib	Cairo University	Egypt
Dipl. Geol. K.W. Töniges	Fernwasserversorgung Rheintal, Sinsheim	France
Dr. U. Siegenthaler	Physikalisches Institut Universität Bern	France/Spain
Dr. R. Basting	Geologisches Institut Universität Heidelberg	Fed. Rep. Germany

Dipl. Phys. M. Bruns	Institut für Umweltphysik Universität Heidelberg	France
Dipl. Phys. B. Kromer	Institut für Umweltphysik Universität Heidelberg	France
Dipl. Phys. I. Levin	Institut für Umweltphysik Universität Heidelberg	Fed. Rep. Germany
Dipl. Phys. J. Rudolph	Institut für Umweltphysik Universität Heidelberg	Norway

This investigation was supported by the Deutsche Forschungsgemeinschaft and the Heidelberger Akademie der Wissenschaften. Dipl. Phys. M. Münnich, Miss A. Ebert and Mrs. K. Pitz took care of the sample preparation for ^{14}C and ^{13}C measurements.

REFERENCES

[1] AMBROGGI, R.P., Water under the Sahara, Sci. Am. **214** (1966) 21.
[2] GISCHLER, C.E., Present and Future Trends in Water Resources Development in the Arab Countries, UNESCO Rep. (August 1976).
[3] BAUMGÄRTNER, A., REICHEL, E., Die Weltwasserbilanz, Oldenbourg, Munich (1975).
[4] KLITZSCH, E., SONNTAG, C., WEISTROFFER, K., El SHAZLY, E.M., Grundwasser der Zentralsahara: Fossile Vorräte, Geol. Rundsch. **65** 1 (1976) 264.
[5] MÜNNICH, K.O., VOGEL, J.C., Untersuchungen an pluvialen Wässern der Ost-Sahara, Geol. Rundsch. **52** (1962) 611.
[6] GONFIANTINI, R., CONRAD, G., FONTES, J.Ch., SAUZAY, G., PAYNE, B.R., "Etude isotopique de la nappe du Continental Intercalaire et de ses relations avec les autres nappes du Sahara", Isotope Techniques in Groundwater Hydrology 1974 (Proc. Symp. Vienna, 1974) **1**, IAEA, Vienna (1974) 227.
[7] SONNTAG, C., NEUREUTHER, P., KALINKE, Chr., MÜNNICH, K.O., KLITZSCH, E., WEISTROFFER, K., Zur Paläoklimatik der Sahara: Kontinentaleffekt im D- und O-18-Gehalt pluvialer Saharawässer, Naturwiss. **63** 10 (1976) 479.
[8] SONNTAG, C., et al., Paläoklimatische Information im Isotopengehalt C-14, datierter Saharawässer: Kontinentaleffekt in D und O-18, to be published in Geol. Rundsch. (1978).
[9] GEYH, M.A., JÄKEL, D., Spätpleistozäne und holozäne Klimageschichte der Sahara aufgrund zugänglicher ^{14}C-Daten, Z. Geomorph. N.F. **18** 1 (1974) 82.
[10] FLOHN, H., Meteorologisches Institut der Universität Bonn, Private communication (1976).

[11] MÜNNICH, K.O., SONNTAG, C., SIEGENTHALER, U., "Lower deuterium excess of fossil Saharian groundwater: an indicator of lower moisture deficit of the marine atmosphere in the past?, in preparation.

[12] CASTANY, G., et al., "Etude par les isotopes du milieu du régime des eaux souterraines dans les aquifères de grandes dimensions", Isotope Techniques in Groundwater Hydrology 1974 (Proc. Symp. Vienna, 1974) 1, IAEA, Vienna (1974) 243.

[13] MAZOR, E., Paleotemperatures and other hydrological parameters deduced from noble gases in groundwaters; Jordan Rift Valley, Israel, Geochim. Cosmochim. Acta 36 (1972) 1321.

DISCUSSION

J. Ch. FONTES (*Chairman*): To which aquifers do the ^{14}C data given in Fig.2 relate?

C. SONNTAG: The isotope data for Algerian groundwater are Mr. Gonfiantini's and yours (see Ref.[6]).

J. Ch. FONTES (*Chairman*): In that case, one must take into account the fact that, unlike those of the eastern Sahara, the western Sahara aquifers are mainly unconfined. One would therefore expect to find higher ^{14}C values in them.

C. SONNTAG: We have considered this problem, and it is discussed in our isotope study on the Murzuq Basin [Ref.4].

J.Ch. FONTES (*Chairman*): From present-day isohyets in southern Tunisia, it appears that precipitation in the eastern erg may come from the Mediterranean. Why do you think that the continental effect can be explained in terms of distance from the Atlantic coast in the past?

C. SONNTAG: I have assumed a small north-south component in the continental effect in Tunisia in the past, but our data from the area are at present insufficient to be able to say more about this.

B.R. PAYNE: I was particularly struck by your Fig.2, which shows an excellent correlation between the stable isotopic composition of natural waters and what you call coastal distance. Since you say in the section of your paper entitled "Continental Effect in Deuterium and Oxygen-18" that the moisture source is the north-north-west, I find some difficulty in agreeing with the figure of 2000 km for Vienna mentioned in your oral presentation. May I ask you how you defined the coastal distance for the different geographical locations?

C. SONNTAG: The coastal distance is tentatively defined along a straight line from Valencia to Vienna.

J. DOWGIALLO: Your study is of great interest for European hydro-geology, and the δD isolines give a very good picture of the continental effect. It is therefore important to know what facts you used for drawing them — what kinds of aquifers were sampled and how many sampling points were used.

C. SONNTAG: In the case of Europe, the isolines are based on about
300 data points obtained from tap-water, the locations of the samples being
homogeneously distributed over the whole area. I can say what kind of aquifer
the samples came from only in few cases.

A.S. ISSAR: Did you collect the samples from the western desert in springs
or artesian wells? The reason for my question is that low deuterium concentrations are found in the water of the Nubian sandstone of the Sinai.

C. SONNTAG: The samples from the western desert of Egypt come mainly
from the artesian groundwater of the Nubian sandstone. I believe that the
Western Drift influence ends in this part of the eastern Sahara.

D.B. SMITH: The interpretation of carbon-14 ages in your paper is made
on the basis of initial input conditions of 85% modern carbon-14. There is not
universal agreement on this method of calculation and it would be interesting
to analyse the results using a correction method such as the WATEQ isotope
programme (see Wigley, Nature 263 (1976) 219—221, and also in press), or
the method described in paper IAEA-SM-228/34[1]. All these methods tend to
produce younger age values, particularly on the old samples, and their application
may clarify the meaning of the age distribution shown in Fig.1 of the paper.

C. SONNTAG: We are not as yet completely convinced of the reliability
of the various correction methods. We therefore prefer to use the simplest way
of interpreting the ^{14}C data.

G. CONRAD: You have found a trough in your ^{14}C age distributions in
Sahara groundwaters at around 16 000 − 20 000 years. From this you deduce
that there must have been a semi-arid period in the Sahara at that time. However,
since there is often mixing, a long residence time and extensive circulation in the
enormous aquifers of the Sahara, do you not think that the climatic change which
you place at 16 000 − 20 000 years could have occurred some thousands of years
earlier?

C. SONNTAG: Although we cannot exclude this possibility, we prefer for
the time being to calculate the ^{14}C groundwater ages on the basis of the piston
flow model, which can be applied particularly well in the case of artesian waters.

G. CONRAD: The climatic conditions of the Sahara vary greatly depending
on latitude and longitude on both sides of the tropic, and I believe that it is
problematic to interpret groundwater movements in accordance with an overall
scheme such as you have adopted. Aquifers are moreover bound to buffer
climatic effects as a result of the mixing which occurs.

C. SONNTAG: At first we were also very surprised that our ^{14}C groundwater
age spectrum seemed to reflect the currently accepted climatic history so well.
The reason may be that our sampling method is in fact a representative one for
purposes of the Sahara groundwater assay.

---

[1] EVANS, G.V. et al., these Proceedings.

Session VII

GEOTHERMAL WATERS – GEOCHEMICAL INTERPRETATION OF ISOTOPIC DATA

Chairman

C.T. RIGHTMIRE
United States of America

RECHERCHES GEOCHIMIQUES SUR LES EAUX THERMALES D'AMELIE-LES-BAINS (PYRENEES ORIENTALES) ET DE PLOMBIERES-LES-BAINS (VOSGES)*

J.C. BAUBRON, B. BOSCH, P. DEGRANGES,
J. HALFON, M. LELEU**, A. MARCE, C. SARCIA
Bureau de recherches géologiques et minières,
Service géologique national,
Orléans, France

Abstract–Résumé

GEOCHEMICAL RESEARCH ON THE THERMAL WATERS OF AMELIE-LES-BAINS
(PYRENEES ORIENTALES) AND PLOMBIERES-LES-BAINS (VOSGES).

A complete geochemical study was carried out on the thermal waters of Amélie-les-Bains and Plombières-les-Bains; chemical analyses were performed for the major elements and trace elements, chromatographic analyses for gases dissolved in the waters and isotopic measurements for ^{3}H, D, ^{18}O, ^{34}S, ^{13}C and ^{40}Ar. The data obtained suggest the following conclusions: (a) The isotopic composition of oxygen and deuterium in the hot waters of the two spas indicates a meteoric origin. (b) At Amélie-les-Bains it is fairly uniform (between -8.9 and $-9.3‰$ for δ^{18}O and between -60.2 and $-61.4‰$ for δD) and independent of the temperature of the emerging water, which varies between 32 and 62°C. These results indicate an absence of mixing in the thermal springs, which is confirmed by the tritium concentrations in the hot waters. Isotopic fractionation of the sulphate-sulphide pair shows a good relationship between Δ^{34}S and the temperature of the springs. This geothermometric interpretation is compared with data provided by chemical geothermometers which show temperatures between 100 and 130°C. The chemical and isotopic compositions of dissolved gases (carbon and argon) in the hot waters are given. (c) At Plombières-les-Bains, mixing of hot and cold waters is clearly demonstrated by the isotopic and chemical data. A good correlation is seen between the isotopic composition of the thermal waters in oxygen, deuterium and carbon as well as the tritium concentration and the temperature of the water at emergence. The Na/K technique for waters which do not appear to be mixed gives an isotopic temperature for the deep reservoir between 145 and 190°C.

RECHERCHES GEOCHIMIQUES SUR LES EAUX THERMALES D'AMELIE-LES-BAINS
(PYRENEES ORIENTALES) ET DE PLOMBIERES-LES-BAINS (VOSGES).

Une étude géochimique complète a été réalisée sur les eaux thermales d'Amélie-les-Bains et de Plombières-les-Bains; on a procédé à des analyses chimiques pour les éléments majeurs et

---

* Cette étude a été réalisée grâce à l'appui financier d'un contrat de recherche de la Commission des Communautés européennes (CCE) et aux fonds propres du Département Minéralogie, Géochimie, Analyses (crédits du Ministère de l'Industrie, du Commerce et de l'Artisanat).

** Centre national de la recherche scientifique, détaché au Bureau de recherches géologiques et minières.

les éléments traces, des analyses chromatographiques de gaz dissous dans les eaux, des mesures isotopiques pour ^{3}H, D, ^{18}O, ^{34}S, ^{13}C, ^{40}Ar. L'ensemble de ces données permet d'aboutir aux conclusions suivantes: a) Dans les deux stations thermales étudiées, la composition isotopique en oxygène et deutérium des eaux chaudes met en évidence leur origine météorique. b) A Amélie-les-Bains, la composition isotopique en oxygène et en deutérium est assez uniforme ($-8,9$ à $-9,3‰$ pour le δ^{18}O et $-60,2$ à $-61,4‰$ pour le δD) et indépendante de la température aux émergences, qui varie de $32°$ à $62°$C. Ce résultat implique une absence de phénomène de mélange dans ces sources thermales. Cette interprétation est confirmée par les concentrations en tritium des eaux chaudes. Le fractionnement isotopique du couple sulfates-sulfures montre une bonne relation entre le Δ^{34}S et la température des sources. L'interprétation géothermométrique est comparée aux données fournies par les géothermomètres chimiques, qui indiquent des températures comprises entre $100°$ et $130°$C. Les compositions chimique et isotopique des gaz dissous (carbone et argon) dans les eaux chaudes sont précisées. c) A Plombières-les-Bains, le mélange entre des eaux chaudes et des eaux froides est clairement démontré par l'examen des données isotopiques et chimiques. On note une bonne corrélation entre la composition isotopique en oxygène, deutérium et carbone ainsi que la concentration en tritium des eaux thermales et leur température à l'émergence. La méthode Na/K pour les eaux qui n'apparaissent pas mélangées donne une température isotopique du réservoir en profondeur comprise entre $145°$ et $190°$C.

1 - INTRODUCTION

        Dans le cadre d'études géothermiques, l'utilisation des isotopes du milieu et plus particulièrement le tritium, l'oxygène 18, le deutérium, le soufre 34 et le carbone 13 permet, avec l'appui des analyses chimiques de préciser certaines données du régime des eaux thermales : en particulier leur origine, leur mode d'alimentation récente ou ancienne, la température atteinte par les eaux en profondeur, les phénomènes de mélange d'eaux, la source profonde ou superficielle des composants de la phase gazeuse dissoute dans les eaux.

        L'application des techniques isotopiques a été réalisée grâce à l'aide de la Commission des Communautés Européennes et les résultats présentés ici concernent deux stations thermales françaises (figure 1) :

        - Amélie-les-Bains au Sud en climat méditerranéen,

        - Plombières-les-Bains au Nord-Est en climat continental.

2 - PRESENTATION DES RESULTATS ET INTERPRETATION DES DONNEES

2.1. Les résultats analytiques

        Les points principaux de la démarche analytique sont présentés tableau I.[1]

        Les résultats concernant les eaux d'Amélie-les-Bains sont présentés tableau II (chimie), tableau III (isotopes), tableau IV (soufre), tableau V (gaz dissous).

---

[1] Les tableaux sont groupés dans l'appendice.

FIG.1.   Carte de situation.

        Les résultats concernant les eaux de Plombières-les-Bains sont
présentés tableau VI (chimie), tableau VII (isotopes), tableau VIII (gaz
dissous).

2.2. Amélie-les-Bains

2.2.1. Cadre géologique

        Les sources étudiées sont situées sur la bordure septentrionale
du massif du Roc de France (1 450 m) constitué de terrains métamorphiques
affectés par l'orogénèse hercynienne. Une faille d'effondrement "dite du
Vallespir" sépare le socle cristallophyllien du Roc de France du bassin
secondaire d'Amélie-les-Bains et permet la remontée des eaux chaudes dans
cette station, dont l'altitude est de 230 m.

BAUBRON et al.

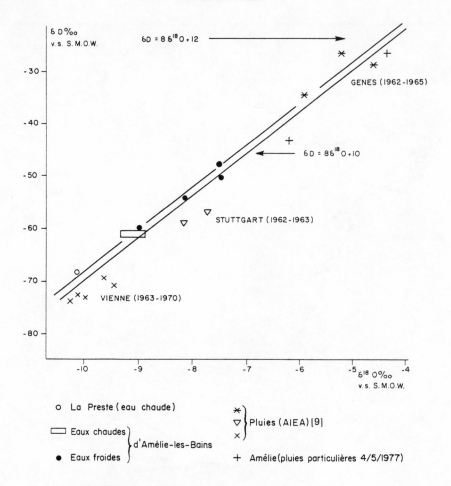

FIG.2. *Amélie-les-Bains − Relation deutérium-oxygène dans les eaux thermales.*

2.2.2. L'origine des eaux

La composition isotopique en oxygène et deutérium des eaux superficielles ou souterraines froides est liée par la relation linéaire suivan-
te (figure 2):

$$\delta D°/_{oo} = 8 \; \delta^{18}O°/_{oo} + 12$$

Cette équation est comparable aux droites des précipitations établies par
H. CRAIG [1] et W. DANSGAARD [2] à l'échelle mondiale.

En ce qui concerne les eaux thermales, leurs teneurs en deutérium
et oxygène 18 sont bien groupées autour des valeurs moyennes
($\delta D°/_{oo}$ SMOW = - 60,94, $\delta^{18}O°/_{oo}$ SMOW = - 9,09) et sont situées sur la droite
des précipitations.

Toutes ces eaux sont donc d'origine météorique et en aucun cas
on ne peut mettre en évidence un effet géothermique marqué par un enrichissement en ^{18}O résultant d'un échange entre l'eau et les roches encaissantes.
L'absence de tritium permet par ailleurs de caractériser ces eaux chaudes
comme anciennes. Les eaux froides superficielles (torrent du Mondony) ou
d'infiltration (Galerie Parès) ont des teneurs en tritium relativement importantes (eaux récentes) et des compositions isotopiques en oxygène et
deutérium différentes de celles des eaux thermales. Ces caractéristiques
isotopiques montrent que les différences de température observées aux émergences des sources thermales (32° à 62°C) ne sont pas dues à des apports
d'eaux superficielles, qui entraîneraient des variations de composition
isotopique des eaux chaudes et la présence plus ou moins importante de
tritium.

2.2.3. La chimie des eaux

Le faciès chimique des eaux d'Amélie-les-Bains est bicarbonaté
calcique pour l'eau froide du Mondony et bicarbonaté sulfuré-sodique pour
les eaux chaudes. La composition chimique des deux types d'eaux est nettement différenciée. L'eau froide de surface est très peu minéralisée, son
pH est voisin de la neutralité et les éléments caractéristiques sont les
teneurs en calcium et magnésium. Les eaux chaudes ont une remarquable uniformité chimique caractérisée par un pH basique, la présence de sulfures,
une augmentation, par rapport à l'eau froide, de la concentration de tous
les anions, de la silice, de certains cations, en particulier le sodium et
plus faiblement le potassium, et des éléments en traces comme le lithium.
L'homogénéité dans le chimisme des eaux chaudes indique une même origine
de ces eaux sans mélange avec des eaux froides. Cette interprétation est
en bonne concordance avec les données isotopiques.

2.2.4. La température atteinte par les eaux en profondeur

On peut, de manière classique, appliquer un certain nombre de méthodes traditionnelles (SiO$_2$, Na/K, Na - K - Ca) ainsi que des méthodes
basées sur les isotopes du soufre ; les résultats obtenus nous ont conduit
à revenir sur ces méthodes, les discuter et en essayer de nouvelles.

2.2.4.1. Les méthodes SiO$_2$, Na/K, Na/K/Ca

Suivant les données de R.O. FOURNIER et A.H. TRUESDELL [3]
ainsi que celles de Y.K. KHARAKA et BARNES [4] pour le quartz, les teneurs
en silice des eaux chaudes conduisent à des températures comprises entre
100° et 120°C.

Suivant les données extrapolées de A.J. ELLIS [5] et celles
de R.O. FOURNIER et A.H. TRUESDELL [3], le rapport Na/K conduit à des températures comprises entre 100° et 130°C.

Par contre, la méthode Na - K - Ca (en utilisant β = 1/3) détermine une température de 180°C, très différente de celle déterminée par les
géothermomètres précédents. Cette anomalie nous a conduit à examiner si
l'eau à l'émergence pouvait être en équilibre avec la calcite : dans cette
hypothèse, les températures théoriques de cet équilibre sont comprises
entre 33 et 76°C, températures voisines de celles mesurées aux émergences.
Cet équilibre est corroboré par la présence de calcite en dépôt aux émergences.

Ainsi toute méthode basée sur la teneur en calcium à l'émergence
ne peut conduire qu'à des résultats sans signification (méthode Na - K - Ca).

2.2.4.2. La teneur en fluor et la solubilité de la fluorine

On peut tenter d'utiliser la solubilité de la fluorine comme géothermomètre ; nous faisons l'hypothèse que le fluor et le calcium proviennent de la solubilisation de la fluorine et à chaque mole de fluor libéré
correspond 0.5 mole de calcium ; en utilisant les données de Y.K. KHARAKA
et BARNES [4], la température en profondeur est alors de 120°C.

2.2.4.3. Les données isotopiques du soufre et la température

La composition isotopique du soufre dans les eaux sulfureuses
d'Amélie-les-Bains a été déterminée sur les espèces réduites (HS⁻) et oxy-
dées (SO₄⁻⁻). Les deux formes de soufre réduit et oxydé sont nettement
différentes du point de vue composition isotopique : les sulfures sont ap-
pauvris tandis que les sulfates sont enrichis. La comparaison de la température d'émergence de chaque source avec le fractionnement Δ entre sulfates
et sulfures dans l'eau prélevée montre de bonnes corrélations et le fractionnement Δ mesuré semble thermodépendant.

Pour essayer d'interpréter cette relation, on peut imaginer qu'en
profondeur les espèces soufrées sous forme gazeuse entrent en réaction suivant l'équilibre proposé par H. SAKAI [6]:

$$H_2S + 2H_2O \rightleftarrows SO_2 + 3H_2.$$

Cet équilibre s'effectue avec un fractionnement isotopique au
niveau du soufre qui varie avec la température et devient nul quand la température est supérieure à 1 000°C.

La mesure de ce fractionnement permet de déterminer une température d'équilibre. Si l'on applique cette démarche aux eaux sulfureuses
d'Amélie, on considère que les espèces soufrées dosées aux émergences sont
isotopiquement comparables à celles en profondeur, et en s'appuyant sur la
courbe de fractionnement isotopique établie par H. SAKAI [6] sur l'équilibre SO₂ - H₂S, on obtient des valeurs thermométriques comprises entre
70° et 115°C pour H₂S et 100° à 145° pour HS⁻.

Dans un second essai d'interprétation, nous utilisons la relation
présentée par B. ROBINSON [7] pour le couple sulfate dissous - hydrogène
sulfuré :

$$\Delta = 5.1 \left(\frac{10^6}{T^2}\right) + 6.3$$
$$(\Delta = \delta SO_4^= - \delta H_2S)$$

Les valeurs obtenues, brutes ou corrigées pour tenir compte du
rapport $\frac{SO_4^=}{\Sigma S}$ sont nettement plus élevées (140 à 250°) (tableau IV).

En comparant les données thermométriques fournies par les géothermomètres chimiques (silice, Na/K et fluorine) de l'ordre de 100° à 130°C
avec celles fournies par le géothermomètre isotopique soufre, on observe
que seul le couple SO₂ - HS⁻ indique des valeurs thermométriques voisines
de celles données par les géothermomètres chimiques adaptés au cas des eaux
thermales d'Amélie-les-Bains.

Ces résultats thermométriques situeraient le réservoir géothermique à une profondeur de l'ordre de 3 500 à 4 500 mètres pour un gradient
géothermique normal de 3°C pour 100 mètres.

2.2.5. L'origine du carbone des carbonates

La composition isotopique du carbone des carbonates dissous est
homogène dans toutes les eaux analysées, aussi bien profondes que superficielles ; les valeurs sont groupées autour de $\delta^{13}C = -9,7‰$; il n'existe
pas de relation entre la température d'émergence et la composition isotopique du carbone des carbonates, argument en faveur d'une seule origine
pour cet élément. Des analyses isotopiques de carbonates effectuées ont
indiqué dans cette partie de la zone axiale des Pyrénées une valeur moyenne de $\delta^{13}C$ aux environs de -9 °/₀₀ vs P.D.B. pour des calcites hydrother-
males en veinules.

2.2.6. Les gaz dissous dans les eaux thermales

Sur des prélèvements d'eaux chaudes réalisés en ampoules de verre
de $1\,000\ cm^3$ sous un vide de 10^{-6} torr, des analyses chimiques et isotopiques
ont été effectuées sur les principaux gaz (Argon, Azote, CO_2, CH_4). La
séparation gaz carbonique - méthane est faite par cryogénie à $-160°C$.
L'argon est séparé et purifié selon la technique utilisée pour les mesures
géochronologiques Potassium/Argon. Les résultats de l'étude chimique et isotopique sont groupés dans les tableaux III et V.

2.2.6.1. L'origine de l'argon et de l'azote dissous

Les eaux thermales présentent une concentration en argon dissous
supérieure à la simple solubilité de l'argon de l'air atmosphérique. De
même, sauf pour l'émergence Arago, le rapport Ar/N_2 des gaz dissous est
supérieur à celui d'une solution d'air dans l'eau.

Ceci peut s'interpréter comme un enrichissement relatif en
argon lors de la circulation de l'eau en profondeur, enrichissement pouvant
être dû à un apport par de l'argon de dégazage de la croûte terrestre, ou
encore provenant de l'altération des minéraux potassiques.

La composition isotopique de l'argon ($^{40}Ar/^{36}Ar$) dissous présentée tableau III varie de 280 ± 2 à 320 ± 2 pour les sources analysées. La
composition isotopique de l'argon de l'air est égale à $295,5 \pm 2$.

S'il n'est pas possible d'interpréter par une relation simple les
variations de la composition isotopique de l'argon, aussi bien en fonction
de la concentration de l'argon que de la température d'émergence, il semble que l'on puisse observer une relation directe entre ces rapports isotopiques et les rapports Ar/N_2 des gaz dissous (figure 3).

L'alignement supérieur, où se place le point représentatif d'une
solution d'air dans l'eau, peut indiquer un système hydrothermal enrichi
en isotope 40 et en argon atmosphérique à partir d'une eau de référence
représentée par la source Arago.

L'ensemble inférieur indique un système complexe où il faut vrai-
semblablement faire intervenir un fractionnement isotopique, un apport d'argon atmosphérique et une mobilisation de l'azote.

2.2.6.2. L'origine du gaz carbonique et du méthane dissous

Si l'on examine le rapport des concentrations en gaz carbonique
et en méthane, on remarque qu'il est sensiblement du même ordre de grandeur
que leur solubilité dans l'eau (<10). Or dans les gaz thermaux, la proportion
du gaz carbonique est généralement plus forte que celle du méthane (≈ 100)

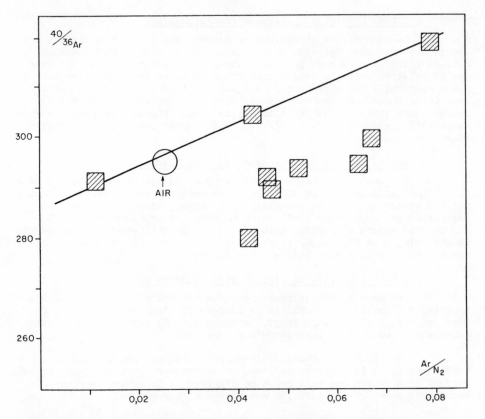

FIG.3. Amélie-les-Bains — Relation entre la composition isotopique en argon et le rapport argon-azote.

en fonction sans doute de l'équilibre qui peut avoir lieu en profondeur entre CO_2, H_2 et CH_4 coexistant avec la vapeur d'eau :

$$CH_4 + 2H_2O \rightleftharpoons CO_2 + 4H_2$$

A Amélie-les-Bains, les premiers résultats montrent que cet équilibre ne semble pas réalisé et par conséquent une interprétation thermométrique de cette réaction déduite de la mesure du fractionnement isotopique du carbone dans le couple gaz carbonique-méthane ne peut être envisagée.

Par contre, à partir de la composition chimique et isotopique du CO_2 et du CH_4 dissous, on peut calculer le carbone 13 initial compris entre - 19 et - 23 $\delta^{13}C°/_{oo}$ vs P.D.B., valeurs qui sont ubiquistes dans les domaines profond et superficiel.

Le couple bicarbonate - CO_2 utilisé comme géothermomètre isotopique conduit à des indications thermométriques basses, voisines de la température aux émergences.

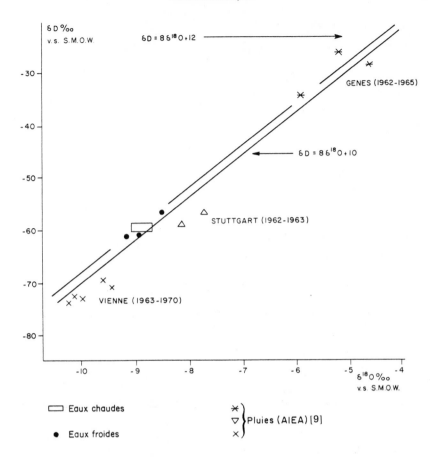

FIG.4. *Plombières-les-Bains — Relation deutérium-oxygène dans les eaux thermales.*

2.3. Plombières-les-Bains

2.3.1. Cadre géologique

Les émergences thermales de Plombières-les-Bains (altitude moyenne 400 m) sont situées en fond de vallée de l'Augronne dans un granite porphyroïde qui en forme les versants entre les alluvions récentes au fond et les grès vosgiens triasiques au sommet. De grands accidents tectoniques affectant la couverture triasique divisent cette région en damiers et correspondent à des failles du socle, en relation avec les émergences des sources chaudes.

2.3.2. L'origine des eaux

La composition isotopique en oxygène et deutérium des eaux froides et chaudes de Plombières vérifie la relation linéaire déterminée à

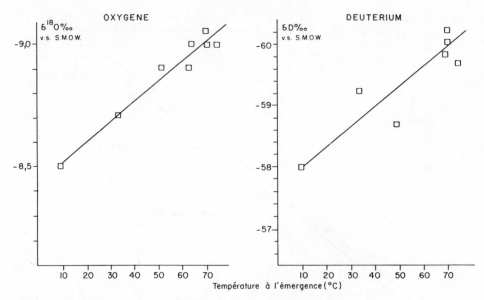

FIG.5. Plombières-les-Bains — Relation entre l'oxygène et le deutérium et la température d'émergence.

Amélie-les-Bains (figure 4) et qui est comparable aux droites des précipitations établies par H. CRAIG [1] et W. DANSGAARD [2] à
l'échelle mondiale. Ce fait prouve clairement l'origine météorique de ces
eaux.

        On note, en outre, une bonne corrélation entre la température
d'émergence des sources et la composition isotopique en oxygène et deutérium de leurs eaux (figure 5). On obtient ainsi une droite de mélange en
proportions variables d'une eau froide (type Alliot) et d'une eau chaude
(type sondage 9). · Cette notion de mélange dans les eaux thermales de
Plombières est confirmée par les teneurs en tritium qui permettent de
préciser du point de vue alimentation trois types d'eaux :

    . *Eaux sans tritium* (< 1 UT)
        Il s'agit uniquement d'eaux chaudes (64° à 71°C) anciennes à
circulation lente dans le granite qui alimentent les sources dites
"impériales" de Jutier, c'est-à-dire Robinet romain, Stanislas, Vauquelin
ainsi que le sondage n° 9 d'aménagement plus récent (1923 - 1924).

    . *Eaux tritiées* (> 30 UT)
        Ce sont des eaux froides récentes qui ont intégré dans leur com-
position du tritium d'origine thermonucléaire donc postérieur à 1952.
On peut ranger dans cette catégorie les eaux froides d'infiltration dans
le fond de la galerie Jutier dont la teneur en tritium (103 UT) est sensiblement plus faible que celles mesurées dans les eaux de ruissellement
collectées par l'Augronne (247 à 268 UT), drain représentatif à l'échelle
du bassin versant de Plombières; les eaux souterraines froides qui circu-
lent dans la couverture sédimentaire triasique et dont la source Alliot
en constitue un exemple.

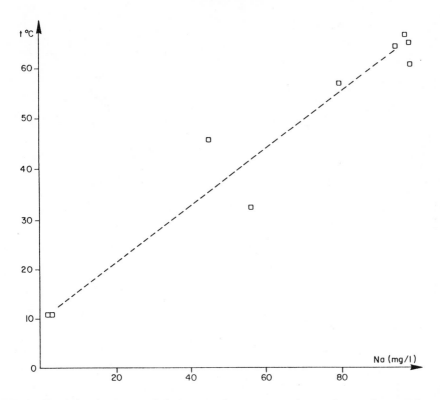

FIG.6. Plombières-les-Bains — Relation entre la teneur en sodium et la température d'émergence.

. Eaux faiblement tritiées (3 < UT < 30)
        Elles résultent de mélanges à différentes concentrations des deux
types précédents d'eaux (sondage n° 9 pour le pôle chaud et Alliot pour le
pôle froid). Le tritium permet d'apprécier la proportion relative de chaque
constituant dans ces mélanges et montre que dans l'ensemble la participation des eaux dites "anciennes" est toujours plus importante que celle
d'eaux "récentes". Sur la base des teneurs en tritium dans les précipitations
mesurées à Orléans. depuis 1967, on peut estimer l'apport d'eaux récentes de
10 à 15 % à la Savonneuse n° 9 et de l'ordre de 5 % à la source Ste Cathe-
rine.

2.3.3. La chimie des eaux

        Les eaux froides de Plombières-les-Bains sont très peu minéralisées,
la moins minéralisée étant représentée par l'eau de surface. : l'Augronne.
La source froide Alliot légèrement plus minéralisée présente un faciès bicarbonaté calcique.

        Les eaux chaudes sont du point de. vue chimique nettement différen-
tes des précédentes par leur minéralisation un peu plus importante et par
leur faciès qui est bicarbonaté sulfaté sodique. Les eaux chaudes sont riches

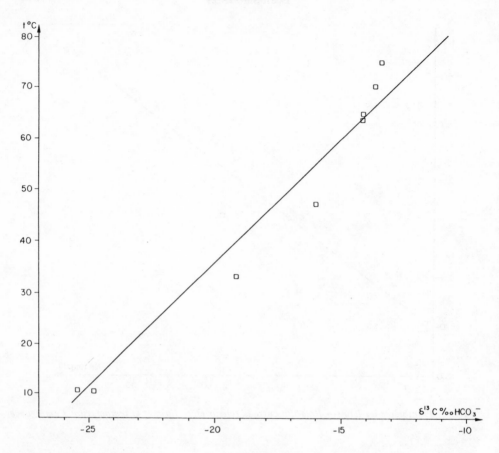

FIG.7.   *Plombières-les-Bains — Relation entre la composition isotopique des bicarbonates et la température d'émergence.*

en un certain nombre d'ions caractéristiques de l'encaissant granitique : silice, fluor, sodium, potassium, lithium, strontium et manganèse. La diminution de la minéralisation en fonction de celle de la température aux émergences peut être expliquée par un modèle de simple mélange avec des eaux froides peu minéralisées (figure 6) mais qui sont susceptibles d'apporter aux eaux chaudes des éléments comme le calcium, le magnésium, les phosphates et les nitrates.

2.3.4. La température atteinte par les eaux en profondeur

L'utilisation des méthodes géothermométriques classiques doit tenir compte à Plombières-les-Bains des deux points suivants :

- présence d'un phénomène de mélange des eaux,
- existence de concrétions aux émergences.

2.3.4.1. Les concrétions

Une étude minéralogique par diffraction X a montré dans les concrétions échantillonnées aux émergences la présence de : calcite, fluorine, quartz, silice amorphe et bassanite.

En raison de ce cortège minéral, il faut considérer avec prudence les indications thermométriques sur la température atteinte par les eaux en profondeur à partir des méthodes basées sur la teneur en silice (130° à 135°C) ou nécessitant la teneur en calcium dont la régulation aux émergences paraît complexe (> 200°C).

2.3.4.2. Détermination de la température par la méthode Na/K

C'est la seule méthode à retenir dans le cas de Plombières-les-Bains et pour les eaux chaudes qui n'apparaissent pas mélangées (sondage n° 9 et Robinet romain). On obtient par ce géothermonètre Na/K des températures du réservoir en profondeur comprise entre 145°C (courbe théorique et extrapolation des données d'A.J. ELLIS [5]) et 190°C (extrapolation des données de J.J. HEMLEY [8]).

2.3.5. L'origine du carbone des carbonates

La composition isotopique du carbone des carbonates dissous présentés tableau VII et figure 7 montre que la source principale du carbone dans les carbonates est d'origine superficielle et correspond à l'oxydation de la matière organique des sols et des roches.

La relation linéaire qui lie la température d'émergence à la composition isotopique en carbone 13 confirme le mélange des eaux de Plombières-les-Bains.

2.3.6. L'origine du soufre des sulfates

Elle a été abordée par l'étude des isotopes du soufre dans les sulfates dissous dans les différentes eaux. Les eaux froides de surface sont caractérisées par de très basses teneurs en sulfates (1 à 2 mg/l) et des compositions isotopiques en soufre légèrement appauvries en soufre lourd par rapport à celles des eaux thermales qui sont assez homogènes et comprises entre 10,7 et 11,8 unités $\delta^{34}S$. Cette homogénéité dans la composition isotopique du soufre des sulfates fait penser à une origine commune de cet élément et sans doute profonde car ces valeurs sont dans la gamme de celles mesurées sur le soufre dans quelques granitoïdes français.

2.3.7. Les gaz dissous dans les eaux thermales

Sur des prélèvements réalisés dans les mêmes conditions opératoires qu'à Amélie-les-Bains, une série de mesures a été réalisée à Plombières-les-Bains sur les gaz dissous dans les eaux chaudes.

2.3.7.1. L'origine de l'argon dissous

L'examen du diagramme de la figure 8 montre d'une part la corrélation température d'émergence/teneur en argon, d'autre part, le phénomène de mélange entre les eaux froides et les eaux chaudes.

Les valeurs isotopiques de l'argon présentées tableau VII sont voisines de celles de l'air.

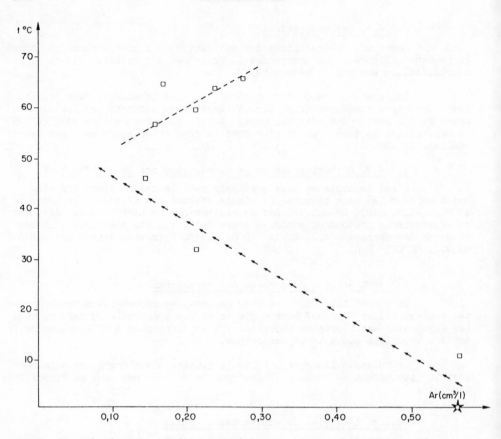

FIG.8. Plombières-les-Bains — Relation entre la teneur en argon et la température d'émergence.

2.3.7.2. L'origine du gaz carbonique et du méthane dissous

        La composition isotopique en carbone de ces deux gaz dissous dans
les sources chaudes du sondage n° 9 et Vauquelin est comparable à celles trouvées dans les bicarbonates des eaux de la rivière l'Augronne, corrélation en
rapport sans doute avec une origine superficielle du carbone. Le couple bicarbonates - CO_2 utilisé comme géothermomètre isotopique donne des indications
de températures voisines de celles des émergences.

3 - CONCLUSIONS

        L'étude des eaux chaudes d'Amélie-les-Bains (Pyrénées Orientales)
et de Plombières-les-Bains (Vosges) met en évidence les points suivants :

        Les eaux thermales d'Amélie-les-Bains sont des eaux d'origine mé-
téoriques réchauffées au cours de leur trajet en profondeur. Ces eaux ne se
mélangent pas avec des eaux superficielles. La température atteinte par les
eaux en profondeur est comprise entre 100° et 130°C ; cette température a

été déterminée par la teneur en silice, la teneur en fluor et le rapport sodium - potassium en tenant compte des variations de répartition des espèces en fonction de la température.

Ces indications thermométriques sont comparables à celles fournies par le géothermomètre soufre pour le couple SO_2-HS^- (100° à 145°C).

On peut montrer que le dépôt de calcite à l'émergence règle la teneur en calcium, ainsi toute méthode thermométrique utilisant la teneur en calcium à l'émergence ne peut conduire qu'à des résultats erronés.

Les eaux thermales de Plombières (Vosges) sont des eaux d'origine météorique réchauffées au cours de leur trajet en profondeur, et mélangées à des degrés divers à des eaux superficielles froides.

L'examen des concrétions aux émergences ne permet, au stade actuel de l'étude de retenir que la méthode du rapport Na/K pour apprécier la température atteinte par les eaux en profondeur ; cette température est comprise entre 145° et 190°C. La température obtenue par le géothermomètre SiO_2 est comprise entre 130 et 135°C, valeur qui doit être considérée comme minimale en raison de la présence de concrétions aux émergences.

Remerciements

Nous sommes heureux de pouvoir remercier le personnel du Département Minéralogie, Géochimie, Analyses du B.R.G.M. : la Section Analyses Monsieur BOUCETTA et ses collègues, la Section Isotopes, Madame MELON, Mademoiselle BOURGEOIS, Monsieur DUMAS, la Section Géochimie, Madame SIMA, dont le travail a permis cette étude.

REFERENCES

[1] CRAIG, H., Isotopic variations in meteoric waters, Science 133 (1961) 1702−03.
[2] DANSGAARD, W., Stable isotopes in precipitation, Tellus 16 4 (1964) 436−68.
[3] FOURNIER, R.O., TRUESDELL, A.H., An empirical Na-K-Ca geothermometer for natural waters, Geochim. Cosmochim. Acta 37 (1973) 1255−75.
[4] KHARAKA, Y.K., BARNES, I., SOLMINEQ: Solution Mineral Equilibrium Computations, US Geol. Survey, Menlo Park, Report PB 215899 (1973).
[5] ELLIS, A.J., «Quantitative interpretation of chemical characteristics of hydrothermal systems», UN Symp. on the Development and Utilization of Geothermal Resources, Pisa 1970, Geothermics 1970, Vol. 2, Part 1 (1970) 516−28.
[6] SAKAI, H., Fractionation of sulfur isotope in nature, Geochim. Cosmochim. Acta 12 (1957) 150−69.
[7] ROBINSON, B., Sulfur isotopic equilibrium during sulfur hydrolysis at high temperature, Earth Planet. Sci. Lett. 18 3 (1973) 443−50.
[8] HEMLEY, J.J., Aqueous Na/K ratios in the systems K_2O, Na_2O, Al_2O_3, SiO_2, H_2O (abstract), 1967 Annual Meeting Geol. Soc. America, New Orleans, Louisiana (1967) 94−95.
[9] INTERNATIONAL ATOMIC ENERGY AGENCY, Environmental Isotope Data Nos 1, 2, 3, 4, 5: World Survey of Isotope Concentration in Precipitation, Technical Reports Series Nos 96, 117, 129, 147, 165, IAEA, Vienna (1969, 1970, 1971, 1973, 1975).

Appendice

TABLEAUX I A VIII

TABLEAU I.   DEMARCHE ANALYTIQUE

Mesure et prélève-ment à l'émergence	Mesures sur le terrain - labo-ratoire mobile	Mesures en labo-ratoire	Résultats présen-tés tableau n° AMELIE ! PLOMBIERES	
t, pH, réserve al-caline, résistivi-té				
- Prélèvement pour dosage	$S^=$, $SO_3^=$, $S_2O_3^=$			
- Prélèvement pour dosages chimiques	Filtration 100 mµ SiO_2 Filtration 10 mµ	anions, cations (maj. + traces) infratraces	II	V_I
- Prélèvement pour dosages isotopi-ques		^{40}Ar ^3H, ^2H, ^{18}O, ^{34}S dans sulfures et sulfates ^{13}C dans CO_2, CH_4, HCO_3^-	III	VII
- Prélèvement pour dosages gaz dissous et libres		gaz dissous et gaz libres dosés par CPG	V	VIII

BAUBRON et al.

TABLEAU II.   DONNEES CHIMIQUES – AMELIE-LES-BAINS

Ba, Al < 1 mg/1
Rb, Sr < 100 mg/m3
Ag, Cd, Co, Cr, Cu, Mn, Mo, Ni, Pb < 1 mg/m3

N° ECHANTILLONS	pH	Li (mg/m^3)	Ca (mg/1)	Mg (mg/1)	Na (mg/1)	K (mg/1)	$CO_3^=$ (mg/1)	HCO_3^- (mg/1)	Cl^- (mg/1)
1 Alcaline	8,67	170	2,4	≤ 0,1	97,0	3,8	12	98	28,0
2 Amelie	8,66	180	2,6	"	95,4	3,7	12	95	28,0
3 Anglada	8,80	180	3,0	"	98,7	3,8	12	101	27,0
4 Arago	8,82	170	2,0	"	96,1	3,8	12	101	28,0
5 Ascensionnelle	8,81	180	2,0	"	96,5	3,8	néant	119	28,0
6 Chomel	8,88	170	2,2	"	96,5	3,7	12	95	28,0
7 Concorde	8,66	180	1,8	"	96,0	3,6	15	92	28,0
8 Galerie	8,82	170	2,2	"	96,5	3,7	15	92	28,0
9 Encome	8,63	170	3,0	"	96,8	3,6	21	79	27,0
10 Fanny	8,67	180	2,0	"	96,7	3,6	21	79	28,0
11 Pares	8,50[a]	180	2,0	"	98,0	3,7	21	82	28,0
12 Pascalone	8,74	180	2,0	"	97,0	3,6	18	85	27,0
13 Petit Escaladou	8,70	180	2,0	"	97,5	3,7	18	85	28,0
14 Petit Monjolet	8,67	180	2,0	"	98,5	3,7	21	85	28,0
15 La Renale	8,94	180	2,2	"	98,2	3,7	21	82	28,0
16 Gros escaladou	8,65	180	2,4	"	97,4	3,6	24	73	28,0
17 Mondony	7,62	10	10,0	3,4	4,4	0,6	néant	52	3,6

[a] pH mesuré à 20°C.

$SO_4^=$ (mg/l)	NO_3^- (mg/l)	NH_4^+ (mg/l)	$PO_4^=$ (mg/l)	B (mg/l)	SiO_2 (mg/l)	F (mg/l)	Zn (mg/m³)	$SO_3^=$ (10^{-4}M/l)	$S_2O_3^=$ (10^{-4}M/l)	HS^- (10^{-4}M/l)	Fe (mg/m³)
36	5,0	0,6	≤0,1	0,3	70	18,00	≤1	0.26	0.74	0.74	10
38	4,5	0,4	"	0,3	73	18,00	12	0.00	0.80	0.27	120
38	5,6	0,6	"	0,3	72	19,00	10	0.22	1.43	0.44	5
38	5,0	0,4	"	0,3	72	18,00	≤1	0.00	1.00	0.78	4
42	3,2	0,4	"	0,3	72	19,00	14	0.00	1.36	0.46	30
42	5,6	0,4	"	0,3	73	18,00	22	0.16	0.95	0.95	40
42	3,8	0,4	"	0,3	74	18,00	6	0.05	1.57	0.85	5
38	6,0	0,4	"	0,3	72	18,00	10	0.17	1.76	0.25	20
38	5,0	0,8	"	0,3	73	18,00	8	0.00	1.68	0.22	10
38	4,7	0,4	"	0,4	71	18,00	7	0.23	1.11	0.55	≤1
38	5,6	0,4	"	0,3	72	18,00	4	0.21	2.72	0.16	7
38	6,0	1,8	0,3	0,3	72	18,00	≤1	0.70	0.79	0.16	16
38	7,0	0,4	≤0,1	0,3	71	18,00	5	0.05	1.01	0.75	10
38	4,7	0,3	"	0,4	72	18,00	6	0.05	1.56	0.66	≤1
38	4,5	0,6	"	0,3	72	18,00	1	0.00	0.75	0.91	"
38	4,5	0,6	"	0,3	71	18,00	6	0.00	1.28	0.54	4
5	≤1,0	0,4	"	≤0,1	13	0,14	≤1				8

TABLEAU III.　DONNEES ISOTOPIQUES — AMELIE-LES-BAINS

PRELEVEMENTS	DATE	Temp. °C	^{3}H UT	$\delta D°/_{oo}$	$\delta^{18}O°/_{oo}$ v. s. S.M.O.W.
Eaux de surface :					
Le Tech à la Preste	1.5.77	7	65 ± 6	-59,9 ± 0,5	-9,0 ± 0,1
Le Mondony à Montalba	2.5.77	10	62 ± 6	-47,7 ± 0,5	-7,5 ± 0,1
Sources froides :					
La Preste Source Reuter	1.5.77	11	70 ± 6	-53,9 ± 0,5	-8,1 ± 0,1
Amélie - Galerie Parès (eaux froides)	29.4.77	32	30 ± 4	-50,3 ± 0,5	-7,5 ± 0,1
Sources thermales : Amélie					
- Alcaline	28.4.77	56	2 ± 1	-61,3 ± 0,5	-9,1 ± 0,1
- Amélie	30.4.77	61	3 ± 1	-61,4 ± 0,5	-9,0 ± 0,1
- Anglada	25.4.77	57	≤ 1	-61,1 ± 0,5	-9,1 ± 0,1
- Arago	26.4.77	57	3 ± 1	-60,9 ± 0,5	-9,3 ± 0,1
- Ascensionnelle	25.4.77	51	≤ 1	-60,7 ± 0,5	-9,0 ± 0,1
- Chomel	26.4.77	46	≤ 1	-60,9 ± 0,5	-9,2 ± 0,1
- Concorde	28.4.77	59	≤ 1	-60,7 ± 0,5	-9,1 ± 0,1
- Galerie	25.4.77	52	≤ 1	-61,0 ± 0,5	-9,2 ± 0,1
- En Côme	28.4.77	56	≤ 1	-60,8 ± 0,5	-9,1 ± 0,1
- Fanny	27.4.77	62	≤ 1	-61,0 ± 0,5	-9,2 ± 0,1
- Parès	28.4.77	56	≤ 1	-60,7 ± 0,5	-9,0 ± 0,1
- Pascalone	30.4.77	50	≤ 1	-61,2 ± 0,5	-9,0 ± 0,1
- Petit Escaladou	28.4.77	57°5	≤ 1	-60,8 ± 0,5	-9,2 ± 0,1
- Petit Monjolet	27.4.77	59	3 ± 1	-61,0 ± 0,5	-9,1 ± 0,1
- La Rénale	03.5.77	32	2 ± 1	-61,3 ± 0,5	-9,0 ± 0,1
- Gros Escaladou	03.5.77	62	≤ 1	-60,2 ± 0,5	-8,9 ± 0,1
La Preste - Appolon	01.5.77	43°5	≤ 1	-68,0 ± 0,5	-10,1 ±0,1
Précipitations					
Amélie Can Malcion	28 au 29.4.77	-	41 ±4	-26,6 ± 0,5	-4,4 ± 0,1
Amélie Can Malcion	3 au 4.5.77	-	49 ±5	-43,1 ± 0,5	-6,2 ± 0,1

δ³⁴S°/₀₀ v. s. C.D. ± 0,2		δ¹³C°/₀₀ v. s. P.D.B.			⁴⁰Ar/³⁶Ar
S(-II)	S(+VI)	HCO₃⁻	CO₂ dissous	CH₄ dissous	Argon dissous
-	+ 9,3	-6,0 ± 0,1	-	-	-
-	+ 9,3	-10,4± 0,1	-	-	-
-	+ 8,3	-11,0 ±0,1	-	-	294
-	-	-	-	-	n.d.
-4,5	+18,9	- 9,3 ±0,1	n.d.	n.d.	280
-3,4	+17,0	- 9,4 ±0,1	-18,2 ±0,1	-27,3 ±0,1	295
-3,6	+18,5	- 9,3 ±0,1	n.d.	n.d.	294
-3,8	+19,2	- 9,6 ±0,1	-20,0 ±0,1	-27,5 ±0,1	291
-4,3	+19,1	- 9,7 ±0,1	n.d.	n.d.	319
-3,7	+19,1	-10,5 ±0,1	n.d.	n.d.	292
-4,3	+19,2	- 9,5 ±0,1	n.d.	n.d.	n.d.
-4,0	+18,6	-10,0 ±0,1	n.d.	n.d.	305
-3,8	+16,3	- 9,5 ±0,1	n.d.	n.d.	284
-2,9	+17,5	- 9,6 ±0,1	-20,3 ±0,1	-29,2 ±0,1	290
-4,0	+19,0	- 9,6 ±0,1	n.d.	n.d.	n.d.
-4,5	+19,6	- 9,8 ±0,1	n.d.	n.d.	311
-4,2	+18,5	- 9,5 ±0,1	n.d.	-27,2 0,1	300
-3,6	+18,0	- 9,7 ±0,1	n.d.	n.d.	n.d.
-3,8	+15,6	- 9,6 ±0,1	n.d.	n.d.	n.d.
-4,5	+18,1	- 9,9 ±0,1	-21,8 ±0,1	-27,4 ±0,1	294
-	+ 6,7	- 9,9 ±0,1	-	-	n.d.
-	-	-	-	-	-
-	-	-	-	-	-

TABLEAU IV. DETERMINATION DE LA TEMPERATURE D'EQUILIBRAGE DES ESPECES REDUITES ET OXYDEES DU SOUFRE – AMELIE-LES-BAINS

SOURCE	Température Emergence	$\Delta SO_4 - H_2S$	Température déterminée par l'équilibre $H_2S + 2H_2O \rightarrow SO_2 + 3H_2$ (SAKAI, 1968)		Température déterminée par la relation de ROBINSON (1973) $SO_4^{=}/H_2S$	
			H_2S	HS^-	Température Brute	Température corrigée
1 Alcaline	56°C	23,4	76°C	100°C	273°C	150°C
2 Amélie	61°C	20,4	103°C	115°	328°	220°
3 Anglada	57°C	22,1	88°	108°	295°	200°
4 Arago	57°C	23,0	79°	102°	179°	150°
5 Ascensionnelle	51°C	23,4	75°	100°	273°	150°
6 Chomel	46°C	22,8	80°	101°	283°	150°
7 Concorde	59°C	23,5	76°	100°	271°	150°
8 Galerie	52°C	22,6	82°	104°	286°	150°
9 Encome	56°C	20,1	105°	135°	335°	220°
10 Fanny	62°C	20,4	103°	131°	328°	220°
11 Parès	59°C	23,0	79°	102°	279°	150°
12 Pascalone	50°C	24,1	71°	95°	262°	140°
13 Petit Escaladou	57,5°C	22,7	82°	104°	284°	150°
14 Petit Monjolet	59°C	21,6	93°	118°	304°	200°
15 La Renale	32°C	19,4	118°	145°	351°	250°
16 Gros Escaladou	62°C	22,6	82°	102°	286°	150°

TABLEAU V. TENEURS EN GAZ DISSOUS – AMELIE-LES-BAINS

SOURCE	CO_2 cm3/l	Ar cm3/l	N_2 cm3/l	CH_4 cm3/l
1 Alcaline	0,119	0,308	7,39	6,98 10^{-2}
2 Amélie	0,153	0,233	3,66	1,86 10^{-2}
4 Arago	0,102	0,280	2,50	2,16 10^{-2}
5a Ascensionnelle	0,134	0,349	4,42	4,61 10^{-2}
5b Ascensionnelle	0,168	0,488	13,00	8,05 10^{-2}
6 Chomel	0,573	0,489	10,70	9,39 10^{-2}
7 Concorde	0,175	0,394	6,38	10,60 10^{-2}
8 Galerie	0,165	0,516	12,00	7,22 10^{-2}
10a Fanny	0,300	0,377	8,85	9,83 10^{-2}
10b Fanny	0,222	0,418	8,84	7,45 10^{-2}
11b Parès	0,149	0,431	6,01	8,83 10^{-2}
13a Petit Escaladou	0,170	0,504	7,57	9,46 10^{-2}
13b Petit Escaladou	0,109	0,380	5,05	6,59 10^{-2}
16 Gros Escaladou	0,301	0,352	6,75	8,31 10^{-2}

TABLEAU VI.   DONNEES CHIMIQUES – PLOMBIERES-LES-BAINS

Ba, Al < 1 mg/l
Cd, Co, Cr, Mo, Ni, Pb < 1 mg/m^3

N° Echantillon	pH t é [a]	Zn (mg/m^3)	Ca (mg/1)	Mg (mg/1)	Na (mg/1)	K (mg/1)	HCO$_3^-$ (mg/1)	Cl$^-$ (mg/1)	SO$_4^=$ (mg/1)	NO$_3^-$ (mg/1)
P1a Sondage 9 (filtré)	8,12	33	5,8	⩽ 0,1	97,5	5,4	119,0	10,5	76	2,2
P1b Sondage 9 (non fil- tré)		3	5,4	⩽ 0,1	97,9	5,3	116,0	14,5	76	1,7
P 2 Robinet romain	8,4	4	5,6	⩽ 0,1	95,0	5,3	113,0	14,0	76	1,3
P 3 Stanislas	8,4	5	6,2	⩽ 0,1	98,8	5,7	113,0	16,0	80	2,0
P 4 Savonneuse 9	7,90	3	10,0	1,6	45,0	3,2	82,0	7,0	38	2,0
P 5 Jutier	7,88	4	13,2	2,0	57,0	5,2	76,0	30,0	46	10,0
P 6 Vauquelin	8,12	8	6,0	⩽ 0,1	95,0	5,4	116,0	11,5	72	1,4
P 7 Ste-Catherine	8,08	3	7,4	0,5	80,0	4,8	104,0	10,5	60	2,0
P 8 Augronne 1	6,2 20°	5	2,4	0,6	2,8	1,2	6,1	5,6	2	2,2
P 9 Augronne 2	6,0 20°	2	2,4	0,6	2,9	1,1	9,2	5,8	⩽ 1	2,0
Alliot	5,3	6	9,2	2,2	2,2	2,0	33,6	1,3	⩽ 1	7,6

[a] Température à l'émergence.

Li (mg/m^3)	NH$_4^+$ (mg/1)	PO$_4^{\equiv}$ (mg/1)	B (mg/1)	SiO$_2$ (mg/1)	F (mg/1)	Rb (mg/m^3)	Sr (mg/m^3)	Ag (mg/m^3)	Cu (mg/m^3)	Fe (mg/m^3)	Mn (mg/m^3)
510	0,4	$\leqslant$ 0,1	< 0,1	103	15,50	$\leqslant$ 100	180	$\leqslant$ 1	$\leqslant$ 1	$\leqslant$ 1	10
500	0,4	$\leqslant$ 0,1	0,2	102	15,50	"	190	$\leqslant$ 1	"	"	8
500	0,6	$\leqslant$ 0,1	0,2	98	15,00	"	170	$\leqslant$ 1	"	"	12
520	0,4	$\leqslant$ 0,1	0,2	99	16,00	"	180	2	8	"	8
220	0,2	$\leqslant$ 0,1	$\leqslant$ 0,1	50	7,20	"	$\leqslant$ 100	$\leqslant$ 1	$\leqslant$ 1	10	$\leqslant$ 1
220	0,4	0,7	$\leqslant$ 0,1	48	7,20	"	$\leqslant$ 100	"	"	$\leqslant$ 1	"
490	0,4	$\leqslant$ 0,1	$\leqslant$ 0,1	102	16,00	"	200	"	"	"	13
400	0,2	$\leqslant$ 0,1	< 0,1	85	14,50	•	$\leqslant$ 100	"	"	"	< 1
$\leqslant$ 10	0,3	0,3	$\leqslant$ 0,1	6	$\leqslant$ 0,10	"	$\leqslant$ 100	"	"	35	27
$\leqslant$ 10	0,4	0,3	$\leqslant$ 0,1	5	$\leqslant$ 0,10	"	$\leqslant$ 100	"	"	37	22
4	$\leqslant$ 0,1	$\leqslant$ 0,1	1	12	1,25	"	40	"	3	15	$\leqslant$ 10

TABLEAU VII.   DONNEES ISOTOPIQUES – PLOMBIERES-LES-BAINS

PRELEVEMENTS	DATE	Temp. °C	^3H UT	δD°/oo v.s. S.M.O.W.	$\delta^{18}O$°/oo	$\delta^{34}S$°/oo v.s. C.D.	$\delta^{13}C$°/oo v.s. P.D.B. HCO_3^-	CO_2 dissous	CH_4 dissous	$^{40}Ar/^{36}Ar$ Argon dissous
Sondage 9	13.06.77	71	<1	-59,7 ± 0,5	- 9,0 ± 0,1	+11,8 ± 0,2	-13,4 ± 0,1	-22,9 ± 0,1	-25,2 ± 0,1	297
Robinet romain	13.06.77	70	<1	-60,0 ± 0,5	- 9,1 ± 0,1	+11,7 ± 0,2	-13,6 ± 0,1	n.d.	n.d.	n.d.
Stanislas	13.06.77	64	<1	-59,8 ± 0,5	-'9,0 ± 0,1	+10,7 ± 0,2	-14,1 ± 0,1	n.d.	n.d.	n.d.
Savonneuse n° 9	14.06.77	47	22 ± 3	-58,7 ± 0,5	- 8,9 ± 0,1	+11,8 ± 0,2	-16,0 ± 0,1	n.d.	n.d.	n.d.
Jutier	14.06.77	33°5	103 ± 8	-59,2 ± 0,5	- 8,7 ± 0,1	+10,8 ± 0,2	-19,3 ± 0,1	n.d.	n.d.	n.d.
Vauquelin	14.06.77	70	<1	-60,2 ± 0,5	- 9,0 ± 0,1	+11,6 ± 0,2	-13,8 ± 0,1	-25,2 ± 0,1	-26,9 ± 0,1	293
Ste Catherine	14.06.77	63	3 ± 1	-59,7 ± 0,5	- 8,9 ± 0,1	+11,5 ± 0,2	-14,1 ± 0,1	n.d.	n.d.	n.d.
L'Augronne 1	15.06.77	11	247 ± 16	-60,8 ± 0,5	- 9,1 ± 0,1	+ 8,8 ± 0,2	-25,5 ± 0,1	n.d.	n.d.	n.d.
L'Augronne 2	15.06.77	11	268 ± 16	-60,5 ± 0,5	- 8,9 ± 0,1	+ 6,7 ± 0,2	-24,9 ± 0,1	n.d.	n.d.	n.d.
Alliot	20.07.77	9°4	40 ± 4	-58,0 ± 0,5	- 8,5 ± 0,1	n.d.	n.d.	n.d.	n.d.	n.d.

TABLEAU VIII. TENEURS EN GAZ DISSOUS – PLOMBIÈRES-LES-BAINS

SOURCE	CO_2 cm3/1	Ar cm3/1	N_2 cm3/1	CH_4 cm3/1	O_2 cm3/1
Sondage 9	0,254	0,274	9,37	$1,07 \ 10^{-2}$	non détecté
Robinet romain	0,294	0,237	9,01	0,92	$54,2 \ 10^{-2}$
Stanislas	0,292	0,212	7,18	0,62	$1,1 \ 10^{-2}$
Savonneuse S9	1,58	0,145	10,30	0,11	$11,2 \ 10^{-2}$
Jutier	1,13	0,212	15,80	non dosé	$301 \ 10^{-2}$
Vauquelin	0,433	0,169	20,60	0,51	$336 \ 10^{-2}$
Sainte Catherine	0,477	0,157	6,82	0,21	$0,91 \ 10^{-2}$
Augronne 1	non dosé	0,563	9,88	0,20	$1570 \ 10^{-2}$
Augronne 2	non dosé	non dosé	non dosé	non dosé	non dosé

DISCUSSION

H. NIELSEN: It is well known that in most hydrothermal springs the sulphide and sulphate are not, thermodynamically, in isotopic equilibrium and that the sulphide-sulphate thermometer fails to give reliable readings, especially in the typical hydrothermal temperature range below, say, 300°C. For this reason a kinetic reaction is assumed whereby the "fractionation factor" would fall within the range given in Table IV. This makes irrelevant any argument about the assumed inorganic process or even about a microbial H_2S production from the primary sulphate content of the waters.

A. MARCE: With regard to the use of the sulphur geothermometer, a theory has been put forward that sulphur species in gaseous form are in the following equilibrium:

$$H_2S + 2 H_2O \leftrightharpoons SO_2 + 3 H_2$$

and that the sulphur species measured at the springs (temperature 32–62°C) are isotopically comparable with those of deep waters. We assumed the $SO_2 - H_2S$ equilibrium and accordingly calculated temperatures between 70 and 145°C. However, it is clear that this sort of calculation is only valid if there is equilibrium.

H. NIELSEN: The SO_4 contents shown in Table II exceed the HS^- contents. The δ values for the spring should therefore be close to the values for SO_4, i.e. in the range above + 15‰, and so it would be interesting to know what the assumed source of the sulphur is.

A. MARCE: You are right, because the oxidized form always represents more than 60% of the total sulphur, except in the Parès spring (56%); however, the δ of the initial sulphur is + 11‰, which makes its origin uncertain.

H. HÜBNER: In Table VII you give $\delta^{13}C$ values for methane and carbon dioxide having a $\delta^{13}C$ difference of about 2‰. I think that if there is isotopic exchange between these two molecules in the presence of water, you will find a greater difference than two δ units at the given temperatures of 71 and 70°C.

A. MARCE: The results seem to show that there is no isotopic exchange between CH_4 and CO_2, which means that this particular geothermometer cannot be used. It should be remembered that these thermal springs are not carbonated.

USE OF ISOTOPIC GEOTHERMOMETERS IN THE LARDERELLO GEOTHERMAL FIELD

C. PANICHI, S. NUTI, P. NOTO
CNR – Istituto Internazionale per le
        Ricerche Geotermiche,
Pisa, Italy

Abstract

USE OF ISOTOPIC GEOTHERMOMETERS IN THE LARDERELLO GEOTHERMAL FIELD.
    Gas samples from productive wells in the Larderello geothermal field have been collected and separated for spectrometric measurements of the isotopic compositions of their individual components. The D/H, $^{18}O/^{16}O$, $^{13}C/^{12}C$ ratios were measured in the H_2O (steam), H_2, CH_4 and CO_2 samples and used to evaluate the equilibration temperature at depth. The $\Delta^{13}(CO_2-CH_4)$, $\Delta^D(CH_4-H_2)$ and $\Delta^D(H_2O_v-H_2)$ geothermometers give base temperatures of 341 ± 37, 314 ± 30 and $254 \pm 25°C$ respectively, which are always higher than the measured temperatures (average $T_m = 216 \pm 25°C$). The $\Delta^{18}(H_2O-CO_2)$ geothermometer generally gives temperature values very similar to those observed at the well-head and appear capable of giving useful information on the physical state of the water (steam or evaporating liquid water) in the surroundings of each well. The H_2O-CH_4 system gives no reasonable temperature values, showing that no equilibrium conditions are reached at depth between the two components. The comparison between the different evaluated temperatures in the same area, along with some chemical and chemico-physical evidence, suggest that the reaction already proposed, i.e. $CO_2 + H_2 = CH_4 + 2H_2O$, is unable to explain the observed isotopic compositions. On the contrary, the water dissociation reaction ($H_2O = H_2 + 1/2O_2$), and the synthesis reaction for methane and carbon dioxide ($C + 2H_2 = CH_4$ and $C + O_2 = CO_2$) seem able to give an appropriate picture of the thermal situation of the geothermal field.

INTRODUCTION

H_2O(steam), CO_2, CH_4 and H_2 are normally present in the high temperature fluids and represent, together with H_2S and N_2, the main components of the gas phase.

Currently a Fischer-Tropsch chemical reaction, such as (1)

$$4H_2 + CO_2 = CH_4 + 2H_2O \qquad (1)$$

provides a generalized pathway for the following isotopic exchange reactions:

$$^{12}CO_2 + {}^{13}CH_4 = {}^{13}CO_2 + {}^{12}CH_4 \qquad \Delta^{13}(CO_2-CH_4)^* \quad (2)$$

$$C^{16}O_2 + H_2^{18}O = CO^{18}O + H_2^{16}O \qquad \Delta^{18}(CO_2-H_2O)^* \quad (3)$$

$$CH_3D + H_2O = CH_4 + HDO \qquad \Delta^D(H_2O-CH_4)^* \quad (4)$$

$$HD + CH_4 = H_2 + CH_3D \qquad \Delta^D(CH_4-H_2)^* \quad (5)$$

$$HD + H_2O = H_2 + HDO \qquad \Delta^D(H_2O-H_2)^* \quad (6)$$

_____

* Δ in this case indicates the isotopic fractionation in the system indicated between brackets.

As the theoretical and/or experimental variations with temperature of the fractionation factors for each reaction are
known, the δ values for each pair measured on the discharged
fluids can be used to evaluate the base temperatures of the
geothermal reservoirs.

Now, if the temperature of the fluids had remained constant
for a period sufficient for all the reactions to reach isotopic equilibria, we would then expect them all to give the
same isotopic temperature. If, however, the temperature of
the fluids has not remained constant as they proceeded towards the point of discharge, the different rates at which
each exchange reaction approaches isotopic equilibrium conditions will be evident in each pair, giving a different isotopic temperature.

Considering vapour phase reactions only, Lyon $\int 5 \mathcal{J}$ suggests
that the rates of reaction for the approach to isotopic equilibrium decrease in this order: $\Delta^D(H_2-H_2O)$, $\Delta^D(H_2-CH_4)$ and
$\Delta^{13}(CO_2-CH_4)$. This sequence could begin with the term
$\Delta^{18}(CO_2-H_2O)$, as shown by the data obtained by Panichi et al.
$\mathcal{L}1\mathcal{J}$ in Larderello geothermal field. Laboratory experimental
data from the New Zealand group $\mathcal{L}2\mathcal{J}$, and field observations,
suggest the following rough estimation of the isotope reac-
tion rates:

$$\Delta^{13}(CH_4-CO_2) \qquad > 5 \text{ years}$$
$$\Delta^D(H_2-H_2O) \qquad < 1 \text{ year}$$
$$\Delta^{18}(CO_2-H_2O) \qquad \text{rapidly}$$

These figures are only indicative because of the approxima-
tions made during the laboratory bomb experiments, in which
a gas phase is only present or in contact with a liquid phase
and no rocks or additional chemical components are added. In
other words, the influence of some rocky components, which
might play an important part in catalyzing these reactions,
is not considered.

Panichi and Gonfiantini $\mathcal{L}3\mathcal{J}$ summarized the evaluated temperatures from several authors in different geothermal fields,
referring to the application of isotopic geothermometers.
The data indicate that the pair CO_2-CH_4 always gives a temperature 50-150°C higher than the measured values, while $\delta^{18}O$
values for CO_2-H_2O and δD values for H_2-H_2O systems give
practically the same T values, and sometimes the same and
sometimes 10-20°C higher from the CH_4-H_2 system.

When different pairs are considered in the evaluation of the
base temperature of a certain geothermal field, the observed
discrepancies are explained in terms of differences in the
reaction rates, assuming that isotopic equilibrium takes place
between the gas pair after their formation, independent of
the chemical reactions governing their actual concentrations.

This hypothesis appears realistic, if we consider that molecular H_2 and H_2O(vapour) seem capable of exchanging their isotope contents and approaching isotopic equilibrium after several months; the oxygen isotopic composition of CO_2 and H_2O suggests that an isotopic exchange also occurs between gas molecules, other than in a liquid phase, via the bicarbonate ion as an intermediate. The generalization of this hypothesis, however, seems to be a critical point because of the lack of re-equilibration observed between molecular CH_4 and CO_2 and the fact that oxygen isotope exchanges between CO_2 and H_2O(vapour) may be permitted to occur by the presence of a small percentage of condensed water occupying the pore space of a vapour-dominated system.

The activation complexes through which the isotopic exchange reactions can proceed are indeed very little understood and quite difficult to define.The generalized form of equation (1) also appears to be of little use in interpreting the isotopic geothermometers.

This paper is aimed at verifying the possibility of explaining the different T values obtained from different geothermometers as being due to different isotopic exchange reactions which are related to several distinct chemical reactions occurring at depth and not necessarily at the same time and in the same place.

PRESENTATION OF THE RESULTS

The fluid components were separated in the field and laboratory following this procedure: a) CO_2 and H_2O(steam) were separated directly at wellhead using liquid air and dry ice-acetone as cooling mixtures, as described by Panichi et al. [1]. In some cases the CO_2 samples were separated from the condensed water using liquid water only as the cooling agent, in order to verify the extent, if any, of the carbon isotopic fractionation, at approximately -30°C, between solid and gaseous CO_2, as required by the previous procedure.

Six different CO_2 samples, collected according to both procedures, indicate that dry-CO_2 is always depleted in ^{13}C by about 1‰ with respect to the wet-CO_2.The fractionation observed moves in the opposite direction from that expected by Grootes et al.[4], who stated that ^{13}C concentrates in the gaseous phase during sublimation at a temperature below zero °C. Despite this discrepancy, the fractionation observed was used to correct the carbon isotopic data utilized for the CH_4-CO_2 geothermometer.

b) the H_2 and CH_4 components were concentrated in the field using 30% NaOH solutions, and then separated in a vacuum line which provides both the oxidation of H_2 and CH_4 on CuO at 380°C and 800°C respectively. The resulting H_2O samples were subsequently reduced on uranium turning at 700°C and

TABLE I. ISOTOPIC COMPOSITIONS OF GEOTHERMAL FLUID FROM THE LARDERELLO AREA AND EVALUATED TEMPERATURES FROM THREE DIFFERENT GEOTHERMOMETERS

Well No.	Sampling date	δD‰(SMOW)		δ13C‰(PDB)		Isotopic Temperatures (°C)			T_m (°C)
		CH_4	H_2	CH_4	CO_2	$\Delta^D_{(CH_4-H_2)}$[a]	$\Delta^D_{(H_2O-H_2)}$[b,d]	$\Delta^{13}_{(CO_2-CH_4)}$[c]	
10	22. 9.77	−154	−433			283	293		230
14	15.11.77	−263	−487			300	228		171
7	16.11.77	−249	−476			318	240		220
55	16.11.77	−	−479			−	238		193
59	12. 1.78	−218	−469			293	248		213
60	13. 1.78	−288	−478			365	238		235
66	17. 4.78		−475	−25.9	+0.5	−	242	276	226
63	14. 4.78	−	−489	−27.0	−1.8	−	227	288	211
65	18. 4.78	−212	−466	−23.8	−0.6	293	250	318	169
62	18. 4.78	−233	−479	−26.7	+0.5	293	239	266	220
64	18. 4.78	−	−451	−24.3	−0.2	−	272	307	258
19		−146	−424	−25.1		290	307	−	228
4	12. 1.78	−218	−439			327	277	−	230
2	7.10.77	−192	−437	−21.5	−2.9	310	283	347	234
40	6.10.77	−190	−410	−24.0	−4.0	357	325	376	235
61	6.10.77	−236	−485	−25.5	−4.4	290	232	360	188

[a] The Eq. used is: $1000 \ln \alpha = -90.888 + 181.269 \times 10^6 T^{-2} - 8.949 \times 10^{12} T^{-4}$ [6].

[b] The Eq. used is: $1000 \ln \alpha = -201.6 + 391.5 \times 10^3 T^{-1} + 12.9 \times 10^6 T^{-2}$ [10].

[c] The Eq. used is: $1000 \ln \alpha = -9.01 + 15.301 \times 10^3 T^{-1} + 2.361 \times 10^6 T^{-2}$ [10].

[d] The deuterium content for the steam samples has been assumed to be equal to −40‰ (V–SMOW) see text.

the resulting CO_2 samples measured directly on the spectrometer.

The standard deviation of the $\delta^{18}O$ is 0.17‰ for water and
0.21‰ for CO_2; the $\delta^{13}C$ measurements deviate by 0.45‰ for
CH_4 and 0.23‰ for CO_2; the δD deviations are roughly evaluated
at 5‰ for CH_4 and 8‰ for the H_2.

The standards used are the PDB for carbon and V-SMOW for
oxygen and hydrogen isotopes.

The data obtained are reported in Tables 1 and 2 which also
indicate the measured and calculated temperatures.

The standard deviation for the isotopic temperatures, due
to the precision of the different measurements, is 5-6°C, 7-
11°C, 8-12°C, 3-5°C for the H_2O-CO_2, H_2-H_2O, H_2-CH_4 and CO_2-CH_4
systems respectively.

DISCUSSION OF THE RESULTS

The isotopic data of Tables 1 and 2 indicate that, once
again, the calculated temperatures differ for the same fluid
when different components are considered and that the measured
temperatures are generally lower than the calculated ones.

Similar results are reported by Lyon [5] for Wairakei field
and by Craig [6] for Yellowstone Park, and have been interpreted in terms of differences in the reaction rates of the isotope exchanges, assuming that the F.T. reaction (1) is the
reaction through which these exchanges effectively occur. However, if the CO_2-CH_4 system indicates the "original" temperature of the formation of a gas phase in the underground, at
which the F.T. is in isotopic and chemical equilibrium, then
some problems arise regarding a possible equilibrium of other
gas components in different parts of the system.Indeed, if the
F.T. represents the pathway for isotopic exchanges, what is
the mechanism by which a successive isotopic exchange is permitted for H_2O-H_2 and CH_4-H_2, while the CO_2-CH_4 remain unaltered?

Let us briefly examine the results obtained in the Larderello
field when applying the different isotopic geothermometers.

$\Delta^{13}(CO_2-CH_4)$ geothermometer

Samples were taken from only 8 wells in order to test this
geothermometer (Table 1), but a more complete picture of the
results obtained in this way in the Larderello area can be attained from the already available data [1], which are in complete agreement with these results.

The total range of the calculated temperatures is 242 to 402°C
with an average value of 341± 37°C; these are considerably
higher than the observed values, the maximum temperature recorded in Larderello area being 305°C.

The calculated figures have been interpreted as the maximum
temperatures reached by the geothermal system below the actual

TABLE II. OXYGEN ISOTOPIC COMPOSITION OF CO_2 AND STEAM SAMPLES AND RELATED ISOTOPIC TEMPERATURES

Well No.	Sampling date	CO_2 $\delta^{18}O$ %o SMOW	H_2O $\delta^{18}O$ %o SMOW	Isotopic Temperature T°C		Well-head Temperature T_m (°C)
				CO_2-H_2O(v) (eq. 1)	CO_2-H_2O(l) (eq. 2)	
1	6.12.76	15.9	-2.4	232	195	231
2	9.12.76	14.5	-4.3	224	188	250
3	10.12.76	17.0	-3.0	212	174	203
4	14.12.76	18.2	-1.9	213	174	234
5	14.12.76	16.5	-5.2	192	152	190
6	16.12.76	14.5	-2.5	250	217	233
7	16.12.76	13.2	-6.3	217	179	219
8	21.12.76	15.5	-2.9	230	193	229
9	22.12.76	13.7	-1.4	277	247	238
10	22.12.76	14.1	-1.7	266	236	232
11	23.12.76	14.3	-2.3	256	223	247
12	9.12.76	16.5	-3.0	216	186	229
13	4.1.77	13.2	-3.4	254	221	222
14	21.4.77	17.1	-6.5	174	134	169
15	21.4.77	17.7	-5.5	178	138	156

15	21.	4.77	17.0	-2.1	224	186	204
16	21.	4.77	15.2	-4.4	216	178	176
17	5.	5.77	15.5	-1.9	247	212	225
18	1.	6.77	10.6	-5.4	264	232	224
19	1.	6.77	15.0	-4.2	221	183	188
20	2.	6.77	14.2	-1.9	263	231	247
21	6.	5.77	15.2	-4.0	221	183	216
22	10.	5.77	14.3	-4.4	238	191	216
23	22.	12.76	12.9	-1.9	284	260	
24	11.	1.77	16.2	-0.3	256	230	192
25	14.	1.77	13.8	-4.3	235	205	194
26	18.	1.77	9.6	-4.5	294	271	209
27	20.	1.77	17.3	-3.7	201	169	157
28	31.	1.77	17.4	-3.9	198	166	180
29	2.	2.77	17.4	-6.0	176	150	157
30	19.	4.77	16.7	-2.0	229	201	218
31	3.	5.77	15.6	-5.4	200	168	176
32	3.	5.77	21.3	-0.3	195	162	225
33	3.	5.77	16.0	-4.8	202	170	199
34	5.	5.77	18.5	-3.3	192	159	156
35	5.	5.77	21.3	-0.4	194	161	204
36	5.	1.77	14.1	-3.2	245	210	205
37	9.	1.77	13.4	-4.1	242	207	203
38	10.	1.77	14.4	-1.6	266	227	240
39	11.	1.77	14.5	-3.1	242	207	209
40	12.	1.77	14.4	-6.4	203	163	190
41	13.	1.77	14.5	-2.5	250	217	235
42	14.	1.77	15.6	-2.1	241	206	217
43	17.	1.77	13.7	-4.0	240	205	217

Table 2(cont.d)

44	18.	1.77	13.1	-2.1	277	247	228
45	19.	1.77	14.8	-2.6	245	209	202
46	20.	1.77	16.8	-2.7	216	188	235
47	20.	1.77	14.6	-3.7	231	202	240
48	21.	1.77	15.9	-3.0	224	187	203
49	25.	1.77	17.0	-2.5	219	182	215
50	26.	1.77	20.6	-3.6	166		180
51	27.	1.77	16.0	-3.3	220	182	235
52	28.	1.77	16.1	-4.3	207	168	203
53	3.	2.77	14.7	-4.3	222	194	241
54	15.	4.77	17.4	-0.5	238	203	197
55	15.	4.77	15.4	-6.4	192	152	190
56	18.	4.77	17.1	-1.7	228	191	210
57	19.	4.77	14.3	-3.5	239	204	222
58	20.	4.77	19.7	-3.5	179	139	174
59	16.	1.78	16.7	-2.8	218	186	213
60	18.	1.78	17.0	-2.1	224	194	235
61	18.	1.78	15.2	-2.7	237	209	230
62	18.	4.78	19.6	-1.1	205	172	
63	14.	4.78	18.1	-2.6	205	172	
64	19.	4.78	15.6	-1.2	252	225	
65	18.	4.78	22.0	+1.0	201	169	
66	17.	4.78	19.2	-0.3	219	188	

Eq. (1): $1000 \ln \alpha = -10.55 + 9.289 \times 10^3/T + 2.651 \times 10^6/T^2$.
Eq. (2): $1000 \ln \alpha = -3.37 + 4.573 \times 10^3/T + 2.708 \times 10^6/T^2$.
Eqs (1) and (2) derived from Bottinga [10].

productive horizons [1]. In this case the carbon isotopic fractionation between CO_2 and CH_4 was considered to be frozen from
the site of gas formation to the wellhead and no re-equilibration is possible because of the relatively short travelling-
time of the fluids. However, the CO_2/CH_4 chemical ratios for
the same wells in the Larderello area indicate "base" temperatures of less than 300°C when an F.T. reaction is considered:
that is, considerably lower than the isotopic values [7].

Furthermore, the chemical analyses always indicate that reasonable equations can be proposed for the synthesis of CO_2
and CH_4 for a temperature interval between 275 and 290°C. That
is:

$$C + 2H_2 = CH_4 \qquad\qquad (7)$$
$$C + O_2 = CO_2 \qquad\qquad (8)$$

Obviously amorphous carbon or graphite must be present in
the geological setting of the region as an independent source
of carbon. We do not know if this is the real situation in all
the existing geothermal systems, but it is certainly true in
the case of Larderello, where graphite has been found scattered throughout several well-cores [8].

In addition, where a liquid phase is present, the hydrolysis
of carbonates, such as that described in equation (9)

$$CaCO_3 + H_2O = CO_2 + Ca(OH)_2 \qquad\qquad (9)$$

can easily occur at a temperature as low as 100°C and may represent an additional source of carbon dioxide [9].

$\Delta^D (H_2-CH_4)$ geothermometer

Twelve wells were sampled and analyzed for their hydrogen
isotopic composition in the CH_4 and H_2 (Table 1). The total
range of the calculated temperatures is 283-365°C and the
average value is 314± 30°C, i.e., about 27°C lower than the
previous geothermometer and 98°C higher than the average measured value (Figure 1).

No indication on the reaction rate is available for this
system, but assuming the original equilibration took place
at the same time as the CO_2-CH_4 pair, through the F.T. reaction, this system must be assumed to have a faster kinetics
than the previous one. The faster kinetics permits the CH_4
and H_2 molecules, during their rise to the surface, to reach
a new re-equilibration state at lower temperatures. This isotopic equilibrium is not frozen, even though the reaction rate
is not fast enough as to permit continuous re-equilibration
until well-bottom conditions are reached.

Equation (7) really provides an appropriate means of evaluating the chemical and isotopic compositions of the H_2 and
CH_4 species in the Larderello geothermal fluid. Using both the

PANICHI et al.

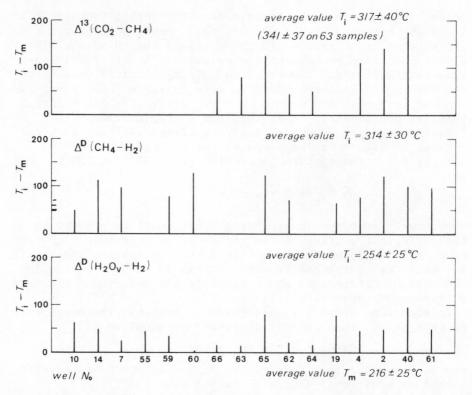

FIG.1. *Differences between measured and isotopic temperatures in some wells in the Larderello geothermal area. The isotopic temperatures are calculated using three different geothermometers; average T values are also given. All temperature values are given in Table 1.*

chemical and isotopic data a deep temperature of   300°C (as an average value) can be calculated.

$\Delta^D (H_2O-H_2)$ geothermometer

The temperatures evaluated using this and the following geothermometers are only approximative, as D/H ratios were not performed on single samples and an average value of $\delta D = -40\%$ was considered as indicative of the condensate steam in Larderello area. This assumption derives from data on D/H ratios obtained previously [1], which indicate almost constant δD values in steam samples having $\delta^{18}O$ values ranging between -5.5 and + 0.5 ‰ . Furthermore, evaluated temperatures in our range of interest change only by 3 to 5°C when the δD values change by ± 5‰ with respect to the average values considered.

The data reported in Table 1 and illustrated in Figure 1 show the excess in the calculated temperatures with respect to the observed ones, the mean value being 254 ± 25°C.

Molecular hydrogen is generally accepted to derive from
the dissociation of water at the geothermal temperature, the
quartz-magnetite-fayalite mineral assemblage being the buffering agent. This mineral association has been found in the Larderello area in several core samples [8] and equation (10)

$$H_2O = H_2 + 1/2 \ O_2 \qquad\qquad (10)$$

appears to exclude the role of the F.T. reaction as controlling
mechanism of the observed isotopic composition.

$\Delta^D (H_2O-CH_4)$ geothermometer

Apart from the limitations mentioned above with regard to
the δD value of the steam samples, this geothermometer seems
incapable of providing reasonable temperature values. The δD
values give $10^3 \ln\alpha$ values ranging from 120 to 200 which
clearly indicate non-equilibrium conditions, when the Bottinga
data [10] are considered.

The lack of equilibrium conditions at depth between H_2O
and CH_4 is in considerable disagreement with the assumption
that the F.T. reaction is an effective agent of the exchange
reactions.

$\Delta^{18} (H_2O-CO_2)$ geothermometer

With the exception of a few cases, the isotopic temperatures
obtained through CO_2-H_2O (vapour) equilibria are equal, or
higher, than the temperature measured at wellhead (Table 2).
As already pointed out [1], this isotopic temperature may
correspond to the actual temperature of the geothermal reservoir when tapped, where CO_2 and steam are in isotopic equilibrium. The difference between the isotopic temperature and
the temperature at wellhead may be a measure of the cooling
undergone by the geothermal fluid on its way up to the surface, the duration of which is too short to bring about any
significant isotopic re-equilibration.

This conclusion may be supported by Fig.2, showing the distribution of the difference between the measured and isotopic
temperatures, as obtained in the 1975 survey and from present
data. This distribution is not normal. The model value corresponds to the isotopic equilibrium between carbon dioxide and
water vapour ($T_m - T_i = 0$), but 69 wells out of 115, i.e.,60%,
definitely exhibit a cooling ($T_m - T_i < 0$) ranging from about 10
to more than 50°C. Only 20 wells (17%) had a wellhead temperature higher than that given by the isotopes; at present we
have no definite explanation for these positive values of
$T_m - T_i$. A partial isotopic re-equilibration at low temperatures, which may have occurred despite our precautions during
the sampling of CO_2 may explain these data. This process,
however, never occurred during our reproducibility tests and,

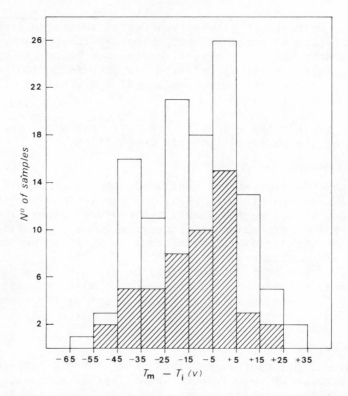

FIG.2. Distribution of $T_m-T_{i(v)}$ values in 110 different wells in the Larderello area. T_m indicates measured temperature and $T_{i(v)}$ the temperature calculated by the $\Delta^{18}(CO_2-H_2O_{steam})$ geothermometer.

although not impossible, seems rather unlikely, considering that both separate collections at each well gave practically the same results.

Another possible explanation may arise from the observation that practically all the wells exhibiting this discrepancy are located in the central area of the field and they represent the largest productive wells of that zone, being characterized by the highest flow-rates and temperatures compared to the surrounding wells. In this situation a certain mixing may occur between fluids having different isotopic compositions of CO_2 and H_2O and simultaneously different temperatures, as a consequence of the widening of the drainage cone of the more productive wells. Now, if a fluid having CO_2 and H_2O in isotopic equilibrium at lower temperature mixes with a fluid equilibrated at higher temperatures, the resulting fluid should simulate an isotopic temperature value which is intermediate between the original values. The actual tempera-

ture of the total fluid ,on the other hand, would remain
practically equal to the highest value if we assume a thermal
re-equilibration of the deep fluid.

Another possible explanation of this uncertainty in the
interpretation may arise from the fact that the actual temperature of the last equilibrium of the gases may have a
quite different value from that reported in Table 2. Indeed,
the wellhead temperature column represents the temperature
values recorded during the last routine measurements before
sampling, and performed by the electric survey team. The time
elapse may be as long as one month. Considering that some
wells show significant fluctuations in repeated measurements,
such as in Well no.2:

Date	8-76	4-77	6-77	8-77	10-77
T°C	253	247	234	231	236

the lack of surface temperature values at sampling time may
result in an erroneous interpretation of the comparison between measured and isotopic temperatures.

In general, the wellhead temperature does not appear to
be the more appropriate value for comparison with the calculated temperatures, considering that the geothermal fluid
undergoes a certain cooling during its flow along the pipe
(up to 15-20°C for a flow-rate of about 3 t/h and from a depth
of about 1000 - 1500 m, and a well diameter of 33 cm [11].
Assuming that inside the pipe the fluid has no time to re-
equilibrate (the actual rate of the fluid can be evaluated as
high as several tens of metres per second), the "true" temperature value to consider is that measured or calculated
at well-bottom.

From the above considerations, the possible evaluation of
the fluid cooling occurring from the tapped reservoir to the
surface, as indicated from the distribution in Fig.2, appears
to be quite approximative. In particular, a cooling of more
than 35°C with respect to the original temperature value appears to be little probable. Some particular physical conditions are possible at depth, where extremely low permeability
in poorly exploited zones may cause the fluids to travel at
very low rates and have sufficient time to cool. In the latter
case a liquid phase may exist instead of a predominant gas
phase, as in the case of a good permeability and strong drainage conditions. Now, if a liquid water exists at the boundary
of the evaporation cone of a certain well, then the oxygen
isotopic composition in the CO_2 and H_2O samples must be interpreted in terms of the isotopic fractionation between carbon
dioxide and liquid water. The temperature values obtained in
this case are reported in Table 2 for all the samples analyzed. The results clearly indicate that when the isotopic temperatures, evaluated by assuming an equilibration between gas

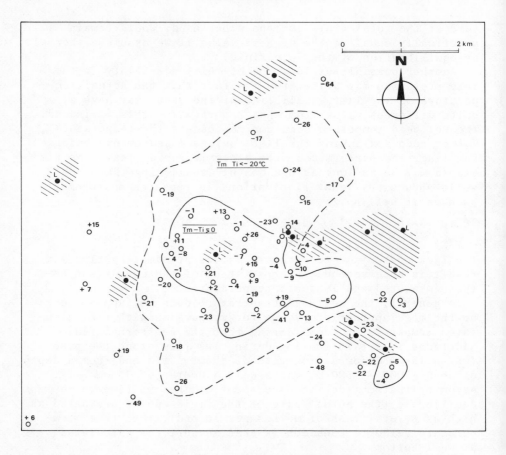

FIG.3. Geographical distribution of $T_m-T_{i(v)}$ values including both data from Table 2 and available data [1]. In the central area (unbroken line) these values are equal to, or higher than, zero. In the shaded zones the black dots represent wells in which equilibrium between carbon dioxide and water is assumed to be reached in presence of a liquid phase.

molecules, is more than 30°C greater than those observed at the surface, a recalculation of the equilibrium temperature for an aqueous phase always gives practically the same values between the isotopic and measured ones.

The contemporary existence of different equilibria (CO_2-$H_2O_{(v)}$ and CO_2-$H_2O_{(l)}$) in the same area seems to be supported by the areal distribution of the isotopic temperature calculated by considering both thermodynamic systems as shown in Fig.3. The black points, which represent equilibrium conditions between CO_2 and liquid water, are generally located at the field boundaries, while the inner part, which is intensely drilled, is characterized by a gas phase in very fast movement and the intermediate wells produce a fluid which has undergone a cooling of about 10-30°C.

If this interpretation is correct, the oxygen isotopic measurements in carbon dioxide and in water not only represent a simple tool for determining the actual temperature at depth in geothermal fields, but also appear to be able to evaluate the thermodynamic conditions of the fluid in the vicinity of the drillhole.

Additional work is required in order to define this point with precision, for example by using the D/H and $^{18}O/^{16}O$ ratios in the condensed steam produced by different wells:the steam produced by an evaporation of a liquid phase should be shown to be isotopically different from that deriving from a vapour phase when the formation temperatures are very similar.

Once again the isotopic fractionation between D and H and ^{18}O and ^{16}O in the liquid-water/water-vapour system very quickly reach equilibrium conditions [12].

CONCLUSIONS

The hydrogen, oxygen and carbon isotopic compositions in the gas components appear capable of recording the temperatures experienced by the geothermal fluids in different parts of the system. Indeed, except for the pair CO_2-CH_4, the equilibration temperatures normally exceed the wellhead temperatures, but the maximum values are only slightly higher than the maximum temperature recorded in a deep unproductive well in the area. The problem arises when interpreting the isotopic temperature values of about 340°C or more calculated using carbon isotopic fractionation between CO_2 and CH_4. This value can be considered realistic if we assume that isotopic equilibrium is reached by these two components. However, a molecular exchange between CO_2 and CH_4 seems practically absent in the hydrothermal conditions, and the only way for equilibration to occur is represented by eq.(1).

As far as the gas phase reduction of carbon dioxide is concerned there are, as yet, relatively few data from which direct conclusions about the mechanism can be drawn. For example, no final unambiguous answer can yet be given to the question of which intermediate is responsible for this reduction reaction. However, some speculation on this point can be made. It is known that the most important stable intermediate of the oxidation of the methane to carbon dioxide is the formaldehyde [13]. Now, it seems reasonable to assume that this compound may be the intermediate of the reverse reaction at high pressure and temperature, and thus an approximation of the F.T. reaction could be written as:

$$CO_2 \xrightarrow[-H_2O]{+H_2} CO \xrightarrow{+H_2} CH_2O \xrightarrow[-H_2O]{+H_2} CH_4$$

In this case the formation rates of carbon oxide and formaldehyde appear to be the regulating factors of the total reaction.

The above sequence provides sufficient steps for the hydrogen and oxygen isotopes in the H_2-H_2O pair (1st step), the hydrogen isotopes in the H_2O-CH_4 pair (final step) and the carbon isotopes in the CO_2-CH_4 pair to reach isotopic equilibrium. We have no reasonable explanation for the absence of any equilibrium conditions between H_2-CH_4 and a possible continuous re-equilibration of the H_2-H_2O and H_2-CH_4 systems, while the carbon isotopes in CO_2 and CH_4 remain unchanged.

These observations lead to the hypothesis that reaction (1) does not occur at depth, and that a series of individual reactions such as nos.(7),(8) and (10) can be responsible for the observed isotopic fractionations. In the latter case the CH_4 molecules never have the opportunity to exchange their isotopic contents with CO_2 and H_2O and the calculated temperatures using these two systems are meaningless.

Additional evidence indirectly supports our hypothesis; Gunter [14] studied the chemical composition of the geothermal gases of Yellowstone Park with respect to their $C_1 - C_4$ hydrocarbon contents, conclusing that "the low abundance of molecular hydrogen, the distribution of alkenes, and the apparent absence of alkynes are taken as additional evidence against the reaction of carbon dioxide and molecular hydrogen as primary source of methane".

ACKNOWLEDGEMENTS

The authors are deeply indebted to Dr.G.C.Ferrara,ENEL-CRG, Pisa and his lab. staff for their invaluable assistance both in the field and laboratory work. Sincere thanks are also extended to Mr.A.Caprai and the other staff of the I.I.R.G. lab for their tireless and competent work and Miss.M.H.Dickson for her revision and typing of the English version. This research was supported by funds from the International Atomic Energy Agency (Contract no.1912/R1/RB) and from the Italian National Research Council (Contract no.412121/92/74804;204121/92/74805).

REFERENCES

[1] PANICHI,C.,FERRARA,G.C., GONFIANTINI,R.,Isotope geothermometry in the Larderello geothermal field,Geothermics, 5,1-4,(1977),81.

[2] HULSTON,J.R.,Isotope work applied to geothermal systems at the Institute of Nuclear Sciences, New Zealand, Geothermics,5, 1-4,(1977),89.

[3] PANICHI,C.,GONFIANTINI,R.,Environmental isotopes in geo-
    thermal studies, Geothermics, 6,3/4,(1978).

[4] GROOTES,P.M.,MOOK,W.G., VOGEL,J.C., Isotopic fractionation
    between gaseous and condensed carbon dioxide, Zeitschr.
    Physik, 221, (1969), 257.

[5] LYON,G.L.,"Geothermal gases", Natural Gases in Marine
    Sediments, (KAPLAN,Ed.), Plenum Press, New York (1974),
    141.

[6] CRAIG,H.,Isotopic temperatures in geothermal systems,
    IAEA Advisory Group on the Application of Nuclear Tech-
    niques to Geothermal Studies, Pisa, unpublished oral com-
    munication, (1975).

[7] D'AMORE,F.,NUTI,S.,Notes on the chemistry of geothermal
    gases, (Proc.2nd Int.Symposium on Water-Rock Interaction)
    3, Strasbourg, (1977),61.

[8] PUXEDDU.M..GIANELLI.G..personal communication.

[9] KISSIN,I.G.,PAKHOMOV,S.I.,The possibility of carbon di-
    oxide generation at depth at moderately high temperatures,
    Akad. Nauk S.S.S.R. Doklady, 174 ,(1967),451.

[10] BOTTINGA,Y.,Calculated fractionation factors for carbon
     and hydrogen isotope exchange in the system calcite-car-
     bon dioxide-graphite-methane-hydrogen-water vapour,Geo-
     chim.cosmochim.Acta, 33, (1969), 49.

[11] RUMI,O.,Determination of the thermodynamic state of water
     vapour in steady adiabatic flow along a vertical pipe
     (well), La Termotecnica, 8(1967).

[12] TRUESDELL,A.H.,FRYE,G.E.,Isotope geochemistry in geother-
     mal reservoir studies, Society of Petroleum Engineers,
     Dallas (1977).

[13] ANTONOVA,I.N.,KUZ'MIN,V.A..MOSHKINA,R.I..NALBANDYAN,A.B..
     NEIMAN,M.B..FEKLISOV.G.I..Investigation of the mechanism
     of methane oxidation by labelled molecules. The mechanism
     of the formation of carbon monoxide, Izv.Akad.Nauk S.S.S.
     R.,(1955),789.

[14] GUNTER,B.D.,C_1-C_4 hydrocarbons in hydrothermal gases,
     Geochim. cosmochim.Acta, 42,(1978),137.

DISCUSSION

J.F. BARKER: We have obtained apparent carbon isotope equilibrium in the
$CH_4 - CO_2$ system over a chromium catalyst in the temperature range 250–350°C
in about 1–2 hours. This emphasizes the difficulty of estimating rates of isotope

exchange reactions in natural systems from laboratory experiments, since from previous rate studies it would seem that more than five years are required for equilibrium to be achieved.

S. NUTI: The aim of our work was not to estimate the rate of isotopic exchange reactions under laboratory conditions in the presence of a particular catalyst. Certainly, carbon isotope equilibrium in the $CH_4 - CO_2$ system is not reached in a short time under natural conditions, as is indicated by the experimental values given in the paper and by those to be found in the literature. Furthermore, studies by Gunter [14] and the temperature values calculated by D'Amore and Nuti [7] on the assumption that the Fischer-Tropsch reaction is valid, support our hypothesis that a molecular exchange between CO_2 and CH_4 seems to be almost non-existent in hydrothermal systems.

P. THEODORSSON: Mr. Nuti showed us a slide which gave some temperatures derived from isotopic exchange between H_2 and H_2O in Iceland. I would like to point out that geochemical methods can be quite as important as isotopic measurements for estimating the base temperature of geothermal areas. In this respect, measurements of silica in geothermal waters in Iceland have given valuable results.

S. NUTI: Apart from the silica geothermometer, other geochemical geothermometers are currently being used in our Institute together with studies of water-rock interactions for estimating the base temperature of geothermal systems. However, since Larderello is a vapour-dominated geothermal field, neither the silica nor the Na-K-Ca geothermometers are used.

HYDRODYNAMIC ASPECTS OF
CARBON-14 GROUNDWATER DATING

M.A. GEYH
Niedersächsisches Landesamt für
    Bodenforschung, Hanover

G. BACKHAUS
Institut für Physikalische Chemie
    der Universität München,
Federal Republic of Germany

Abstract

HYDRODYNAMIC ASPECTS OF CARBON-14 GROUNDWATER DATING.
    The influence of man-made hydraulic disturbances on the ^{14}C ages of groundwater from confined aquifers is examined, also taking into account ^{14}C diffusion, which has an effect on ^{14}C ages only if the hydrostatic pressure in the lower, confined aquifer is not more than 0.5 m higher than that in the upper, unconfined aquifer. If the water head of the lower aquifer exceeds this value, the ^{14}C ages of the confined groundwater are reliable. If the water head is lower, the ^{14}C water ages rapidly approach values of a few thousand years, which no longer reflect the history of the groundwater regeneration. With regard to the palaeohydrogeological situation in Central Europe and the Central Sahara during the last 40 000 years, the ^{14}C ages of Holocene groundwater, and the duration of the preceding hiatus of the groundwater regeneration during the last glacial period, can be determined reliably. ^{14}C ages older than that are too small in many cases; thus, groundwater velocities derived from such data are too great. Recently operations were started to use the groundwater from confined aquifers associated with rates for lowering the water table at 0.1−0.5 m/a that result in a rapid decrease in the ^{14}C ages determined for these aquifers, delayed for one or two decades after the beginning of the withdrawal. The ^{3}H level and the chemical content of the groundwater may also be changed after the same delay period. Changes of this kind can be used to estimate the hydraulic properties of the aquifer system. In conclusion, an interpretation of the ^{14}C content of the groundwater from confined aquifers in terms of its age is only possible if the water head of the confined aquifer has not been lower than that of the upper aquifer for even a relatively short period.

INTRODUCTION

Discussion on the reliability of ^{14}C groundwater dating was revived when Sonntag discussed ^{14}C diffusion in aquifer systems again [1]. ^{14}C diffusion occurs between young groundwater with a high ^{14}C concentration in an unconfined shallow aquifer and old groundwater with a low ^{14}C concentration in a lower confined aquifer. This ^{14}C diffusion reduces the ^{14}C age of the groundwater in the deeper aquifer. This is also true if a separating layer of low permeability (aquiclude) is present.

It is felt that the ^{14}C diffusion effect can help to explain the apparently too low ^{14}C ages of confined groundwater samples from a large area of the Central Sahara [1]. However, the method has failed when applied to other areas such as that south of Bad Hersfeld in the Federal Republic of Germany. Here, an aquifer in sheet dolomite of about 25 m thickness and 5–10% porosity is covered with layers of Upper Zechstein clay (8–10 m) and of crumbly shale (25–40 m) with an estimated water conductivity $K = 10^{-8}$ m/s $= 0.3$ m/a and an assumed water content of 30%. The ^{14}C water ages to be expected, owing to ^{14}C diffusion, should be less than 20 000 years, but most of the results exceed 30 000 years.

During discussions on this observation with Dr. Andres, Bayerisches Landesamt für Wasserwirtschaft, Munich, the question arose regarding whether ^{14}C diffusion is also effective in the case of confined groundwater. Our deliberations resulted in a theoretical study, whose practical results for the interpretation of ^{14}C ages of groundwater from confined aquifers are discussed here.

1.   MODEL CONSTRUCTION OF A CONFINED AQUIFER

The isolines of hydrostatic potential in an isotropic aquifer are perpendicular to the direction of the groundwater flow. As can be seen from the pattern of the potential lines in an unconfined aquifer (Fig.1), at distances far enough from the regeneration area the groundwater of the deeper parts of the aquifer has a higher hydrostatic pressure than that of the upper parts. This is also true if an aquiclude separates two aquifers.

Let us assume that a confined aquifer, CA, is separated from the unconfined aquifer, UA, by an aquiclude, AC, with a thickness of H_1 and a pore volume of P_1 (Fig.2). The confined aquifer is considered as a closed box with a thickness of H_2 and a pore volume of P_2. The groundwater in it should be well-mixed, at the latest at the moment of pumping. The difference Δ in hydrostatic pressure between the two aquifers CA and UA shall be positive if the lower aquifer has a higher hydrostatic pressure than the upper one, i.e. there is an upwards flow. If not otherwise mentioned, within the regeneration area of the confined aquifer and the unconfined aquifer there is no aquiclude. Thus, the initial ^{14}C concentrations C_0 of both aquifers are equal. Moreover, that from the upper aquifer should be constant over the time range for which the dating is done. This is the result of rapid groundwater regeneration in the confined aquifer.

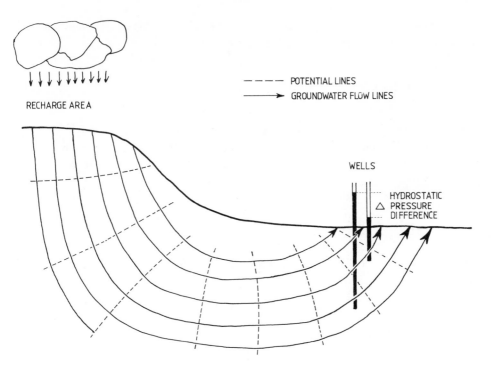

FIG.1. *Scheme of the hydrostatic pressure distribution in an unconfined aquifer and presentation of the groundwater flow lines and the potential lines.*

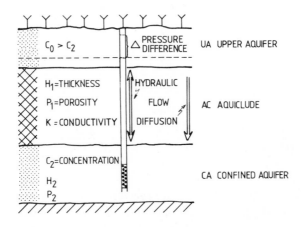

FIG.2. *Scheme of an aquifer system consisting of a shallow unconfined aquifer separated from a deeper confined aquifer by an aquiclude.*

If the ^{14}C concentration in the groundwater C_2^+ changes with increasing age only as a result of radioactive decay, then

$$A = \lambda \ln C_0 / C_2^+ = \frac{1}{T} \ln C_0 / C_2^+ \qquad (1)$$

where A is the age of the groundwater and λ is the decay constant ($T_{1/2} = \ln 2/\lambda$ = 5568 a for ^{14}C and 12.26 a for tritium). If C_2 is influenced by hydrochemical processes [2], isotope exchange [3], diffusion or other effects, A is the apparent age.

2.   DIFFUSION MODEL

Diffusion of ^{14}C between the upper aquifer with a high ^{14}C concentration and the lower aquifer with a low ^{14}C concentration, is described for $\Delta = 0$ by Fick's Second Law:

$$\frac{\partial C}{\partial t} = D \frac{\partial^2 C}{\partial x^2} - \lambda C \qquad (2)$$

Sonntag [1] has used a value of $D = 3.8 \times 10^{-2}$ m^2/a for the diffusion constant for tritium and radiocarbon. A variation of 50% in D does not drastically change the results described below. If the thickness of the aquiclude is constant, the horizontal groundwater velocity need not be considered since this term drops out when Eq.(2) is solved. Integration of Eq.(2) yields

$$C_2(t) = C_0 \frac{Z+1}{Z + e^{\lambda t}} \qquad (3)$$

with

$$Z = \frac{P_1}{P_2 H_2^*} \left[\frac{e^{\lambda t} - 1}{\sinh H_1^*} + \frac{2H_1^*}{\pi^2} \sum_{n=1}^{\infty} \frac{(-1)^n (1 - e^{-n^2(\pi/H_1^*)^2 \lambda t})}{n^2 + (H_1^*/\pi)^2} \right]$$

$H_1^* = H_i / \sqrt{DT}$ is defined as the relative relaxation distance. $\sqrt{DT}$ is the relaxation distance in which the tracer concentration decreases by a factor of $e = 2.72$. This distance is 17.4 m and 0.82 m for ^{14}C and tritium, respectively.

Equation (3) is equivalent to Sonntag's equation when $C_2^+ \ll C_0$. The results are seen in Fig.3.

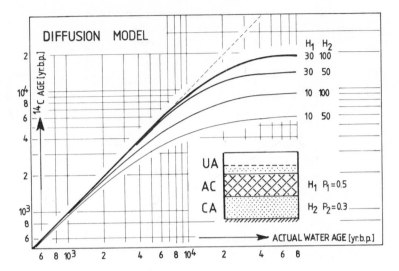

FIG.3.   Non-steady-state relationship between apparent ^{14}C groundwater ages and actual ^{14}C ages in a confined aquifer owing to ^{14}C diffusion.

3.   HYDRAULIC MODEL

If there is a hydrostatic pressure difference Δ between aquifers UA and CA, there is a vertical groundwater movement according to Darcy's Law. Its velocity is given by

$$V = \frac{K\Delta}{H_1 P_1} \tag{4}$$

If Δ is positive, water flows upwards from CA by definition, and the ^{14}C concentration C_2 does not change. In the reverse case, surface water flows through the aquiclude into CA. The ^{14}C concentration decreases along the way by radioactive decay and at the boundary, AC/CA, reaches a value of

$$C_2 = C_0 \, e^{\lambda H_1^2 P_1 /(K\Delta)} \tag{5}$$

With this, the differential equation, also taking into consideration the hydraulics

$$\frac{\partial C}{\partial t} = -V \frac{\partial C}{\partial t} - \lambda C \tag{6}$$

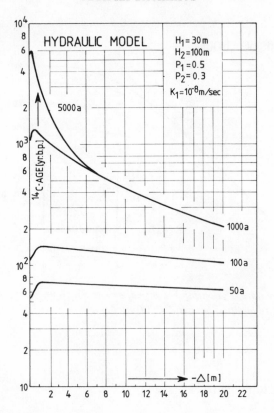

FIG.4. Changes of the ^{14}C ages of confined groundwater owing to a hydraulically forced admixture of sub-surface water through the aquiclude.

has the solution

$$C_2(\Delta) = C_0 e^{-\lambda^* t} \left[1 + \frac{V}{H_2 P_2 \lambda^*} e^{\lambda H_1 / V(e^{\lambda^* t} - 1)} \right] \tag{7}$$

with $\lambda^* = \lambda + V/(H_2 P_2)$.

^{14}C ages calculated from Eqs (7) and (1) are given in Fig.4 as a function of the hydrostatic pressure differences. When $\Delta > -0.5$ m, water from the upper aquifer, aged while it was in the aquiclude, causes a reduction of the ^{14}C concentration in the deeper aquifer and an apparent increase in the ^{14}C age. At greater negative hydrostatic pressure differences, the apparent ^{14}C ages in the lower aquifer decrease, approaching zero at $\Delta = -\infty$. This behaviour can be explained since, within this interval of Δ, V is large compared with the average diffusion velocity. This is almost always the case.

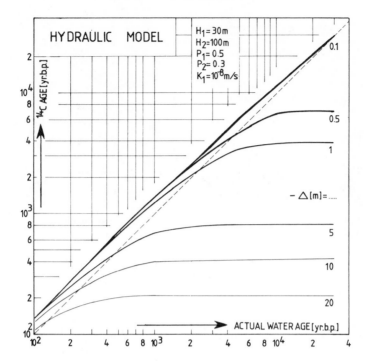

FIG.5.   Relationship between apparent ^{14}C ages and actual ^{14}C ages owing to the hydraulic effect.

In Fig.5 the apparent ages are plotted above the actual ages. It can be seen that the hydraulic effect is greater for high ^{14}C water ages than for low ones. But, even for Δ values as low as -0.5 m, the apparent ^{14}C groundwater ages approach values of less than 10 000 years, independent of the actual ages, despite an aquiclude thickness of 30 m clay.

Many of the ^{14}C ages of groundwater samples from confined aquifers in northern Germany are constant over wide areas and may therefore be interpreted as the result of a steady-state admixture of young groundwater from shallow aquifers.

4.   COMBINED HYDRAULIC-DIFFUSION MODEL

A mathematical treatment combining the hydraulic and diffusion effects leads to the following differential equation:

$$\frac{\partial C}{\partial t} = D \frac{\partial^2 C}{\partial x^2} - V \frac{\partial C}{\partial t} - \lambda C \tag{8}$$

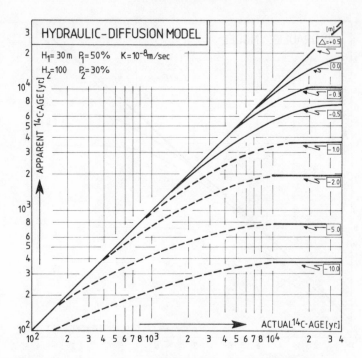

FIG.6.   Relationship between apparent ^{14}C ages and actual ^{14}C ages owing to the hydraulic-
diffusion effect.

In solving Eq.(8) the main difficulty is due to the choice of initial and boundary
conditions. The fact must be taken into account that, according to the hydraulic
effect, the ^{14}C concentration of the water crossing the boundary AC/CA is
determined by the hydrostatic pressure difference only. But, according to the
diffusion effect, it depends on the ^{14}C end concentration in the confined aquifer.
    The solution of Eq.(8) is too complicated to be presented here. However,
Eq.(3) is a special solution for $\Delta = 0$ [m]. The results of the general solution
are shown in Fig.6. For hydrostatic pressure differences of nearly zero and high
water ages, only the diffusion effect plays a role, leading to a significant reduction of the ^{14}C age of the water. The hydraulic countercurrent resulting from
the hydrostatic pressure differences of only a few decimetres is sufficient to
counterbalance the diffusion effect. For larger negative hydrostatic pressure
differences, the inflow of young surface water to the confined aquifer will be
so great that ^{14}C ages approach steady-state values that are much too small and
can no longer be interpreted in terms of water ages. Thus, reliable ^{14}C water
ages, which are not influenced by diffusion and hydraulic effects, can be expected
only from confined groundwater which had been under high hydrostatic pressure
during its presence in CA.

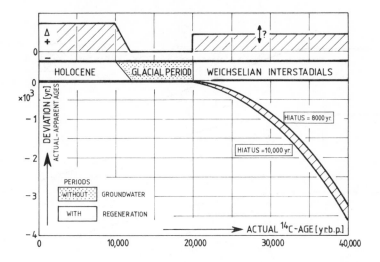

FIG.7. Deviation between apparent ^{14}C ages and actual ^{14}C ages from confined groundwater from Central Europe owing to the influence of the ^{14}C diffusion effect during the last glacial period.

5. HYDRAULIC AND DIFFUSION EFFECTS AND THE ^{14}C GROUNDWATER TIME SCALE

Regarding the palaeohydrogeological situation in Central Europe during the last 40 000 years, Δ and C_0 must be considered as variable. According to findings by Andres and Geyh [4, 5], the rate of groundwater regeneration during the Weichselian Interstadials up to 20 000 BP and after 12 000–10 000 BP during the Holocene, was large enough to keep the hydrostatic pressure of the confined aquifers high. Consequently, the diffusion effect could not become effective and the ^{14}C time scale remained unchanged. However, during the last glacial period, the earth's surface was sealed by permafrost and groundwater regeneration interrupted. In this situation, Δ became zero and C_0 decreased by radioactive decay for at least 8000 years to a maximum of 10 000 years and ^{14}C diffusion changed the horizontal ^{14}C age pattern to such a degree that it becomes measurable. The effect on the ^{14}C ages of confined groundwater from our model aquifer can be seen in Fig.7. The ages of Holocene groundwater, and the duration of the hiatus in the groundwater regeneration, are correctly reflected by the corresponding ^{14}C ages. But the ^{14}C ages of groundwater generated during the last interstadial periods turn out to be too small by up to 10%. Thus, the corresponding groundwater velocities derived from ^{14}C data are too great.

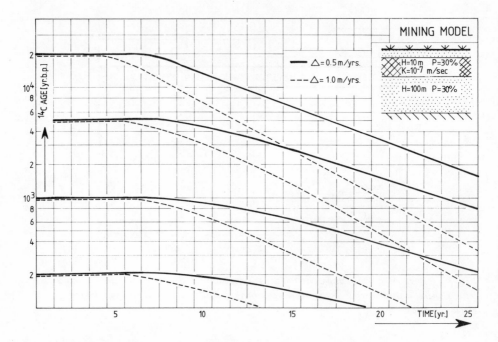

FIG.8. *Changes of the ^{14}C ages in a confined aquifer owing to groundwater mining as a function of the duration of the groundwater withdrawal and different depletion rates of the hydrostatic pressure.*

The palaeohydrogeological situation in the Central Sahara is less well known. It is commonly agreed that groundwater regeneration occurred previous to 20 000 years ago and between 10 000 and 2000 BP. However, the geomorphological and hydrogeological conceptions for the period between 20 000 and 10 000 BP are contradictory [1, 6]. Interpretations range from arid to humid climate. According to conclusions drawn from the situation in Central Europe, differences between apparent and actual water ages may be expected only for interstadial waters in the Sahara, but hardly to the extent estimated by Sonntag [1].

6.   GROUNDWATER MINING MODEL

The increasing use of confined groundwater by man has led to a significant lowering of the groundwater table in many areas. Rates of up to 1 m/a have been observed. In the heavily populated and industrialized area between Mannheim and Heidelberg, for instance, the water-table depression in the confined aquifer is already as much as 13 m [7]. From this and similar examples, the question arose whether and how quickly such hydraulic disturbances are reflected by the isotopic and chemical content of the water from the lower aquifer owing to an admixture of water from the surface and/or a shallow aquifer.

Let us assume that, before the hydraulic disturbance, the water in the lower aquifer was under higher hydrostatic pressure than in the upper aquifer. Then, the groundwater flowed upwards and its ^{14}C concentration decreased in the same direction. When the hydraulic disturbance started at time t_0, the hydrostatic pressure difference began to decrease at a rate Δ_p. As soon as Δ became $\leqslant 0$, the increasingly rapid flow of old water from the aquiclude into the deeper aquifer CA increased the age of the groundwater there. When finally the young water from the upper aquifer reached the boundary AC/CA, the ^{14}C age of the water in the deeper aquifer started to decrease abruptly, the tritium content began to increase and the chemical content may also change.

The situation of the above-mentioned Mannheim-Heidelberg aquifer is shown in Fig.8. The initial slow increase in the ^{14}C ages lasted for several years. But the ^{14}C ages began to decrease rapidly within only seven years when the water-table depression reached -3.5 m. Over a ten-year period, wells whose water had previously had ^{14}C ages of 100, 5000, 10 000 and 20 000 years later showed ^{14}C ages of 500, 2100, 3500 and 4800 years, respectively.

Such remarkable decreases of ^{14}C ages caused by hydraulic disturbances make it possible to determine values for hydraulic properties of an aquifer. In addition, it shows that mining of confined groundwater resources may also change the chemical content. This means that the effect of groundwater mining must be considered in the future when making any hydrogeological interpretations with radioactive or stable isotope data, chemical data or geothermal records.

ACKNOWLEDGEMENTS

Financial support by the University of Munich is kindly acknowledged. Special thanks are due to Professor Dr. Hahn, Hanover, who gave very valuable comments to our work.

REFERENCES

[1] KLITZSCH, E., SONNTAG, C., WEISTROFFER, K., EL SHAZLY, E.M., Grundwasser der Zentralsahara: Fossile Vorräte Geol. Rundsch. 65 (1976) 264.
[2] WIGLEY, T.M.L., Carbon-14 dating of groundwater from closed and open systems, Water Res. Res. 11 2 (1975) 324.
[3] MÜNNICH, K.O., Isotopen-Datierung von Grundwasser, Naturwissenschaften 55 4 (1968) 158.

[4] ANDRES, G., GEYH, M.A., Untersuchungen über den Grundwasserhaushalt im über-
    deckten Sandsteinkeuper mit Hilfe von [14]C- und ^{3}H-Wasseranalysen, Wasserwirtschaft 8
    (1970) 259.
[5] GEYH, M.A., Basic studies in hydrology and ^{14}C and ^{3}H measurements, 24th Int. Geol.
    Congr. Montreal (1972), Section 11, 227.
[6] GEYH, M.A., JÄKEL, D., The climate of the Sahara during the Late Pleistocene and
    Holocene on the basis of available radiocarbon data, Natural Res. Dev. 6 (1977) 65.
[7] ARMBRUSTER, J., JOACHIM, H., LAMPRECHT, K., VILLINGER, E., Grenzen der
    Grundwassernutzung im Rhein-Neckar-Raum (Baden-Württemberg), Z. dt. Geol. Ges. 128
    (1977) 263.

DISCUSSION

B.Th. VERHAGEN: In Fig.8 you show a decrease in radiocarbon ages during exploitation. Does this mean that the aquifer was leaking?

M.A. GEYH: No, the figure relates to a formerly confined aquifer separated by a 10-m clay aquiclude of constant thickness.

P. FRITZ: Is it correct to assume that the models as you present them are valid only for geochemically inactive systems and, if so, how realistic are they? Geochemical reactions will alter the water not only when it is diffusing through the aquiclude but also after its arrival in the confined aquifer.

M.A. GEYH: In my presentation I was considering only inactive geo-chemical systems. However, the theory is a fundamental one and could easily be applied to geochemically active systems — for example to isotopic exchange between bicarbonate and aquifer rocks.

J.F. BARKER: Does your model take the water and ^{14}C present in the aquiclude into account?

M.A. GEYH: Yes, it does.

J.F. BARKER: With thick aquicludes, would it be reasonable to assume that the only significant amounts of water entering the confined aquifer would be aquiclude water?

M.A. GEYH: Groundwater from the aquiclude can enter the confined aquifer only if the hydrostatic pressure difference is negative. At low absolute values the water flow from the shallow aquifer is so insignificant that it can be defined as aquiclude water. However, with large negative pressure differences, or where groundwater mining is practised, I assume that there might well be entries of shallow groundwater, possibly influenced only by the rock chemistry of the aquiclude.

P. FRITZ: From a strictly practical point of view, how valuable is your mining model? The usual assumption seems to be that fracture flow is more important, even in clays. This is certainly the case in a situation in central Canada where leakage through 20–30 m of lake clay occurs because of the exploitation of an underlying carbonate aquifer. Here we believe we can show that fracture flow is transmitting the water from the unconfined aquifer to the deep aquifer and that diffusion is only important because the geochemistry of the water flowing through the fractures is affected by its "exchange" with the water in the clay matrix.

M.A. GEYH: Your comment sums up the whole basis of the mining model. Vertical flow through the aquiclude is considered quantitatively and is found to be considerable. If, for instance, isotope exchange also occurs, this can be taken into account by a corrected half-life value. The mode of correction was described by Münnich in 1968.

D.B. SMITH: Figure 1, which illustrates an unconfined isotropic aquifer, is somewhat misleading, partly because of the compressed horizontal scale. Recharge would normally occur over the whole area where the water table is below ground level. Discharge would occur where the water table intersects the ground owing to slope or to an aquiclude. Pressure differences with depth near the output act along the flow direction and, unless the aquifer were inhomogeneous and included aquicludes, they would not provide a potential for lateral mixing across the flow lines which is the impression given by Fig.1.

M.A. GEYH: Thank you for your comment. If your recommendation had been made before we drew Fig.1 we certainly would have made our point more clearly. However, this mode of presentation of flow lines and equipotential lines was taken from the hydrodynamics literature. I agree with you that mixing does not occur under the circumstances shown in the figure.

URANIUM ISOTOPE DISEQUILIBRIUM IN GROUNDWATERS OF SOUTHEASTERN NEW MEXICO AND IMPLICATIONS REGARDING AGE-DATING OF WATERS

G.E. BARR, S.J. LAMBERT
Sandia Laboratories,
Albuquerque, New Mexico

J.A. CARTER
Oak Ridge National Laboratory,
Oak Ridge, Tennessee,
United States of America

Abstract

URANIUM ISOTOPE DISEQUILIBRIUM IN GROUNDWATERS OF SOUTHEASTERN NEW MEXICO AND IMPLICATIONS REGARDING AGE-DATING OF WATERS.

Regional exploration of bedded salt for a radioactive waste repository in the Delaware Basin (Permian, New Mexico, USA) included boreholes into the evaporites and associated rocks. One such hole, ERDA No.6, encountered an accumulation of saturated $NaCl$-Na_2SO_4 brine accompanied by H_2S-rich gas. This fluid, and fluids from other boreholes elsewhere in the area, have been characterized geochemically according to solute content, $^{18}O/^{16}O$ and D/H ratios and natural actinide content. Deviations from the equilibrium $^{234}U/^{238}U$ activity ratio (α) of 1.0 were found in all water samples. These deviations are used to affirm the isolation of ERDA No.6 and to establish bounds on the age of the ERDA No.6 fluid. A mathematical model for the age of ERDA No.6 water intrusion was formulated in terms of the following variables: initial α-value (α_0) of the brine precursor waters, zero-order kinetic rate constants of leaching of ^{234}Th and ^{238}U from the rocks, and degree of leaching of ^{234}Th from the rocks. The model allows calculation of α-values of brines as a function of time, leach rate and α_0. Various combinations of α_0's, leach fractions and leach rates indicate that the leach rate must be low. If no leaching is assumed and the α_0 is the highest known α from the nearby Capitan Reef ($\alpha_0 = 5.2$ at present), the fluid is older than 880 000 years. ERDA No.6 brine has undergone more profound rock/fluid interactions, reflected in its solutes and stable isotopes, compared with younger meteoric Capitan waters.

1. INTRODUCTION

1.1 History

Regional exploration of bedded salt deposits for a radio-active waste repository in the Delaware Basin (Permian, New Mexico) included boreholes into the evaporites and associated

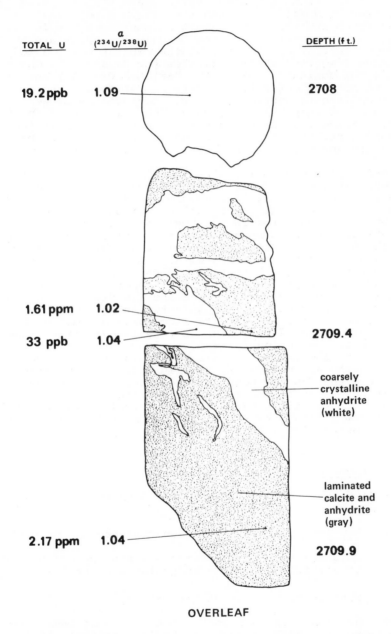

TOTAL U	a $(^{234}U/^{238}U)$	DEPTH (f t.)
19.2 ppb	1.09	2708
1.61 ppm	1.02	
33 ppb	1.04	2709.4
		coarsely crystalline anhydrite (white)
		laminated calcite and anhydrite (gray)
2.17 ppm	1.04	2709.9

OVERLEAF

Plate 1 and Overleaf. Core fragments from Castile amhydrite serving as host rock for the ERDA No.6 brine reservoir. Depths of origin, uranium contents and $^{234}U/^{238}U$ ratios are given in the overleaf for various parts of the core fragments.

rocks. These rocks contained various amounts of fluids, already described [1]. Although the halite deposits of the
Ochoan Epoch are the prime target of consideration for waste
storage, the older rocks of the Guadalupian Epoch are included
in the present study, because of the Capitan reef, a body of
cavernous limestone which encircles the Basin. The Capitan not
only is the dominant rock in the Carlsbad Caverns, but also
supplies potable water to the city of Carlsbad. Therefore, it
is appropriate to consider possible connections between the
Capitan and accumulations of water found in pockets in the evaporites. Brine and gas pockets are known to occur in the
Carlsbad district potash mines and in the Castile Formation
(anhydrite and halite) underlying the Salado Formation.

    One exploration hole, ERDA No. 6, drew national attention
[2], when it encountered an accumulation of saturated NaCl
brine laced with Na_2SO_4, accompanied by H_2S-rich gas at 826 m
below the surface. Discovery of this artesian fluid, together
with bedding-plane dips in the host Castile Formation exceeding 70°, led to abandonment of that immediate site. Geophysical investigations showed that this and other occurrences of
brine and gas were associated with closed structure contours
drawn for the Castile Formation. These structures, restricted
to a narrow strip of evaporites near the reef have been termed
"salt anticlines" [3]. Core from the fluid-producing zone of
ERDA No. 6 consisted of gray laminated anhydrite whose fractures
were filled with massive crystalline white anhydrite (Plate 1).
While fluid accumulations are not rare in evaporites, the possible use of the region for radioactive waste disposal requires
attention to the origin of this particular type of accumulation. In the extreme cases, the fluid originates directly from
water from the Capitan Reef, or the water was entrapped in the
Castile Formation at the time of evaporite deposition in the
Permian. In the first case, the water would have an age little
older than Capitan waters. In the second case, the water would
be in excess of 230 million years old.

1.2   Implications

    Indeed, the very presence of the ERDA No. 6 accumulation
of fluid poses several questions:

    1. How long has the water been confined in the pocket?

    2. Does the time of the water's intrusion correspond
        to any significant known geologic events?

    3. What is the source of the water?

    4. How long has this pocket been disconnected from
        nearby aquifers such as the Capitan Reef, if it
        ever was connected?

    5. What are the limits to rates of influx (if any) or
        recharge and discharge (if any) of this fluid
        pocket?

6. To what degree does the ERDA No. 6 occurrence (or
    any other in the area) represent a flowing system?

7. Which radiometric clock should be used to age-date
    the water and what should be taken as initial condi-
    tions for age-dating?

2. ANALYTICAL APPROACH

2.1 General Geochemistry of Groundwaters

A number of groundwaters from the Delaware Basin have been
characterized in terms of their general geochemistry with respect to their host rocks [1]. In terms of solutes, the waters
from below the fresh-saline interface in the Capitan were similar to each other. In addition the shallow (≤ 300 m depth)
waters were similar to one another and all deep (≥ 1000 m depth)
waters were similar to one another. The ERDA No. 6 water was
unique in its solute content, indicative of pronounced disequilibrium between solution and host rock and also between solution
and associated gas phase.

In terms of D/H and $^{18}O/^{16}O$ ratios, Capitan and shallow
waters all are similar and are of meteoric origin. (Carlsbad
Caverns waters are also of meteoric origin, but are not related
to other waters in the Capitan.) Deep water stable isotope
ratios are similar to one another, and are not characteris-
tically meteoric. ERDA No. 6 has uniquely non-meteoric isotopic
ratios among those of Delaware Basin waters. No waters were
found which represent original evaporite mother-liquor.

2.2 Experimental Procedures

In spite of the evidence that this fluid has experienced
strong   interactions with rock   [1], radioisotopes were examined to establish that the ERDA No. 6 fluid was not of relatively recent origin. Plutonium concentration was determined by
isotope dilution employing a spike of ^{244}Pu. The actual measurement was made on a three-stage thermal emission solid source
mass spectrometer equipped with pulse counting for ion detection. The instrument has a very high abundance sensitivity
(10^7) and low background, allowing for the determination of
precise isotopic ratios on small samples amounting to 0.1 to
1 nanograms of plutonium. The plutonium concentration of the
ERDA No. 6 brine was found to be less than 10^{-15} grams per
gram, corresponding to the lower limit of analytical detect-
ability. The absence of plutonium suggests the brine has been
isolated at least since 1945, a date corresponding to the first
atmospheric detonation of nuclear weapons.

Carbon 14 was measured in the CO_2, which amounted to 55%
of the total gas collected from ERDA No. 6. Carbon 14 content
was counted in a gas proportional detector whose efficiency was
calibrated with the NBS oxalate standard. The result was the
same as the background, which was about 2 counts per minute
in a 1.5 liter detector at 2 atmospheres pressure. From this
it was concluded either that the brine is older than about

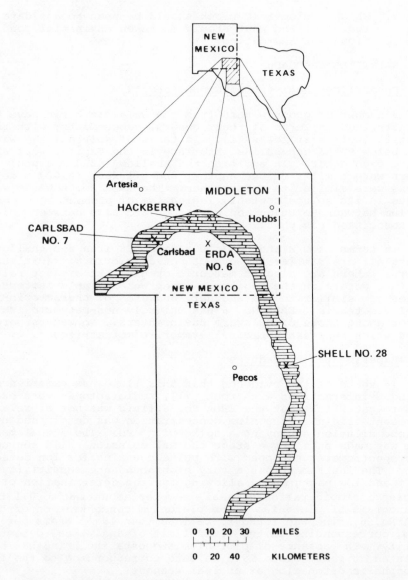

FIG.1.   Map of a portion of the Delaware Basin (southeast New Mexico, west Texas) showing locations of holes whose fluids were sampled for $^{234}U/^{238}U$ disequilibrium age-dating.   The brick-pattern shows the surface projection and outcrop of the Capitan Reef limestone.

30,000 years or possibly that carbonates in the host rock have diluted the carbon in the system with non-radioactive carbon.

The disequilibrium of the activity of ^{234}U relative to ^{238}U in natural waters has been attractive as an indicator of the age of groundwaters [4]. ^{234}U and ^{238}U are related by the decay scheme:

$$_{92}^{238}U \xrightarrow[4.5 \times 10^9 a]{\alpha} {}_{90}^{234}Th \xrightarrow[24d]{\beta-} {}_{91}^{234}Pa \xrightarrow[1\ min]{\beta} {}_{92}^{234}U \qquad (1)$$

It has been proposed that excess ^{234}U builds up in confined waters as an aging effect as a result of the alpha-recoil of ^{234}Th, which ejects ^{234}Th out of the host rock crystal lattice into the water [5,6]. From a knowledge of decay constants and initial activity ratio α_0 ($\alpha = {}^{234}U/{}^{238}U$ activity ratio), time of confinement is calculated. It cannot be assumed, however, that the earth's natural abundance ratio, $\alpha = 1.0$, is a priori valid for a confined rock-water system. The water carried in with it an initial uranium concentration and original activity ratio, α_0. The water ultimately comes to thermodynamic equilibrium with its host rock and may leach or even deposit ^{234}U and ^{234}Th. Uranium concentrations and α-ratios of rocks and waters in this study were determined by isotope dilution mass spectrometry, described above. Figure 1 shows the locations of principal water samples of this study. Table I shows uranium concentrations and values of α for various waters.

3. RESULTS AND DISCUSSION

The overleaf of Plate 1 shows the variations of U content and α in anhydrite of rock core taken from the zone which produced fluid in ERDA No. 6 (826 m, 2709 ft depth). Table I results show that in ERDA No. 6, both U concentration and α vary with time, until U concentration falls asymptotically to about 2 parts in 10^9, and α rises asymptotically to about 1.35. In the early part of a several-hour flow test, water samples were found to be contaminated with drill mud and spallations from metal pipe. By 1400 hours not only had the uranium reached steady state values, but the iron concentration had fallen from 260 parts in 10^6 (1130 hours) to a steady value of 5 parts in 10^6 (1400 hours), indicating the flushing out of drilling-introduced contaminants. Thus, the later samples are probably representative of the fluid accumulation.

The total variation in α among all water samples is relatively small (1.2 to 5.2). In contrast, absolute uranium content in the waters of Table I varies by more than two orders of magnitude. Similarly, rock material illustrated in Plate 1 shows a wide variation in U content: 1 to 2 parts in 10^6 for "original" laminated gray anhydrite down to 20 to 30 parts in 10^9 in the secondary white anhydrite fracture filling only a few centimeters away.

TABLE I

URANIUM CONCENTRATIONS AND ISOTOPIC RATIOS IN DELAWARE BASIN WATERS

Sample	U Concentration	α ($^{234}U/^{238}U$)	$\delta^{18}O$(SMOW) ‰	δD(SMOW) ‰
Shell	0.60 parts in 10^9	2.75	-7.7	-56
Middleton	0.54 parts in 10^9	1.81	-7.5	-55
Hackberry	0.02 parts in 10^9	1.22±0.05	-6.5	-46
Carlsbad 7	0.05 parts in 10^9	5.14	-7.9	-54
ERDA 6 1140 h	89 parts in 10^9	1.11±0.04		
ERDA 6 1212 h	4.8 parts in 10^9	1.26±0.05		
ERDA 6 1445 h	2.14 parts in 10^9	1.37±0.07	+10.3	0
ERDA 6 1520 h	1.88 parts in 10^9	1.33±0.07		
ERDA 6 Drill Mud	67.2 parts in 10^9	1.19		
Brine Lake (Mud Ingredient)	56.3 parts in 10^9	2.04		

4. APPLICATION OF THE URANIUM ISOTOPE DISEQUILIBRIUM MODEL

   If N_2 is the amount of ^{234}U present in a phase, and N_1 is the amount of ^{238}U, the change in ^{234}U content with time in the brine is

$$\left(\frac{dN_2}{dt}\right)_{brine} = \lambda_1 N_1 - \lambda_2 N_2 + Q_1 + Q_2, \qquad (2)$$

in which Q_1 is the rate of release of ^{234}Th (the parent of ^{234}U) from the rock, Q_2 is the rate of release of ^{234}U from the rock and λ_1 and λ_2 are decay constants for ^{238}U and ^{234}U, respectively. Release rates Q_1 and Q_2 can be modeled in the fashion of Kigoshi [5], who studied the release of ^{234}Th from zircon powder. He gave the accumulated activity of ^{234}Th in solution after time t as

$$(\lambda_{Th} N_{Th})_{brine} = 1/4 \ L \ S\rho \ (N_{238} \ \lambda_1)_{rock} \ (1 - e^{-\lambda_{Th}t}), \qquad (3)$$

in which λ_{Th} is for ^{234}Th, L is the recoil distance of ^{234}Th in decay scheme (1) above, S is the surface area of parent rock, ρ is its density, and N_{238} is expressed as the number of ^{238}U atoms per gram of rock. Thus, the activity of ^{234}Th in the brine is related to the activity of ^{238}U in the rock by a proportionality constant, and, of course, the time-dependent term. When $t \gg 24$ days, (3) can be approximated as

$$(\lambda_{Th} N_{Th})_{brine} \approx f_1 \cdot (\lambda_1 N_1)_{rock}, \qquad (4)$$

in which f_1 is now the leach fraction of ^{234}Th, incorporating 1/4 L Sρ in (3). In general, ^{238}U activities in rock and coexisting brine will not be equal, but will be related by a distribution coefficient r, which is theoretically an equilibrium constant, neglecting the kinetics of dissolution and precipitation:

$$(\lambda_1 N_1)_{rock} = r \ (\lambda_1 N_1)_{brine} \qquad (5)$$

In actual fact, the time required to achieve chemical equilibrium between rock and brine might be very long relative to the half-life of ^{234}Th. Q_1 now becomes

$$Q_1 = r \cdot f_1 \cdot (\lambda_1 N_1)_{brine} \qquad (6)$$

Similarly, the amount of ^{234}U released is dependent on the decay of its precursor ^{238}U, and on the equilibrium ratio r, and on the leach fraction for ^{234}U, f_2:

$$Q_2 = r \cdot f_2 \cdot (\lambda_1 N_1)_{brine} \qquad (7)$$

In addition, some fraction, f_3, of the ^{238}U will be leached from the rock,

$$\lambda_1 N_1 \Rightarrow r \cdot f_3 \cdot (\lambda_1 N_1)_{brine} + (\lambda_1 N_1) \qquad (8)$$

Now, equation (2) becomes:

$$\left(\frac{dN_2}{dt}\right)_{brine} = (\lambda_1 N_1)_{brine} \left[1 + r (f_1 + f_2 + f_3)\right] - \lambda_2 N_2 \quad (9)$$

If f is defined as a composite leach fraction $(f_1 + f_2 + f_3)$, the solution to (9) is:

$$N_2(t) = \frac{\lambda_1}{\lambda_2 - \lambda_1} (1+fr)N_1(t) + N_1(t=0) \frac{\lambda_2 N_2(t=0)}{\lambda_1 N_1(t=0)} \frac{\lambda_1}{\lambda_2} (1+fr) e^{-\lambda_2 t}$$

$$(10)$$

which allows us to express the solution in terms of $\alpha = \frac{\lambda_2 N_2}{\lambda_1 N_1}$:

$$\alpha_{brine} = \frac{\lambda_2}{\lambda_2 - \lambda_1} (1 + fr) + \left[\alpha_o - \frac{\lambda_2}{\lambda_2 - \lambda_1} (1 + fr)\right] e^{-(\lambda_2 - \lambda_1)t}$$

$$(11)$$

and since $\lambda_2 \gg \lambda_1$

$$\alpha_{brine} \approx 1 + fr + \left[\alpha_o - 1 + fr\right] e^{-\lambda_2 t} \approx \alpha_b \quad (12)$$

The corresponding equation for the rock is:

$$\left(\frac{dN_2}{dt}\right)_{rock} = \lambda_1 (1 - f) (N_1)_{rock} - \lambda_2 (N_2)_{rock} \quad (13)$$

Since for a closed system, the $^{234}U/^{238}U$ equilibrium activity ratio is 1.000, the solution to (13) is:

$$N_2(t)_{rock} = \frac{\lambda_1}{\lambda_2 - \lambda_1} (1 - f)N_1(t)_{rock} + \frac{\lambda_1}{\lambda_2} N_1(0)_{rock} e^{-\lambda_2 t}$$

$$- \frac{\lambda_1}{\lambda_2 - \lambda_1} N_1(0)_{rock} (1 - f) e^{-\lambda_2 t} \quad (14)$$

which in terms of α is

$$\alpha_{rock} = \frac{\lambda_2}{\lambda_2 - \lambda_1} (1 - f) + e^{-(\lambda_2 - \lambda_1)t} - \frac{\lambda_2}{\lambda_2 - \lambda_1} (1 - f) e^{-(\lambda_2 - \lambda_1)t}$$

$$(15)$$

The approximation $\lambda_2 \gg \lambda_1$ reduces the equation (15) to

$$\alpha_{rock} \approx 1 - f + f e^{-\lambda_2 t} \approx \alpha_r \quad (16)$$

Let us now assume that the geochemical conditions (i.e. U content and α) observed in the ERDA No. 6 rock-fluid assemblage represent equilibrium conditions. This requires the mutual consistency of equations (12) and (16), and that

$$f = \frac{(\alpha_o - 1) (\alpha_r - 1)}{\alpha_b - \alpha_o + r (\alpha_r - 1)} . \quad (17)$$

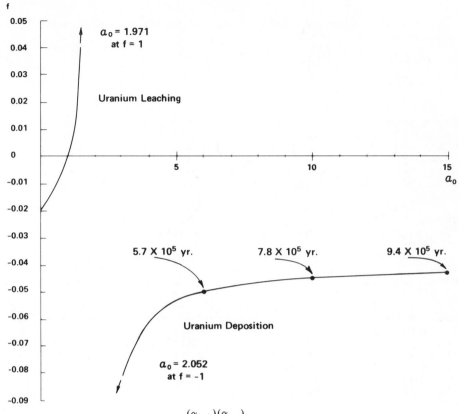

FIG.2.   Graph of the equation $f = \dfrac{(\alpha_{0-1})(\alpha_{r-1})}{\alpha_b - \alpha_0 + r(\alpha_{r-1})}$ for the ERDA No.6 assemblage of
rock and brine, in which $r = 16.42$, $\alpha_r = 1.04$, $\alpha_b = 1.35$. Uranium disequilibrium ages for
the brine are shown according to the equation

$$t = \frac{\ln\left(\dfrac{\alpha_b - 1 - fr}{\alpha_0 - 1 - fr}\right)}{-\lambda_2}$$

Note that the ERDA No.6 assemblage is indicative of uranium deposition (f < 0) rather than
uranium leaching.

Anhydrite found in the veins is the most recently formed
phase in the core, and using the U values from 2709.4 white
anhydrite (U = 33 parts in 10^9, α = 1.04; Plate 1 overleaf),
and the averages of the last two ERDA No. 6 brines in Table I,
(U = 2.01 parts in 10^9, α = 1.35) equation (17) gives rise to
the plot in Figure 2. Note the prominent singularity at
α_0 = 2.01, and that $\alpha_0 \leq 1.971$ at f = 1 (corresponding to 100%
leaching, and $\alpha_0 \geq 2.052$ at f = -1 (corresponding to all the
uranium in the rock having precipitated from the brine).

Solving equation (12) for t, we obtain an expression for the age of the brine:

$$t = \frac{\ln \left(\dfrac{\alpha_b - 1 - fr}{\alpha_0 - 1 - fr} \right)}{- \lambda_2}. \tag{18}$$

For values of $\alpha_0 \leq 1.971$, negative ages result. Only for values of $\alpha_0 \geq 2.052$ are realistic ages obtained. This result implies that there has been minimal leaching from the rock. Furthermore, it suggests rather than precipitation of uranium has occurred from brine to rock which is consistent with reducing conditions implied by the presence of H_2S.

Clearly, the amount of precipitation which has occurred (negative leaching), is related to the original α at the time of closure of the rock-fluid system.

Values of α for nearby waters in the Capitan limestone (Table I) which by proximity afford the most likely ultimate source of water for the ERDA No. 6 brine pocket, mostly have α values in excess of 1.35. In addition, the ERDA No. 6 brine is not saturated in $CaSO_4$ [1], and is not likely to have been responsible for the precipitation of substanial amounts of vein anhydrite. Furthermore, the limited volume of the brine pocket [7] could not realistically be expected to dissolve and reprecipitate anhydrite in veins to the extent observed in the core. Therefore, the α_0 of the ERDA No. 6 brine must have been larger than 2.052, with very little interaction with the reservoir rock.

5.   MODEL AGES BASED ON NO LEACHING

Values of α and α_0 will in this model allow an age determination of a water to be made. The highest value thus far determined is 14.2, the result of isotope dilution mass spectrometry on a sample from Israel supplied by J. Kronfeld. Its α value was 10.1 ± 1.6 by alpha-spectroscopy [6]. In addition, the occurrence of waters elsewhere in the Delaware Basin (Table I) suggests that a reasonable α_0 value might be 5.14 (in the Capitan limestone), the highest thus far found in the Basin. If the ERDA No. 6 brine represents water escaped from the Capitan, and its α_0 was on the order of 5, Figure 2 shows that its interaction with the ERDA No. 6 reservoir rock (fractured Castile anhydrite) precipitated less than 6% of the rock's uranium from the brine.

Neither total concentration nor isotopic composition of uranium in the ERDA No. 6 brine was inherited through interaction with the Capitan limestone, for the ERDA No. 6 brine contains substantially more uranium than the Capitan waters (Table I) presently do. Likewise, uniformity in the stable isotope composition of Capitan waters [1] shows that the peculiar stable isotope composition of ERDA No. 6 water cannot have arisen by interaction with the Capitan limestone. Composition of stable and unstable isotopes in ERDA No. 6 brine

might have arisen from interaction between water and rocks encountered by the water as it moved from the Capitan to its
ERDA No. 6 environment.

In its simplest form, equation (18), under conditions of
no interaction between ERDA No. 6 brine and its reservoir rock,
reduces to:

$$t = \frac{\ln \left(\frac{\alpha_b - 1}{\alpha_o - 1} \right)}{- \lambda_2} \tag{19}$$

which very nearly corresponds to the combination of the solutions to the equations

$$\frac{dN_1}{dt} = - \lambda_1 N_1$$

$$\frac{dN_2}{dt} = - \lambda_2 N_2 + \lambda_1 N_1$$

for which

$$t \approx \frac{\ln \left(\frac{\alpha_b - 1}{\alpha_o - 1} \right)}{- \lambda_2 + \lambda_1} \tag{20}$$

The above conclusions allow limits to be assigned to the age
of confinement of the ERDA No. 6 brine, and also to ages of
waters in the Capitan. Equation (2) differs from equation (19)
only by the approximation that

$$\frac{\lambda_2}{\lambda_2 - \lambda_1} \approx 1,$$

since $\lambda_2 = 2.806 \times 10^{-6}$ a^{-1} for ^{234}U

and $\lambda_1 = 1.537 \times 10^{-10}$ a^{-1} for ^{238}U.

Figure 3 shows the family of curves obtained from equation (20), using observed values of α from Table I, and various
values for α_o. The model ages indicated on Figure 2 for the
ERDA No. 6 brine are less than those in Figure 3 because 4 to
5 percent interaction between brine and rock (uranium precipitation) involves loss of uranium from solution by means other
than radioactive decay. There is no evidence for any degree
of chemical equilibrium between ERDA No. 6 rock and fluid. If
we realistically limit f (Figure 2) to have a value between
-0.02 and -0.05, the minimum age of the ERDA No. 6 occurrence
is 570,000 years ($\alpha_o = 6$). If in actual fact, as the data of
Kronfeld et al (1975) suggest, maximum α_o values are universally in
the range 10 to 15, the age is between 800,000 and 1,000,000
years. According to the no-interaction model, (Figure 3), water
could have escaped from the Capitan ($\alpha_o = 5.14$) into the brine
pocket 880,000 years ago. The highest α_o value (14.2) confirmed
by isotope dilution mass spectrometry gives an age of 1,300,000
years.

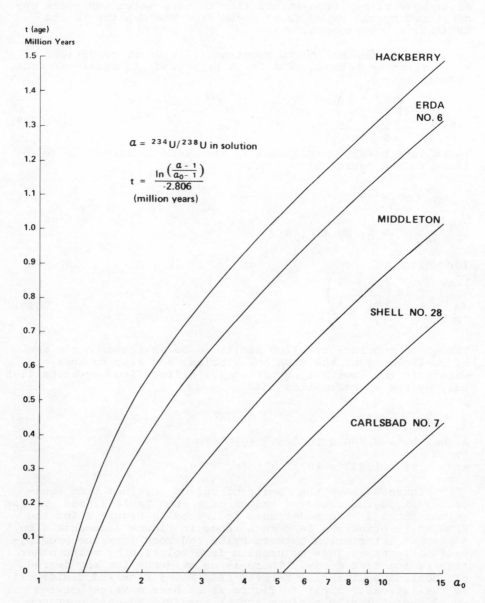

t (age)
Million Years

$$a = {}^{234}U/{}^{238}U \text{ in solution}$$

$$t = \frac{\ln\left(\frac{a-1}{a_0-1}\right)}{-2.806}$$
(million years)

HACKBERRY

ERDA NO. 6

MIDDLETON

SHELL NO. 28

CARLSBAD NO. 7

a_0

FIG.3. Uranium disequilibrium ages as a function of original $^{234}U/^{238}$ ratio, for Deleware Basin groundwaters. These curves are derived from a model involving no exchange of uranium between rock and water.

If the Carlsbad area (near the Pecos River) is a major
recharge area for much of the Capitan reef, and if Carlsbad
water (α_0 = 5.14) is the basis for age-dating other waters in
the Capitan, ages between 300,000 and 1,100,000 years are obtained. The maximum α_0 (14.2) would imply that Capitan waters
are at least 400,000 years old.

6. IMPLICATIONS AND CONCLUSIONS

A mathematical model based upon analytical data has showed
that the ERDA No. 6 occurrence of brine can be age-dated by the
uranium-disequilibrium method. Combinations of leach fractions
and ages could be derived, and the interaction between rock and
fluid was indicated to be minimal. If the brine pocket was once
connected to the Capitan Reef, the most productive aquifer in
the region, such connection was severed at least 500,000 years
ago, and probably more than 900,000 years ago. The brine pocket
has been stagnant ever since, and there is little evidence to
indicate that chemical equilibrium has been established among
the solid, liquid and gaseous phases involved in the brine
pocket.

Even though the α value of the ERDA No. 6 brine was close
to those of nearby meteoric waters taken from the northern apex
of the Capitan Reef, these α values are far removed from those
of more remote meteorically-derived saline Capitan waters from
the east and west arms of the reef (Carlsbad No. 7 and Shell
No. 28). In fact, the remote waters appear to be younger than
apex waters (Hackberry and Middleton), implying that groundwater flow in the reef is indeed toward the apex and recharge
is in the east and west part of the reef.

Fresh and saline Capitan Reef waters have retained their
meteoric D/H and $^{18}O/^{16}O$ ratios, even though the ages of some
of them are comparable to that of ERDA No. 6. Although reef
apex waters and ERDA No. 6 brine are radiometrically similar,
the solutes and stable isotopes of ERDA No. 6 reflect a more
profound rock/fluid interaction than Capitan waters have experienced. This interaction, however, is more likely to have
taken place in rocks between the Capitan and the brine pocket
than in the brine pocket itself.

The ERDA No. 6 reservoir rock has not replenished ^{234}U
to the brine in spite of long contact. Furthermore, uranium
mobility into the white recrystallized anhydrite from the gray
laminated anhydrite was shown to be very low in this anoxic
environment.

Dating by the uranium-disequilibrium method is not necessarily dependent on a closed system, and bounds on ages can
be assigned. Requirements for $^{234}U/^{238}U$ dating include: α
values of fluid and rock, leach rate, and knowledge of a value
for original activity ratio, α_0. The α value for the rock is
necessary to evaluate the degree of rock-fluid disequilibrium.
The leach rate might be negligible even in a closed system. A
unique solution to an age-dating problem, however, requires
knowledge of initial conditions, just as in the case of the
well-established uranium-lead and carbon 14 age-dating techniques.

ACKNOWLEDGEMENT

We thank Dennis Powers for arranging for some of the
samples used in this study. We also thank Felton Bingham
and James Krumhansl for criticism of the manuscript.

This work was supported by the United States Department
of Energy through the Sandia Corporation.

We are grateful to J. Kronfeld who provided us one of
his samples for verification of his high-α result by mass
spectrometry.

REFERENCES

[1]   Lambert, Steven J., "The geochemistry of Delaware Basin
          groundwaters" (New Mexico Bur. Mines and Min. Res.
          Circ. 159, in press, 1978).

[2]   Boffey, Philip M., Radioactive waste site search gets
          into deep water, Science 190 (1975) 361.

[3]   Anderson, Roger Y., and Powers, Dennis W., "Salt anti-
          clines in the Castile-Salado evaporite sequence,
          northern Delaware Basin, New Mexico" (New Mexico
          Bur. Mines and Min. Res. Circ. 159, in press, 1978).

[4]   Osmond, J. K., and Cowart, J. B., The theory and uses
          of natural uranium isotopic variations in hydrology,
          Atomic Energy Review 14 (1976) 621.

[5]   Kigoshi, Kunihiko, Alpha-recoil thorium-234: dissolution
          into water and the uranium-234/uranium-238 disequi-
          librium in nature, Science 173 (1971) 47.

[6]   Kronfeld, J., Gradsztajn, E., Müller, H. W., Radin, J.,
          Yaniv, A., Zach, R., Excess ^{234}U: an aging effect
          in confined waters, Earth Planetary Sci. Letters
          27 (1975) 342.

[7]   Griswold, George G., Site selection and evaluation studies
          of the Waste Isolation Pilot Plant (WIPP), Los
          Medaños, Eddy County, New Mexico, Sandia Laboratories,
          SAND77-0946 (1977).

DISCUSSION

H. ZOJER: Do you have any experience of leaching processes involving
uranium isotopes, not only in solid rocks but also in sediments?

G.E. BARR: If you mean unconsolidated sediments, such as present oceanic
or lacustrine sediments, we have no such experience. In another programme at
Sandia, R. Dosch has been studying some of the leaching properties of sea-bed
sediments in connection with the possibility of deep-sea disposal of nuclear
wastes, but I do not know what stage this experiment has reached.

CARBON-14 MEASUREMENTS IN AQUIFERS WITH METHANE

J.F. BARKER, P. FRITZ
Department of Earth Sciences,
University of Waterloo,
Waterloo, Ontario

R.M. BROWN
Chalk River Nuclear Laboratories,
Atomic Energy of Canada Limited,
Chalk River, Ontario,
Canada

Abstract

CARBON-14 MEASUREMENTS IN AQUIFERS WITH METHANE.

A survey of various groundwater systems indicates that methane is a common trace constituent and occasionally a major carbon species in groundwaters. Thermocatalytic methane had $\delta^{13}C_{CH_4} > -45\%$ and microbially produced or biogenic methane had $\delta^{13}C_{CH_4} < -45\%$. Groundwaters containing significant biogenic methane had abnormally heavy $\delta^{13}C$ values for the inorganic carbon. Thermocatalytic methane had no apparent effect on the inorganic carbon. Because methanogenesis seriously affects the carbon isotope geochemistry of groundwaters, the correction of raw ^{14}C ages of affected groundwaters must consider these effects. Conceptual models are developed which adjust the ^{14}C activity of the groundwater for the effects of methanogenesis and for the dilution of carbon present during infiltration by simple dissolution of rock carbonate. These preliminary models are applied to groundwaters from the Alliston sand aquifer where methanogenesis has affected most samples. In this system, methanogenic bacteria using organic matter present in the aquifer matrix as substrate have added inorganic carbon to the groundwater which has initiated further carbonate rock dissolution. These processes have diluted the inorganic carbon ^{14}C activity. The adjusted groundwater ages can be explained in terms of the complex hydrogeology of this aquifer, but also indicate that these conceptual models must be more rigorously tested to evaluate their appropriateness.

INTRODUCTION

Radiocarbon dating of groundwaters has become a widely used technique in the assessment of hydrologic activity of subsurface flow systems. As more subsurface environments are investigated and the need to generalize the interpretation of ^{14}C dates increases, the limitations of the technique become evident. A major problem is the adjustment of the measured groundwater ^{14}C activity to take into account carbonate dissolution and precipitation under closed-system conditions. Wigley [1,2], Fontes and Garnier [3] and Reardon and Fritz [4] have applied open- and closed-system carbonate dissolution/precipitation models to adjust ^{14}C ages in groundwaters.

661

Although the addition of carbon from organic sources has been
recognized in the Carizzo sand [5] and New Jersey deltaic aquifers [6],
techniques for adjusting ^{14}C ages for these additions are at an early
stage of development. Plummer [7] included consideration of organic
matter oxidation during sulphate reduction in a study of mass transfer
along a flow path in the Flordan aquifer. Calculations indicated that
the flow system appeared to become open to CO_2 well down the flow path
where sulphate reduction was occuring. Mass transfer equations for
lignite oxidation during sulphate reduction were used to calculate the
mass of CO_2 added to the groundwater and its ^{14}C activity was assumed to
be 0% modern. This paper deals with the influence methane-producing
bacteria exert on the carbon isotope geochemistry of groundwaters, and
the implication of this process on ^{14}C dating of the affected groundwaters.

THE ORIGIN AND OCCURRENCE OF METHANE IN GROUNDWATER

Methane-producing bacteria (methanogens) have been recognized
in environments where associated bacteria maintain a low Eh and also
produce methanogenic substrates. Methanogens utilize a limited range of
carbon sources such as CO_2, CO, methanol, formate, and acetate. In
anaerobic environments, fermentation of complex organic molecules leads
to the production of biogenic CH_4, usually CO_2 and simpler organics in the
presence of methanogens. However, all methanogens can grow on CO_2 + H_2
as sole carbon and energy source, which emphasizes their considerable
biosynthetic powers. It is commonly observed that methanogenesis occurs
only in sulphate-depleted environments ([8,9] for example). CH_4 production
is inhibited by sulphate-reducing bacteria apparently because methanogens
cannot compete with sulphate reducers for H_2, a strong reducing agent,
usually provided by interspecies H_2 transfer mechanisms [10]. However,
in this study, groundwaters containing >1mM biogenic CH_4 and >0.1 mM
sulphate were encountered.

Methane is also generated in the thermal decomposition of organic matter during burial and diagenesis. This non-bacteriological methane
is termed thermocatalytic, recognizing the probable importance of heat
and catalysts in its formation. The groundwater systems sampled in this
study, because of their low temperature (<10°C) and shallow depths,
should be free of thermocatalytic methane, except where leakage from
underlying thermocatalytic natural gas deposits or storage/distribution
facilities has occurred.

Hazardous concentrations of methane in groundwaters associated
with landfills [11], with natural gas deposits and storage/distribution
systems [12,13], and with normal biological activities in aquifers
[14,13] have been reported. The distribution of methane in the groundwater environment is poorly documented. Dyck et al. [15] surveyed about
1700 water wells, mainly in crystalline rocks, in eastern maritime
Canada, and found only 14 contained more than 0.2 cm^3 CH_4/litre. Barker
[16] surveyed 300 wells and piezometers completed in various unconsolidated Pleistocene deposits, and carbonate and lignite bedrock aquifers
in Canada and the United States and found about 80 containing more than
0.2 cm^3 CH_4/litre. The presence of more organic matter in the latter
aquifers may be one reason for the increased methane concentrations
observed.

CARBON-13 IN METHANE AND AQUEOUS CARBONATE IN GROUNDWATER

Carbon isotope ratio measurements on methane provide one tool useful in defining the origin of methane. Natural methane exhibits a wide range in $^{13}C/^{12}C$ ratio, with measured $\delta^{13}C_{CH_4}$* values from 0 to -90‰. Thermocatalytic methane dominates the heavier (more positive) values and biogenic methane dominates the lighter (more negative) values. Carbon isotope ratios in the range of overlap, -38‰ to -65‰, do not uniquely define the origin of methane and integration of other geochemical and hydrogeological information is required for its identification.

In groundwaters, the identification of more than traces of higher hydrocarbons has been used as evidence of the thermocatalytic origin of the associated methane [12]. The lack of associated C_{2+} compounds does not uniquely define methane as biogenic because such "dry" natural gas mixtures are commonly formed by thermocatalytic processes [17]. The lack of likely sources of thermocatalytic methane, such as in igneous rock flow systems, makes a stronger case for a biogenic origin of methane found in such groundwaters.

In the groundwaters sampled in this study, biogenic methane always had $\delta^{13}C_{CH_4}$ values < -45‰ and thermocatalytic methane had $\delta^{13}C_{CH_4}$ values > -45‰. Of 30 groundwater samples containing sufficient methane for isotopic analysis, only 4 contained thermocatalytic methane. Thermocatalytic methane was associated with groundwater having "normal" total inorganic carbon (TIC) $\delta^{13}C$ values i.e. values similar to those observed in the absence of CH_4. Significant concentrations of biogenic methane were associated with groundwaters whose TIC was anomalously enriched in ^{13}C - $\delta^{13}C_{TIC}$ values up to +15‰ were encountered. This indicates a large isotopic fractionation factor exists for CO_2 and CH_4 produced during methanogenesis. This interpretation is supported by calculated equilibrium fractionation factors [18], and by fractionation factors measured in experiments with pure-cultures of methanogens [19]. Anomalously heavy $\delta^{13}C_{TIC}$ values are another indicator of biogenic methane production in groundwater systems.

The influence methanogenic bacteria have on $\delta^{13}C_{TIC}$ values in groundwater is evident in two shallow flow systems in unconsolidated Quaternary sandy deposits on the Canadian Shield. The Kenora site is in the Experimental Lakes Area operated by Fisheries and Environment Canada in NW Ontario and the Chalk River site in the Perch Lake Basin at the Chalk River Nuclear Laboratories about 200 km NW of Ottawa. Table I presents TIC and CH_4 concentrations and $\delta^{13}C_{TIC}$ values representative of the groundwaters sampled. The biogenic methane in Kenora samples had $\delta^{13}C_{CH_4}$ = -74.9‰ and in Chalk River (22-74) $\delta^{13}C_{CH_4}$ = < -50‰. The complete results and interpretation of sampling in these areas is given by Fritz et al. [20]. The general trend of increasing $\delta^{13}C_{TIC}$ with depth is interpreted as being the result of dissolution of minor amounts of marine carbonate fragments in these sandy deposits.

---

* Carbon isotope ratio measuremetns are reported in the standard delta (δ) notation, relative to the PDB standard, in units of per mille (‰). The compound or phase for which the measurement was made is subscripted.

TABLE I. METHANE AND TOTAL INORGANIC CARBON CONCENTRATIONS AND $\delta^{13}C_{TIC}$ VALUES FROM KENORA AND FROM CHALK RIVER, ONTARIO

The number of replicate determinations are in brackets

	PIEZOMETER	DEPTH (m)	METHANE (μMOL./ℓ)	TIC (μMOL./ℓ)	$\delta^{13}C_{TIC}$ (°/oo)
KENORA	P55-4.5	1.5	5	--	-23.9
	P55-33	11	3	--	-16.4
	P80-18	6	<1	--	-14.7
	P85-4.7	1.4	490	3800	-7.1
	P86-8	2.6	570–1310 (2)	--	+10.3
	P87-14	4.6	600–1270 (2)	--	+11.3 (2)
CHALK RIVER	4-74	8.4	<0.3	850	-19.1 to -20.0 (7)
	25-74	9.3	1 to 3 (3)	860 to 940 (3)	-13.5 to -16.4 (7)
	24-74	12.2	3 to 11 (3)	1060 to 1150 (3)	-14.6 to -15.6 (5)
	22-74	22.8	19 to 180 (4)	920 to 1120 (4)	-4.8 to -15.6 (5)

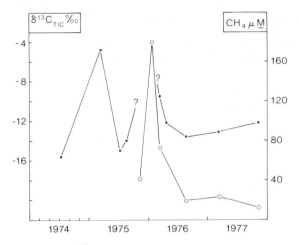

FIG.1. *Temporal variation of* $\delta^{13}C_{TIC}$ *and CH conc. in groundwater from 23 m depth, Perch Lake Basin, Chalk River, Ontario.* (● = $\delta^{13}C_{TIC}$, ○ = CH_4 *conc.)*

It is clear that when methane concentration in the groundwaters is significant when compared to the TIC concentration, the $\delta^{13}C_{TIC}$ values increased even more. As well, the ranges in $\delta^{13}C_{TIC}$ values in samples with methane taken over the 4 year sampling period are much larger than in samples unaffected by methanogenesis. Figure 1 illustrates the effect of variable methanogenic activity on $\delta^{13}C_{TIC}$ values from piezometer 22-74 at the Chalk River site. With increased methanogenic activity the TIC is enriched in ^{13}C. TIC concentrations decreased slightly from 1.12 mM to 0.92 mM as CH_4 concentration increased, suggesting inorganic carbon could be a significant carbon source. The precise mechanism responsible for these large, temporal variations cannot as yet be defined.

Another well, completed in sand and gravel at about 76 m near the north shore of Lake Erie in Ontario produces water containing more than 5 mM of biogenic CH_4. The water was sampled on 6 occasions over a two year period. Concentrations of TIC and CH_4 varied over the ranges 5.0 to 6.1 mM and 2.0 to 6.3 mM respectively. $\delta^{13}C_{CH_4}$ values ranged only from -59.1 to -60.2‰, but $\delta^{13}C_{TIC}$ values ranged from +16.8 to +10.8‰ --a much larger variability.

These data demonstrate that methanogenesis in groundwaters can have a pronounced effect on TIC concentrations and $\delta^{13}C_{TIC}$ values, especially significant in producing anomalously heavier $\delta^{13}C_{TIC}$ values. This addition of carbon to the TIC of groundwaters is temporally variable. The influence methanogenic bacteria exert on the carbon isotope distribution of groundwater TIC must be taken into account when ^{14}C dating of the groundwater so affected is attempted.

CORRECTION OF RAW ^{14}C AGES OF GROUNDWATERS AFFECTED BY METHANOGENESIS

To calculate the mean ^{14}C age of groundwaters affected by methanogenesis two steps are usually necessary:

1) an adjustment for the dilution of carbon present during infiltration by dissolution of rock carbonate
2) an adjustment for the effects of methanogenesis on groundwater composition.

The first adjustment can be made using the stable carbon isotope ^{13}C as a label to distinguish carbonate present in infiltrating water from carbonate added by dissolution of marine carbonates. This stable isotope correction method has evolved from the work of Ingerson and Pearson [21] and is incorporated in the computer-modelling programme of Reardon and Fritz [4]. The adjustment for methanogenesis effects must take into account both the direct microbial influence on the TIC of the groundwater during methanogenesis and the carbonate dissolution or precipitation resulting from this change in TIC. Unfortuantely, this adjustment is in an early stage of development and must be based on numerous assumptions.

Three possible sources of carbon for the methanogens must be considered because the utilization of each source leads to different adjustment models. Since these bacteria can utilize CO_2, the carbonate species in the groundwater itself are a possible substrate. In this situation the ^{14}C activity of the CH_4 and residual TIC should differ only by the isotope fractionation factor between CH_4 and TIC. This is complicated by the possible precipitation of calcite as CO_2 or H_2CO_3 is removed biologically. This model may be applied in groundwater systems where increased methane concentrations are paralleled by decreasing TIC concentrations and where ^{14}C activity in CH_4 and TIC are similar. If methane and CO_2 are co-produced by methanogens from organic matter recharged with the groundwater then ^{14}C activity of the CH_4 would define the age of the groundwater after the relatively minor carbon isotope fractionation between organic matter and CH_4 is considered. Unfortunately, the large concentrations of CH_4 encountered in some groundwaters and the usually much lower organic carbon concentration in recharge waters suggests this model is not often appropriate. A more likely situation would be co-production of CH_4 and CO_2 by methanogens whose major carbon source is organic material leached from the aquifer materials. The increase pCO_2 (as H_2CO_3 aqueous) creates a groundwater capable of dissolving more carbonate. Adjustment of ^{14}C ages for addition of methanogenic CO_2 and resulting carbonate dissolution are outlined in the following section.

Corrections for the Effects of Methanogenesis

These corrections cannot be made without recourse to certain assumptions based on conceptual models for methanogenesis. One approach assumes that a sampled groundwater had specific water chemistry and δ^{13}C values before methanogenesis and that the actual, sampled groundwater evolved from this specific parent by addition of methanogenic H_2CO_3 and carbonate dissolution/precipitation reactions. This approach does not consider other possible inputs of biogenic CO_2, such as occurs

during sulphate reduction. This evolution brought about by methanogenesis
is assumed to have occurred at 0 years BP (i.e. 1950).

The chemistry and $\delta^{13}C_{TIC}$ values for the parent water are
obtained from a groundwater relatively free of methane from an environment
similar to that being studied, preferably from the same aquifer. These
values are subscripted as P such as $\delta^{13}C_{TICP}$ or $^{m}Ca_P$. The measured
values of the actual sampled groundwater are subscripted as S, such as
$\delta^{13}C_{TICS}$ or $^{m}Ca_S$. The purpose of this adjustment is to calculate the
^{14}C activity of the parent water's TIC ($^{14}C_{TICP}$)--the ^{14}C activity of the
sampled groundwater before methanogenesis occurred. It is assumed that
methanogenic bacteria add aqueous CO_2 (H_2CO_3-b) to this parent water.
This biogenic H_2CO_3 has a molal concentration represented as $m_{H_2CO_3-b}$,
a $\delta^{13}C$ value as $\delta^{13}C_{H_2CO_3-b}$ and a ^{14}C activity as $^{14}C_{H_2CO_3-b}$. This
addition of H_2CO_3-b creates an aggressive groundwater capable of
carbonate dissolution. Simple carbonate dissolution or incongruent
dissolution of dolomite with precipitation of calcite may occur. The
first possibility is considered in Case 1, the second in Case 2.

In case 1, the ^{14}C activity of the parent is calculated from
the carbon isotope mass balance (CIMB):

$$^{14}C_{TICP} = \frac{^{14}C_{TICS}\, m_{TICS} - {}^{14}C_{H_2CO_3-b}\, m_{H_2CO_3-b} - {}^{14}C_{carb}\, m_{carb}}{m_{TICP}}$$

The molality and ^{14}C activity of the actual groundwater sample (m_{TICS} and
$^{14}C_{TICS}$) are measured values. The TIC molality of the parent (m_{TICP}) is
obtained from the analysis of the groundwater selected as an appropriate
parent. The molalities of methanogenic carbonate (mH_2CO_3-b) and rock
carbonate (m_{carb}) added during methanogenesis can be calculated from the
following carbon and carbon isotope mass balances:

$$m_{H_2CO_3-b} + m_{carb} = m_{TICS} - m_{TICP}$$

$$\delta^{13}C_{H_2CO_3-b}\, m_{H_2CO_3-b} + \delta^{13}C_{carb}\, m_{carb} = \delta^{13}C_{TICS}\, m_{TICS}$$

$$-\delta^{13}C_{TICP}\, m_{TICP}$$

These equations specify that the increased TIC molality and the change
in $\delta^{13}C$ values between the parent and the sampled groundwaters are due
to addition of methanogenic H_2CO_3-b and carbonate dissolution. In the
last equation, $\delta^{13}C_{TICS}$, m_{TICS},$\delta^{13}C_{TICP}$ and m_{TICP} are obtained from the
sampled and selected parent groundwater analyses. $\delta^{13}C_{carb}$ is assumed
to be 0:3‰, a typical value for marine carbonates. A value of 0.0‰
is normally selected for $\delta^{13}C_{carb}$, but for this preliminary treatment a
non-zero value was selected to keep all these terms in the calculations.

A value for $\delta^{13}C_{H_2CO_3-b}$ is also needed in this last equation.
$\delta^{13}C_{H_2CO_3-b}$ cannot be measured, so it is necessary to assume possible
values. Three assumptions were made in order to produce a range of
$\delta^{13}C_{H_2CO_3-b}$ values:

1) case 1-A: $\delta^{13}C_{H_2CO_3-b} = \delta^{13}C_{CH_4} + \varepsilon CO_2-CH_4$. This assumes
isotopic equilibrium between methanogenic CO_2 and CH_4 and that
$\delta^{13}C_{CH_4 aqueous} = \delta^{13}C_{CH_4 gas}$. The small fractionation factors
between $CO_2 gas$ and $H_2CO_3 aqueous$ are ignored.

2) case 1-B: $\delta^{13}C_{H_2CO_3-b} = \dfrac{\delta^{13}C_{org}m_{org} - \delta^{13}C_{CH_4S}m_{CH_4S} - \delta^{13}C_{TOCS}m_{TOCS}}{m_{H_2CO_3-b}}$
where $\delta^{13}C_{org}$ = -25‰, an average value for the organic matter and
$m_{org} = m_{CH_4S} + m_{TOCS} + m_{H_2CO_3-b}$. This assumes that wood breaks down
completely to produce methane, carbon dioxide and organic compounds
now found in the groundwater.

3) case 1-C: $\delta^{13}C_{H_2CO_3-b} = 2\delta^{13}C_{org} - \delta^{13}C_{CH_4S}$. This assumes that
$\varepsilon_{H_2CO_3-b-org} = \varepsilon_{org-CH_4S}$ and is considered to provide a more positive
$\delta^{13}C_{H_2CO_3-b}$ value than the other two cases.

Now there are three values for $\delta^{13}C_{H_2CO_3-b}$, from cases 1-a, 1-b, and 1-c,
and so there are three solutions for the carbon istope mass balance
equations yielding three sets of values of $m_{H_2CO_3-b}$ and m_{carb}.

   The only remaining unknowns in the initial CIMB are ^{14}C
activities of the dissolving carbonate ($^{14}C_{carb}$) and of the methanogenic
CO_2 ($^{14}C_{H_2CO_3-b}$). The carbonate can be assumed to be old enough so that
$^{14}C_{carb}$ = 0.0% mod. The $^{14}C_{H_2CO_3-b}$ can be calculated using the assumption
that fractionation factors between CH_4 and H_2CO_3-b for ^{14}C are twice those
for ^{14}C. Then

$$^{14}C_{H_2CO_3-b} = {}^{14}C_{CH_4S}(1 - 2(\delta^{13}C_{CH_4S} - \delta^{13}C_{H_2CO_3-b})10^{-3})$$

The ^{14}C activity of the groundwater prior to methanogenesis ($^{14}C_{TICP}$) can
now be calculated with the initial CIMB. Three values may be obtained
because three distinct values for $\delta^{13}C_{H_2CO_3-b}$ were calculated.

   For case 2, where dolomite $(CaMg)_{.5}CO_3$ dissolution and
calcite, $CaCO_3$, precipitation accompany methanogenesis, a more complicated
CIMB equation is used to calculate the ^{14}C activity of the sampled
groundwater with these effects removed ($^{14}C_{TICP}$):

$$C_{TICP} = \frac{{}^{14}C_{TICS}m_{TICS} + {}^{14}C_{calc}m_{calc} - {}^{14}C_{dol}m_{dol} - {}^{14}C_{H_2CO_3-b}m_{H_2CO_3-b}}{m_{TICP}}$$

Again, the values of $^{14}C_{TICS}$ and m_{TICS} are obtained from the analysis of
the sampled groundwater and m_{TICP} from the analysis of the selected
parent groundwater.

   The dolomite dissolved is assumed to be $Ca_{.5}Mg_{.5}CO_3$ and the
molality of carbonate added to the groundwater by do omite dissolution
(m_{dol}) is calculated from the increase in Mg^{2+} molality from the selected
parent to the sampled groundwater assuming that ion exchange of Na^+ for
Mg^{2+} has also occurred:

$$m_{dol} = (m_{Mg_S} - m_{Mg_P} + \tfrac{1}{2}(m_{Na_S} - m_{Cl_S})) \times 2$$

Any excess of Na^+ over Cl^- is assumed to result from ion exchange with Mg^{2+}. This represents an extreme case, since Mg^{2+}-Na^+ exchange is not common in carbonate aquifers. The molality of carbonate lost due to calcite precipitation (m_{calc}) is calculated from the decrease in Ca^{2+} molality from parent to sampled groundwater taking into account Ca^{2+} addition from dolomite dissolution:

$$m_{calc} = m\,Ca_p - mCa_s + \frac{1}{2}\,m_{dol}$$

The ^{14}C activity of the dolomite ($^{14}C_{dol}$) can again be taken as 0% mod. The ^{14}C activity of the precipitated calcite ($^{14}C_{calc}$) must be specified also. The simple assumption is made that:

$$^{14}C_{calc} = {}^{14}C_{TICS} + {}^{14}C_{H_2O_3} - b$$

The molality of methanogenic CO_2 ($m_{H_2CO_3-b}$) added can be calculated from the carbonate mass balance:

$$m_{H_2CO_3-b} = m_{TICS} - m_{TICP} + m_{calc} - m_{dol}$$

The $\delta^{13}C_{H_2CO_3-b}$ value is calculated using the following CIMB:

$$\delta^{13}C_{H_2CO_3-b} = \frac{\delta^{13}C_{TICS}m_{TICS} - \delta^{13}C_{TICP}\,m_{TICP} + \delta^{13}C_{calc}m_{calc} - \delta^{13}C_{dol}m_{dol}}{m_{H_2CO_3-b}}$$

where all values other than $\delta^{13}C_{calc}$ and $\delta^{13}C_{dol}$ have been previously calculated. $\delta^{13}C_{calc}$ is obtained by an assumption:

$$\delta^{13}C_{calc} = \frac{\delta^{13}C_{TICP} + \delta^{13}C_{TICS}}{2}$$

$\delta^{13}C_{dol}$ is assumed to be +0.3‰, a typical value for marine dolomites. The calculated $\delta^{13}C_{H_2CO_3-b}$ value is used in the calculation of $^{14}C_{H_2CO_3-b}$ with the relationship previously used in case 1:

$$^{14}C_{H_2CO_3-b} = {}^{14}C_{CH_4S}\left[1 - 2(\delta^{13}C_{CH_4S} - \delta^{13}C_{H_2CO_3-b})10^{-3}\right]$$

Now the case 2 parent ^{14}C activity ($^{14}C_{TICP}$) can be calculated using the original CIMB equation for ^{14}C.

In both case 1 and case 2, the chemistry of this parent groundwater can also be calculated taking into account the mass transfers outlined above. However, it is assumed that pH is not controlled by these methanogenic reactions alone and therefore the pH of the parent water is assumed to be constant during these reactions. This assumption is necessary as calculated pH values based solely on consideration of the above reactions and the sampled groundwater chemistry yield unreasonable pH values for the parent water. The oversimplification of the methanogenesis reactions, the neglect of sulphate reduction and the simplified treatment of carbonate-mineral/water interactions are major limitations in this preliminary attempt to correct ^{14}C ages for methanogenic effects.

TABLE II. CHEMISTRY OF GROUNDWATERS FROM THE UPPER NOTTAWASAGA DRAINAGE BASIN, ONTARIO
(*Concentrations are mM*)

WELL No.	AQUIFER	DEPTH (m)	Ca	Mg	Na	Cl	SO_4	TIC	TOC	CH_4	SiO_2	Fe	pH	Eh (mV)	T (°C)
1	ALLISTON SAND	76	2.33	1.54	4.18	3.02	0.069	9.69	0.180	3.50	0.163	0.073	7.85	+59	11
2	ALLISTON SAND	65+	2.27	1.36	2.75	1.39	0.048	9.17	0.450	2.04		0.073	7.58		7
3	ALLISTON SAND	59+	2.29	1.54	2.61	1.41	0.041	9.20	0.090	1.10		0.048	7.58		6
4	ALLISTON SAND	77	0.941	1.08	1.24	1.33	0.044	4.11	0.150	0.320	0.196	0.002	8.10	+65	10
5	ALLISTON SAND	82	0.721	0.876	0.961	0.465	0.001	3.91	0.130	0.090	0.146	0.004	8.28	+35	10
6	UNNAMED SAND	13	1.25	1.03	0.757	0.536	0.259	4.27	0.070	0.008		0.002	7.75	+160	8
7	ALLISTON SAND	65	0.659	0.617	3.34	0.638	0.271	5.34	0.950	2.05		0.011	8.08		7
8	ALLISTON SAND	78	1.05	1.05	4.59	1.72	0.047	7.11	0.230	3.61		0.003	7.56	+44	11

TABLE III. ^{13}C AND ^{14}C ANALYSES OF GROUNDWATERS FROM THE UPPER NOTTAWASAGA DRAINAGE BASIN, ONTARIO

WELL No.	TIC		TOC	CH_4	
	$\delta^{13}C_{TIC}$ ($^o/oo$)	^{14}C % MOD	$\delta^{13}C_{TOC}$ ($^o/oo$)	$\delta^{13}C_{CH_4}$ ($^o/oo$)	^{14}C % MOD
1	+2.9	5.90	−25.2	−71.0	4.40
2	+2.3	12.05	−26.7	−67.7	8.49
3	+2.2	21.80	−27.9	−66.3	
4	−11.4	39.31	−20.4	−72.3	
5	−11.9	31.91	−21.1	−61.8	
6	−12.7	51.67	−17.1		
7	−0.3	7.98	−23.5	−73.1	2.46
8	+1.8	3.01	−16.1	−71.5	3.02

Correction of Dilution of Recharged Carbonate by Rock Dissolution Before Methanogenesis

The calculations outlined above adjust the ^{14}C content of the sampled groundwater TIC for methanogenesis and associated carbonate dissolution/precipitation. However, before an actual age can be calculated, the $^{14}C_{TIC}$ must be adjusted for carbonate mineral contributions which occur before methanogenesis.

From the numerous techniques available for making this adjustment, the modelling technique of Reardon and Fritz [4] was selected. A subroutine ISOTOP, written by Reardon and Fritz [4] to be incorporated with WATEQF, a water-analysis treatment program [22] was used to model the evolution of groundwater from initial open-system conditions to the calculated parent of the actual groundwater sample. In this subroutine a variety of initial open-system pH, pCO_2 and water chemistry conditions is stipulated. The open-system conditions are assumed to occur in the soil zone with soil-CO_2 values of $\delta^{13}C$= -24‰, ^{14}C = 100% modern, and temperature of 5°C. This open-system groundwater is assumed to evolve under closed-system conditions via carbonate dissolution to the sampled or calculated parent groundwater values. The carbonate dissolving is assumed to be marine ($\delta^{13}C$ = +0.3‰) and free of ^{14}C. The evolution is calculated by adding carbonate to the initial open-system TIC concentration until the TIC concentration of the modelled groundwater attains the sampled or parent groundwater value. A $\delta^{13}C$ of the CO_2 gas phase in equilibrium with each modelled groundwater is then computed and compared to the actual groundwater's value. If these values are similar (±1-2‰)

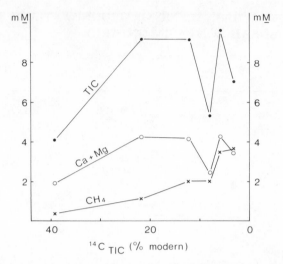

FIG.2. Trends for TIC, CH₄ and total Ca+Mg (Ca+Mg+½ (Na-Cl)) conc. with decreasing ¹⁴C_TIC activity in the Alliston sand aquifer.

then the model represents a possible simulation of the evolutionary
path of the actual groundwater. A ¹⁴C age is computed for each
modelled groundwater. When δ¹³C values of modelled and actual groundwaters CO₂ gas phase agree and non-negative ¹⁴C ages are computed, the
calculated ¹⁴C age represents a possible length of time since the
groundwater was recharged.

        When this model is applied to the parent groundwater, previously
adjusted to remove the effects of methanogenesis, a possible ¹⁴C age
of the actual, sampled groundwater is obtained. For groundwaters free
of the effects of methanogenesis, the ISOTOP modelling may be applied
directly.

        These adjustments were applied to groundwaters from the
Alliston sand aquifer, Ontario, in an attempt to provide ¹⁴C dates for
groundwater from this aquifer.

¹⁴C DATING OF GROUNDWATER FROM THE ALLISTON SAND AQUIFER, UPPER
NOTTAWASAGA DRAINAGE BASIN, ONTARIO, CANADA

        The Upper Nottawasaga Drainage Basin is underlain by thick
(up to 150 m) unconsolidated Pleistocene deposits overlying a carbonate/
shale bedrock. The Alliston sand aquifer, a deep, confined aquifer in
sandy lacustrine deposits underlies the eastern portion of the basin
and is probably part of a deep lacustrine formation that extends south
towards Lake Ontario and north towards Georgian Bay, Lake Huron. The
aquifer is of variable thickness (3-13 m) [24] and interbedded or
replaced by silts and clays. It is overlain by >35 m of clays and silts
and occasionally by more permeable deposits. Flowing wells in this

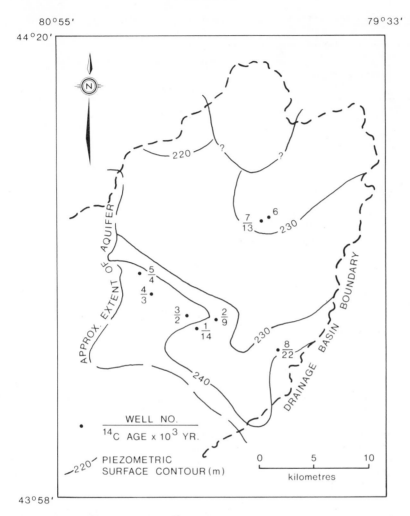

FIG.3.   Representative corrected ^{14}C ages and the piezometric surface of the Alliston sand aquifer, Ontario.

aquifer are common. Recharge is thought to be from the west and/or south through more sandy deposits. Carbon-14 dating of groundwater in this aquifer was expected to help assess the rate and source of recharge and the possible extent of groundwater "mining" in the Alliston sand aquifer.

Pertinent chemical and isotopic information for the eight samples from this drainage basin are presented in Tables II and III. Analytical procedures will be described elsewhere. pH, Eh, and temperature were measured at the well-head. Methane is a common problem in water from the Alliston sand aquifer and a biogenic origin for it is suggested

TABLE IV. CORRECTED ^{14}C GROUNDWATER AGES

WELL No.	$^{14}C_{TIC}$ % MOD	CORRECTED GROUNDWATER AGE ($\times 10^3$ Years)				
		ISOTOP ONLY	ISOTOP WITH			
			CASE 1-A	CASE 1-B	CASE 1-C	CASE 2
1	5.90	--	--	12.7-16.1	12.9-15.3	--
2	12.05	--	--	--	7.7-11.1	--
3	21.80	--	--	--	0.6-2.8a	--
4	39.31	1.1-4.2	--	--	--	--
5	31.9	3.3-5.1	--	--	--	--
6	51.67	0.2-2.0	--	--	--	--
7	7.98	future	--	--	--	11.2-14.6
8	3.01	--	--	--	22.9-26.3	18.0-21.4

a - $^{14}C_{CH_4}$ assumed 6.0% MOD.

on the basis of the very light $\delta^{13}C_{CH_4}$ values and abnormally ^{13}C-enriched TIC. The source of the methane appears to be organic matter buried in the lacustrine material, but a contribution from inorganic carbon reduction cannot be dismissed. Recharge organic carbon is an unlikely source of the high concentrations of methane observed. Figure 2 illustrates the decrease in $^{14}C_{TIC}$ activity is roughly proportional to the increase in CH_4, the increase in TIC and, to a lesser extent, the increase in Ca + Mg corrected for ion exchange. Dissolution of carbonate minerals (Ca + Mg) does not account for the total increase in TIC, and so the addition of methanogenic CO_2 is proposed to account for part of both the TIC increase and the decrease in $^{14}C_{TIC}$ activity.

The previously mentioned conceptual models were applied to obtain possible "corrected" ^{14}C dates for groundwater in the Alliston sand aquifer (Fig.3). Samples were considered to have evolved from Well #5 via methanogenes and water-carbonate interaction. Corrected ^{14}C ages are presented in Table IV.   Blanks (--) appear in Table IV where ISOTOP was unable to model the sampled or the calculated parent water composition or where methanogenesis correction was not appropriate (Wells 5 and 6). ISOTOP corrections alone failed for samples with $\delta^{13}C_{TIC} > 0.3$‰; that is, for samples seriously effected by methanogeneis. The future age obtained for well 7 also reflects the effects of methanogenesis. Case 1-a and 1-b commonly failed because the calculated $m_{H_2CO_3-b}$ added to the groundwater was found to be in excess of the increase in TIC from parent to sampled groundwater. For well 4, the minor increase in TIC from the assumed parent well 5 is insufficient for the calculated addition of H_2CO_3-b and

dissolved carbonate. Only wells 7 and 8 have chemical differences from
the parent well 5 which could be interpreted in terms of the incongruent
dissolution model (case 2). The ^{14}C age ranges result from the matrix
of assumed open-system pH-pCO$_2$ values used in the subroutine ISOTOP.

These corrected ^{14}C groundwater ages must be younger than the
aquifer materials. Since the methane is considered to be generated from
organic aquifer material, the age of this material will be similar to
the age of the methane. From the ^{14}C data for methane listed in Table 3,
the aquifer materials' age is probably 20,000-30,000 years B.P. Corrected
^{14}C ages beyond this limit would reflect either problems with the
conceptual model calculations or leakage of older groundwaters into this
aquifer. Although well 8 yields a corrected ^{14}C age of 23,000-26,000
years B.P. (case 1-b) which is within the age range of the aquifer, the
age of the aquifer material calculated from the ^{14}C$_{CH_4}$ activity in that
sample is 27000 years. It is concluded that all the corrected ^{14}C ages
for groundwaters listed in Table IV are possible.

In all cases the corrected ^{14}C ages of groundwaters are considerably less than the raw ^{14}C ages. The trend of decreasing ^{14}C$_{TIC}$
activity with increasing methane concentration shown in Fig. 2 is generally
maintained when corrected ^{14}C$_{TIC}$ activities are considered. That is,
the older corrected ages are still associated with waters seriously
effected by methanogenesis. This could reflect the slow production of
methane in these aquifers. However, it must be emphasized that these
corrected ages only represent possible groundwater ages and are based on
preliminary conceptual models of groundwater evolution.

CONCLUSIONS.

Methane in groundwater can have two origins - biogenic and
thermocatalytic. Thermocatalytic methane has no apparent effect on
groundwater geochemistry, but methane-producing bacteria serious affect
the carbon geochemistry of groundwaters and so complicate ^{14}C dating
of affected groundwaters. Even in hydrogeological systems on the Canadian
Shield, methanogenesis occasionally appears to strongly affect the
carbon geochemistry of groundwaters. The most obvious manifestation of
this process is increased δ^{13}C$_{TIC}$ values. This influence appears to be
temporally variable and the ability of methanogens to utilize inorganic
and various organic substrates further complicates matters.

Groundwaters from the Alliston sand aquifer seriously affected
by methanogenesis have received input of TIC in excess of that supplied
by carbonate dissolution. This extra TIC was attributed to CO$_2$ co-
produced with methane by methanogens from old, but not ^{14}C-free, organic
matter from the aquifer matrix. Some of the decrease in ^{14}C$_{TIC}$ activity
observed in affected groundwaters reflects this input of older methanogenic CO$_2$ and the associated dissolution of ^{14}C-free carbonates.

The corrected ^{14}C ages for these groundwaters, calculated using
simple conceptual models for groundwater evolution which include methanogenic processes, appear reasonable in light of the limited hydrogeological
data available. The Alliston sand aquifer appears to be rather complex
in light of the proximity of groundwaters of very different corrected ^{14}C
ages. Certainly some of this complexity could have resulted from

inappropriate assumptions in the ^{14}C age correction models. The conceptual models must be more rigorously tested in hydrogeologically well-understood flow systems to permit better quantification of the ^{14}C correction procedures.

Acknowledgements

        Assistance in the field studies was provided by the Manitoba Water Resources Branch, University of North Dakota, Freshwater Institute of Environment Canada, the Chalk River Nuclear Laboratories of Atomic Energy of Canada Limited, municipal agencies in Waterloo Region and Haldimand-Norfolk Region and in other municipalities, and by graduate students at the University of Waterloo. Financial assistance was provided by the Inland Waters Directorate, Fisheries and Environment Canada.

        Personnel of the Isotope Laboratory, University of Waterloo assisted with analyses. Particular thanks must go to R. Drimmie, J. Catterall, and D. Killion. The helpful comments of C. I. Mayfield, E. J. Rearon, and J. A. Cherry in the course of this research are acknowledged.

REFERENCES

[1]   WIGLEY, T. M. L., "Carbon-14 dating of groundwater from closed and open systems", Water Resour. Res. 11 (1975) 324.

[2]   WIGLEY, T. M. L., "Effect of mineral precipitation on isotopic composition and ^{14}C dating of groundwater", Nature 262 (1976) 219.

[3]   FONTES, J. C. and GARNIER, M. J., "Correction des activites apparantes en ^{14}C du carbon dissous: estimation de la vitesse des eaux in nappes captives", C. R. Reunie Annu. Sci. Terre, Soc. Geol. Fr., 1976, Abstr.

[4]   REARDON, E. J. and FRITZ, P., "Computer modelling of groundwater ^{13}C and ^{14}C isotope compositions", J. Hydrol. 36 (1978) 201.

[5]   PEARSON, F. J., Jr. and WHITE, D. E., "Carbon-14 ages and flow rates of water in Carizzo Sand, Atascosa County, Texas, Water Resour. Res. 3 (1967) 251.

[6]   WINOGRAD, I. J. and FARLEKAS, G. M., "Problems in ^{14}C dating of water from aquifers of deltaic origin", Isotope Techniques in Groundwater Hydrology 1974 (Proc. Symp. Vienna, 1974) 1, IAEA, Vienna (1975) 69.

[7]   PLUMMER, L. N., "Defining reactions and mass transfer in part of the Floridan aquifer", Water Resour. Res. 13 (1977) 801.

[8]   NISSENBAUM, A., PRESLEY, B. J. and KAPLAN, I. R., "Early
       diageneses in a reducing fjord, Saanich Inlet, British Columbia
       - I. Chemical and isotopic changes in major components of
       interstitial water",Geochim. Cosmochim. Acta 36 (1972) 1007.

[9]   BRYANT, M. P., CAMPBELL, L. L., REDDY, C. A. and CRABILL, M R.,
       "Growth of Desulfovibrio in lactate or ethanol media low in
       sulfate in association with H_2-utilizing methogenic bacteria,"
       Appl. Environ. Microbiol. 33 (1977) 1162.

[10]   OREMLAND, R. S. and TAYLOR, B. F., "Sulfate reduction and
       methanogenesis in marine sediments", Geochim. Cosmochim. Acta
       42 (1978) 209.

[11]   HUGHES, G. M., LANDON, R. A. and FARVuLDEN, R. N., "Hydrogeology
       of Solid Waste Disposal Sites in Northeastern Illinois", U.S.
       Environ. Protection Agency, Rept. SW-12d (1971) 154p.

[12]   COLEMAN, D. D., MEENTS, W. F., LIU, C-L and KEOGH, R. A.,
       "Isotopic Identification of leakage gas from underground storage
       reservoirs - a progress report", Ill. State Geol. Survey, Ill.
       Petrol. 111 (1977) 10p.

[13]   FRITZ, P. and BARKER, J. F., "Methane in Groundwater, Progress
       report, Inland Waters Directorate, Fisheries and Environment
       Canada (1976) 20p.

[14]   COLEMAN, D. D., "Isotopic Characterization of Illinois natural
       gas", Thesis, University of Illinois, Urbana-Champaign, Illinois,
       1976.

[15]   DYCK, W., CHATTERJEE, A. K., GEMMELL, D. E. and MURRICANE, K.,
       "Water well trace element reconnaissance, eastern maritime Canada",
       J. Geochem. Explor. 6 (1976) 139.

[16]   BARKER, J. F., "Carbon isotope geochemistry of methane in ground-
       water", Thesis, in preparation.

[17]   FUEX, A. N., "The use of stable carbon isotopes in hydrocarbon
       exploration", J. Geochem. Explor. 7 (1977) 155.

[18]   RICHET, P., BOTTINGA, Y. and JAVOY, M.,"A review of hydrogen,
       carbon, nitrogen, oxygen, sulphur, and cnlorine stable isotope
       fractionation among gaseous molecules", Ann. Rev. Earth Planet.
       Sci. 5 (1977) 65.

[19]   GAMES, L. M. and HAYES, J. M., "On the mechanicsm of CO_2 and CH_4
       production in natural anaerobic environments", Environmental
       Biogeochemistry (Nriagu, J. O., Ed.), Ann Arbor Science, Ann
       Arbor (1976)1 5.

[20]   FRITZ, P., REARDON, E. J., BARKER, J. F., BROWN, R. M., CHERRY,
       J. A., KILLEY, R. W. D. and McNAUGHTON, D., "The carbon-
       isotope geochemistry of a small groundwater system in Northeastern
       Ontario", submitted to Water Resour. Res., Nov. 1977.

[21]   INGERSON, E. and PEARSON, F. J. Jr., "Estimation of age and rate
       of motion of ground water by the C^{14} method,"Recent Researchers
       in the Fields of Hydrosphere, Atmosphere, and Nuclear Geochemistry,
       Maruzen, Tokyo (1964) 263.

[22]   PLUMMER, N. L., JONES, B. F. and TRUESDELL, A. H., WATEQF-F FORTRAN
       IV Version of WATEQ.   U.S. Geol. Surv., Water Res. Invest.,
       76-13 (1976) 61p.

[23]   SIBUL, U. and CHOO-YING, A. V., "Water Resources of the Upper
       Nottawasaga River drainage basin", Ont. Water Resour. Comm.,
       Water Resour. Rept. 3 (1971) 128p.

DISCUSSION

A. ZUBER: In a study which we have just finished, on the origin of methane appearing in a zinc and lead mine, we found $\delta^{13}C$ values for both CO_2 and CH_4 typical of biogenic production of methane. Our measurements indicated that this methane was produced by lignosulphate sewage discharged into the ground from a paper factor. The shift in $\delta^{13}C$ and ^{14}C values for dissolved bicarbonate was more or less in accordance with the figures reported by Mr. Barker.

M.A. GEYH: Mr. Barker's paper is based on an excellent study, the preliminary results of which were presented two years ago by P. Fritz in Amsterdam. This work stimulated us to search for methane in groundwater, mainly in confined aquifers. About 40 aquifers throughout the Federal Republic of Germany were selected, from which at least 10 ^{14}C groundwater dates have been taken. We analysed 150 samples, more than 90% of which showed methane concentrations of less than 2 $\mu mol/l$. This would correspond to a ^{14}C age correction of about 10 years. Two samples with methane contents of 1 mmol/l gave reasonable ^{14}C groundwater ages, as the methane was of thermocatalytic origin. Our conclusion is that in general methane has a minor effect where the correction of ^{14}C groundwater ages is concerned.

J.F. BARKER: Our survey of 300 groundwaters may be biased in favour of groundwaters containing methane, and your survey is in agreement with that of Dyck and co-workers (Ref.[15] of my paper). It must be stressed, however, that where methanogenesis occurs the interpretation of ^{14}C activity by standard methods is inappropriate. This problem can only be handled properly when either methane concentrations and $\delta^{13}C_{CH_4}$ or chemical parameters such as alkalinity and $\delta^{13}C_{TIC}$ are measured. This shows how necessary it is to collect rather complete chemical and isotopic data in ^{14}C studies on groundwater.

SOME PROBLEMS IN THE INTERPRETATION OF ISOTOPE MEASUREMENTS IN UNITED KINGDOM AQUIFERS

G.V. EVANS, R.L. OTLET
United Kingdom Atomic Energy Authority,
AERE Harwell, Oxon

R.A. DOWNING, R.A. MONKHOUSE, G. RAE
Central Water Planning Unit,
Reading, Berks,
United Kingdom

Abstract

SOME PROBLEMS IN THE INTERPRETATION OF ISOTOPE MEASUREMENTS IN UNITED KINGDOM AQUIFERS.

Corrections are discussed which allow for the dilution of biogenic carbon in the determination of groundwater age by ^{14}C measurements. A reappraisal is made of the procedure used in earlier work on groundwaters from the Lincolnshire Limestone, which allowed for diluting during an incongruent stage of dissolution. A new correction formula is derived for groundwaters which have been affected by both congruent and incongruent dissolution. The significance of such corrections in relation to other limitations imposed by the mixing of groundwaters of different ages is also discussed. Some of the practical problems encountered are illustrated from studies of groundwaters from the Carboniferous Limestone and Lower Greensand. In the Lower Greensand the problem is further complicated by uncertainty about the $\delta^{13}C$ value for the rock carbonate as both fresh-water and marine carbonate cements appear to be present. Corroboration of groundwater ages have been sought from D/H and $^{18}O/^{16}O$ relationships. Equations are derived from the relationship between D/H and $^{18}O/^{16}O$ for rainfall based on data for a number of European stations. These are compared with that for groundwaters covering a wide age range. A relationship between $^{18}O/^{16}O$ in rainfall and surface air temperatures for northwest Europe is also derived and is then used in conjunction with known climatic conditions at the time of origin of the groundwaters to investigate a possible relationship between $^{18}O/^{16}O$ and groundwater age.

1. INTRODUCTION

The use of isotope measurements of ^{3}H, ^{14}C, $^{13}C/^{12}C$, $^{18}O/^{16}O$ and D/H in groundwater investigations has been the subject of considerable research in recent years. Carbon-14, in particular, has been invaluable in establishing flow patterns and age estimates of groundwater in two important British aquifers, the Chalk of the London Basin [1,2] and the Lincolnshire Limestone [3].

Following on from these studies, consideration is given here to three main aspects of isotope data interpretation relevant to a further understanding of groundwater behaviour. These are:-

(i)   the derivation of groundwater age from residual ^{14}C in dissolved carbonates,

(ii)   the analysis and interpretation of results of ^{14}C dating in two problematical applications and

(iii) the relationship between stable isotope measurements and groundwater age determinations in conjunction with an analysis of published data on stable isotopes in rainfall particularly applicable to the UK.

2.   THE DERIVATION OF GROUNDWATER AGE FROM RESIDUAL ^{14}C IN DISSOLVED CARBONATES

The derivation of age based simply upon the decay of the radiogenic component is inevitably complicated by processes occurring within the aquifer such as,

(a)   the dilution by non-radiogenic carbonate,
(b)   the mixing of waters of different ages and
(c)   chemical reactions between the water and the aquifer.

Only the first two will be specifically discussed in this paper. For the first various correction procedures have been reported [4]. One technique which has been widely used is based on an assessment of the simple dilution of biogenic carbon by rock carbon from measurements of the ^{13}C/^{12}C ratio of the total dissolved carbonates in the groundwaters [5]. This technique was applied in the initial studies of the London Basin. A more detailed analysis [6] allowed for isotopic fractionation during precipitation of carbonate in situations where this also contributes to the change in the isotopic ratio, ^{13}C/^{12}C, observed in the total dissolved carbonates.

Two distinct stages of dilution of ^{14}C were thus identified, firstly, congruent solution, involving the simple dissolution of bed-rock carbonate by the dissolved biogenic carbon dioxide and, secondly, incongruent solution, involving continuous solution and reprecipitation of carbonate minerals [1]. The problem in this second stage is that ^{14}C is lost during precipitation and the amount cannot be assessed merely by assuming a simple dilution effect in the ^{13}C/^{12}C ratio of the resulting solution. Allowances have to be made for both ^{14}C losses and isotopic fractionation during repeated sequences of precipitation and dissolution. Ignoring this can lead to a considerable overestimate of the corrected age [6].

A technique allowing for incongruent solution was employed in the study of groundwater from the Lincolnshire Limestone referred to above. The technique involved an iterative procedure which, by means of a computer program, calculated, step by step, the mass balance changes and its effect on the ^{13}C/^{12}C ratios in multiple precipitation and dissolution stages. A subsequent examination of the procedure showed that the computer iterations were unnecessary since the whole process could in fact be represented by a single step equation. However, close examination of the basic assumptions of the procedure has led to a revised formula which more correctly allows for both ^{13}C and ^{14}C fractionation effects. The full derivation of the revised equation, given below, is in Appendix I.

TABLE I. COMPARISON OF DATING RESULTS WITH DIFFERENT
CORRECTION PROCEDURES FOR GROUNDWATER FROM THE
LINCOLNSHIRE LIMESTONE

Site	$^{14}C*$ raw data (% modern)	$\delta^{13}C*$ (‰)	Corrected Ages (% modern or 10^3 years)		
			(1)**	(2)**	Diff
Tallington	45.9	-11.7	92%	90.2	-1.8%
Kates Bridge	60.0	-12.6	114%	109%	-5%
Langtoft	48.6	-10.3	109%	109%	0%
Baston Fen (i)	1.9	- 2.8	14.9K	15.2K	0.3K
(ii)	2.6	- 2.8	12.4K	12.6K	0.2K
Six Score Farm	0.4	- 2.2	25.7K	26.0K	0.3K
Tongue End Farm	1.1	- 2.0	16.5K	16.8K	0.3K
Cuckoo Bridge (i)	2.4	- 1.8	9.0K	9.5K	0.5K
(ii)	2.2	- 1.8	9.8K	10.2K	0.4K

Notes

* Data from [3], qv for further isotope measurements data

**Column (1) data from [3]
Column (2) recalculated using the correction formula of this
paper (Appendix II):

$$ t = 8267 \ln \frac{50}{^{14}C_s} \left(\frac{K - \delta^{13}C_s}{K - \delta^{13}C_i} \right) \quad 1 + \frac{\varepsilon_{13}}{1000} $$

where $K = \delta^{13}C_{rock} - \varepsilon_{13}$

Constants used are: ε_{13} = + 2.42‰
$\delta^{13}C_{rock}$ = + 2.35‰
$\delta^{13}C_i$ = - 11.5‰

$$t = 8267 \ln \frac{50}{{}^{14}C_s} \left[\frac{K - \delta^{13}C_s}{K - \delta^{13}C_i} \right]^{(1 + \frac{\varepsilon_{13}}{1000})}$$

where   t is groundwater age (years)

K is $\delta^{13}C_r - \varepsilon_{13}$

ε_{13} is the fractionation factor for ${}^{13}C$ (ie $\delta^{13}C_p - \delta^{13}C_s$)

subscript   s refers to sample, r to the rock, p to precipitate

and   i to the initial state of the incongruent dissolution.

Table I shows the data for the Lincolnshire Limestone recalculated using this more precise equation and it is noted that in comparison with the originally published data the difference is minimal. This conclusion is in agreement with that of Wigley et al [7] who, in a comprehensive study of the evaluation of ${}^{14}C$ and ${}^{13}C$ corrections for natural waters, have derived general expressions which additionally allow for changes in the carbonate chemistry.

The differences between these more recent attempts to make precise allowances for carbonate dissolution are perhaps of diminishing importance when the second problem, (b), that of mixing, is more closely examined.

In some cases, where a proportion of the water may be modern enough to contain tritium, an approximate allowance can be made from inferences regarding the observed tritium levels. However, this can easily be shown to be an undue simplification as it takes no account of contributions of relatively modern waters which contain no tritium (ie those originating before 1954).

The implications of mixing are discussed in more detail in Appendix II and the results in Figs. (1) and (2) illustrate the extent of the problem for even a simple two-source model.

The importance of these considerations is to stress that a balance must be maintained between using and developing highly sophisticated calculation techniques for one process (eg dissolution) and ignoring other more obvious problems which can cause much larger errors.

This is illustrated by the examples considered in the next section which are concerned with groundwaters from less easily defined aquifer conditions and which indicate some of the uncertainties that can arise.

3.   THE ANALYSIS AND INTERPRETATION OF RESULTS OF ${}^{14}C$ DATING IN TWO
       PROBLEMATICAL APPLICATIONS

3.1   The Carboniferous Limestone

The Carboniferous Limestone is a hard karstic limestone with a very low matrix porosity (about 1%). Water movement is almost entirely by means of fissures, see for example [8]. From the point of view of ${}^{14}C$ interpretation it has proved to be a difficult aquifer to investigate because of both geochemical and hydrological problems.

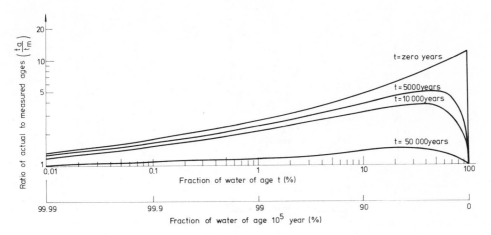

FIG.1.   Effect of mixing of groundwaters on ^{14}C dating in a simple two-source model.

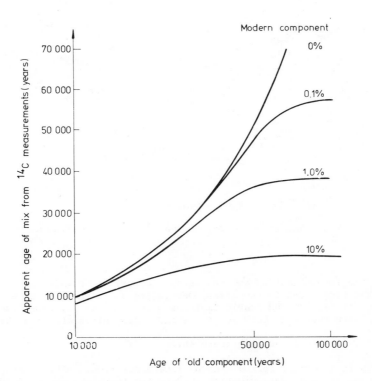

FIG.2.   Effect of the presence of 'modern' water on the ^{14}C dates of a two-component mix of groundwaters.

FIG.3.   Outcrops of aquifers and general location of sampling sites.

In the investigation reported in this paper, groundwater samples were taken from sites in two areas, Derbyshire and Bath (Fig. 3) and analysed for ^3H, ^{14}C and stable isotopes. The samples all came from thermal water sources. Details of the individual sites are given below.

3.1.1 St. Ann's Well, Buxton (Derbyshire)

This is a thermal spring, known since Roman times. The water temperature is relatively constant at 28°C and the flow, 850 m³/d, does not respond directly to rainfall. The water is of the calcium bicarbonate type and it is saturated with respect to calcium carbonate.

3.1.2 Stoney Middleton (Derbyshire)

This is also a thermal spring. Its temperature when sampled was 18°C and the rate of flow at least 130 m^3/d, but there are no long-term records to indicate whether the temperature or flow varies with time or rainfall. The water issues from the massive limestone and, as at Buxton, it is a calcium bicarbonate water saturated with respect to calcium carbonate.

3.1.3 Kings Spring (Bath)

The thermal waters at Bath have also been known since Roman times and have been extensively exploited for therapeutic purposes. There are three thermal springs but only the Kings Spring was sampled. The temperature of the water does not fluctuate significantly. Over the last 200 years the recorded extremes are 45 and 48.5°C [9]. The total flow of the thermal water is about 1600 m^3/d and it does not respond to short-term variations in rainfall. The springs issue from marls in the Lower Lias along a fault but the thermal water almost certainly issues from the Carboniferous Limestone at a depth of about 100 m and passes through Triassic marls and the Lias to the surface. The water is believed to be of meteoric origin having entered the limestone outcrops to the south and west of Bath and been heated as it descended to considerable depths in the syncline of the Bristol Coalfield [9]. At present there is insufficient evidence to define precisely the deep flow path of the water although Andrews and Wood [10] have suggested, on the basis of the radium and radon contents of the water that circulation into the Old Red Sandstone occurs below the Carboniferous Limestone. The radioactivity could, however, be derived from uraniferous horizons at the base of the Quartzitic Sandstone group of the Millstone Grit [11]. The water is of the calcium sulphate type and is saturated with respect to calcium carbonate.

3.1.4 Results and interpretation

The results of the isotope analyses are given in Table II, Section a. Calculation of the ^{14}C age, which is also given, was made using the correction formula already mentioned (Section 2). The equation requires the δ^{13}C value of the rock and the value of the bicarbonate in the solution on input to the incongruent stage. Values for the rock in the areas investigated were not available but since it is a marine deposit it would be reasonable to expect values of approximately 0‰. Results of three analyses of the Carboniferous Limestone from a deep borehole at Ashton Park, Table III, support this, giving a mean δ^{13}C of -0.05‰. Thus, taking the biogenic δ^{13}C as -26‰, the value for input to the incongruent stage is about -13‰. The measured solution values at all sites were more positive than this and incongruent dissolution was assumed.

Referring to Table III (b), an obvious difficulty arises in the interpretation if the flow path extends into the Old Red Sandstone for the few δ^{13}C values available for this formation (mean -10.7‰) are more consistent with freshwater carbonate cement (a not unexpected result since the rocks are of continental origin).

The remaining parameter required to correct for incongruent dissolution is the step fractionation factor ε_{13}. Being dependent on the prevailing pH and temperature [7] it is a value which can only be approximately derived from the values observed in the issuing groundwater. A value was derived from reference 7 assuming a water temperature 10°C higher than measured to be more appropriate to the true value at depth.

TABLE II. TRITIUM, ^{14}C AND STABLE ISOTOPE MEASUREMENTS IN GROUNDWATERS FROM THE CARBONIFEROUS LIMESTONE AND LOWER GREENSAND

Site and Aquifer	Site* No	Tritium (TU)	$\delta^{18}O$ (‰)	δD (‰)	$\delta^{13}C$ (‰)	^{14}C raw data (% modern)	^{14}C age (years)
(a) Carboniferous Limestone:							
(i) Bath:							
Kings Spring	1	5.9	-7.4	-46	-2.8	0.9	Indeterminate
(ii) Derbyshire:							
Stoney Middleton	2	10	-8.3	-52	-9.1	37.3	Modern
St. Ann's Well, Buxton	3	4.5	-7.9	-49	-8.6	20.0	3,300
(b) Lower Greensand:							
(i) Kent:							
Ryarsh Pumping Stn	8	4.2	-7.3	-47	-15.3	71.2	Modern
Luddesdown Pumping Stn	9	0.1	-7.4	-42	-14.1	39.0	2,700
Northfleet Pumping Stn	10	3.1	-7.5	-46	-13.3	32.2	3,800
Bowaters	11	2.3	-8.6	-56	-9.6	2.8	21,300
(ii) Cambridgeshire:							
Kingston Pumping Stn	4	2.5	-7.7	-48	-12.6	37.3	2,200
Lords Bridge Pumping Stn	5	0.4	-8.1	-51	-12.5	21.0	6,800
Harston Pumping Stn	6	2.7	-8.9	-60	-12.7	2.0	26,400
Great Shelford	7	1.7	-8.8	-58	-12.6	8.6	14,300
(iii) Slough:							
Industrial Estate	12	0	-8.5	-53	-11.3	1.3	29,000

*See Figs 3, 4 and 5

TABLE III. $\delta^{13}C$ VALUES FOR ROCK SAMPLES

Site and Aquifer	Depth (m)	$\delta^{13}C$ (‰)
(a) Ashton Park Borehole:		
Carboniferous Limestone	452	− 0.88
Carboniferous Limestone	548	− 2.29
Carboniferous Limestone	656	+ 3.00
(b) Hamswell Borehole:		
Old Red Sandstone	103	− 9.82
Old Red Sandstone	150	−11.21
Old Red Sandstone	255	− 8.4
		−10.7
Old Red Sandstone	290	−12.2
Old Red Sandstone	292	−11.9
(c) Warlingham Borehole:		
Folkestone Beds	229	−17.9
		−16.6
Hythe Beds	290	− 0.08
		+ 0.76
Hythe Beds	299	+ 0.56
		+ 0.34
Hythe Beds	311	− 9.27
		− 9.31

In the case of the Kings Spring water (Bath) the mean age is 11.5K years, assuming $\delta^{13}C_{rock}$ is zero, ε_{13} is + 2‰ and the incongruent dissolution input value is −13‰. However, if the actual extremes of the measured $\delta^{13}C_{rock}$ values (Table III, a) are considered, it can be shown that large variations about this mean age estimate are obtained, ranging from indeterminate to 22.5K years, the first result being due to the solution $\delta^{13}C$ value of −2.8‰ exceeding the limiting condition of

$$\delta^{13}C_{rock} - \varepsilon_{13}$$

(see Appendix I, Equation 5).

This thoroughly unsatisfactory situation leaves no option but to give an indeterminate result for Kings Spring.

The occurrence of tritium indicates the presence of fractions of modern water in all the Carboniferous Limestone samples. It is, there-fore, inferred that they are all mixes of waters derived from different

depths in the aquifer, or, in the case of Bath, possibly the deposits
above the aquifer. As established earlier any attempt to correct for
simple mixing from the tritium values is misleading if the mix also
includes significant quantities of relatively modern, although earlier
than 1954, waters. Consequently, no attempt has been made to resolve
anomolies by correction of the ^{14}C ages from the tritium measurements.
As explained in Appendix II the presence of any modern ^{14}C could greatly
reduce the apparent age estimates. Thus the apparent results of "modern"
for Stoney Middleton and 3.3K years for Buxton may in reality represent
~100% and at least 13% respectively of very modern water mixed with a
second component of very old origin (say > 10^5 years).

A possible explanation for the water movement and chemistry at
Derbyshire and Bath can be postulated. As the water descends in the
limestone the temperature increases, some precipitation of carbonate
occurs and the water becomes saturated with respect to calcium carbonate.
At greater depths incongruent solution occurs giving the enrichment in
δ^{13}C observed particularly at Bath. This process appears to have
progressed to a lesser extent in Derbyshire, but there is the possibility
that its effect is masked by dilution of the thermal water by younger
components. A further effect which could render the δ^{13}C value more
positive is the loss of carbon dioxide with consequent precipitation of
calcium carbonate [12] as the water rises to the surface. The magnitude
of this effect is unknown but it may not be very significant.

The interpretation that the thermal component in the Derbyshire
waters is several thousand years old is credible but no definite
conclusions can be drawn about the water from Bath. Although the top of
the Carboniferous Limestone below Bath is only 100 m below the surface,
the temperature of the water indicates it was derived from at least
1000 m and geological considerations suggest the depth could be of the
order of 2000 m. The limestone is overlain by the Keuper Marl and the
Lias, relatively impermeable deposits, but possibly sufficiently
permeable to contribute a small amount of post-1953 water, though not
enough to cool the thermal water significantly. Consideration is given
in Section 4 to the possible use of ^{18}O/^{16}O and D/H to provide further
insight into the true age of the waters at Bath.

3.2 Lower Greensand

The Lower Greensand is a Lower Cretaceous sandstone aquifer that
crops out around the periphery of the London Basin (Fig. 3). It is
a marine deposit laid down on the flanks of an ancient massif referred
to as the London Platform. The aquifer provides appreciable amounts of
groundwater to wells drilled both in the outcrop and down-gradient below
more recent deposits where the flow is confined.

Wells were sampled along two lines running in general down-gradient
directions in the confined area. One line was south-east of London in
Kent (Fig. 4, hereafter referred to as the Kent section) and the other
was north-west of London in Cambridgeshire (Fig. 5, referred to as the
Cambridge section); a sample was also collected at Slough (Fig. 3).
Previously the age of water in the aquifer had been determined at a
number of sites to the west of London, including Slough [13].

South of London the Lower Greensand sequence comprises in order of
increasing age: Folkestone Beds, about 50 m thick; Sandgate Beds, about
5 m; Hythe Beds, about 30 m; Atherfield Clay, about 15 m. The Sandgate

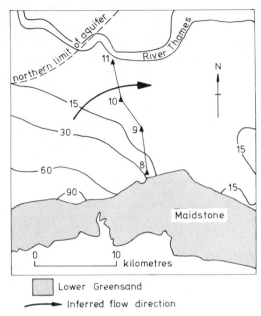

FIG.4.   Location of sites along the Kent section and groundwater levels in the lower Greensand
(sites shown as ▲ with numbers as given in Table II: rest water levels in metres above sea level,
based on hydrogeological map of Kent published by the Institute of Geological Sciences
in 1970).

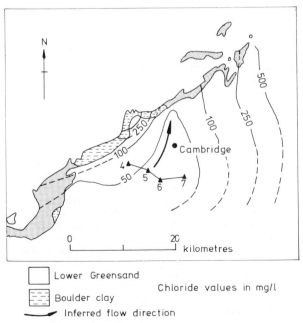

FIG.5.   Location of sites along the Cambridge section and isochlors of groundwater in the
lower Greensand (sites shown as ▲ with numbers as given in Table II).

Beds are variable in character but in the outcrop to the south of the
Kent line of section they tend to be argillaceous and the Folkestone and
Hythe beds are the main aquifers. The Hythe Beds consist of alternating
layers of sandy limestone and poorly cemented calcareous sandstone; they
die out about 3 to 5 km north of the outcrop. The Folkestone Beds are
mainly unconsolidated sands.

Along the Cambridgeshire outcrop the divisions just discussed are not
recognised and the formation consists of a series of sandy deposits about
10 m thick at outcrop but thickening to over 20 m some 12 km from the
outcrop. Lithologically the deposits are fine to coarse-grained, cross-
bedded sands with varying amounts of glauconite.

In the past, the Cambridgeshire outcrop was covered by boulder clay
and the area close to the sampled sites is still so covered, although
elsewhere erosion has removed the clay from extensive areas. The boulder
clay tends to be chalky and contains numerous small fragments of chalk
and larger boulders including both chalk and limestone [14].

The groundwaters along the Kent outcrop are calcium bicarbonate waters
but because of the unconsolidated nature of the aquifer the total ionic
content is relatively low and the waters are under-saturated with respect
to calcium carbonate. Along the line of the section, calcium bicarbonate
waters occur for a considerable distance below the overlying confining
bed but the waters eventually pass into a sodium bicarbonate type; the
waters from all the sites sampled except the extreme northern site
(Bowaters) are of the calcium bicarbonate type.

The quality of water in the Lower Greensand in Cambridgeshire is
strongly influenced by the boulder clay [15]. Water in the aquifer
below the boulder clay has a high total ionic content due to sulphate
and chloride as well as other constituents. In and near the outcrop
where the boulder clay is present, calcium sulphate-bicarbonate waters
occur but these pass into sodium bicarbonate or sodium sulphate-
bicarbonate types down-gradient.

3.2.1 Results and interpretation

The results of the isotope analyses of groundwaters from the Lower
Greensand are given in Table II, b. In the derivation of the corrected
^{14}C ages which are also given, the main problem is to understand the
chemical processes within the aquifer in relation to variations of the
$\delta^{13}C$ of the rock. The magnitude of the observed $\delta^{13}C$ values of the
solution strongly suggests that only congruent solution of carbonate
is involved. Thus, the age correction formula (Equation 2, Appendix I)
originally proposed by Ingerson and Pearson [5] is required. For this,
the unknown parameter is the $\delta^{13}C$ of the rock. Table III, c lists
determinations carried out on 4 samples from various depths in a borehole
which penetrated the aquifer below overlying deposits in the London Basin.
These are the only results available from a total of 12 core samples from
the Lower Greensand. Because of the unconsolidated nature of the formation
most of the samples did not contain carbonate and reliable values for the
$\delta^{13}C$ of the rock were only obtained for the four samples listed; two
measurements were made on each of these samples. Since the Lower Greensand is regarded as a shallow water marine deposit, any original carbonate
in the aquifer would be expected to have a $\delta^{13}C$ value close to zero.

However, the results of these very limited data are equivocal; two are close to values to be expected for marine carbonates but two are more typical of freshwater carbonates.

Clearly this information is far too limited to draw firm conclusions but, taken at its face value, parts of both the Folkestone Beds and the Hythe Beds appear to contain a freshwater carbonate cement and the inference is that they were cemented either at the time of formation of the deposit, or very soon afterwards during an aerial stage probably of short duration. Alternatively the carbonate may have been precipitated more recently from circulating meteoric groundwaters. Of these two possibilities the former would seem more likely, for the Lower Greensand in Kent was deposited near a land mass and does contain plant debris and lignite. Aerial stages of short duration do not seem incompatible with such a scenario. Further studies will be required to reconcile the conflict but in the present context, because the Lower Greensand is recognised as a marine deposit, it has been assumed that the carbonate in the aquifer is predominantly of marine origin with a value of $\delta^{13}C$ close to zero.

3.2.1.1 Kent section

The $\delta^{13}C$ values of the water samples along the Kent section (Table II, b, i) are listed in order of increasing distance from the outcrop. They are seen to become less negative in a down-gradient direction.

On the assumption that the carbonate is of marine origin the North-fleet water represents the end of the congruent stage. It follows that the Ryarsh and Luddesdown samples suggest the carbonate/carbon dioxide reaction has not been completed. The $\delta^{13}C$ value of -9.6‰ for the Bowaters sample is probably due to ion exchange which allows increasing solution of carbonate from the aquifer. (This is also likely to be the case for the Slough sample but to a much lesser extent.) Along the Kent section the age is seen to increase down-gradient to about 21K years but there is a marked discontinuity between the Northfleet and Bowaters sites. The values imply a mean velocity of about 4 m/year between the outcrop and Northfleet and the outcrop and Luddesdown but only about 0.25 m/year between Northfleet and Bowaters. The sharp reduction in velocity is improbable and a more likely explanation is that the main flow path is diverted to an outlet further east and escapes by upward leakage into the Chalk and ultimately the Thames. There is an interface between this more recent water and the older water represented by the sample from the Bowaters site.

3.2.1.2 Cambridge section

The line of the Cambridge section was selected to cross part of the aquifer where the chloride content initially decreases down-gradient before rising again [15]. The chloride ions derived from the boulder clay act as a tracer indicating the down-gradient flow path. The chloride content of water in the aquifer below the boulder clay is generally over 100 mg/ℓ but does exceed 250 mg/ℓ and locally 500 mg/ℓ. This declines to about 25 mg/ℓ in the confined area before increasing steadily to more than 500 mg/ℓ. The implication is that there are two distinct waters which mix along the line of minimum chloride content. The relatively high chloride water, moving down-gradient from the sub-outcrop below the boulder clay, is diluted by low chloride water flowing north-east from the natural recharge area further south. Although the form of the

groundwater levels accepted at present does not support this view, data
are limited and an important point may be that the Lower Greensand is
much thicker along a south-west to north-east line passing through the
area of minimum chloride values. The high chloride water further down-
gradient may be a connate water or more likely represents a high chloride
water that originated before the last interglacial when the Lower
Greensand outcrop was probably more extensively covered with boulder clay.

The aquifer in this region is overlain by boulder clay and the
source of bicarbonate in the water is believed to be carbonate present
in the clay. The $\delta^{13}C$ of the water samples range from -12.5‰ to -12.7‰
and assuming this represents the end of the congruent stage it implies
a $\delta^{13}C$ value for the carbonate of zero which is typical of a marine calcium
carbonate, the most likely source in this case.

Referring to Table II, b, ii, the two calcium-rich waters give
measured ages of 2.2K and 6.8K years and the sodium-rich types 14.3K and
26.4K years; of the sodium-rich waters the younger water is farther down-
gradient implying non-homogeneous conditions in the aquifer and its
attendant consequences on the age determinations. The ages of the
calcium-rich waters indicate a velocity of 2 m/year near the outcrop and,
as along the Kent section, there is a marked decrease in velocity further
down-gradient; between Lords Bridge and Harston the apparent flow rate
is about 0.25 m/year. This could be partly due to the increase in
thickness of the sandstone but because of the chloride content the over-
riding reason must be a gradual mixing with an older water showing that
the discontinuity is real. The older water is likely to have a zero ^{14}C
content and, if so, the actual mean age of the water mix is greater than
that measured.

3.2.1.3 Slough

The sample from the well at Slough has a ^{14}C content approaching
the limit of detection but this low value implies that the age is at
least 29K years which is in keeping with the position of the site near
the axis of the London Basin.

No significant levels of tritium were measured in any Lower Greensand samples. The implication is that no immediately modern components
are present, but the reservations already mentioned regarding pre-1953,
but still modern, components may still apply. It may well be that
preferred flow routes exist through the aquifer due to heterogeneities
arising from conditions existing at the time the rock was deposited. As
before significant under-estimations of the mean actual age would occur
if this were the case.

The above discussion has assumed that the $\delta^{13}C$ of the Lower Greensand
is zero but if the presence of some freshwater carbonate gave a negative
mean rock value then the ages in Table II would be reduced. The effect
would be more significant for younger waters.

The marked discontinuity in the age of the groundwaters in Kent and
in Cambridgeshire may infer an extended period of no infiltration. The
major glacial advance of the Devensian stage of the Pleistocene occurred
between about 14K and 24K years BP. Although the Lower Greensand outcrop
was not covered by ice at this time, it must have been affected by permafrost which would have prevented infiltration. Thus waters with a mean

TABLE IV. DERIVED RELATIONSHIPS BETWEEN δD AND δ^{18}O FOR
RAINFALL AT SOME EUROPEAN STATIONS

Site	Slope	Intercept	Correlation coefficient
Gibraltar	4.2	-6.4	0.72
Isfjord Radio, Norway	6.0	-8.9	0.93
Liege, Belgium	6.5	-3.2	0.96
Lista, Norway	6.9	-0.3	0.93
Reykjavik, Iceland	7.0	-0.9	0.91
Stuttgart, FDR	7.5	+1.5	0.99
Taastrup, Denmark	7.3	+3.3	0.97
Valentia, Eire	6.1	-2.8	0.85
Vienna, Austria	6.5	-7.9	0.99
All above stations	7.1	+0.8	0.97
Maritime Europe*	6.9	-0.3	0.95

*Includes Isfjord Radio, Liege, Lista, Reykjavik, Taastrup and
Valentia. The relations have been derived from raw data taken from
World Survey of Isotope Concentration in Precipitation, 1953-1971,
in Technical Reports Series Nos 96, 117, 129, 147 and 165,
International Atomic Energy Agency, Vienna.

age in excess of about 14K years must have originated as rain during an
interstadial prior to the last major glacial advance. The post-glacial
waters appear to be slowly displacing the pre-glacial waters from the
aquifer but the extent of this is limited to the areas affected by modern
flow paths. The relationship between climate and recharge is discussed
further in the next section.

4.   RELATIONSHIP OF STABLE ISOTOPE ANALYSES AND GROUNDWATER AGE

Measurements of the stable isotopes of hydrogen and oxygen were
carried out on all groundwater samples in an attempt to provide possible
corroboration of dating interpretations and to broaden the understanding
of past recharge mechanisms. For example, information on the climatic
history of a water can be obtained from the stable isotope ratios D/H and
$^{18}O/^{16}O$ [16]. Also the relationship between the isotopic ratios for
deuterium and Oxygen-18 in precipitation based on a global precipitation
survey [17] is:

$$\delta D = 8 \, \delta^{18}O + 10$$

and provides an aid to the identification of meteoric water [18]. A
similar equation was reported for northern hemisphere continental
stations including a number from Europe [16].

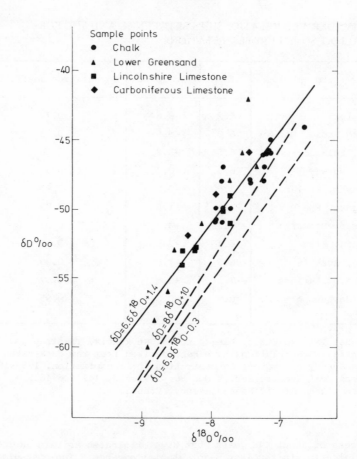

FIG.6.   Relationship between δD and δ^{18}O (δD = 6.6 δ^{18}O + 1.4 is for groundwater;
δD = 6.9 δ^{18}O − 0.3 is for rainfall over maritime Europe).

4.1   Specific data analyses and interpretation

4.1.1 $^{18}O/^{16}O$ and D/H relationship

A study was made of published data from a number of rainfall collection stations in Europe. The relationships between $^{18}O/^{16}O$ and D/H for rain at these stations were derived from an analysis of the monthly values (Table IV). It is seen that considerable deviations from the accepted Craig relationship [17] given above are observed for individual stations. An analysis of the data from selected stations, marked in the table as 'maritime Europe', gives the relation:

$$\delta D = 6.9 \ \delta^{18}O - 0.3$$

This is probably the best representation for recent rainfall over Britain and it is of note that it is very similar to that derived from the combined results of groundwater samples from the Chalk [19], Lincolnshire Limestone [3] and the Lower Greensand and Carboniferous Limestone given in this paper, Fig. 6, viz,

$$\delta D = 6.6 \, \delta^{18}O + 1.4$$

It is also worth observing that the relationship for groundwaters younger than 10K years, when the climate was temperate and similar to that of today, is virtually identical to that given above for all the groundwater samples.

The agreement between the relationships for rainfall over 'maritime' Europe and the groundwaters, if not simply fortuitous because of such a small groundwater sample, implies that at no time was rainfall significantly affected by evaporation before recharge. Agreement also suggests that the thermal waters of the Carboniferous Limestone have not been modified by isotopic exchange and, if so, supports the view that they are meteoric waters which have not penetrated to great depths.

The reliability of such deductions depends upon the constancy of the $^{18}O/^{16}O$ and D/H relationship for rainfall over Britain during this period. This might not be so, as climatic circulation patterns may have differed at various times in the last 10K years and during the late Pleistocene [20].

4.1.2 $^{18}O/^{16}O$ and temperature relations

The widely accepted relation between $^{18}O/^{16}O$ in rainfall and temperature is

$$\delta^{18}O = 0.69t - 13.6 \, (‰)$$

where t is the mean air temperature in $^{\circ}C$. The result, derived empirically by Dansgaard [16], is based on data for the annual means of $\delta^{18}O$ in precipitation and surface air temperature at many stations, including North Atlantic coastal and Greenland ice-cap stations. The data covered the temperature range $-30^{\circ}C$ to about $+11^{\circ}C$. Using the groundwater analyses for $^{18}O/^{16}O$, this equation gives higher temperature values than would have been expected to exist when the groundwaters originated.

A new study has helped to clarify this apparent anomaly. The relationships between both the mean annual $\delta^{18}O$ of rainfall and surface air temperature and between the mean $\delta^{18}O$ of winter rainfall and surface air temperature were examined for data published for the five north-west European stations: Groningen, Isfjord Radio, Lista, Reykjavik and Valentia with the following results:

For monthly $\delta^{18}O$ values:

$\delta^{18}O = 0.23t - 8.62$ (correlation coefficient 0.77, range of monthly data, $\delta^{18}O$ -18.9 to $-2.2‰$ and temperature $-20^{\circ}C$ to $18^{\circ}C$)

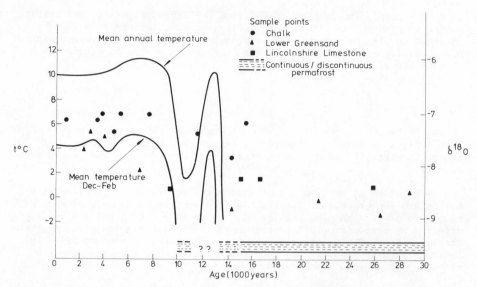

FIG. 7.   Relationship between $\delta^{18}O$, surface temperature (based on $\delta^{18}O = 0.23t - 8.62$) and age of groundwater.

For mean winter (October to March) $\delta^{18}O$ values:

$\delta^{18}O = 0.22t - 8.26$ ·(correlation coefficient 0.78, range of monthly data, $\delta^{18}O$ -18.9 to -3.3‰ and temperature -20°C to 14°C)

Both these equations give more compatible lower surface air temperatures than those derived from Dansgaard's relationship for the time when the groundwaters originated as rainfall. Before discussing the relative merits of the equations, it is necessary to summarise briefly the climatic conditions during the period relevant to any discussion of the ages of the groundwater samples.

4.1.3 Summary of climatic conditions over the last 70K years

The period of interest includes the last stage of the Pleistocene, referred to in Britain as the Devensian, which commenced about 70K years BP. It included three cold phases separated by interstadials when the climate was milder and at times approached that of the present day. The first interstadial, the Chelford Interstadial, occurred about 60K years BP. It was succeeded by the second cold phase from about 55K to 45K years BP, the complex Upton Warren Interstadial from 45K to 25K years BP and the last major cold phase from 25K to 14.5K years BP; the latter is generally believed to be the only truly glacial phase in Britain during the Devensian.

The boundary between the Devensian and the succeeding post-glacial Flandrian stage is taken as 10K years BP. In the period between 14.5K and 10K years BP the climate was very cold but it did include one

interstadial, between 13.5K and 11K years BP, referred to as the
Windermere Interstadial, when climatic conditions were temperate. The
post-glacial Flandrian stage has had a temperate climate which attained
a climatic optimum about 7K years BP. For much of the early part of the
Flandrian the climate was somewhat warmer than today. Deductions about
past climatic conditions are difficult to draw and are open to conjecture.
Nevertheless, as a general guide, the mean annual temperature and the
mean December-February temperature are shown in Fig. 7. This is based
mainly on the work of Lamb et al [21], Coope [22], West [23] and Lamb
[20].

     Of particular interest from the groundwater point of view is the
length of time during the Devensian when permafrost existed. The
hydrological conditions in areas affected by permafrost are not completely
known, but although water can move through frozen ground, the recharge of
aquifers from infiltration is very significantly reduced [24, 25].

     It is known that Britain was extensively glaciated only during the
final cold phase of the Devensian but the ice front did not reach the
aquifer outcrops discussed in this paper. However, climatic conditions
over the outcrops during all three cold phases are believed to have been
severe enough for the formation of continuous permafrost [26]. The
climate of the Chelford Interstadial has been compared to that currently
found in southern Finland between 60°N and 65°N [22] with a mean summer
temperature somewhat lower than central England today; the mean annual
temperature was possibly between 2°C and -3°C [27]. Watson [26] has
suggested that discontinuous permafrost may have prevailed.

     The climate during much of the period represented by the Upton
Warren Interstadial was arctic; tundra and permafrost occurred, except
between about 44K and 42K years BP, when a temperate climate prevailed with
summer temperatures higher than today and winter temperatures only a
little lower [22].

     During the final cold phase continuous permafrost probably continued
until about 15K years BP [26]. Thereafter it became discontinuous and during
the Windermere Interstadial temperate conditions were attained. After
the interstadial, tundra was re-established possibly with discontinuous
permafrost which persisted until the final amelioration of the climate
at the beginning of the Flandrian.

4.1.4 Discussion of relationship between climate and recharge

     When the information in the previous section is considered from the
point of view of natural recharge of groundwater, it would seem that
recharge could only occur during relatively limited periods of the Chelford
Interstadial, the early part of the Upton Warren Interstadial and the period
after the retreat of the ice about 14.5K years BP, following the last major
glaciation. During limited periods the climate was similar to that of
today and winter infiltration probably occurred. At other times, when
conditions were transitional between arctic and temperate, seasonally
frozen ground would have prevented infiltration in the winter and the
main infiltration periods were likely to have been in the spring and
autumn, as, for example, in Finland today.

     It is important to consider whether the mean annual air temperature
at the time of recharge, as derived from Oxygen-18 values of the groundwater samples, can be related to the climatic variations in the

Flandrian and the Devensian. As has been pointed out previously [2],
the surprising feature is the small range of the $\delta^{18}O$ values for the
groundwater during the last 30K years since it only extends from about
-7‰ to -9‰.

During the last 8K years, when temperate conditions prevailed, the
main season for groundwater recharge was the winter six months from
October to March. The average air temperature at the time of recharge
would have been expected to be about 6°C but applying Dansgaard's
equation to the Oxygen-18 data implies that the recharge water originated
at a temperature of about 9°C, only slightly below the annual mean.
However, if the equations derived for north-west European stations are
applied, the $\delta^{18}O$ values generally give temperatures for the time of
origin of groundwaters younger than 8K years, that are within the range
to be expected for recharge by rainfall in the six months, October to
March. If application of Dansgaard's equation is correct, a possible
explanation is that part of the water recharging the aquifer in the winter
originated as summer rainfall, was concentration by evaporation in the soil
zone and then displaced by the subsequent winter rainfall. It would tend
to make the infiltration less depleted in Oxygen-18. This is at variance
with the previously stated observation that rainfall is not significantly
affected by evaporation before recharge. The uncertainty could possibly
be resolved by measurements of $\delta^{18}O$ for rainfall over the outcrops of the
aquifers, together with the determination for $\delta^{18}O$ values for soil water
throughout the year, and for water in the unsaturated zone.

Dansgaard's equation suggests the pre-8K year old groundwater
originated at a temperature only 1 to 2°C lower than post-8K year old
waters but the equations for north-west European stations suggest
temperatures ranging from 6°C to -2°C for waters older than 8K years,
with the temperature decreasing with increasing age of the water as
shown in Fig. 7.

The temperatures deduced for the time of origin of groundwaters
older than 8K years from both the Dansgaard and the north-west European
equations are high for the climatic conditions that existed at the time.
However, because of permafrost, samples older than about 15K years must
contain a component derived from the early part of the Upton Warren
Interstadial or the Chelford Interstadial or, in view of the short
period of these "infiltration windows", possibly the Ipswichian Interglacial. The equations relating $\delta^{18}O$ to surface temperatures for
north-west European stations imply that waters older than 15K years
originated at temperatures between -2°C and +2°C and this conforms with
a view that infiltration occurred during interstadial or interglacial
periods. At times infiltration would have been possible in the winter
but at other times when conditions were colder, it is likely to have
occurred in the spring and autumn because of frozen ground in the winter;
a component may have been derived from melting winter snow which would
be reflected in the $\delta^{18}O$ of the groundwater. Some of the waters from
the period 9K to 15K years BP are probably also mixes and include a
component from earlier warm periods.

The equations relating $\delta^{18}O$ to surface temperatures for north-west
European stations appear to give a better interpretation of the
temperature at the time of infiltration, than application of Dansgaard's
equation. But the conclusions drawn about air temperature at the time
of origin of the groundwater depends upon the empirical equations
relating air temperature and $\delta^{18}O$ of modern rainfall applying equally

IAEA-SM-228/34   699

well in the past. As already mentioned atmospheric circulation patterns
may have differed.

There is an apparent general linear relationship between $\delta^{18}O$ and
the age of the groundwater over the last 30K years although, of course,
this is not altogether real as $\delta^{18}O$ reflects climatic conditions, which
varied during the Flandrian and late Devensian periods. Nevertheless,
while bearing this in mind, the relationship can be used to assist in
the interpretation of the data. For example, it would appear on the
basis of $\delta^{18}O$ that the mean age of the Kings Spring at Bath is unlikely
to exceed 16K years.

The relationship between $\delta^{18}O$ and groundwater age can be crudely
applied in situations where some doubt exists about the reliability of
the age determination because of, for example, uncertainty about the
$\delta^{13}C$ value of the rock carbonate, a situation existing in the Lower
Greensand.

5.   CONCLUSIONS

1.   A simplified formula has been developed for use in the correction
of ^{14}C measurements when the groundwater has been affected by incongruent
dissolution. Its application has been compared with previously published
corrected data.

2.   Greater attention has to be given to the nature and origin of aquifers
and their subsequent geological history in so far as this relates to
deposition of carbonates and geochemical corrections of ^{14}C dating.
Further analyses of the $\delta^{13}C$ of the rock will throw much light on the
origin of the carbonate in aquifers and may provide evidence for some
re-interpretations of geological histories. In Britain, the continental
Old Red Sandstone contains freshwater carbonates and the present work
indicates that this may also be true of part of the Lower Greensand. It
could also apply to the Permo-Triassic sandstones, one of the most
important aquifers in Britain.

3.   An approximate relationship between the age of a groundwater and its
$^{18}O/^{16}O$ and D/H contents has been examined for groundwaters in Britain.
The preliminary indication is that a relationship exists which may be of
great value in the corroboration of ^{14}C dating measurements. Furthermore
the interpretation of surface temperatures at the time of origin of
groundwaters from their Oxygen-18 and Deuterium contents can provide a
useful insight into past climates especially if applied on a regional
scale.

4.   The effect of mixed waters on the determination of mean age by
radioactive isotope measurements can be large, particularly where the
old water component is of a great age.

5.   Overall it is essential that dating measurements are supported by
hydrogeological information for a correct interpretation to be made. The
effect of mixing can be significantly greater than effects of different
geochemical models and care should be taken that the use of highly
sophisticated mathematical procedures does not in itself imply accuracy
of interpretation. The dangers of mixing and contamination will be
greatly magnified when the dating of very old waters becomes possible
by methods now being developed for ^{14}C enrichment and the use of high
energy accelerators for ^{14}C measurements.

6.   The greater attention now being given to the prevention of groundwater pollution increases the importance of being able to obtain a
reliable mean age of a groundwater especially in fissured rocks. An
apparent age of several thousand years for an outcrop water or near-
outcrop water does not necessarily indicate that it contains no recent
components. Similarly a mixed water cannot be reliably dated because
varying proportions of different components could given the same
apparent age.

ACKNOWLEDGEMENTS

    This paper is published with the permission of the Directors of the
Central Water Planning Unit and the United Kingdom Atomic Energy
Research Establishment, Harwell. The authors would like to thank
Dr. T.M.L. Wigley for providing a pre-publication copy of his paper [7]
and Mr. D.B. Smith for helpful discussions.

Appendix I

DERIVATION OF AGE FROM CARBON ISOTOPE MEASUREMENTS

    Dissolution of carbonates by dissolved carbon dioxide produced in
the   organic   soil layer, and hence of biogenic origin, is represented
primarily by the equation

$$(CaCO_3)_{rock} + (H_2CO_3)_{biogenic} \rightleftharpoons Ca(HCO_3)_2 \text{ solution}$$

If in a given situation the fraction of carbon originating from biogenic
carbon is f, the mean value for the ^{13}C isotopes incorporated in solution
($\delta^{13}C_x$) is given by the balance equation

$$\delta^{13}C_x = f \, \delta^{13}C_b + (1 - f) \, \delta^{13}C_r$$

and hence

$$f = \frac{\delta^{13}C_x - \delta^{13}C_r}{\delta^{13}C_b - \delta^{13}C_r} \qquad \qquad \dots (1)$$

    The subscripts b, r and x refer to the biogenic, rock and solution
components respectively and $\delta^{13}C$ has the usual meaning, viz

$$\delta^{13}C = \left[\frac{(^{13}C/^{12}C)}{(^{13}C/^{12}C)_{standard}} - 1 \right] \times 1000$$

    In the initial dissolution stage it is generally assumed that the
carbonate dissolves from the parent rock and no precipitation occurs.
This is termed congruent dissolution. If it is the only process which
need be considered, the age of the sample is given by:

$$t = 8267 \ln \frac{A_o}{A_t} f$$

where A_o is the original biogenic ^{14}C level
and A_t is the measured level in the solution t years later.

Taking the modern level as 100% and expressing the measured level ^{14}C in terms of percent modern,

$$t = 8267 \ln \left(\frac{100}{^{14}C} \right) \left[\frac{\delta^{13}C_x - \delta^{13}C_r}{\delta^{13}C_b - \delta^{13}C_r} \right] \qquad \ldots(2)$$

Wigley [6] has shown that where repeated precipitation and dissolution occurs such simple representation of balance is inapplicable. Correction must be made for the combined effects of isotopic fractionation occurring in the precipitating step, $Ca(HCO_3)_2 \rightarrow H_2CO_3 + CaCO_3$, and dissolution into the modified solution components.

This process is known as incongruent dissolution. A balance equation for any elementary step during incongruent dissolution can be evolved by consideration of the change which occurs in the isotope ratios in solution, eg $d(^{13}C/^{12}C)$ and the separate amounts of ^{12}C in

(a) the element of solution (S)
(b) the quantity precipitating out of the element (P)
and (c) the quantity dissolving into the element (R)

This elemental relationship is represented by the model below:

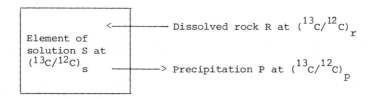

Thus, for a small change $(d(^{13}C/^{12}C)_s)$ of the $^{13}C/^{12}C$ composition of the solution,

$$S \, d(^{13}C/^{12}C)_s = R \, (^{13}C/^{12}C)_r - P \, (^{13}C/^{12}C)_p$$

Subscripts s, r and p refer to solution, dissolving rock and precipitating carbonates respectively.

If the quantity being precipitated P is exactly balanced by the quantity being dissolved R, ie R = P, the total ^{12}C content is unaltered by the step.

Within this first condition,

$$d\left(^{13}c/^{12}c\right)_s = \frac{P}{S}\left[\left(^{13}c/^{12}c\right)_r - \left(^{13}c/^{12}c\right)_p\right] \qquad \ldots (3)$$

$$= \frac{P}{1000S}\left[\delta^{13}c_r - \delta^{13}c_p\right]\left[^{13}c/^{12}c\right]_{std}$$

where subscript std refers to an agreed standard.

It is seen that no further change in the isotopic ratio in solution $(^{13}c/^{12}c)$ can proceed once the composition of the rock precipitate becomes the same as the composition of the rock, ie,

$$\delta^{13}c_r = \delta^{13}c_p \qquad \ldots (4)$$

In the usual isotopic terminology we may derive the step fractionation factor as

$$\epsilon_n = \left[\frac{\left(^nc/^{12}c\right)_p}{\left(^nc/^{12}c\right)_s} - 1\right] \times 1000 \ (\text{‰})$$

(subscript n being the mass number of the isotope in question).

This second limiting condition, Equation 4 above, is thus

$$\delta^{13}c_s = \delta^{13}c_r - \epsilon_{13} \qquad \ldots (5)$$

Continuing from Equation 3,

$$d\left(^{13}c/^{12}c\right)_s = \frac{P}{S}\left[\left(^{13}c/^{12}c\right)_r - (1 + \frac{\epsilon_{13}}{1000})\left(^{13}c/^{12}c\right)_s\right] \qquad \ldots (6)$$

Dissolution of ^{12}c and ^{13}c occurs because of its presence in the rock. This is not true of ^{14}c, except for the extremely small amount in the precipitate, but it is assumed that this does not redissolve. The same relationship therefore cannot be used to describe both ^{13}c and ^{14}c isotope behaviour.

The relationship of ^{14}c is:

$$d\left(^{14}c/^{12}c\right)_s = -\frac{P}{S}(1 + \frac{\epsilon_{14}}{1000})\left(^{14}c/^{12}c\right)_s \qquad \ldots (7)$$

Combining Equations 6 and 7 and integrating between the solution states corresponding to the start of incongruent dissolution (i) and the sample condition (s).

$$-\int_{i}^{s} \frac{d\left[\dfrac{^{14}C}{^{12}C}\right]_{s}}{\left[1 + \dfrac{\varepsilon_{14}}{1000}\right]\left[\dfrac{^{14}C}{^{12}C}\right]_{s}} = \int_{i}^{s} \frac{d\left[\dfrac{^{13}C}{^{12}C}\right]_{s}}{\left[\dfrac{^{13}C}{^{12}C}\right]_{r} - \left[1 + \dfrac{\varepsilon_{13}}{1000}\right]\left[\dfrac{^{13}C}{^{12}C}\right]_{s}}$$

$$\frac{1}{\left[1 + \dfrac{\varepsilon_{14}}{1000}\right]} \ln \left\{ \frac{\left[\dfrac{^{14}C}{^{12}C}\right]_{s}}{\left[\dfrac{^{14}C}{^{12}C}\right]_{i}} \right\} = \frac{1}{\left[1 + \dfrac{\varepsilon_{13}}{1000}\right]} \ln \left\{ \frac{\left[\dfrac{^{13}C}{^{12}C}\right]_{r} - \left[1 + \dfrac{\varepsilon_{13}}{1000}\right]\left[\dfrac{^{13}C}{^{12}C}\right]_{s}}{\left[\dfrac{^{13}C}{^{12}C}\right]_{r} - \left[1 + \dfrac{\varepsilon_{13}}{1000}\right]\left[\dfrac{^{13}C}{^{12}C}\right]_{i}} \right\}$$

Hence

$$^{14}C_{o} = {}^{14}C_{s} \left\{ \frac{\left[\dfrac{^{13}C}{^{12}C}\right]_{r} - \left[1 + \dfrac{\varepsilon_{13}}{1000}\right]\left[\dfrac{^{13}C}{^{12}C}\right]_{i}}{\left[\dfrac{^{13}C}{^{12}C}\right]_{r} - \left[1 + \dfrac{\varepsilon_{13}}{1000}\right]\left[\dfrac{^{13}C}{^{12}C}\right]_{s}} \right\}^{\dfrac{\left[1 + \dfrac{\varepsilon_{14}}{1000}\right]}{\left[1 + \dfrac{\varepsilon_{13}}{1000}\right]}} \qquad \ldots (8)$$

This relationship when expressed in terms of $\delta^{13}C\text{‰}$ becomes

$$^{14}C_{i} \simeq {}^{14}C_{s} \left\{ \frac{\delta^{13}C_{r} - \varepsilon_{13} - \delta^{13}C_{i}}{\delta^{13}C_{r} - \varepsilon_{13} - \delta^{13}C_{s}} \right\}^{\left[1 + \dfrac{\varepsilon_{13}}{1000}\right]} \qquad \ldots (9)$$

The dating of waters affected by both congruent and incongruent stages of dissolution can therefore be derived by combining Equations 1, 2 and 9. When $^{14}C = {}^{14}C_{i}$, $f = \tfrac{1}{2}$ and $\delta^{13}C_{x}$ has the specific value of $\delta^{13}C_{i}$, the solution value on input to the incongruent stage, then

$$t = 8267 \ln \frac{50}{^{14}C_{s}} \left[\frac{K - \delta^{13}C_{s}}{K - \delta^{13}C_{i}} \right]^{\left[1 + \dfrac{\varepsilon_{13}}{1000}\right]}$$

where $K = \delta^{13}C_{r} - \varepsilon_{13}$

The use of this expression eliminates the need for digital computation [3] and takes account of the differences in the precipitation processes of the ^{13}C and ^{14}C isotopes.

Appendix II

EFFECT OF MIXING ON GROUNDWATER DATING

The dating of groundwater by measurement of radioactive isotope concentration implies that the water is of a single 'age' (eg from a confined and isolated aquifer or one subject to piston-type flow). If the groundwater consists of components of different ages, as for example in a mixed flow system, errors can result in the determination of the mean age by isotope measurements. The relative errors increase with the mean age of the water and may be considerably greater than those due to differences in the geochemical correction procedures adopted. A simple example consisting of a mix of two groundwaters A and B of different ages t_A and t_B is given to illustrate the possible significance of error from this source. If fraction x of B is present with (1-x) of A, the mean concentration of isotope in the mix is

$$C_{mean} = (1-x) \; C_A + xC_B$$

where C_A and C_B are the respective isotope concentrations in A and B, and the actual mean age is

$$t_{actual} = (1-x) \; t_A + xt_B$$

The measured age by a radioactive isotope method would be

$$t = \alpha \ln \frac{100\beta}{C_{mean}}$$

where C_{mean} is the mean concentration of the isotope (% modern) and α and β are constants dependent on the halflife of the isotope and the geochemical changes of the water.

$$t = 8267 \ln \frac{100}{(1-x) \; C_A + xC_B}$$

The effect of mixing water of different ages with water of 100K years is shown in Fig. 1 and the marked effect of the presence of modern water on the measured ages of old waters is seen in Fig. 2. These examples, although relevant to those in the main text also forcibly show the dangers, due to contamination by younger waters, in the interpretation of groundwater age when the age range of the ^{14}C method is extended (eg by enrichment of ^{14}C in the laboratory or by the use of high energy accelerators for ^{14}C measurements). For example, groundwater with a

measured age of say 57K years may be water of 100K years with a mix of
only 0.1% of modern water. The result is little changed even if the
modern component is 5K years old.

The problem of mixing applies in particular to all fissured aquifers.
For example, the Chalk and the Lincolnshire Limestone contain a
significant post-1953 component when the apparent age is less than 5K
years implying older but still relatively modern waters. This does not
occur in those sandstones where flow is primarily intergranular as in
the Lower Greensand. Therefore fissure aquifers are more likely to
produce underestimates of the mean age of the mixed waters than such
sandstones.

REFERENCES

[1] SMITH, D.B., OTLET, R.L., DOWNING, R.A., MONKHOUSE, R.A., PEARSON, F.J.
    Stable carbon and oxygen isotope ratios of groundwaters from the
    Chalk and Lincolnshire Limestone. Nature, 257 (1975) 783–784.

[2] SMITH, D.B., DOWNING, R.A., MONKHOUSE, R.A., OTLET, R.L., PEARSON, F.J.
    The age of groundwater in the Chalk of the London Basin. Wat. Rcs.
    Res., 12 (1976) 392–404.

[3] DOWNING, R.A., SMITH, D.B., PEARSON, F.J., MONKHOUSE, R.A., OTLET, R.L.
    The age of groundwater in the Lincolnshire Limestone, England and its
    relevance to the flow mechanism. J. Hydrol., 33 (1977) 201–216.

[4] PEARSON, F.J., HANSHAW, B.B. Sources of dissolved carbonate species
    in groundwater and their effects on Carbon-14 dating. Proc. Symp.
    Isotope Hydrology, IAEA, Vienna (1970) 271–286.

[5] INGERSON, E., PEARSON, F.J. Estimation of age and rate of movement
    of groundwater by the ^{14}C method. Recent Researches in the Fields of
    Hydrosphere, Atmospheric and Nuclear Chemistry, Maruzen, Tokyo (1964)
    263–283.

[6] WIGELY, T.M.L. Effect of mineral precipitation on isotopic
    composition and ^{14}C dating of groundwater. Nature, 263 (1976) 219–
    221.

[7] WIGLEY, T.M.L., PLUMMER, L.N., PEARSON, F.J. Mass transfer and
    carbon isotope evolution in natural water systems. Geochim.
    Cosmochim Acta (in press).

[8] ATKINSON, T.C. Diffuse flow and conduit flow in limestone terrain
    in the Mendip Hills, Somerset (Great Britain). J. Hydrol., 35
    (1977) 93–110.

[9] EDMUNDS, W.M., TAYLOR, B.J., DOWNING, R.A. Mineral and thermal
    waters of the United Kingdom. 23rd Int. Geol. Cong., 18 (1969) 139–
    158.

[10] ANDREWS, J.N., WOOD, D.F. Mechanism of radon release in rock
     matrices and entry into groundwaters. Trans. Instn. Min. Metall.
     B., 81 (1972) 198–209.

[11] KELLAWAY, G.A. The Geological Survey Ashton Park Borehole and its
     bearing on the geology of the Bristol district. Bull. Geol. Surv.
     G.B., 27 (1967) 49–153.

[12] DEINES, P., LANGMUIR, D., HARMON, R.S. Stable carbon isotope ratios
     and the existence of a gas phase in the evolution of carbonate
     groundwaters. Geochim. Cosmochim Acta, 38 (1974) 1147–1164.

[13] MATHER, J.D., GRAY, D.A., ALLEN, R.A., SMITH, D.B. Groundwater
     recharge in the Lower Greensand of the London Basin - results of
     tritium and Carbon-14 determinations. Q. Jl. Engng. Geol., 6
     (1973) 141–152.

[14] EDMONDS, E.A., DINHAM, C.H.   Geology of the country around Huntingdon
     and Biggleswade.   Mem. Geol. Surv. G.B., HMSO, Lond., (1965) 90 pp.
[15] MONKHOUSE, R.A.   An assessment of the groundwater resources of the
     Lower Greensand in the Cambridge-Bedford region.   Water Resources
     Board (1974) 37 pp.
[16] DANSGAARD, W.   Stable isotopes in precipitation.   Tellus, 16 (1964)
     436-468.
[17] CRAIG, H.   Isotopic standards for carbon and oxygen correction
     factors for mass-spectrometric analysis of carbon dioxide.   Geochim.
     Cosmochim Acta, 12 (1957) 133-149.
[18] PAYNE, B.R., HALEVY, E.   Introduction.   Guidebook on nuclear
     techniques in hydrology.   Tech. Rep. Series 91, IAEA (1968) 1-20.
[19] DOWNING, R.A., PEARSON, F.J., SMITH, D.B.   The flow mechanism in
     the Chalk based on radioisotope analyses of groundwater in the
     London Basin.   Submitted to J. Hydrol. (1978).
[20] LAMB, H.H.   Climate: Present, past and future.   Climatic History and
     the Future, Methuen & Co Ltd, London, 2 (1977) 835 pp.
[21] LAMB, H.H., LEWIS, R.P.W., WOODROFFE, A.   Atmospheric circulation and
     mean climatic variables between 8000 and 0 BC: Meteorological
     evidence.   World Climate from 8000 to 0 BC, Roy. Met. Soc., Lond.,
     (1966) 174-215.
[22] COOPE, G.R.   Fossil coleopteran assemblages as sensitive indicators
     of climatic changes during the Devensian (last) cold state.   Phil.
     Trans. R. Soc., Lond., B., 280 (1977) 313-340.
[23] WEST, R.G.   Early and middle Devensian flor and vegetation.   Phil.
     Trans. R. Soc., Lond., B., 280 (1977) 229-242.
[24] HARLAN, R.L.   Dynamics of water movement in permafrost: A Review.
     Permafrost Hydrology, Can. Nat. Comm. Int. Hyd. Decade (1974) 69-77.
[25] VAN EVERDINGEN, R.O.   Groundwater in permafrost regions of Canada.
     Permafrost Hydrology, Can. Nat. Comm. Int. Hyd. Decade (1974) 83-93.
[26] WATSON, E.   The periglacial environment of Great Britain during the
     Devensian.   Phil. Trans. R. Soc., Lond., B., 280 (1977) 183-198.
[27] SIMPSON, I.M., WEST, R.G.   On the stratigraphy and palaeobotany of
     a late Pleistocene organic deposit at Chelford, Cheshire.   New
     Phytol., 57 (1958) 239-250.

DISCUSSION

B.R. PAYNE: In your paper you compare the relationship of deuterium and ^{18}O in groundwater samples from a number of different sites in the United Kingdom with the relationship for precipitation at a number of selected stations described as "Maritime Europe" in Table IV and Section 4.1.1 of the paper. May I ask whether this latter relationship was defined on the basis of weighted or unweighted values?

R.A. DOWNING: It was based on unweighted values, but in fact no significant differences are observed when the data are weighted for rainfall. Our objective was to derive an acceptable $\delta^{18}O/\delta D$ relationship for rainfall over the United Kingdom. Our equation gives the relationship for $\delta^{18}O$ and δD for all samples analysed from the six stations we included in "Maritime Europe", the data being considered together to represent a single weather zone.

B.R. PAYNE: As you say in Section 4.1.1, the similarity in the relationships may be fortuitous. I should like to draw your attention to the fact that

92% of the variations in the δD and $\delta^{18}O$ of the samples described in your
paper can be explained by a relationship which has a slope not very different
from eight.

R.A. DOWNING: Although one cannot rule out the possibility that the
agreement is fortuitous, it is nevertheless striking and we believe that it shows
that the rainfall was not greatly affected by evaporation before recharge. The
equation for the groundwater data is found by the method of least squares
which all the available data fit.

P. FRITZ: Could you please tell us what the slope of the meteoric waterline
is for individual stations you sampled? In other words, does the slope you
give in Section 4.1.1 of your text represent a family of lines with higher slopes
but changing deuterium excess?

R.A. DOWNING: The $\delta^{18}O/\delta D$ relationships for the various stations we
examined are given in Table IV. It is interesting to note that for all the slopes
values of less than eight were found, and values less than ten for all the intercepts.

P. FRITZ: I would also like to comment on the question of permafrost
and recharge through it. First, very little is known about the distribution of
permafrost and its existence during the last glaciation. Secondly, permafrost
does not necessarily entail the existence of a continuous ice seal but is defined
as a temperature regime. Therefore, some groundwater recharge may well occur
and studies are being done to evaluate this possibility. As a matter of fact,
active groundwater systems with local isotopic compositions do occur in the
Arctic. In addition, we have found tritium at depths exceeding 10 m in areas
where soil-ice was present — which again shows the permeability of frozen
soil. (See Michel, F. and Fritz, P., Proc. Third Int. Conf. Permafrost, Edmonton,
Canada, 1978.)

R.A. DOWNING: With regard to permafrost, we have to interpret conditions
in the United Kingdom during the Pleistocene from modern analogues in the
Arctic. Data in the literature (for example Refs [24] and [25]) indicate that,
in general, infiltration is very limited where permafrost exists. If "infiltration
windows" did exist in the UK would you not expect very depleted ^{18}O values
in the groundwater?

P. FRITZ: No, low ^{18}O contents are only found in deep soil-ice, and all
active groundwater systems have ^{18}O contents similar or identical to those
found in precipitation. Differences exist only where regional flow systems
discharge, as for example in the thermal springs which are found in many parts
of the Canadian Arctic. Typical modern $\delta^{18}O$ groundwater values lie between
-21 and $-23\%_{oo}$, whereas the deep soil-ice values may be as low as $-35\%_{oo}$
and are typically close to $-32\%_{oo}$.

R.A. DOWNING: In fact the $\delta^{18}O$ values in the groundwater imply an
origin from rainfall under climatic conditions not so very different from those
of today.

M.A. GEYH: Concerning the question whether permafrost interrupted groundwater recharge during the last glacial period or not, I would like to draw attention to the increasing number of observations indicating that the frequency distributions of ^{14}C groundwater ages of confined aquifers in Central Europe show a definite gap between about 20 000 and 12 000 years B.P. ago. This can only be explained by a drastic reduction in groundwater regeneration during this period. The causes might be interruption of rainwater seepage by permafrost and/or reduction in rainfall rates. The same conclusion can be drawn from the frequency distribution of approximately 300 ^{14}C dates obtained from stalagmites taken from various caves in Central Europe: these show the same gap.

W.G. MOOK: I wonder what data you used to derive the annual temperature curve in your Fig.7. Great caution is required in this respect, because many studies on climatic variations employ very different temperature schemes from those which we are looking for in isotope studies (for instance, pollen analyses

generally show minimum average July temperatures rather than mean annual temperatures). It is also important to be careful when applying known phenomenological $\delta^{18}O$ versus temperature relations. To use the relation formulated by Dansgaard (see Section 4.1.2 of the paper) is a serious over-simplification. In my opinion, use of the seasonal temperature dependence of $\delta^{18}O$ might also be wrong. In some stations no such dependence exists at all.

R.A. DOWNING: The mean annual and winter temperature curves were derived from an examination of the literature (particularly Refs [20–23]). The curves are intended merely to indicate general changes in temperature. The relationship between $\delta^{18}O$ and temperature formulated by Dansgaard does not appear to give a good indication of the temperature at the time of infiltration of the groundwater. The equation we have derived from published data gives a more acceptable interpretation.

R.G. THOMAS: A complication which arises in temperature analysis of waters 10 000 to 12 000 years old is the fact that large amounts of meltwater were available for recharge in some places for short periods of perhaps only hundreds of years, and the meltwater may not have been representative of rainwater of the time. When analysing this factor it would be necessary to recognize the possibility that a specific sample locality might have had such a history.

R.A. DOWNING: I agree. Melting ice formed under different climatic conditions must have recharged groundwaters at the end of the last glacial phase. The melting would have occurred relatively slowly and the effects are likely to have been concentrated along drainage channels. In the fifth paragraph of Section 4.1.4 we make the point that the annual snow melt in the spring could have influenced the [18]O content of the groundwater.

Session VIII

GROUNDWATER RECHARGE STUDIES –
UNSATURATED ZONE

Chairman

D.B. SMITH
United Kingdom

CHANGES IN FORMATION GAS COMPOSITION AND ISOTOPE CONTENT AS INDICATORS OF UNSATURATED-ZONE CHEMICAL REACTIONS RELATED TO RECHARGE EVENTS

C.T. RIGHTMIRE
US Geological Survey,
Idaho Falls, Idaho,
United States of America

Abstract

CHANGES IN FORMATION GAS COMPOSITION AND ISOTOPE CONTENT AS INDICATORS OF UNSATURATED-ZONE CHEMICAL REACTIONS RELATED TO RECHARGE EVENTS.

The constituent composition and carbon and oxygen isotopes for CO_2 in unsaturated-zone gases have been utilized to provide an independent means of determining what chemical reactions occurred during an extended artificial recharge experiment on the High Plains of Texas. Gas samples were collected from a sample point at a depth of 16 metres beneath a waterspreading basin to see what effect recharging water had on the unsaturated-zone atmosphere. Two different systems, one open and one closed, were sampled during this study. The ^{13}C, ^{18}O, P_{CO_2}, and N/Ar data for gas collected during a 1975 experiment show the influence of unsaturated-zone gases being displaced by recharging waters, the trend towards equilibration with dissolved atmospheric gases as the sampling point is submerged by rising waters, the onset of a reducing environment and, following the termination of recharge, the slow recovery of the gas composition to pre-recharge conditions. The ^{13}C content of the CO_2 in the gases varied from -18.6% for the pre-recharge formation gas, to -10.1% for gases approaching equilibrium with dissolved atmospheric gases, to -38.1% for a CO_2 produced from methane generated in a reducing environment. The log P_{CO_2} for the above samples varied from -2.10 to -3.02 to -4.18 atmospheres, respectively. Recovery to pre-recharge unsaturated-zone atmospheric conditions took many months following the termination of recharge.

INTRODUCTION

To be able to predict the chemistry of a ground water, it is essential to understand the chemical reactions that influence the water. The initial reactions which strongly influence ground-water chemistry occur in the unsaturated zone. Information about the reactions affecting waters percolating through the unsaturated zone will aid in our ability to predict changes in ground-water chemistry.

This study utilizes the constituent composition of soil gas and the stable isotope content of the CO_2 in that gas, in conjunction with hydrology, mineralogy, and water chemistry, to determine what chemical

reactions occur during and after artificial recharge of large quantities
of water, over relatively long periods of time, through a thick unsaturated
zone. The gas and stable isotope relationships during one recharge experiment are compared with water chemistry from an earlier experiment under
nearly identical environment conditions to determine whether the same
chemical reactions could have been predicted.

PREVIOUS WORK

Several artificial recharge experiments have been conducted on the
High Plains of West Texas to study the potential for recharging the
Ogallala aquifer by spreading basins. This work was conducted by the
U. S. Geological Survey's High Plains Artificial Recharge Research
Project, located at Lubbock, Texas. The initial experiment was conducted
in 1972 with the geochemical interpretation being presented by Wood and
Signor [1], Wood and Bassett [2], and Wood and Brown [3]. During this
experiment, water samples for chemical analyses were collected from
porous cups in the unsaturated zone at depths of 0.6, 1.9, 7.9, 16.2,
22.1, and 32.8 meters (m) below the bottom and about 3-m from the edge of
the spreading basins.

Physical-chemical reactions that occur in the unsaturated zone during
recharge may lead to changes in pore geometry and hence to alteration of
the infiltration rate through the unsaturated zone. Changes in geometry
leading to a decrease in effective porosity can result from expansion of
clay lattice, hydration of certain minerals, clay dispersion, precipitation of mineral species, and trapped gases [1]. These changes, therefore,
decrease the rate of infiltration of recharging water. Porosity can be
increased by solution of sediment-rock framework, gases, or organic
material.

The 1972 experiment was conducted in a spreading basin from which the
vegetation and approximately 0.5-m of surface sediment had been removed.
Experimental data led to the conclusion that the chemistry of the
recharging waters was most strongly influenced by 1) cation exchange,
2) anion exchange, 3) mineral solution, 4) adsorption, 5) desorption, and
6) redox reactions. To these chemical reactions should be added the
influence of organic matter present or introduced into the system.

The second recharge experiment was conducted in 1975 in a vegetated
spreading basin directly north of the one used in 1972. This basin was
slightly enlarged (by approximately 3 percent) to include a series of gas-
sampling screens installed at known depths. It was felt that information
on the changes in formation gas composition and isotopic content would aid
in our understanding of unsaturated-zone chemical reactions during
recharge.

RECHARGE FACILITY

The recharge facility utilized for the 1975 experiment, located near
the Lubbock Regional Airport (Fig. 1), is a basin of approximately 0.41
hectare formed by building a 1-m-high berm around the area. The water
inlet was controlled by a float-activated motorized valve which maintained
the water level at 0.6±0.05-m. The water level was continuously monitored
and the inflow was recorded by a totalizing meter that was read daily [1].

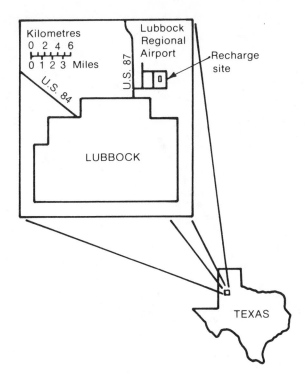

FIG.1. Map showing location of the recharge site.

The geologic section consists of Amarillo series soils developed on the top of approximately 7-m of Holocene eolian sediments which overlie the Ogallala Formation of late Miocene age. These soils are developed to depths of approximately 2.5-m and contain measurable organic carbon throughout. The organic-carbon content of this type of soil ranges from a low of 0.06 percent at 2.0 to 2.75-m to a high of 0.70 percent at 0.6 to 1.27-cm [4].

The Ogallala Formation is dominantly a reddish-brown fine sand with abundant iron-oxyhydroxide-coated grains. Secondary calcium carbonate accumulations (caliche) are found throughout the unsaturated section, both as nodular layers and as interparticle cement. Amorphous silica is commonly present in zones rich in secondary calcium carbonate. Clay minerals in the system are dominantly mixed-layer illite and montmorillonite, with kaolinite and well-formed illite also present. Clay-size carbonate particles dominate the finer-sized fractions in many parts of the section.

The vegetation and approximately 0.5-m of surface sediment were removed before the 1972 experiment but were left undisturbed during the 1975 recharge experiment. It is felt that the P_{CO_2} of the soil atmosphere

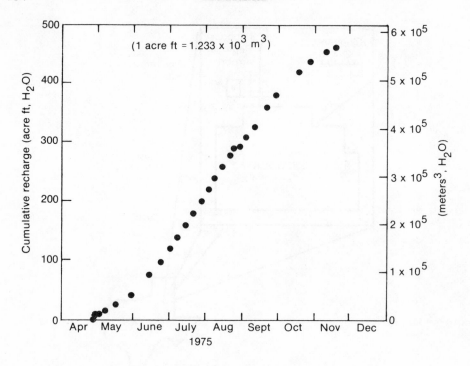

FIG.2. *Plot of cumulative recharge for the 1975 recharge experiment.*

and the organic carbon in the deeper surface sediment in these semi-arid, moderately vegetated areas were the same in both basins even though the vegetation had been removed from the basin for the 1972 experiment.

HYDROLOGY OF THE RECHARGE SITE

A brief discussion of a generalized sequence of hydrologic events in this recharge experiment will aid in the understanding of some of the complexities of sampling in the unsaturated zone and of interpreting the data derived from these samples.

Figure 2 shows a plot of cumulative input to the recharge basin during the 1975 experiment. This plot shows that the rate of infiltration increases following the initial input of water, attains a fairly constant rate, and then decreases toward the end of the experiment. The following is a brief chronology of the development of sub-basin hydrology as it relates to the gas sampling:

Water, first introduced into the north spreading basin on April 28, 1975 (Fig. 3a), entered a sedimentary medium containing about 17 to 18 percent water [5]. According to E. P. Weeks [6], the 16.2-m sampling screen may or may not have been submerged by water from the developing

a)

Late April pre-recharge and early-recharge conditions.

b)

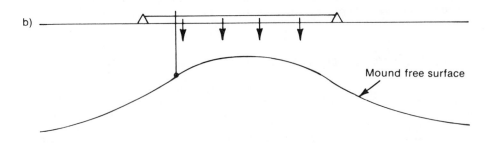

May 14 — Water beginning to mound at 16.2 meter level, aerobic conditions.

c)

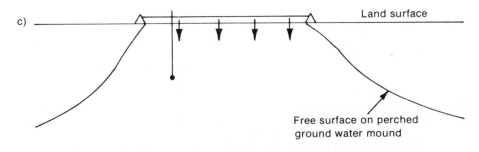

June 1 — Ground water mound above 16.2 meter level, aerobic conditions.

FIG.3. Development of the groundwater mound.

ground-water mound on May 14 (Fig. 3b); channel flow was evidently the
principal mechanism of infiltration. At the time of sampling on June 5,
1975, the 16.2-m screen was submerged by the ground-water mound. Conditions were still aerobic at that time (Fig. 3c). By September 4, 1975
the sampling screen was still submerged, but anaerobic conditions had been
established in the shallow sub-basin sediment (Fig. 4a). A very slight
decrease in infiltration rate may be observed on Fig. 2. During the June
and September sampling, the screen and a portion of the sampling pipe
contained water.

With the reduction of infiltration rate following the onset of reducing
conditions in the shallow sub-basin sediments, a slow dewatering of the
unsaturated zone occurred as the mound declined. Unsaturated flow replaced
saturated flow as the means of water movement. Formation gases migrated
laterally to fill the voids as the mound declined (Fig. 4b). Recharge was
terminated on November 17, 1975, and as the basin bottom dried, atmospheric
gases migrated vertically downward as the ground-water mound declined
(Fig. 4c). Gas sampling continued until January 10, 1977, when the gas
composition was observed to have stabilized.

SAMPLING TECHNIQUES AND ANALYTICAL PROCEDURES

The technique for sampling unsaturated zone waters involves the
utilization of buried porous ceramic cups and is described by Wood [7]
and Wood and Signor [1]. Samples were collected daily during the initial
phases of the 1972 recharge experiment, then weekly as the recharge
progressed. Analyses of pH, alkalinity, and specific conductance were
carried out in the field. Because the 1975 recharge experiment was conducted in an undisturbed area directly adjacent to the 1972 site and
because, as explained above, it was felt that the 1972 chemical samples
could be utilized to represent the chemistry of the 1975 experiment, no
water samples were collected for chemical analysis [8].

Since the water sampling technique requires the application of a
vacuum to the porous cup in order to overcome soil-moisture tension, the
possibility arises that the carbonate chemistry of the water was disturbed
and that the supersaturation with respect to calcite observed in almost
all the samples is apparent and not real with respect to carbonate
minerals.

The gas samples were collected by a technique described by Rightmire and
Hanshaw [9] in which metering valves control the flow of gas into a large
evacuated sample cylinder. By controlling the flow rate it is possible
to minimize the disruption of the gas-liquid equilibrium and therefore to
obtain a sample representative of the unsaturated zone. The samples were
collected at flow rates of approximately $2\,cm^3/min$ over a period of 24 hours
or longer. Sampling over this long period minimizes the influence of
diurnal changes in gas composition or isotope content.

The installation from which the gas samples were collected was
installed to permit the determination of vertical permeability by
fluctuations in formation gas pressures relative to atmospheric pressure
[10]. A hole was augered to the water table and isolated screens were set
next to more permeable units in the unsaturated zone, backfilled, and

a)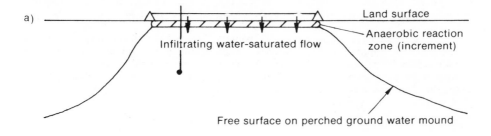

Land surface

Anaerobic reaction zone (increment)

Infiltrating water-saturated flow

Free surface on perched ground water mound

September 1 - Mound height about maximum, anaerobic conditions incipient

b)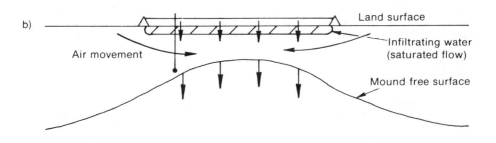

Land surface

Air movement

Infiltrating water (saturated flow)

Mound free surface

November 1 - Mound declining, substantial clogging in anaerobic zone, air moving from beyond perimeter to beneath basin

c)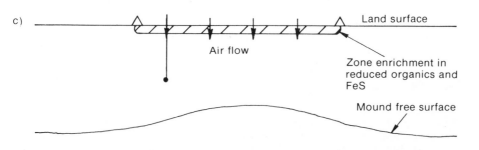

Land surface

Air flow

Zone enrichment in reduced organics and FeS

Mound free surface

December 1 - Mound continues to decline, recharge ended, airflow vertically downward

FIG.4.   Dewatering of the sub-basin unsaturated zone.

cemented to eliminate any vertical interconnection. Two screens at
(16 and 30-m) were initially selected for gas sampling, but evidence of a
valve leakage necessitated the abandoning of the 30-m screen.

The gas samples were returned to the laboratory where an aliquot of
whole gas was separated for chromatographic analysis and the CO_2 component
was separated by passing the air slowly through a succession of
liquid nitrogen traps. The ^{13}C and ^{18}O of the CO_2 were then analysed
using an isotope ratio mass spectrometer. These values are reported in
the standard δ notation, in parts per mil ($^O/oo$):

$$(^O/oo) = \left[\frac{R_{spl} - R_{STND}}{R_{STND}} \right] \times 100$$

where $R = {}^{13}C/{}^{12}C$ or ${}^{18}O/{}^{16}O$. The carbon isotope values are reported
relative to the PDB standard [11] and the oxygen values relative to
Standard Mean Ocean Water (SMOW) standard [12].

RELATIONSHIP OF WATER CHEMISTRY TO GAS CHEMISTRY

In view of the facts that water samples for chemical analysis were
collected only during the 1972 recharge experiment and that gas samples
were collected only during the 1975 experiment, only a cursory discussion
of the relationship between the two is justified. It is felt that, since
the two basins are within 50 feet of each other and exhibit identical
hydrologic, mineralogic, and geochemical features, it should be feasible
to compare the gas chemistry from the 1975 experiment to the water
chemistry from the 1972 experiment based on the cumulative water inflow
for the two experiments. A brief discussion of the water-chemistry
changes observed in the shallow sub-basin provides a basis for validating
interpretations made on the data from the analysis of the gases,
especially as related to the presence of reducing conditions.

WATER CHEMISTRY

The recharge water for both experiments was obtained from Lake
Meredith, an impoundment of the Canadian River north of Amarillo, Texas,
and piped more than 100 miles to the recharge site. The water is a
sodium chloride-sodium sulfate water with a dissolved solid concentration
of 1,100-1,200 mg/l. The concentration of suspended solids was usually
in the range of 3 to 6 mg/l, and only rarely exceeded 10 mg/l. The
average chemistry of the recharge water is given in Table I. The
calculated log P_{CO_2} for this water, -3.29, is slightly higher than
atmospheric.

This discussion will concentrate on the chemistry of water samples
collected from the 0.6-m porous cup, those showing the greatest chemical
fluctuations. Samples were also collected from a porous cup at 16.2-m,
but the interval was much greater and the sampling was discontinued prior
to the appearance of evidence of reducing conditions.

TABLE I. CHEMISTRY OF AVERAGE RECHARGE WATER
(mg/l, except specific conductance and pH)

SiO	0.8	HCO	203	pH	8.50
Ca	62	CO	2.8	Spec. Cond.	1656
Mg	26	SO	277	(µmho)	
Na	264	Cl	284		
K	7.4	F	0.7		

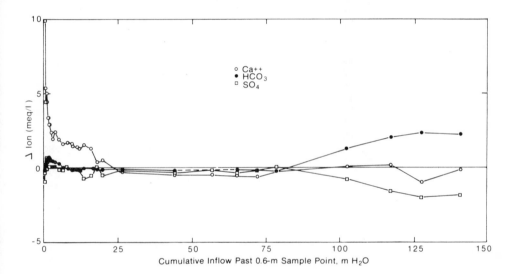

FIG.5. *Fluctuation of calcium, bicarbonate, and sulphate with cumulative inflow past the 0.6-m sampling point.*

Figure 5 shows differences (Δ ion) between average recharge water and the water collected from the 0.6-m sampling point for calcium, bicarbonate, and sulfate. Presenting the data in this fashion shows changes from the input water. This plot shows significant variations during the 1972 recharge experiment which are discussed at length in other studies [1, 2, 3]. The plot also shows the onset of sulfate reduction at about the time at which the cumulative inflow had reached 75-m of water. The pH and the calcium, bicarbonate, and sulfate concentrations for samples from the 0.6-m porous cup are given in Table II.

RIGHTMIRE

TABLE II. ION CONCENTRATIONS FOR CALCIUM, BICARBONATE, AND SULFATE (IN mg/l) AND pH FOR WATERS COLLECTED FROM 0.6-METRE POROUS CUP DURING 1972 RECHARGE EXPERIMENT

DATE	pH	Ca^{++}	HCO_3^-	$SO_4^=$
Apr 13	7.80	260	189	230
14	8.20	170	209	260
15	7.72	150	220	310
16	7.72	160	230	280
17	7.83	150	238	270
18	7.90	130	239	280
19	7.61	130	227	280
20	7.71	120	239	280
21	7.72	120	238	280
22	7.92	110	232	280
24	7.84	100	230	280
26	7.92	110	224	280
29	7.90	100	217	270
May 2	7.80	94	205	270
5	7.82	96	201	280
9	7.90	94	190	270
10	7.90	91	193	270
13	8.10	90	197	270
15	8.02	88	192	270
20	8.15	94	198	240
24	8.10	88	200	250
28	8.32	69	195	280
June 1	8.35	72	194	250
13	8.50	56	197	270
July 5	8.24	52	193	260
21	8.15	53	----	270
31	8.25	50	194	260
Aug 7	8.10	49	194	270
14	8.25	56	188	280
Sep 11	7.79	64	281	240
Oct 2	7.15	66	329	200
18	7.90	42	344	180
Nov 15	7.90	59	337	190

PRESENTATION OF DATA

Gas samples were collected from the 16-m sampling screen at various times before, during, and after the 1975 artificial recharge test. Sampling was continued after the termination of recharge to determine the length of time required for the formation atmosphere to recover to a stable composition. Samples were analysed for whole-gas-constituent composition by gas chromatography and for carbon and oxygen stable-isotope content of separated CO_2 by mass spectrometry.

The first gas sample for this study was collected on April 16, 1975 (#1), twelve days prior to the introduction of water into the north spreading basin on April 28, 1975. This sample is used to establish pre-recharge formation-gas composition and stable-isotope content. Gas samples also were collected on April 30 (#3), May 14 (#5), June 5 (#7), and September 4, 1975 (#9), as water was being applied to the basin. The sample of November 20, 1975 (#11) was collected three days after the termination of recharge and slightly after the water level of the mound had declined below the 16-m screen.

Samples were collected as dewatering continued and the formation atmosphere recovered toward pre-recharge conditions on December 29, 1975 (#13); February 3 (#15), April 21 (#17), June 1 (#19), July 18 (#21), and October 5, 1976 (#23); and January 10, 1977 (#25). The last two samples are identical in P_{CO_2}, P_{O_2}, and $\delta^{13}C$, suggesting that the system had arrived at a stable state.

The whole-gas composition and the isotope content of the CO_2 from these gases are given in Table III and Table IV respectively. Figures 6 and 7 show the relationships of the ^{13}C content of the CO_2 to the P_{CO_2}, and of the ^{18}O content of the CO_2 to the P_{O_2}, respectively. Two gas samples collected from the 30-m screen prior to its abandonment (collected April 16, 1975 (#2) and April 30, 1975 (#4)), yield $\delta^{13}C_{PDB}$ values of -16.1 and -14.1 $^{o}/oo$ respectively. The $\delta^{13}C_{PDB}$ of the total inorganic carbon in the recharge water is -0.9 $^{o}/oo$. Formation gas samples collected by the author at various depths beneath the recharge basin in 1970 and 1971 yield $\delta^{13}C$ ranging from -14.3 $^{o}/oo$ to -23.5 $^{o}/oo$ with a log P_{CO_2} range of -1.78 to -2.93. The results and their variation with depth are shown in Table V. These data show significant changes in $\delta^{13}C$ with depth, indicating that either enrichment or depletion in ^{13}C content may occur at a fairly constant P_{CO_2}.

DISCUSSION

Three separate gaseous environments were sampled during this study, and significant chemical changes can be expected during the transition from one environment to another. The three principal environments are:

1) The pre-recharge environment, with gases in, or approaching, an equilibrium with the atmosphere above and with the water in the aquifer system below. The mechanisms controlling the development of these equilibrium conditions are gaseous diffusion in the unsaturated zone and exchange across the liquid-gas interface;

2) The recharge environment, with interstitial gases not in equilibrium or at best in metastable equilibrium with water infiltrating through the unsaturated zone. The principal mechanisms controlling the approach to equilibrium in this environment include (a) the flow regime of the infiltrating water (saturated or unsaturated flow), (b) the velocity of the water, (c) the dissolved gas content of the water, and (d) exchange across the liquid-gas interface; and

TABLE III. CONSTITUENT COMPOSITION OF FORMATION GASES FROM
16.2-METRE SAMPLING SCREEN

No. on Figures	Date		$-\log P_{N_2}$	$-\log P_{O_2}$	$-\log P_{CO_2}$	$-\log P_{Ar}$	N/Ar
Standard	Air		0.107	0.68	3.48	2.03	83.75
1	16 Apr.	75	0.088	0.79	2.10	1.94	70.92
	28 Apr.	75			Flow initiated		
3	30 Apr.	75	0.098	0.74	2.07	1.96	72.99
5	14 May	75	0.087	0.79	2.08	1.95	72.99
7	5 June	75	0.106	0.69	3.12	2.02	81.96
9	4 Sept.	75	0.101	0.71	3.02	2.00	81.97
	17 Nov.	75			Flow initiated		
11	20 Nov.	75	0.017	1.93	4.18	1.87	69.66
13	29 Dec.	75	0.017	1.71	4.30	1.86	69.44
15	3 Feb.	76	0.011	3.00	4.40	1.85	68.97
17	21 Apr.	76	0.007	3.70	2.79	1.88	74.63
19	1 June	76	0.013	2.16	2.05	1.89	75.34
21	18 July	76	0.009	2.38	2.49	1.88	74.30
23	5 Oct.	76	0.114	0.66	3.47	2.02	80.54
25	10 Jan.	77	0.115	0.66	3.46	2.01	78.52

3) The post-recharge environment which yields gases again approaching
   equilibrium with the atmosphere above and the water in the aquifer
   system below. The principal mechanisms controlling the approach
   to equilibrium in this environment are the dewatering rate of the
   unsaturated zone, the oxidation of sulfide minerals and organic
   material, and gaseous diffusion and convection in the unsaturated
   zone.

An additional complexity, as mentioned in the section on site
hydrology, is that during at least two sampling intervals, June 5, 1975
(#7) and September 4, 1975 (#9), the screen and pipe were partly filled

TABLE IV. P_{CO_2}, P_{O_2}, AND STABLE ISOTOPE CONTENT OF CO_2 FROM THE 16.2-METRE SAMPLING SCREEN

No. on Figures	Date	$\log P_{O_2}$	$\delta^{18}O_{SMOW}$	$\log P_{CO_2}$	$\delta^{13}C_{PDB}$
1	16 Apr. 75	−0.79	+35.10	−2.10	−18.61
	28 Apr. 75		Flow initiated		
3	30 Apr. 75	−0.74	+32.86	−2.07	−18.09
5	14 May 75	−0.79	+36.77	−2.08	−14.90
7	5 June 75	−0.69	+34.99	−3.12	−12.49
9	4 Sept. 75	−0.71	+43.30	−3.02	−10.13
	17 Nov. 75		Flow terminated		
11	20 Nov. 75	−1.93	+37.58	−4.18	−38.09
13	29 Dec. 75	−1.71	+29.31	−4.30	−27.60
15	3 Feb. 76	<−3.00	+23.99	−4.40	−30.88
17	21 Apr. 76	<−3.70	+37.09	−2.79	−15.47
19	1 June 76	−2.16	+37.21	−2.05	−13.19
21	18 July 76	−2.36	+30.10	−2.49	−12.22
23	5 Oct. 76	−0.664	+27.08	−3.47	−9.86
25	10 Jan. 77	−0.662	+32.69	−3.46	−9.42

Recharge Water $\delta^{13}C_{PDB}$ $(C_{TOT}) = -0.94$ °/oo

$\log P_{CO_2}$ (calc) = −3.29

with water, in effect leading to the sampling of a closed system. It is possible that during one other sampling interval, May 14, 1975 (#5), we were sampling a closed system but one with a larger gas reservoir. This reservoir could have been isolated by the rising ground-water mound below, and the channel flow above and laterally adjacent, to the sampling screen.

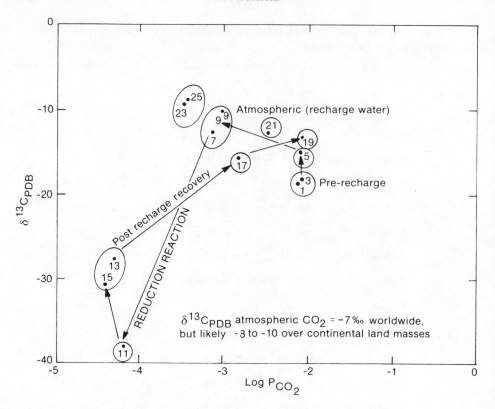

FIG.6.   Relationship of ^{13}C content of CO_2 to P_{CO_2} of formation gas.

Table IV shows significant changes in both ^{13}C content and P_{CO_2} of the gases sampled. The April 16, 1975 sample (#1), collected prior to the initial application of water to the spreading basin on April 28, yields both the P_{CO_2} and the $\delta^{13}C$ that one would expect for an undisturbed semi-arid grassland environment (log P_{CO_2} = -2.10 and $\delta^{13}C$ = -18.6 $^{o}/oo$), and $\delta^{18}O$ (CO_2) = +35.1 $^{o}/oo$). Previous analysis of formation gas collected in 1970 from the recharge site at the 16-m depth yield a log P_{CO_2} of -2.09, and $\delta^{13}C_{PDB}$ of -14.5 $^{o}/oo$, and a $\delta^{18}O_{SMOW}$ (CO_2) of +38.1 $^{o}/oo$.
Other samples collected at various depths over a three-year period (1970-1973) show variations in ^{13}C content at the 26-m depth of up to 8 per mil (Table V). The P_{CO_2} remains fairly constant for two years; data are unavailable for the third year. It is not known how the ^{13}C content of the gases from the 16-m depth varies, but it is not unreasonable to assume that the variation would be similar to that at 26-m.

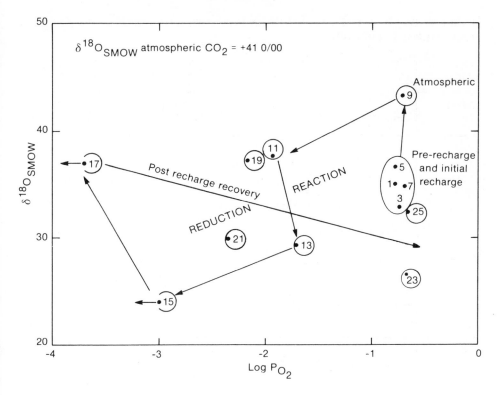

FIG.7.   Relationship of ^{18}O content of CO_2 to P_{O_2} of formation gas.

Gas sample #3 was collected after the initiation of recharge but before a significant quantity of water reached the 16-m gas-sampling depth; it yielded $\delta^{13}C$, $\delta^{18}O$ (CO_2), and P_{CO_2} values approximately those of the pre-recharge gases.

The May 14 gas sample (#5) was collected at the time the surface of the ground-water mound approached the sample screen from below. This sample shows an enrichment in ^{13}C while maintaining the same P_{CO_2} (Fig. 6, Table IV). The $\delta^{13}C$ values reported for the two gas samples collected from the 30-m screen are both enriched in ^{13}C relative to the corresponding gases from the 16-m screen yet have approximately the same P_{CO_2}. The enrichment in ^{13}C while maintaining the same P_{CO_2} is likely the result of upward displacement of the deeper gas by the rising ground-water mound. The whole-gas composition (Table III) and ^{18}O content (Table IV and Fig. 7) show only minimal change from pre-recharge values.

TABLE V. P_{CO_2} AND $\delta^{13}C$ VARIATIONS IN FORMATION GAS WITH DEPTH AND TIME, AIRPORT SPREADING SITE

Depth (m)	August 24, 1970			June 4-5, 1971		
	$\log P_{CO_2}$	$\delta^{13}C$	$\delta^{18}O$ (CO_2)	$\log P_{CO_2}$	$\delta^{13}C$	$\delta^{18}O$ (CO_2)
10	-2.16	-14.3	+37.4	-2.21	-15.7	+37.0
16	-2.09	-14.5	+38.1	Insufficient Sample		
22	Insufficient Sample			-2.20	-23.5	+38.3
26	-2.31	-16.5	+34.8	-2.23	-21.9	+34.8
35	-1.78	-18.4	+38.7	-2.93	-21.9	+43.2

Depth (m)	June 20-21, 1972			August, 1973		
	$\log P_{CO_2}$	$\delta^{13}C$	$\delta^{18}O$ (CO_2)	$\log P_{CO_2}$	$\delta^{13}C$	$\delta^{18}O$ (CO_2)
10	-2.21	Sample Lost		Not Sampled		
16	Not Sampled			Not Sampled		
22	-2.23	Sample Lost		Not Sampled		
26	Not Sampled			N.M.	-13.5	N.M.
35	Not Sampled			N.M.	-14.1	N.M.

By June 5 (Sample #7) the screen and a portion of pipe had been below the surface of the ground-water mound for several weeks. At this time interstitial water flow was in excess of 2.5-m per day. At this velocity the water was not likely to attain chemical equilibrium with the rock through which it was flowing [3]. The gas sample collected on that date has, with the exception of CO_2, the chemical constituents of atmospheric gas (Table III) and an isotopic composition approaching that of the atmosphere. Sampling procedures had prevented the introduction of atmospheric gas into the sampling tube. The calculated $\log P_{CO_2}$ of the

recharge water in the basin was -3.29, while that of the sample, measured
by gas chromatograph, was -3.31; volumetric determinations yield a sample
log P_{CO_2} of -3.12.

The ^{13}C content of the CO_2 of this sample is -12.49 ^{o}/oo (the calculated $\delta^{13}C$ for CO_2 gas in isotopic equilibrium with water with an initial
$\delta^{13}C$ of -0.94 ^{o}/oo is -9.40 ^{o}/oo), suggesting that the carbon-isotopic
composition was approaching equilibrium with water that was in equilibrium
with atmospheric gases. The presence in the sample pipe of residual CO_2
depleted in ^{13}C could have caused the difference between calculated and
observed ^{13}C. The technique of sampling waters from depth using a vacuum
makes it impossible to attempt meaningful interpretations on calculated
P_{CO_2} values from porous-cup water samples. The whole-gas composition,
the velocity of the water moving through the system, and the trend of
$\delta^{13}C$ all suggest that the sample collected on June 5 represents a gas of
atmospheric composition which outgassed from the rapidly recharging water
as it moved past the sampling screen. By this time approximately 10.5-m
of water (cumulative inflow) would have passed the screen.

Wood and Bassett [2] state that a significant decrease in infiltration
rate in the 1972 experiment occurred on August 15, 1972. This corresponds
to a cumulative inflow of 80-m of water. This decrease in infiltration
rate is attributed to the growth of anaerobic bacteria. A similar
decrease in infiltration rate was observed in the 1975 recharge experiment
on August 26, 1975 and was attributed to the onset of reducing conditions
in the spreading basin [13]. The onset of reducing conditions in the 1972
experiment is indicated in the water samples from the 0.6-m sampling
point by an increase in HCO_3^- and a decrease in SO_4 (Table II and Fig. 5).
However, neither chemical sampling in the 1972 experiment nor recharge in
the 1975 experiment was maintained long enough to identify the onset of
reducing conditions in the water from the 16-m sampling point.

The September 4, 1975 gas sample (#9) was collected following the onset
of anaerobic conditions in the basin water before water corresponding to
reducing conditions reached the 16-m sampling point. The whole-gas
analysis shows some change in the constituent composition, but the gas
composition is very similar to that of the June 5 sample (#7) and,
therefore, to atmospheric gas. The P_{CO_2} is slightly higher than that of
June 5 (#7), (log P_{CO_2} = -3.02 vs. -3.12 for June), suggesting a slight
increase in the solubility of CO_2 in the recharge water with a slight
decrease in temperature. The $\delta^{13}C$ of the September 4 sample (#9) is
-10.13 ^{o}/oo, quite close to the value of -9.40 ^{o}/oo predicted for CO_2 in
isotopic equilibrium with water with an initial $\delta^{13}C$ (C_{TOT}) of -0.94 ^{o}/oo.
The $\delta^{18}O_{SMOW}$ of the CO_2 is +43.30 ^{o}/oo, close to the value for atmospheric
CO_2 of +41 ^{o}/oo. If contamination from a leaking valve were causing the
approach to atmospheric-gas composition and isotope content, one would
expect the P_{CO_2} to decrease as the ^{13}C content approached atmospheric
values, not to increase. The carbon-isotope value and the gas composition
for the September 4 sample suggest continued outgassing of the water in
the pipe following the drop in total pressure brought about by the
sampling of June 5, but no comparison with the water chemistry from the
1972 experiment is possible because of irregularity of sampling at the
16.2-m depth in the later stages of the test.

The sample collected on November 20, 1975 (#11) shows significant changes in gas composition and stable-isotope content. At this time the water surface of the ground-water mound had only recently dropped below the 16.2-m level, restoring an open, unsaturated system. The gas-composition sample should then be representative of a mixture of the last gas outgassed from the water while the screen and pipe were submerged and of the first gas migrating into the unsaturated zone left by the dewatering of the mound. The log P_{CO_2} decreased to -4.18, less than atmospheric, and the log P_{O_2} decreased to -1.93. This change in gas composition is the first evidence of reducing conditions observed at the 16-m sampling depth. The water chemistry for this depth in the 1972 experiment shows no evidence of the reducing conditions observed in the 0.6-m waters in late August and early September 1972.

The ^{13}C content of the CO_2 decreases dramatically to a $\delta^{13}C = -38.1$ $^o/oo$ (Table IV and Fig. 6), indicating the presence in the CO_2 of carbon generally considered derived from reduced carbon species, in particular, from methane. The equilibrium carbon-isotope fractionation factor between methane and carbon dioxide at 20°C is approximately 1.070 [14]. Therefore, if the CO_2 seen in this sample were formed under reducing conditions in equilibrium with methane, it would be expected to be enriched in ^{13}C. This enrichment is not observed. It is, therefore, felt that the CO_2 observed, while in very small quantities, is derived from the oxidation of methane. No methane was detected on chromatographic analysis of the whole gas.

The nitrogen-cycle data may aid in our understanding of the reduction process. As reduction proceeded after August 26, any nitrate would have been reduced to nitrogen gas and the N/Ar would have increased. This did not occur because of minimal nitrate in the system. Later, as observed in sample #11, as reduction continued nitrogen gas was reduced to NH_4^+, which was absorbed on clays, and the N/Ar ratio decreased.

The following hypothesis may explain the observed ^{13}C and ^{18}O content of the CO_2 as well as the CO_2 and O_2 partial pressures, but should not be considered as the only mechanism for removing CO_2 and O_2 and generating the observed ^{13}C and ^{18}O contents:

Under anaerobic conditions present in the basin bottom after August 26, oxygen would not enter the subsurface with the infiltrating water. According to Morris and Stumm [15], once conditions favorable for the reduction of $SO_4^=$ to HS^- are developed, the conditions for the reduction of CO_2 to CH_4 could develop almost simultaneously. If a series of reactions postulated by Berner [16] is used as an example, CO_2 generated by the fermentation of sugar could be used as a source for production of CH_4:

I. $C_6H_{12}O_6$ ——— Bacteria I ——— $2C_2H_5OH + CO_2$

II. $CO_2 + 2C_2H_5OH$ —— Bacteria II ——— $CH_4 + 2CH_3OOH$

Methane and HS^- are generated by bacterial reduction in the shallow sediments in the bottom of the basin. The HS^- is precipitated almost instantly as finely disseminated iron sulfide. Part of the methane remains in the shallow sediment as trapped gas bubbles or escapes to the water in the basin through the sediment-water interface. Part of the

methane dissolves and is carried downward by recharging water. Since
methane is fairly stable, only a small part of it would be oxidized to CO_2
if it encountered a source of residual oxygen. It is conceivable that
methane outgassing from the recharging water into the sample pipe could be
oxidized by the small quantities of oxygen remaining to produce the
observed CO_2. The gas in the pipe would likely not have been mixed significantly with the convective gases that were following the declining
mound surface.

At the time when the surface of the perched-water mound declined below
the 16.2-m sampling screen, the recovery begins. Actually, very little
dewatering occurs while recharge is in progress even though the infiltration rate has been decreased by plugging. As the free surface of the
ground-water mound declines, the voids left behind must be filled with
air. This is accomplished by the inflow of gas, initially from the perimeter of the spreading basin as recharge is still in progress (Fig. 4b).
A small amount of the gas filling the voids may be derived from outgassing
of the water in the mound due to slight partial pressure differences.
When recharge is terminated and the basin drains, vertical downward air-
flow occurs through the bottom of the basin as the mound continues to
recede (Fig. 4c).

The first true recovery-phase sample collected on December 29 (#13),
has an extremely low P_{CO_2} (log P_{CO_2} = -4.30) and a log P_{O_2} of -1.71
slightly increased from the previous sample and is somewhat enriched in
^{13}C ($\delta^{13}C$ = -27.6 $^{o}/oo$). The $\delta^{13}C$ is still suggestive of the methane
influence. The February 3, 1976 (#15) sample is similar to the previous
one but with a log P_{O_2} of -3.00. The $\delta^{13}C$ = -30.88 $^{o}/oo$ is similar but
not identical to the December sample. The P_{O_2} has decreased still
farther in the April 1976 sample but the P_{CO_2} has started to recover,
increasing to $10^{-2.79}$ (a normal soil-CO_2 value). The $\delta^{13}C$ = -15.47 $^{o}/oo$
is also a normal soil CO_2 value. The N/Ar ratio is also increasing toward
pre-recharge formation-gas and atmospheric values due to the oxidation of
NH_4^+ sorbed on clays and to the introduction of gases of an atmospheric
N/Ar ratio.

With the exception of its very low-oxygen content, the formation gas
attained pre-recharge conditions by June 1, 1976 (#19). At this time the
P_{CO_2}, $\delta^{13}C$ and N/Ar ratio returned to their pre-recharge values (Fig. 6).
Up to an additional 126 days were required before the oxygen content of
the gases approached normal pre-recharge values. During this period the
P_{CO_2} and $\delta^{13}C$ fluctuated as the data observed in Table V. The last two
samples, those for October 5, 1976 (#23) and January 10, 1977 (#25),
yielded atmospheric values for all parameters.

The following hypothesis explains, in a qualitative fashion, the
observations of the recovery phase of the formation atmosphere:

When the basin dries, atmospheric gas moves downward through the
reduced zone. The small amount of residual methane and a large amount of
finely disseminated iron sulfide are oxidized, generating small amounts
of CO_2 and removing the oxygen. Following the oxidation of the iron
sulfide, the organic material formed and deposited during recharge is

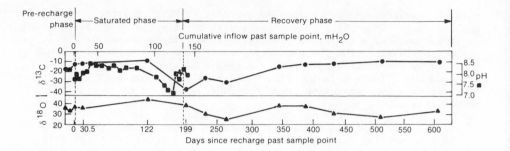

FIG. 8.   *Changes in $\delta^{13}C_{PDB}$, $\delta^{18}O_{SMOW}$, and pH with both cumulative inflow past 16.2-m sample point in metres of water (top scale) and days since recharge past 16.2-m sample point.*

oxidized. The results of this hypothesis are: 1) a reduction of O_2 to a very low level, 2) the same N/Ar ratio as atmospheric gases, 3) an initial CO_2 depleted in ^{13}C, 4) lack of CO_2 in the initial gas because of O_2 consumption by oxidation of iron sulfide. The FeS is oxidized preferentially before carbon because of its finely dispersed state. 5) The CO_2 generated after oxidation of FeS would be of normal soil-gas isotope content.

Plots of isotope content versus constituent composition (Figs. 6 and 7) show general trends that indicate which chemical reactions may be occurring. The plot for ^{13}C vs P_{CO_2} is most dramatic in displaying the conditions under which the gas was sampled: 1) the pre-recharge phase, 2) equilibration with infiltrating water, 3) reducing environment, and 4) post-recharge recovery.

Figure 8 shows the general relationship of $\delta^{13}C_{PDB}$ (CO_2) and $\delta^{18}O_{SMOW}$ (CO_2) with time for the entire recharge experiment. Included for reference are pH measurements for waters from the 0.6-m sampling point for the 1972 experiment. The effects of saturated flow and of the reducing conditions on all three parameters are apparent, as is the lag in evidence of those conditions in the carbon isotopes behind the pH, and in the oxygen isotopes behind the carbon. Again it must be noted that, while the waters from the 16.2-m porous cup showed no evidence of reducing conditions, sampling at that depth was irregular late in the recharge test.

CONCLUSIONS

This study points out that formation gas composition and the stable-isotope content of the CO_2 of that gas give an independent means of ascertaining chemical reactions occurring during long-term recharge. It also raises several points which must be considered in natural and artificial recharge studies, in dewatering studies, and in any study where reducing conditions may occur.

Studies of the constituent chemistry and the isotopic composition of gases in the unsaturated zone give an indication of the chemical reactions that occurred in recharging waters, even though indications in water chemistry alone may be masked by later reactions. Whereas the chemical analysis of waters from the 0.6-m sampling site shows marked changes in water chemistry with the amount of water through the systems, these changes are masked by other reactions by the time the water reaches the 16-m level. Gas samples show changes in both constituent composition and isotopic composition which can be used to infer the chemical reactions that have occurred in the water.

The length of time required for the unsaturated zone to recover following the termination of recharge is significantly greater than originally expected. Once reducing conditions are established and a significant quantity of oxidizable material (reduced mineral phases or organic material) have developed, large quantities of oxygen will be consumed in the shallow subsurface. Thus substantial time will elapse before more than minimal quantities of O_2 reach any significant depth. The length of time required for the unsaturated atmosphere to recover to pre-saturation conditions is in part a function of the extent to which reducing conditions prevailed. If these conditions were prevelant in the entire zone which was temporarily saturated, then all the gases must be brought in from outside the reduced environment. The potential for consuming the oxygen as the gas moves through organic-rich, FeS-rich sediment is very high. The generation of CO_2 as O_2 is being consumed would lead to the lag in O_2 recovery behind that of CO_2 even if most of the CO_2 escaped into the atmosphere.

ACKNOWLEDGEMENTS

I would like to thank Messrs. E. P. Weeks, F. J. Pearson, W. W. Wood, R. L. Bassett, D. C. Thorstenson, and others too numerous to mention for their support, discussion, and criticism of this study. The samples were collected by W. W. Wood and D. L. Redinger. Isotope samples were prepared and analysed by the author and Ms. J. C. Woodward. Chromatographic analyses were conducted by Mr. D. W. Fisher who significantly influenced my thinking on the whole gas part of the work.

REFERENCES

[1]   WOOD, W. W. SIGNOR, D. C., Geochemical factors affecting artificial ground water recharge in the unsaturated zone, Transactions ASAE 18 (1975) 677.

[2]   WOOD, W. W., BASSETT, R. L., Water quality changes related to the development of anaerobic conditions during artificial recharge, Water Resources Research 11 (1975) 553.

[3]   WOOD, W. W., BROWN, R. F., Experimental Geochemistry and Solute Transport in an Artificial Recharge Experiment, in preparation.

[4]   U. S. Department of Agriculture, Soil Survey Lab Data and Description for some soils of Texas, SCS, USDA, Soil Survey Inv. Rept. #3 (1976) 89.

[5]   KEYS, W. S., BROWN, R. F., The use of well logging in recharge studies of the Ogallala Formation in West Texas, U. S. Geological Survey, Prof. Paper 750-B (1971) B270-B277.

[6]   WEEKS, E. P., private communication, 1977.

[7]   WOOD, W. W., A technique using porous cups for water sampling
       at any depth in the unsaturated zone, Water Resource Research
       9 (1973) 486.

[8]   WOOD, W. W., private communication, 1975.

[9]   RIGHTMIRE, C. T., HANSHAW, B. B., Relationship between the carbon
       isotope composition of soil CO_2 and dissolved carbonate species in
       ground water, Water Resources Research 9 (1973) 958.

[10]   WEEKS, E. P. Field determination of vertical permeability to air in
       the unsaturated zone, U. S. Geological Survey Open-File Report
       77-346 (1977) 92.

[11]   CRAIG, H., Isotopic standards for carbon and oxygen and correction
       factors for mass-spectrometric analysis of carbon dioxide: Geochem,
       Cosmochim. Acta 12 (1957) 133.

[12]   CRAIG, H., Standard for reporting concentrations of deuterium and
       oxygen-18 in natural waters, Science 133 (1961) 1833.

[13]   WOOD, W. W., private communication, 1976.

[14]   BOTTINGA, Y., Calculated fractionation factors for carbon and
       hydrogen isotope exchange in the system calcite-CO_2-graphite-
       methane-hydrogen and water vapor, Geochem, Cosmochim. Acta 33
       (1969) 49.

[15]   MORRIS, J. C., STUMM, W., Redox equilibria and measurements of
       potentials in the aquatic environment in Equilibrium concepts in
       Natural Water Systems, Advances in Chemistry Series 67 (1967) 270.

[16]   BERNER, R. A., Principles of chemical sedimentology, McGraw-Hill,
       New York (1971).

DISCUSSION

D.B. SMITH *(Chairman):* The main data in your paper relate to changes
under artificial recharge conditions. Have you any further data on the range of
annual variations in $\delta^{13}C$ which may occur in the unsaturated zone?

C.T. RIGHTMIRE: A short paper I have written, which is in press and is
to be published in Water Resources Research, entitled "Seasonal variations in
P_{CO_2} and ^{13}C content of soil atmosphere", shows significant seasonal variation
in both P_{CO_2} and ^{13}C content in the shallow unsaturated zone. The $\delta^{13}C$
approaches atmospheric value during the winter, as does the P_{CO_2}. During the
growing season the P_{CO_2} increases significantly and the $\delta^{13}C$ approaches $-22\%_{oo}$,
an expected value for the humid environment sampled.

TRITIUM PROFILES IN
KALAHARI SANDS AS A MEASURE
OF RAIN-WATER RECHARGE

B.Th. VERHAGEN, P.E. SMITH, I. McGEORGE
Nuclear Physics Research Unit,
University of the Witwatersrand,
Johannesburg

Z. DZIEMBOWSKI
Division of Geohydrology,
Department of Water Affairs,
Pretoria,
South Africa

Abstract

TRITIUM PROFILES IN KALAHARI SANDS AS A MEASURE OF RAIN-WATER RECHARGE.

This paper attempts to relate recharge measurements in the Kalahari by tritium profiles in the unsaturated zone to isotopic, hydrochemical and hydrological data from an underlying, semi-confined aquifer. Auger holes into the sand cover were drilled along a line of experimental deeper holes penetrating the saturated zone. A further line of auger holes was drilled into the dune sand cover of a control area. Variable moisture contents, apparently independent of grain-size distribution and indicating transients, are observed in the different profiles. ^{3}H and ^{18}O measurements on the moisture contents allow for the identification of the 1962/63 bomb tritium rise and successive drier and wetter periods. Infiltration, or potential recharge as a percentage of infiltration, was found to be strongly dependent on the annual rainfall. The distribution of ^{14}C, ^{13}C, ^{3}H and chemistry in the shallower of two underlying aquifers leads to the consideration of three possible mechanisms of recharge. Arguments favouring vertical recharge are presented, which lead to possible extrapolations into the sand-covered areas of the Kalahari in general.

1. INTRODUCTION

The question of rain recharge to ground waters in the Kalahari Thirstland has been the topic of a number of early studies such as by Dixey [1], Martin [2], and Boocock and van Straten [3]. Some of these authors were of the opinion that such recharge does not occur and recently Smit [4] concluded that the thickness of the tertiary to recent Kalahari Beds deposits imposes a limit to the direct recharge. Isotopic studies of the ground waters themselves (Mazor et al [5], Verhagen et al [6]) concluded that such recharge is actually occurring over wide areas. This paper presents an attempt at reconciling, in a limited area, the preliminary information obtained from infiltration studies in the sand cover using the tritium profile method and the hydrological, hydrochemical and isotopic information from the immediately underlying ground waters.

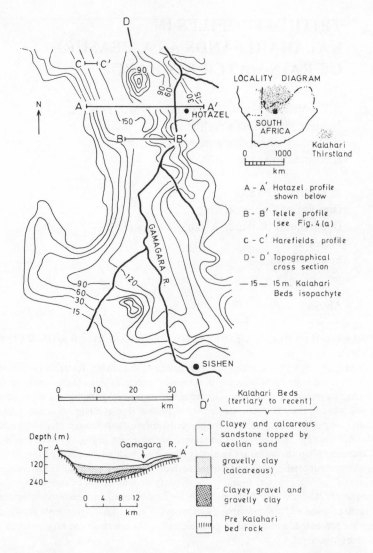

FIG.1. The Gamagara River catchment showing the increasing thickness of Kalahari Beds to the north and the location of sections through these beds.

2.   PHYSIOGRAPHY, CLIMATE AND GEOLOGY

        The Kalahari (see insert, Fig. 1) is a largely flat, sand-covered area, stretching northwards from the northern Cape Province in South Africa through central and western Botswana as well as the eastern parts of SWA/Namibia. The climate for the greater part of the region is semi-arid (Köppen classification) with an annual (summer) rainfall varying from 200-500mm. In the extreme south-western part desert-like conditions prevail. For the rest, except where over-grazing has degraded the plant cover, the area is well-vegetated, alternating between grassland and shrub country to bushes and tall (>15 m) trees.

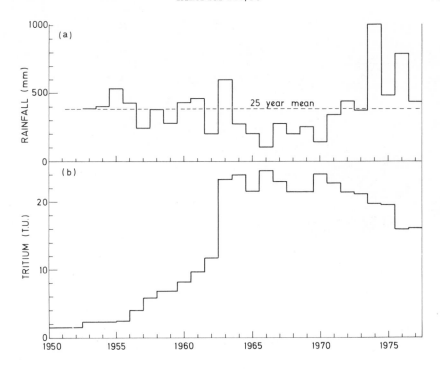

FIG.2.   (a) Seasonal precipitation at a rainfall station 5 km to the south of the Telele section B-B'. (b) Tritium 'input' curve decay corrected to 1978, based on measurements from Pretoria station and weighted for local precipitation.

The area under study is on the southern edge of the Kalahari, within the catchment of the Gamagara river (see Fig. 1). It has a mean annual precipatation of 330mm. The river is ephemeral. Only during the large rainfall event of 1973/74 (see Fig. 2a) did the river flow substantially over its entire length for several months.

Sub-Kalahari rock outcrops occur to the south, east and west of the valley. The base of the valley has been incised by glacial action, leaving deposits of tillites on the floor rocks. The remaining northwards dipping depression has been filled by increasing thicknesses of Kalahari Beds, consisting of basal gravels, clays, gravels and calcretes (for description see Verhagen et al [4]) and topped almost everywhere by sands which are at least in part aeolian (see general geological section A-A',Fig. 1). It is through the deeper part of the trough that the main sub-surface, northwards drainage in the valley is thought to occur.

3.   RECHARGE STUDIES

Geohydrological and chemical observations in the southern and eastern outcrop areas have led Smit [4] to estimate rainwater recharge at around 3% to the hard-rock aquifers. However, two main factors made him pessimistic about recharge through the Kalahari Beds:

1) the loss of infiltrated rainwater by evapotranspiration and

2) the sharply deepening rest levels for Kalahari Beds thicknesses greater than 15m.

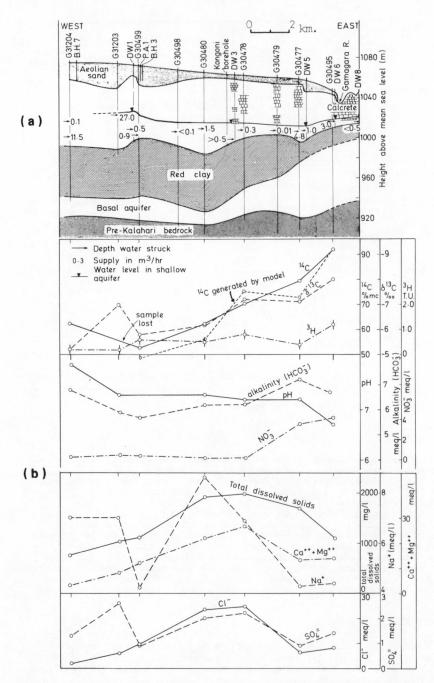

FIG.3. (a) Telele profile B-B' showing depth of water struck, driller's estimates of yield for the shallow aquifer and borehole locations. Water levels in the shallow aquifer can only be measured in the DW series boreholes. (b) Total dissolved solids, pH, major cations and anions as well as ^3H, $\delta^{13}C$, ^{14}C and model generated ^{14}C data for the G series boreholes shown in (a).

Smit regards 15m as the limiting thickness for direct rainwater recharge, in general agreement with Martin's [2] estimate of 6m. During 1974-1976, the Department of Water Affairs drilled a series of 10 boreholes through the Kalahari Beds, in order to investigate the shape of, and deposits in, the trough as well as the different ground water horizons encountered. The boreholes in this profile (B-B', Fig. 1), are indicated by G number in Fig. 3(a). In the light of the results from this drilling programme, which suggested the suitability of the terrain, augering was undertaken in the sand cover for infiltration studies and further deep holes drilled to investigate the shallow aquifer. (DW-Series). By way of comparison further auger drilling was undertaken in a somewhat different setting, 30km to the north west (Profile C-C' Fig. 1).

4.   SAMPLE COLLECTION AND PROCESSING

4.1   Chemical and Isotopic Analysis

Chemical data were obtained by standard analytical techniques. Total alkalinity, conductivity, pH and temperature are determined in the field at the time of sampling. Isotopic data processing is discussed elsewhere (Verhagen et al [7]).

4.2   Sand Sampling

Sampling for moisture profiles was performed exclusively in the sand cover. Hand augers were initially employed, but later a modified core barrelling technique was developed, using a standard diamond drilling rig. The top and bottom fractions of each "core" were discarded and the middle section only packed into hermetically sealed and dried 2 litre jars.

Apart from the top 10 - 50cm the sand is invariably moist and coherent. With careful drilling, a minimum of caving occurs, the holes remaining intact, even when uncased for several months. Depths are usually limited to 20 metres or by drill refusal: when calcretes, coarse gravels or boulders are encountered.

4.3   Moisture Extraction

Moisture is distilled directly from the jars under vacuum at about $100^{\circ}C$ to complete dryness (about 8 hours). Recovery of water (mass collected divided by loss of weight of sample jar) is 98% on the average. Aliquots of 200-250 ml are made up by combining 2 or more samples. These are electrolysed 7x to increase the tritium concentration before counting. Some sample fractions were retained for stable isotope determinations.

Sand moisture content is determined as weight %. As a field density is very difficult to determine, the density is taken as the maximum value of the well packed material in the sampling jar.

4.4   Grain Size Analysis

Preparation and sieve analysis of 100g aliquots of dried sand was carried out according to Folk [8]. Only the graphic mean grain size is here given. It is defined as 1/3 ($\emptyset16 + \emptyset50 + \emptyset84$), where $\emptyset = - \log_2 d$ and d = particle diameter in mm. $\emptyset16$, $\emptyset50$ and $\emptyset84$ are $\emptyset$ values corresponding to the accumulated weight % of the retained fractions.

5.   TRITIUM INPUT FUNCTION

The tritium profiles have to be discussed in relation to the "input curve" given in Fig. 2b. This curve has been constructed on the basis of tritium data for Pretoria from IAEA world survey of isotope concentration in precipitation [19], unpublished data from this laboratory and interpolations for periods over which no data are available. The tritium values have been decay-corrected for 1978. The Pretoria station $(25^{\circ}50'S)$ is taken as representative for the region in question $(27^{\circ}15'S)$, as both are in continental climatic zones. As the absolute rainfall and rainfall distribution for the study area differ substantially from Pretoria, monthly data have, as far as possible, been weighted for rainfall in the area.

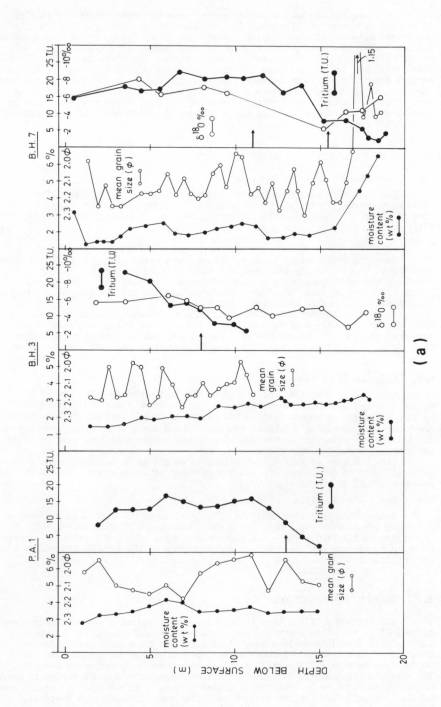

(a)

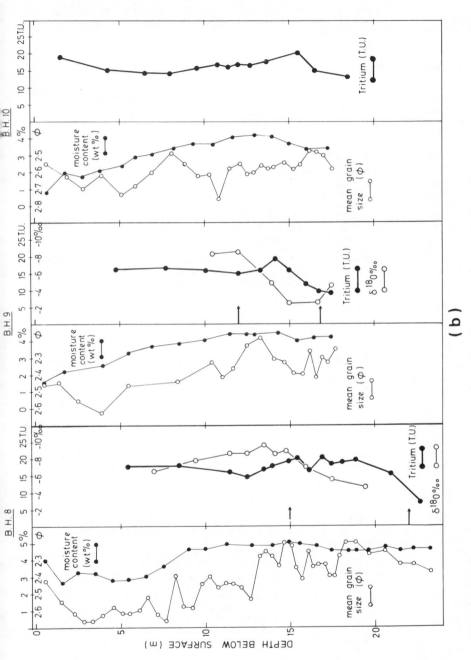

FIG.4. (a)/(b) Mean grain size, moisture content, ³H and δ¹⁸O concentrations for sand profiles from sections B-B' and C-C'.
Upper and lower arrows on the profiles represent estimated 1973–1978 and 1963–1973 input, respectively.

(b)

It is clear from this curve that the only prominent feature at present remaining from the Southern hemisphere bomb tritium peak is the sharp change in concentrations associated with the 1962/63 rainy period. It is this feature, which we refer to as the "cut-off", which will act as the basic parameter in the interpretation of the profiles.

6.   TRITIUM PROFILES

Some of the tritium profiles obtained from sand-augering are shown in Fig. 4a,b. Given along with the tritium results are some ^{18}O-measurements on the extracted water samples, the weight % moisture contents and the mean grain size of the sand samples.

6.1   Profiles from Section"B"

The three profiles here presented are from different sites along section "B" on the farm Telele (Fig. 1)

PA1 is a hand-augered hole sunk in August 1977. BH3 and BH7 were sunk in February 1978. BH3 was located beneath a large camel thorn tree (Acacia giraffae) and was the second attempt at drilling, the first being stopped by a major root at a depth of 7.5 metres. PA1 was drilled in a grassy clearing amongst major trees and shrubs; BH7 in an open savannah-type area, with no major trees within hundreds of metres. PA1 shows a clear cut-off of tritium concentrations between 11 and 15 metres. On the other hand, BH7 suggests considerably deeper penetration, the cut-off ranging from 12 - 18 metres. Tritium concentrations in BH3 drop sharply between 5 and 11 metres. This suggests a much lower infiltration rate than in the other two profiles.

6.2   Profiles in Section"C"

The area on the farm Harefields (Fig. 1) is characterised by a very deep sand cover, with east-west striking dunes. These dunes have been fixed by quite abundant vegetation. The three profiles (Fig. 4b) for this section were obtained from sites with differing vegetal cover running along the trough between two dune ridges.

BH8 was sunk on the edge of a large overgrazed clearing, some 50m away from major trees, BH9 amongst major shrubs and bushes but well away from trees, and BH10 amongst, and within 10 metres from, major trees. The drilling was done on three consecutive days during February 1978.

All three profiles suggest a considerably deeper penetration of bomb tritium than was observed in section B. BH8, in particular, exhibits   a bomb tritium cut-off region between 19 and 23 metres. As opposed to the auger holes from section "B", which were all terminated by the occurrence of either calcrete bands or gravels, section"C" holes continued in pure sand cover to their ultimate maximum depths.

7.   MOISTURE CONTENTS AND MEAN GRAIN-SIZE

The most striking feature exhibited by the moisture content profiles is the marked difference between the two sections. The % moisture in section"C" is considerably higher than observed for section"B", and the distribution in the former three profiles is very similar. Considerable variation exists between the moisture contents in different profiles in section"B", with extremely low values observed at some depths in BH7. Individual moisture profiles in both sections show rough correlations with the mean grain-size profiles.

However, particular grain size ranges correspond to quite different moisture contents both within each section and between the sections, even within individual profiles. It is therefore possible that part of the moisture variation down the profiles is determined by transients. Apart from individual variations near the surface, the highly similar profiles in BH's 8, 9 and 10, suggest a common downward-moving moisture wave.

The relatively smaller grain sizes observed at shallower depths in BH's 8, 9 and 10 is taken to be associated with the dune environment. At greater depths, the grain sizes approach the values observed in section "B".

As BH7 and BH's 8, 9 and 10 were sunk within days of each other late in the rainy season the moisture contents of the profiles suggests that the environment of section C is considerably more promising for infiltration than section B. This is supported by the observed depth of penetration of bomb tritium.

8.   OXYGEN-18 PROFILES

Oxygen-18 measurements were done on sample material from the moisture extraction which remained after the tritium measurements had been performed. In view of the high recovery of moisture from the soil samples (98%), no appreciable change in stable isotope composition of the resulting water should be expected. The results are shown in Fig. 4(a,b). A strong similarity is observed between the ^{18}O profiles of BH7, 8 and 9. Relatively light isotopic values are found in the upper part of the profiles, with a rather abrupt increase to much heavier values, some distance above the tritium cut-off. The O-18 profile in BH3 shows some similarities but at much attenuated amplitude, with somewhat higher values above the tritium cut-off and slightly heavier below.

The rainfall record of the area (Fig. 2a), shows a period of below average rainfall from 1964-1971. The following period, with higher than average rainfall, includes 1974, with nearly three times the annual mean. The period of drought was preceded by a decade of average rainfall, the year of arrival of the bomb tritium peak (1963) being well above average.

We interpret the ^{18}O data in the light of this rainfall history. The light values (-6.5‰ to -9‰) are associated with the post-1973/74 season, the heavier values with the low rainfall period and the inflection at the bottom of the profile we ascribe to the higher rainfall season preceding the drought. This interpretation is supported by the time scale predicted by the tritium input curve (Fig. 2b), and reflected in the tritium profiles.

The different behaviour of the ^{18}O profile in BH3, drilled underneath a tree, seems to support the much attenuated infiltration suggested by the shallow cut-off of the tritium profile, due to rainfall interception. This aspect of the recharge study, so important to an understanding of the losses from the unsaturated zone, require generalisation in the Kalahari area and further detailed investigation.

9.   INTERPRETATION OF PROFILES

Tritium profiles measured earlier in the central and western Kalahari by Verhagen [9] as well as reported by Bredenkamp et al. [10] for southern African environments with slightly higher rainfall have shown that bomb tritium had penetrated to depths of 4-6 metres by 1973/74.

The profiles reported here indicate that, during the subsequent 4 years, the bomb tritium front has been displaced down to 16-23 metres in the southern Kalahari sands. Although by now subjected to considerable dispersion, the bomb tritium cut-off and the ^{18}O variations associated with the variable rainfall periods between 1963 and 1973 have remained largely preserved. This is so in spite of the rapid downward displacement of up to 15 metres. These facts are taken to justify the use of the piston or plug flow concept in analysing the profiles; the rise above 10 T.U. at present is taken as the best time marker on the tritium profile (1962/63).

No attempt will be made to reconstruct or model the infiltration on the basis of estimates of recharge effectiveness and/or losses. As will be pointed out later, the loss factors from this environment can be appreciable, even from depths greater than the penetration of the tritium profile at present and as yet quite impossible to handle quantitatively.

From the corresponding values on the ^{18}O and ^{3}H profiles, arrows are drawn in Fig 4(a,b) for BH7, BH8 and BH9, indicating the estimated 1962/63 and 1973/74 positions, respectively. For BH3 and PA1, only the 1962/63 marker can be identified. The ^{18}O profile in BH3 is too undifferentiated to discern the 1973/74 position. No ^{18}O data are available for PA1 and BH10.

The total precipitation over the two periods 1962/63 - 1973/74 and 1973/74 - 1977/78 is, respectively, 2800 mm and 3500 mm. The moisture balance was drawn

TABLE I   PERCENTAGE INFILTRATION FOR HIGH AND LOW RAINFALL
PERIODS IN DIFFERENT PROFILES

Profile	% infiltration	
	1973/74 - 1977/78	1962/63 - 1973/74
BH 9	17.1	12.1
BH 8	26.6	14.5
BH 7	10.6	5.5
BH 3	3.5	
PA 1	11.4	

up for the different profiles and % infiltration calculated on the basis of
moisture accretion over a specified period.   The results for the different
profiles are given in Table I.

Several features can be observed from these results.

a)   In each of the profiles in which the differentiation of the rainfall
periods has been possible, a marked contrast is seen between the percentage
infiltration of the pre 1973 and post 1973 periods.   The ratio of the
percentages for BH7, 8 and 9 has a mean value of 1.7±0.3.
b)   The % infiltration found for the 1962/63 - 1973/74 period in the
different profiles is in general agreement with recharge estimates for the
southern African environment in general: up to 14% estimated by Enslin [11],
from 4%-10% by Jennings [12] and estimates ranging from 5% - 10% based on earlier
tritium profile studies by Bredenkamp [10] and Verhagen [9].

c)   Considerably higher percentage infiltration for the section C sands
than found in section B.   The particular dune environment and sand grain size
might be determining factors.
d)   Very low percentage infiltration in BH3, close to the tree.

10.   OBSERVATIONS ON THE SATURATED ZONE

It has been established that infiltration into the sand cover occurs,
that it can be quantified using the bomb tritium profile and resolved by the
^{18}O profile.   This infiltration in the sand cover can be called potential
recharge.
The question arises as to whether evidence of actual recharge to the
water table can be found.   Although no tree roots were observed in any of the
auger holes sunk, except when deliberately drilling beneath a large tree, it
is known that phreatophytes tap perched water tables tens of metres deep.
In fact, large trees are ubiquitous in the Kalahari and seem, from the evidence
of BH3, to adversely affect infiltration in their immediate environment.
Their water demands are to be quantitatively investigated and may be as
high as 0,5 - 1 m^3/day [13].
Hydrologic, isotope and chemical observations were done on a series
of deep boreholes sunk along the Telele section 'B'.   The geological section
is shown in Fig 3a, along with borehole positions.
Two water horizons were encountered: the upper is semi-confined in
clayey gravel, the lower, below the red clay band, is confined and similarly
in clayey gravel.   The depth to water struck and the standing water level of
the shallow aquifer is indicated for each of the holes as well as driller's
estimates of borehole yields.   Samples were obtained from the top aquifer
before the holes were drilled through the clay and cased off.
No long term rest level observations are as yet available and no proper
pump tests conducted.   The estimated yields give the only indication of the
aquifer characteristics at present.   The inferred transmissivities are poor,
except for the gravel around G31203 and G31204,where the yields are about two
orders of magnitude higher.   This is thought to be the main northwards ground
water drainage channel down the valley.

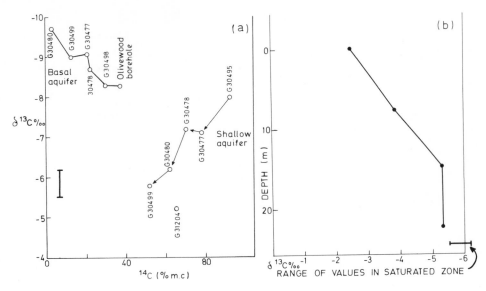

FIG.5. (a) $\delta^{13}C$ versus ^{14}C for the shallow and basal aquifers shown in Fig.3(a). The opposing
trends of the two sets of data show the separation of the two samples. (b) $\delta^{13}C$ values from
increasing depths in a calcrete profile above the water table compared with a range of values
from aquifer materials.

        Also shown in Fig 3b are the profiles of the major ion concentrations,
and pH, as well as ^3H, ^{14}C and ^{13}C concentrations for the upper aquifer, plotted
against borehole position.   The individual stable isotope concentrations are
not plotted, being almost identical, with $\delta D = -38.9 \pm 1.5\%$ and $\delta^{18}O =$
$5.8 \pm 0.4\%$ for the group of boreholes.
        The lower aquifer appears to be hydrologically quite distinct from the
upper aquifer and shows quite different carbon isotopic behaviour.   Radiocarbon
concentrations are plotted against $\delta^{13}C$ for the two aquifers in Fig 5(a).
Two opposing trends are found: the shallow aquifer exhibits a trend consistent
with carbonate exchange.   The deep aquifer shows a mixing behaviour presumably
between an old, low $\delta^{13}C$ water, derived from the bedrock and a younger,
higher $\delta^{13}C$ component, presumably from the Kalahari Beds.   The lower aquifer
which, in addition, has a higher piezometric level than the upper, is therefore
left out of consideration in this discussion on infiltration.

        The ^{14}C profile shows steadily decreasing concentrations away from the
river bed.   The tritium concentrations, with the exception of the first sample
next to the river, are not much above the measureable limit. $\delta^{13}C$ values
become gradually more positive along the profile.   In addition, the total
dissolved solids in the water increase with a shift in chemical type (see
Piper diagram, Fig 6) showing a reversal in trends at G30499.   Chemical and
isotopic compositions alter sharply at G31203 and G31204, the two boreholes
thought to be located in the higher transmissive zone.
        We discuss three possible mechanisms which could account for the
observed profiles.

1.   Flow away from and at roughly right angles to the river, along the borehole
profile.   The increasing ^{13}C concentrations appear to be produced by exchange
with the carbonate component of the clays and gravels in the aquifer which

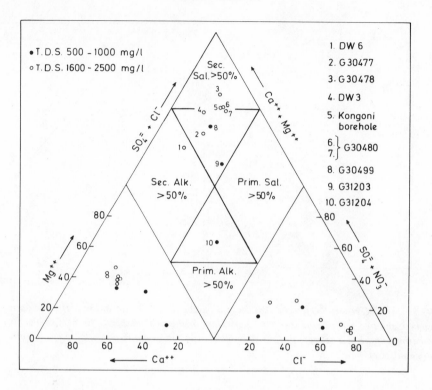

FIG.6. Piper diagram with chemical analyses of samples from the section B-B'. The samples have been numbered from east to west, i.e. from the Gamagara River to the high yielding trough, Fig.7(b).

should have a higher value of $\delta^{13}C$. Increasing salt content appears to develop along the profile, the trend reversing with dilution and exchange with water in the high transmissive (high yielding) zone.

2.   Infiltration all along the river bed. This would produce, as a result of the northwards gradient in ground water levels (Fig 7a) the flow pattern schematically shown in Fig 7b. The different flow times could produce the ^{14}C gradient observed along the profile and the increasing exchange times could account for the gradual increase in ^{13}C concentration. The chemical profile could also be consistent with this mechanism.

3.   Vertical infiltration through the unsaturated zone. This mechanism will be discussed in the light of supporting evidence and of objections that can be raised against mechanisms 1 and 2.

    a)   Chemical, as well as isotopic, exchange processes are unlikely to occur within an established aquifer. This has been discussed in the context of carbonate chemistry cf. Thilo and Münnich [14] and is supported by Smit [4] in numerous observations of the chemical behaviour of ground waters in this environment.

    b)   The type of chemical evolution seen in the groundwaters, especially the parallel growth of chlorides and sulphates as well as the major cations, suggesting disequilibrium conditions.

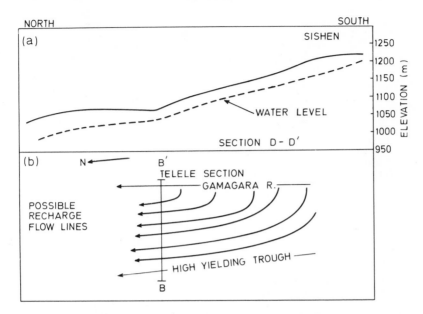

FIG.7. (a). Section D-D' of Fig.1 showing the water level in relation to the falling land surface along a north-south section. (b) Diagrammatic representation of infiltrating flow lines of possible recharge from the Gamagara River into the shallow aquifer shown on section B-B' of Fig.3(a).

c) Tests done on the aquifer materials obtained from drill cores, could not produce the major ions at sufficient concentration in solution to explain the observed chemistry.

d) Values of $\delta^{13}C$ measured in a variety of aquifer materials ranged from -5,2‰ to -6,5‰ i.e. close to equilibrium with the ground waters.

e) Tritium profiles have produced proof of the downward movement of rain-recharge in the overlying sand and the addition of substantial amounts of water during recent years. Even with severe and deep evapotranspirative loss mechanisms operating and evidence that this C-C' section exhibits relatively poor infiltration, some of the soil moisture should occasionally move downwards beyond the sand cover.

f) The calcretes encountered in the drill cores were highly weathered and interbedded with sands. Massive losses of drilling fluid were experienced. Using a single core barrel, only 30% of a 24 metre core was recovered. The calcrete showed numerous solution channels, weathered surfaces and many rounded calcrete fragments were observed.

g) $\delta^{13}C$ values for the calcrete (see Fig 5b) are more positive than the aquifer materials and show strong evidence of exchange and redeposition.

h) Some increase in salinity with depth was observed in the sand profiles. It is possible that the vegetal water demand can enrich the major ions in the soil moisture during dryer periods to the levels which are observed in the saturated zone.

i) Tritium concentrations are observed in the shallow aquifer which may be just significantly above 0. This would not be possible if either of the mechanisms 1 or 2 would be operative.

We therefore conclude that, on the balance of the above arguments, the direct infiltration mechanism is the most probable. This would make the ground waters in the shallow aquifer recent on the radiocarbon time-scale. In the following section we propose a mechanism which would produce the observed values.

11. RADIOCARBON LEVELS PRODUCED IN THE UNSATURATED ZONE

Many schemes have been proposed by which the initial concentration of ^{14}C in ground waters can be reconstructed. The models of Mook [15] and Wallick [16] were tried. Mook's model was not found to be applicable as the exchange sequence proposed by him does not hold for intermittent movement through a discontinuous vadose zone. Wallick's chemical/isotopic model, although designed for a similar environment, is difficult to apply because of the high chloride contents of the waters.

A modified dilution model, along the lines of Pearson and Hanshaw [17] was therefore tried, based partially on circumstancial evidence in this particular case. The shallow water in borehole G30495 has values of 91.7% m.c. and -8.0‰ for ^{14}C and $\delta^{13}C$ respectively, with an accompanying 1.2 T.U. This relatively low tritium, high radiocarbon combination could be the result of biological 'pumping' action as proposed by Münnich [18]. It also serves as an indicator that very little recharge from the 1973/74 flood in the Gamagara river has reached this relatively shallow water table.

The soil CO_2 was found to lie in the range $\delta^{13}C = -15‰$ to $-18‰$. Let us assume that 90% m.c., -8‰ and pH <7 are the values established in the ground waters where they leave the humic (root) zone, as is suggested by borehole G30495. The water now enters the calcrete zone and the dissolved carbon exchanges with secondary lime. From the data in Fig 5(b), we take a mean value for $\delta^{13}C = -3.5‰$ as well as for $^{14}C=5\%$ m.c. for the calcrete. Such exchange will be possible, as the calcrete surfaces are clearly weathering and therefore out of equilibrium with the infiltrating water. The ^{13}C gradient and finite ^{14}C concentrations confirm this and the exchange will produce the observed $\delta^{13}C$ values when:

$$a\delta_R + b\delta_c = \delta_s \qquad (1)$$

$$a + b = 1 \qquad (2)$$

where a, b and δ_R, δ_c are the proportions and $\delta^{13}C$ values for the recharge bicarbonate and calcrete carbonate respectively and δ_s refers to bicarbonate of water in the saturated zone. The values of a and b can be evaluated from eqns. 1 and 2 for each value of δ_s and the ^{14}C concentration A_s in the water can be calculated from A_R and A_c, the ^{14}C concentrations in the recharge and calcrete respectively:

$$aA_R + bA_c = A_s \qquad (3)$$

The calculated values of A_s are shown in Fig 3(b). They generally approximate the observed radiocarbon concentrations, and demonstrate the possibility of generating the observed values by an exchange mechanism in the calcrete of the vadose zone. The increasing ^{13}C values along the profile and the resulting decrease in ^{14}C concentration, could be generated by increasing thicknesses of limey material and increasing exchange times. The model is not taken to apply to the water in G31203 and G31204, which belong to a different hydrological regime.

12. CHEMICAL EVOLUTION OF THE GROUND WATERS

The chemical composition of the shallow aquifer ground waters are plotted on a Piper diagram in Fig 6. The group of high (1600 - 2000) TDS ground waters in Fig 3(b), plot in a relatively tight group in the 'secondary salinity' sector of the diagram. Three of the others (G30477, DW6 and

G30499), at lower TDS, plot somewhat lower down on the diagram, whilst the two analyses for the high-yielding holes G31203 and G31204 plot well away from the two other groups.   The latter two results confirm the difference in the flow regimes assumed earlier and set these waters apart in the arguments regarding local development of the water chemistry and isotopic content.

The high TDS chemistry cases are taken to represent the chemical type that might be produced by direct infiltration through ($\sim$ 40 m) calcretized vadose zone, involving secondary dissolution and exchange by the water. The lower TDS waters are interpreted as transition cases, where these waters are partially diluted by fresher waters or appear as chemically partially developed cases.

13.   SUMMARY AND CONCLUSIONS

Tritium profiles in Kalahari sands have produced evidence of much accelerated infiltration due to a succession of high rainfall seasons. Although dispersed, the 1962/63 bomb tritium cut-off is still clearly discernable.   The position of moisture derived from a sequence of average, below average and much above average rainy periods is visible in variable 18O concentrations in the soil moisture.

This separation of the tritium profile shows the contrast between the percentage infiltration for low and high rainfall periods.   For the former, the % infiltration into the sands conforms to estimated recharge rates for other southern African environments;   over the last four years it appears to have been at the average 1.7 times higher.

Infiltration into the sands must be regarded as potential recharge to the aquifers.   Loss mechanisms can operate both at shallower depths (trees and lesser vegetation) and from deeper levels (phreatophytes).   The profiles prove that trees exert a considerable influence on their immediate environment.

No clear evidence was seen of recharge to the shallow aquifer at the Telele section (B-B') except for the fact that ^{14}C concentrations suggest slow but steady turnover of water.

Of the different mechanisms proposed to account for the ^{13}C concentrations, direct rain recharge movement through the calcretes seems the most plausible and a simple carbonate exchange model can account for the observed ^{14}C values. The tritium concentrations are as yet too close to the minimum detectable level to allow for a clear distinction between the mechanisms.

The direct recharge mechanism would furthermore type the chemical composition of the waters of higher salinity as that which could be expected from rain recharge through the Kalahari Beds (sands and calcretes).   This would in turn raise the possibility that the salination of ground waters, which in the Kalahari Beds can rise to >10000 mgl^{-1}, can be a function of direct rain recharge.

Although bomb ^{14}C and ^{3}H have been observed in many shallower Kalahari ground waters [5,6] and massive infiltration has occurred into the sand cover during the past four years, the bomb components have not yet reached the deeper water tables.   The next few years are likely to see the infiltration mechanisms decided by the arrival or otherwise of the bomb components.

Further investigations of the saturated zone would seem most fruitful along section C-C', where considerably higher infiltration rates are observed.

ACKNOWLEDGEMENTS

The research project of which the work described in this paper forms a part, has been undertaken under contract to the South African Water Research Commission; their generous financial support for this project is gratefully acknowledged.

The authors would like to express their gratitude to members of the Division of Geohydrology, Department of Water Affairs, and in particular Mr. J.R. Vegter, for their close cooperation and encouragement; also to the Geological Survey and Dr. P.J. Smit, who initiated much of the geohydrological research in the area.

Finally they are indebted to the University of the Witwatersrand for the indispensable facilities and financial backing; Professor J.P.F. Sellschop, Director of the Nuclear Physics Research Unit for his support and constant encouragement and last but far from least, the hard-working staff of the Environmental Isotope Laboratory: Mr. E. Ndhlovu, Mrs. Z. Richter, Mr. M. Mchunu and Mr. F. Chikane as well as Mr. I. McGeorge for field work.

REFERENCES

[1]   Dixey, F.   Water supply problems in arid regions in the British colonies and Southern Africa (1956) Colon. Geol. Min. Res. 6, 307-325

[2]   Martin, H. Hydrology and water balance of some regions covered by Kalahari Sands in South West Africa.   Inter-African Conf. on Hydrology Nairobi (1960)

[3]   Boocock, C. and O.J. van Straten.   Notes on the geology and hydrology of the central Kalahari region, Bechuanaland Protectorate (1962) Trans. Geol. Soc. S. Afr. 65, 125-171

[4]   Smit, P.J., Die Geohidrologie in die opvanggebied van die Molopo-rivier in die noordelike Kalahari (1977).   Ph.D. thesis. University of the Orange Free State, Bloemfontein

[5]   Mazor, E., B. Th. Verhagen, J.P.F. Sellschop., M.T. Jones, N.E. Robins, L. Hutton and C.M.H. Jennings.   Northern Kalahari ground waters: Hydrologic, isotopic and chemical studies at Orapa, Botswana (1977) Jnl. of Hydrology. 34, 203-234

[6]   Verhagen, B.Th, E. Mazor and J.P.F. Sellschop Radiocarbon and tritium evidence for direct rain recharge to ground waters in the northern Kalahari (1975) Nature 249, 643-644

[7]   Verhagen, B.Th., P.E. Smith, J. Dziembowski and J. Oosthuizen An environmental isotope study of a major dewatering operation at Sishen mine, Northern Cape Province.   These proceedings.

[8]   Folk, R.L., Petrology of sedimentary rocks.1968. University of Texas.

[9]   Verhagen, B.Th. Environmental Isotopes in ground waters - some case studies in southern Africa.   SARCCUS - Meeting of specialists in underground water (1975) Bulawayo.

[10]   Bredenkamp, D.B., J.M. Schutte, G.J. du Toit. Recharge of a dolomitic aquifer as determined from tritium profiles. Isotope techniques in groundwater hydrology 1974.   IAEA, Vienna, 1974. 73

[11]   Enslin, J.F. Die grondwater potensiaal van Suid-Afrika. Convention: Water for the Future Pretoria (1970)

[12]   Jennings, C.M.H. The Hydrogeology of Botswana. PhD. Thesis. University of Natal, Pietermaritzburg (1974).

[13]   Bate, G.C. Private communication.

[14]   Thilo, L. and K.O. Münnich.   Reliability of carbon-14 dating of groundwater: effect of carbonate exchange. Isotope Hydrology 1970. I.A.E.A. Vienna 1970. 259

[15]   Mook, W.G. The dissolution-exchange model for dating groundwater with ^{14}C.   Interpretation of environmental isotope and chemical data in groundwater hydrology I.A.E.A. Vienna 1976. 213.

[16]   Wallick, E.I. Isotopic and chemical considerations in radiocarbon dating of groundwater within the semi-arid Tucson Basin, Arizona. Interpretation of environmental isotope data in groundwater hydrology. I.A.E.A. Vienna 1976. 195

[17]   Pearson, F.J., B.B. Hanshaw.   Sources of dissolved carbonate
        species in groundwater and their effects on carbon-14 dating.
        Isotope Hydrology 1970. I.A.E.A. Vienna 1970. 271.

[18]   Münnich, K.O., W. Roether, L. Thilo.   Dating of groundwater with
        tritium and 14C.   Isotopes in Hydrology. I.A.E.A. Vienna 1967. 305

[19]   Environmental Isotope Data Nos. 1-5. Technical Report series No.
        165. I.A.E.A. Vienna

DISCUSSION

T. DINÇER: Did you measure the sand to determine the temperature profile?

P.E. SMITH: No, we did not.

T. DINÇER: I think that the increase in moisture with depth shown in your Fig.4 is more likely to be related to the moisture transport mechanism in the sand than to a transient moisture wave.

P.E. SMITH: The moisture content does not necessarily always increase with depth. With the first profile, in many cases the moisture content remains relatively stable through most of the hole; it may increase, for instance, where there is a discontinuity in grain size. The second profile shows a high moisture content which is independent of grain size; it was this that induced us to consider the possibility of a moisture wave, since the profile was taken after heavy rain.

T. DINÇER: There are studies which indicate that thermal gradients could be the main force behind movements of soil moisture in the upper layers of soils, especially where the soil-moisture values are generally low.

P.E. SMITH: We will have to look into this question in more detail.

A.H. BATH: I was very interested in your description of this work, which is essential if environmental isotopes are to be used for evaluating recharge in arid zone aquifers. I would like to ask you whether grain-size distribution rather than mean grain size might not be the parameter to which moisture content should be related?

P.E. SMITH: Mean grain size was the only parameter that showed any correlation at all with moisture content.

A.H. BATH: You relate the observed saturated zone salinity to direct vertical recharge by means of a gradient of increasing salinity in the unsaturated zone. How do you reconcile this concentration in the unsaturated zone with the apparent lack of evaporative modification of $\delta^{18}O$?

P.E. SMITH: We believe that the salinity is caused by vegetation, since in dry periods there is a concentration of salts around roots as a result of osmotic processes. The ^{18}O and deuterium values are fairly constant throughout the profile.

A.H. BATH: With such profiles of low moisture content and apparently high potential moisture retention capacity, it is essential to establish at what depth in the profile there ceases to be any chance of upward flux of moisture either as vapour or liquid, and below which only downward movement occurs. It is only below this level that the tritium profile can be used confidently to determine whether small but nevertheless significant amounts of recharge are occurring or not. So far, very little is known about the physical mechanism which would help us to understand the behaviour of isotopes in such cases.

P.E. SMITH: I quite agree with you, and this problem will have to be considered.

K.E. WHITE: My question relates to the downward movement of water indicated by the ^{18}O and ^3H fronts, especially where the results for borehole 9 are concerned. In section 9 of the paper you state that "... during the subsequent four years the bomb tritium front has been displaced down to 16−23 metres ...". This is equivalent to a displacement of 10 to 17 m in four years, i.e. between 2.5 and 4.2 m/a. There appears to be a good correlation of fronts for borehole 9, possibly better than that indicated by your arrows set at arbitrary concentrations. The two fronts relating to fallout and rainfall, separated by about ten years in time, are clearly separated by only a few metres in depth. This suggests a percolation rate which is an order of magnitude less, i.e. about 0.3 m/a or about 0.5 m/a on the basis of your arrow criteria. How do you explain this apparent anomaly? Could there be other processes involved as well as simple percolation, e.g. could there be a difference in the physical behaviour of water molecules containing ^{18}O and ^3H in certain sand or rock strata − or posssibly different percolation rates of rains at different times of the year with very different tritium concentrations?

P.E. SMITH: The rainfall in the area takes the form of showers and so there are different amounts of rainfall in different places. Even when boreholes are very close together they may show different infiltration rates, the rainfall being far greater in one than in another.

B.Th. VERHAGEN: A comment on the profiles. Up to 1973 we observed tritium profiles showing a cut-off at 4−6 metres. The major recharge events since 1973/74 seem to have lowered this cut-off to the values seen in Fig.4. The ^{18}O values observed appear to support this conclusion. The markers in borehole 9 are about 5 m apart.

P. FRITZ: Since you have ^{14}C on calcretes found at depth, could it be said that the calcrete formation is related to present-day water movements? In other words, what do you know about calcite saturation in the water samples collected?

B.Th. VERHAGEN: The ^{13}C and ^{14}C measurements on the calcrete core were done on bulk samples. The ^{13}C gradient in the calcrete seems to reflect long-term redeposition. We would expect the surface ^{14}C of the calcrete to be considerably higher, i.e. a ^{14}C gradient should exist. We regard these particular calcretes as being the result of an earlier deposition, related to the calcretes observed at Sishen (see paper IAEA-SM-228/1)[1]. The $\delta^{13}C$ values of up to about $-2‰$ indicate that the sources may be atmospheric CO_2 and dolomite, and that the calcretes may be related to a major pluvial period. The profile of $\delta^{13}C$ with depth suggests weathering by groundwater movement up to the present time. We have not looked into the question of carbonate chemistry but we do intend to do this.

---

[1] VERHAGEN, B.Th., et al., these Proceedings.

NEW TECHNIQUE OF IN-SITU SOIL-MOISTURE SAMPLING FOR ENVIRONMENTAL ISOTOPE ANALYSIS APPLIED AT PILAT SAND DUNE NEAR BORDEAUX

HETP modelling of bomb tritium propagation in the unsaturated and saturated zones

G. THOMA, N. ESSER, C. SONNTAG, W. WEISS, J. RUDOLPH
Institut für Umweltphysik,
Universität Heidelberg,
Federal Republic of Germany

P. LEVEQUE
Laboratoire de radiogéologie et de
mécanique des roches,
Université de Bordeaux, France

Abstract

NEW TECHNIQUE OF IN-SITU SOIL-MOISTURE SAMPLING FOR ENVIRONMENTAL ISOTOPE ANALYSIS APPLIED AT PILAT SAND DUNE NEAR BORDEAUX – HETP MODELLING OF BOMB TRITIUM PROPAGATION IN THE UNSATURATED AND SATURATED ZONES.

A new soil-air suction method with soil-water vapour adsorption by a 4-Å molecular sieve provides soil-moisture samples from various depths for environmental isotope analysis and yields soil temperature profiles. A field tritium tracer experiment shows that this in-situ sampling method has an isotope profile resolution of about 5–10 cm only. Application of this method in the Pilat sand dune (Bordeaux/France) yielded deuterium and tritium profiles down to 25 m depth. Bomb tritium measurements of monthly lysimeter percolate samples available since 1961 show that the tritium response has a mean delay of five months in the case of a sand lysimeter and of 2.5 years for a loess loam lysimeter. A simple HETP model simulates the layered downward movement of soil water and the longitudinal dispersion in the lysimeters. Field capacity and evapotranspiration taken as open parameters yield tritium concentration values of the lysimeters' percolate which agree well with the experimental results. Based on local meteorological data the HETP model applied to tritium tracer experiments in the unsaturated zone yields in addition an individual prediction of the momentary tracer position and of the soil-moisture distribution. This prediction can be checked experimentally at selected intervals by coring.

INTRODUCTION

Various geoscience disciplines pay increasing attention to transport and mixing processes in the unsaturated zone. Special attention is directed towards the downward movement of the soil water which acts as a carrier for most substances deposited on the ground. Environmental tritium and tritium tracer experiments [1–6] proved this water movement to be layered for homogeneous soil strata and moderate precipitation pattern. This has also been shown by bomb tritium vertical profiles of soil moisture [7–10], where the tritium time variation in rain of the preceding years was found to be reproduced in the unsaturated zone. In principle, both methods can give reliable local groundwater recharge data. Recharge is equal to the amount of soil water within the soil-core section, the displacement of which is given by the displacement of a tracer or bomb tritium peak. Since soil-core sampling costs increase greatly with core length, the mean recharge estimates obtained by this technique are rather short-term. By using simple hand augers in tracer experiments, cores 2 m long, in some cases a few metres, can be obtained. For environmental tritium measurements in soil moisture a greater effort in core sampling is necessary. On the other hand, under Central European climate conditions the prominent bomb tritium peak of the 1963/64 precipitation is found below 10 m depth, depending on the field capacity of the individual soil. In many cases it has already passed into the saturated zone. In arid zones, however, these high bomb tritium contents are expected to be still preserved in the uppermost soil or sand-dune layers, if any groundwater recharge occurs. Indeed, distinct tritium and stable isotope profiles were observed in the Saudi Arabian Desert by Dinçer et al. [9]. Such isotope profiles are expected to be suitable tools for studying the recharge and evaporation pattern of arid zone soils, which is of particular interest for current projects in arid zone hydrology [11]. For this purpose we have developed a new in-situ soil-moisture sampling technique that reaches about 30 m deep under favourable conditions (dune sand). Bomb tritium as a hydrological tracer has been applied to lysimeters [12], springs [13] and rivers [14–17] by many authors.

This paper also describes the bomb tritium response of a sand and of a loam lysimeter, based on measured tritium values of monthly percolate samples available since 1961. The measured concentration values and the lysimeter percolation rates are explained by a simple HETP model with field capacity and evapotranspiration as open parameters. The same model is applied to predict water movement in tritium-tagged natural soils.

1. THE SUCTION METHOD OF IN-SITU SOIL-MOISTURE SAMPLING

The basic idea of this new technique is simple. A probe, comprising a standard 1-inch steel tubing, previously used for shallow groundwater sampling [10],

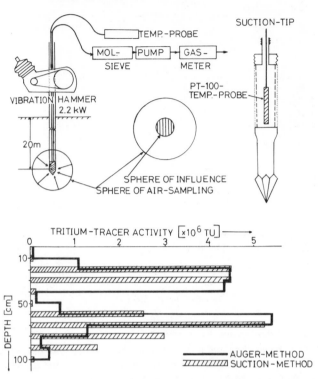

PRINCIPLE OF SUCTION-METHOD

FIG.1. Resolution comparison: Suction vs auger method.

300 litres of soil air are pumped, corresponding to a soil sphere of 1 m dia. However, a rapid exchange between the streaming gas and the liquid phase results in the water sampled originating only from the vicinity of the suction tip (sphere of influence dia. < 10 cm). This gives a depth resolution of a few centimetres, as seen from a double-peak tritium tracer experiment in sandy soil.

is rammed down to the depth selected in the unsaturated soil zone. A vibration hammer[1] is used.(Fig.1). The suction tip of the probe is connected to a pumping system by a plastic hose running inside the probe. The small pump[2] operated from a car battery via a 12-V d.c./220-V a.c. chopper/transformer pumps the moisture saturated soil air at a rate of, e.g., 30 litres/min. The water vapour delivered is adsorbed in a molecular sieve (4 Å) bed. The loaded molecular sieve samples are stored in standard glass bottles. The soil moisture is later re-extracted quantitatively in the laboratory by heating to 400°C in a vacuum system and freezing the released water in a dry-ice cold trap (max. sample loss < 1‰).

---

[1]   Made by Zremb, Gdansk (Poland).
[2]   Reciprotor, Copenhagen (Denmark).

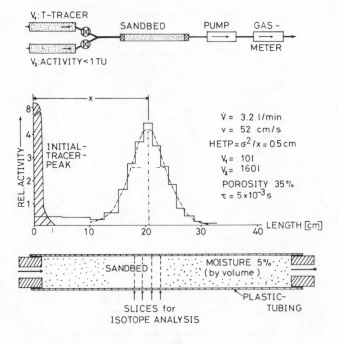

FIG.2. *Vapour/soil-moisture exchange experiment.*

   *Gaussian-approximated tritium tracer activity observed in a sand bed, after passage of
10 litres tritium-tagged humid air and subsequently 160 litres tritium-free air. The relaxation
time τ between gas and liquid phase is calculated to be about 5 ms.*

At 10°C soil-temperature 1 m³ soil air sampled yields about 10 cm³ water.
Normally, 3-cm³ sample size is sufficient for isotope analysis, i.e. the pumping
time per sample is about 10 min. Then the probe proceeds to the next deeper
position. The size of the soil-moisture samples can in principle be measured by
the weight difference of the loaded and unloaded molecular sieve. Sample size and
pumped soil-air volume provide information about the soil temperature at the
suction tip, an amount which can be checked with the soil temperature measured
directly by a Pt-100 temperature probe of the suction tip and a digital display unit.
Exact soil-temperature data is indispensable for stable isotope measurements of
soil moisture, since the stable isotope data of the moisture samples must be
corrected for equilibrium isotope fractionation between the collected soil-water
vapour and the liquid soil water.

   The depth resolution of suction technique is surprisingly high. This has been
shown in a field experiment where tritium point injections were made at two
different depths. Soil-water profiles have then been sampled by both the suction
probe and the standard hand auger technique. The results of this tritium double-
peak experiment are shown in Fig.1.

2. MOLECULAR EXCHANGE BETWEEN VAPOUR AND LIQUID PHASE

At first glance the high resolving power of the suction method is surprising. If about 1 m³ soil air is pumped to obtain a 10-cm³ moisture sample, the air sampled originates from a sphere of about 1 m radius around the suction tip, if a porosity of 0.25 is assumed. On the other hand, the vapour component of the soil air obviously represents a sphere of influence of a few centimetres radius only, yielding the observed high resolution. This paradox is due to the fact that vapour/liquid molecular exchange in moist soil is extremely rapid, with a relaxation length of a few millimetres only, at the air-flow velocities used. This has been shown by the following laboratory experiment (Fig.2).

Air, saturated with vapour, alternately tritium-tagged and free of tritium is passed at constant flow-rate through a sand bed with about 5 vol.% moisture content. This sand bed is contained in a standard plastic hose of 1.8 cm inner diameter. Initially about 10 litres of air with tritiated moisture and afterwards 160 litres of air with tritium-free moisture were passed through the sand bed, which was then cut into small sections of about 1 cm long.

The tritium activity of the moisture extracted from the sections was measured in a scintillation counting system. In Fig.2 the tritium distribution along the sand bed of one of these experiments is shown. Originally, the moisture in the sand had been tritium-free. The Gaussian-type tritium peak observed after the whole experiment, at a position 20 cm from the front end, is the result of a chromatographic exchange between vapour moving and the liquid phase at rest. The tritiated vapour contained in the first 10 litres of air was totally absorbed by the liquid moisture at the front end of the sand bed initially. The subsequent flushing with air containing tritium-free moisture shifts the peak to the position observed. Analysis of the tracer distribution yields a mean exchange time of about 5 ms. With this relaxation time the relaxation length in the field experiment (lower flow velocity) is negligibly reduced to a few millimetres. Therefore, the sphere of influence is just the soil volume containing the amount of moisture collected by pumping (sphere radius about 5 cm under the conditions of the dune experiment).

3. WATER ADSORPTION PROPERTIES OF MOLECULAR SIEVE

The water adsorption properties of the molecular sieve have been studied in great detail. Moist air of well-defined isotopic composition was passed through molecular sieve beds at various flow-rates. The water content of the molecular sieve shows an error-function-type distribution from the one end to the other, the gradient of which decreases with increasing flow-rate. The isotopic fractionation of molecular sieve adsorption at saturation level is similar to that of condensation at equilibrium. At the rear end of the molecular sieve column, however,

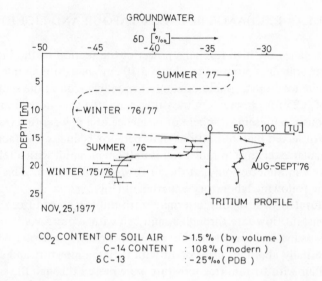

FIG.3. *Pilat sand-dune isotopic profile.*
Preliminary deuterium and tritium profile, obtained by the suction method. The shape of the deuterium profile is probably due to seasonal variations of deuterium in precipitation, which are preserved in the unsaturated zone under these conditions. The tritium peak of 95 TU at 15.5 m depth was also observed in the rain of August − September 1976.

the air is so much depleted in water vapour that the absorption by the dry sieve is a one-way process and kinetic preference of the lighter water molecules is clearly observed. On the whole, the molecular sieve has proved to be sufficiently reliable to sample vapour even for stable isotope analysis. Its memory is about 3% and can be corrected for.

4. TEST EXPERIMENT IN THE PILAT DUNE: FUTURE PROGRAMME

In November 1977 the suction method was tested at the Pilat dune, a shifting dune 110 m high, 2 km long and about 250 m wide, near Bordeaux, France. Unfortunately, owing to very bad weather conditions at that time (storm with heavy rainfall), only one profile of soil-moisture samples between 15 and 25 m deep could be collected.

The deuterium and tritium data of this depth section is presented in Fig.3. This isotope data is believed to be due to winter 1975/76, and summer 1976 precipitation [18, 19]. This assumption is supported by the occurrence of an isolated tritium peak of 95 TU which was also observed in rain of August− September 1976 at various European collection stations. This tritium peak in precipitation was probably due to tritium release by nuclear industry. If our tentative interpretation is correct, this would mean a yearly groundwater recharge through this dune of about 800 mm.

The Pilat dune results are only preliminary. A more detailed study is planned for this year (1978). We are, however, also prepared to do sand-dune studies in the arid zone where soil moisture of the pre-bomb time should be found within a 30-m depth range. Field work is planned for the desert of Saudi Arabia, in co-operation with T. Dinçer [9], and the Food and Agricultural Organization, Riyadh, Saudi Arabia.

5. BOMB TRITIUM RESPONSE OF LYSIMETERS SINCE 1961

The tritium concentration of about 450 monthly samples of precipitation and the lysimeter percolate since 1961 were measured at two lysimeter stations:

(a) The lysimeter E1a at Limburgerhof[3] (8°25'E, 49°30'N near Ludwigshafen) is filled with sand (depth 1 m, area 1 m^2), supported by a gravel layer.

(b) Lysimeter B at Sindorf[4] (6°45'E, 50'N near Cologne) consists of an upper loess layer of 40 cm thickness, followed by 110 cm loam and a gravel pack of 150 cm thickness.

Both lysimeters are grass covered and not weighable.

Figure 4(a) shows the tritium response of the sand lysimeter Limburgerhof, together with the tritium time variation of precipitation for this locality. The prominent tritium peaks of 1963/64 can be easily identified, showing a delay of five months behind the bomb tritium in rain, which corresponds to a moisture displacement velocity of v = 2.4 m/a.

There is no obvious broadening of the input histogram by dispersion, which means that monthly steps are too coarse to resolve the pattern caused by dispersion. In the case of the loam lysimeter Sindorf the peaks of 63/64 are not resolved, they form a broad maximum with a delay of about 2.5 years. Some irregular points in 1963 indicating a direct rain contribution of a few per cent may be due to passage of heavy rainfall through soil cracks.

A second sand lysimeter at Limburgerhof and a second loam lysimeter at Sindorf show a response very similar to the corresponding examples presented here.

---

[3] Lysimeter station: Agricultural Research Institute, BASF Ludwigshafen.

[4] Lysimeter station: Ministry of Food, Agriculture and Forestry, Düsseldorf.
For deuterium measurements of seepage water from both lysimeters see Ref.[21].

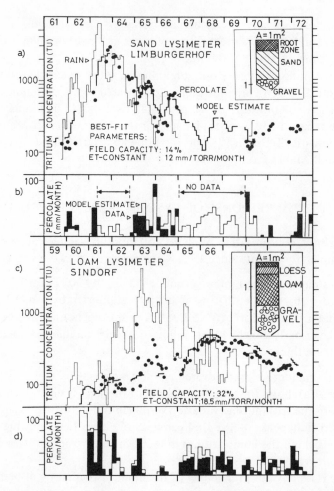

FIG.4.(a) *Measured tritium concentration at Limburgerhof lysimeter station (in TU, logarithmic scale, monthly values 1961-66, 69-72), lysimeter percolate (●), precipitation (thin line histogram), HETP model best parameter fit (thick line histogram). The tritium response is delayed by about 5 months (moisture displacement velocity: 2.4 m/a). For HETPs to be chosen smaller than 10 cm the model prediction becomes independent of the HETP. This seems to be obvious since the monthly percolate samples represent the amount of water contained in a layer of about 20 cm thickness. The model tritium histogram is interrupted when the model does not predict percolation water at the lysimeter outlet.*

(b) Bi-monthly values of lysimeter percolate (Limburgerhof) at top of dark area; measured values (1960-61, 63-64, 68-72), top of light area: HETP model estimate with the same parameters as in (a).

(c) Measured tritium concentrations at Sindorf, 1961-70. The bomb tritium input characteristic is smoothed out to a broad maximum with a delay of 2.5 years. Model parameters are fitted to the tritium response of 1964-70. Some irregular points in 1963, indicating a direct rain contribution of a few per cent, may be due to passage of heavy rainfall through soil cracks, and cannot be explained by the model.

(d) Bi-monthly values of lysimeter percolate (Sindorf) 1960-70.

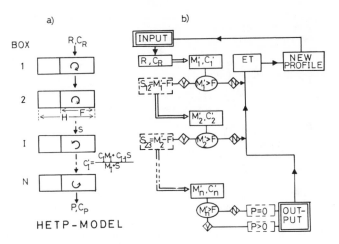

BOX

HETP-MODEL

FIG.5. *Schematic sketch of the HETP model.*

(a) The unsaturated zone is described as a series of internally well-mixed boxes of height H and field capacity F. The model works as follows. The individual monthly precipitation R(t) with tritium activity $c_R(t)$ is added to the amount m_1 of water in the uppermost soil box whereby its tritium content is changed accordingly to c_1'. If $m_1 + R$ exceeds the amount of water corresponding to the field capacity F, the excess $S = m_1' - F$, after complete mixing, is transferred to the box below. Isotope fractionation effects are neglected throughout. The sample procedure goes on from box to box, transferring water exceeding field capacity until the last box is reached, where the excess water P (if $m_n' > F$) leaves the lysimeter outlet. Its tritium concentration c_p is the result of multiple mixing.

(b) Flow chart. Double-lined arrows indicate fluxes of infiltrated precipitation (R) and percolating water (S). After each time iteration step ($\Delta t = 1$ month) the evapotranspiration loss, calculated from Eq.(1), is removed from the uppermost box.

6. SIMULATION MODEL OF WATER MOVEMENT IN THE UNSATURATED ZONE: APPLICATION TO LYSIMETER RESPONSE

The bomb tritium response of both lysimeters can be simulated by a simple one-dimension model, which describes the unsaturated zone as a series of internally well-mixed layers (boxes or 'plates') — Fig.5. The plate height H (see [20]) and the field capacity F of the soil are chosen as open-model parameters, controlling longitudinal dispersion (cf. HETP = Height Equivalent to a Theoretical Plate in chromatography) and average tracer travel velocity, respectively. The model works as follows:

Starting from a soil-moisture profile, m_i (i = box number), and a tritium profile, c_i, the individual monthly precipitation, R(t), with a tritium activity, $c_R(t)$, is added to the amount of water in the uppermost soil box (new water content m_1'),

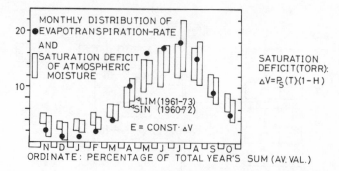

FIG.6. Seasonal variation of evapotranspiration and saturation deficit in atmospheric moisture. Evapotranspiration for north and middle Germany, F.R. (denoted by full dots (●), [22] and saturation deficit (vertical bars representing ±1σ range of values observed during around 1960 to 1973 at both lysimeter stations). Both sets of data (dots and bars) show approximate equivalence and are therefore assumed to be proportional. (Eq.(1)).

whereby its tritium content is changed accordingly to c_1'. If $m_1' < F$, no further moisture movement occurs; otherwise the quantity $S = m_1' - F$ with the tritium concentration c_1' is mixed into the following box. This procedure is repeated from box to box and either leads to a stop of percolation in the nth box or to water release at the lysimeter outlet with the tritium concentration c_n' as the result of multiple mixing. Between these calculating loops which simulate infiltration and seepage, the evapotranspiration is simulated by removing water from the uppermost box at a rate E to be discussed in Section 7.

7. ESTIMATING EVAPOTRANSPIRATION

To evaluate monthly evapotranspiration rates E(t) a very crude approximation was used. Figure 6 shows the monthly distribution of the average annual evapotranspiration, determined for northern and central Germany, F.R. [22], compared with the monthly variation of the saturation deficit ΔV in atmospheric moisture at at both lysimeter stations. Both sets of data show approximate equivalence and we we can therefore assume

$$E(t) \sim \Delta V(t) \tag{1}$$

with

$$\Delta V = P_S(T) \cdot (1-h)$$

TABLE I. LONG-TERM ANNUAL AVERAGES OF PRECIPITATION (R),
PERCOLATE (P), WHOLE YEAR's SUM OF SATURATION DEFICIT IN
ATMOSPHERIC MOISTURE ($\Sigma \Delta V$), CALCULATED EVAPOTRANSPIRATION
(E) AND EVAPOTRANSPIRATION (ET-) CONSTANT K FOR BOTH
LYSIMETERS

Lysimeter	Soil	R (mm)	P (mm)	$\Sigma \Delta V$ (torr)	E (mm)	K (mm/month per torr)
Limburgerhof	Sand	569	233	28	336	12
Sindorf	Loam	670	175	26.8	495	18.5

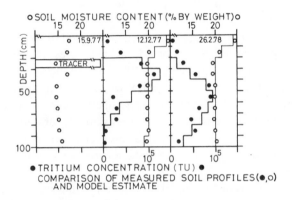

FIG.7. *Measured tracer (●) and soil-moisture content (○) profiles of loamy sand in a field experiment at Bammental (20 km southeast of Heidelberg) and an estimate due to the HETP model (solid lines), as calculated from daily data of precipitation, air temperature and saturation deficit in atmospheric moisture. Model parameters are the field capacity, the evapotranspiration constant and the in-situ density of the dry soil.*

where

$E(t)$　= monthly evapotranspiration rate

$P_S(T)$ = saturation vapour pressure (torr) of atmospheric moisture at the average temperature of the proper month

h　　= average relative air humidity (%) of the proper month

The proportionality constant (to be called ET-constant K), determined by the condition that the long-term integral of Eq.(1) (November to October over one or more years) must yield the long-term evapotranspiration (lysimeter water balance). The numerical values known experimentally for each individual lysimeter are given in Table I.

Evapotranspiration formulas such as Penman's formula, the Bowen ratio method or other methods [23] could not be applied, since the necessary meteorological data such as soil heat flow or net radiation were not monitored at either of the lysimeter stations.

In the case of the sand lysimeter the best model fit (Fig.4(a) (b)) was obtained for a field capacity F = 14% by volume and an evapotranspiration constant K = 12 mm/month per torr. The corresponding values of the loam lysimeter (Fig.4(c) (d)) are F = 32% and K = 18.5 mm/month per torr. The evapotranspiration constants correspond to an evaporation transfer velocity of w = 0.45 and 0.7 cm/s, respectively (for comparison of these values with values of w for evaporation from open-water surfaces see Ref.[24]).

The HETP model described can also be used to predict the displacement of the tritium tracer peak, the development of the soil-moisture profile, and thus the groundwater recharge in tritium tracer studies [1]. Figure 7 shows a first example of such a model prediction compared with the measured profiles. In this case the field capacity, varying with depth, is based on auger cores. The exact in-situ soil density, coming into play by the necessity of transforming soil-moisture wt% into vol.% can be adjusted from winter core data, where evapotranspiration is almost negligible. From summer core data it is then hoped to obtain more reliable values of the evapotranspiration constant K, which presumably, at a higher level of approximation, depends on season as well as on type of soil and plant cover. This procedure is at present being tested at several locations with different types of soil and plant cover.

Another application of the model described lies in modelling the tritium response in spring and river waters. The modelling of data from continuous sets of such samples, available from 1960 onwards, is at present under way in this laboratory.

ACKNOWLEDGEMENTS

The authors would like to thank Mrs. Christel Junghans, who did the stable isotope measurements. Thanks are due to Hugo and Horst Koch, Limburgerhof, and H. Diehl, Habbelrath-Horrem and the Staatliches Amt für Wasser- und Abfallwirtschaft, Aachen, for providing the lysimeter water samples and the lysimeter data. The Minister für Ernährung, Landwirtschaft und Forsten, Düsseldorf, gave the permission for publishing the data of the Sindorf lysimeter. This work was supported by the Deutsche Forschungsgemeinschaft and the Akademie der Wissenschaften, Heidelberg.

REFERENCES

[1] MÜNNICH, K.O., SONNTAG, C., "Tritium-Markierung", in Methoden zur Bestimmung der Grundwasserneubildungsrate" Geol. Jahrb., Reihe C **19** (1977) 22.

[2] ROETHER, W., Tritium und Kohlenstoff-14 in Wasserkreislauf, Z. Dtsch. Geol. Ges. Sonderh. Hydrogeol. Hydrogeochem. (1970) 183.

[3] SMITH, D. B., WEARN, P.L., "Water movement in the unsaturated zone of high and low permeability strata by measuring natural tritium", Isotope Hydrology (Proc. Symp. Vienna 1970) IAEA, Vienna (1970) 73.

[4] ZIMMERMANN, U., EHHALT, D., MÜNNICH, K.O., "Soil water movement and evapotranspiration: Changes in the isotopic composition of water", Isotopes in Hydrology (Proc. Symp. Vienna 1967), IAEA, Vienna (1966) 567.

[5] DATTA, P. S., GOEL, P. S., RAMA, F.A., Sc., SANGAL, S.P., "Groundwater recharge in Western Uttar Pradesh," in Proc. Indian Acad. Sci. **78** Sec 'A' 1 (1973).

[6] GOEL, P. S., DATTA, P.S., TANWAR, B.S., Measurements of vertical recharge to groundwater in Haryana State (India) using tritium tracer, Nord. Hydrol. **8** (1977) 211.

[7] MÜNNICH, K.O., ROETHER, W., THILO, L., "Dating of groundwater with tritium and C-14", Isotopes in Hydrology (Proc. Symp. Vienna, 1966) IAEA, Vienna (1967) 305.

[8] ATAKAN, Y., ROETHER, W., MÜNNICH, K.O., "The Sandhausen shallow-groundwater tritium experiment", Isotope Techniques in Groundwater Hydrology 1974 (Proc. Symp. Vienna 1974), IAEA, Vienna (1974) 21.

[9] DINÇER, T., AL-MUGRIN, A., ZIMMERMANN, U., Study of the infiltration and recharge through the sand dunes in arid zones with special reference to the stable isotopes and thermonuclear tritium, J. Hydrol. **23** (1974) 79.

[10] ATAKAN, Y., "Bomb tritium hydrology of a sandy unconfined aquifer", Ph.D. diss. (English version), University of Heidelberg, Heidelberg 1972 (unpublished).

[11] SONNTAG, C., Paläoklimatische Information im Isotopengehalt C-14 datierter Saharawässer: Kontinentaleffekt im Deuterium und Sauerstoff-18, to be published in Geol. Rundsch. (1978).

[12] SAUZAY, G., "Interpretation of Lysimeter Data", Technical Report, IAEA, Vienna (1974).

[13] PRZEWLOCKI, K., YURTSEVER, Y., "Some conceptual mathematical models and digital simulation approach in the use of tracers in hydrological systems", Isotope Techniques in Groundwater Hydrology 1974 (Proc. Symp. Vienna, 1974), IAEA, Vienna (1974) 425.

[14] WEISS, W., ROETHER, W., Tritium run-off of the Rhine River 1961-1973, Dtsch. Gewässerkundl. Mitt. **19** 1 (1975) 1.

[15] ROETHER, W., "Estimating the tritium input to groundwater from wine samples: Groundwater and direct run-off contribution to Central European surface waters", Isotopes in Hydrology (Proc. Symp., Vienna 1966), IAEA, Vienna (1967) 73.

[16] BROWN, R.M., "Distribution of hydrogen isotopes in Canadian waters", Isotope Hydrology (Proc. Symp. Vienna, 1970), IAEA, Vienna (1970) 3.

[17] ERIKSSON, E., Atmospheric tritium as a tool for the study of certain hydrologic aspects of river basins, Tellus **15** 3 (1963) 303.

[18] EHHALT, D., KNOTT, K., Kinetische Isotopentrennung bei der Verdampfung von Wasser, Tellus **17** 3 (1965) 389.

[19] CRAIG, H., Isotopic variations in meteoric waters, Science **133** 3465 (1961) 1702.

[20] ZIMMERMANN, U., MÜNNICH, K.O., ROETHER, W., Downward Movement of Soil Moisture Traced by Means of Hydrogen Isotopes, Geophysical Monograph No.2, Am. Geophys. Union, Washington, D.C. (1967).

[21] EICHLER, R., Deuterium — Isotopengeochemie des Grund- und Oberflächenwassers, Geol. Rundsch. **55** (1965) 144.

[22] KELLER, R., Gewässer und Wasserhaushalt des Festlandes, Verlag Haude und Spener, Berlin (1961).

[23] GRANT, D.R., Comparison of evaporation measurements using different methods, Q.J.R. Soc. **101** (1975) 543.

[24] MÜNNICH, K.O., CLARKE, W.B., FISCHER, K.H., FLOTHMANN, D., KROMER, B., ROETHER, W., SIEGENTHALER, U., TOP, Z., WEISS, W., "Gas exchange and evaporation studies in a circular wind tunnel, continuous Radon-222 measurements at sea, and tritium/helium-3 measurements in a lake", Paper presented at NATO Symp., Turbulent Fluxes Through the Sea, Wave Dynamics and Prediction, Marseille, 12-16 September 1977, (to be published by Plenum Press New York).

DISCUSSION

R.L. OTLET: I am interested in the values of 108% for ^{14}C and $-25\permil$(PDB) for $\delta^{13}C$ you give in Fig.3. In the associated slide of the study area shown in your oral presentation there appeared to be no plant cover which might explain the $\delta^{13}C$ value. Also, 108% ^{14}C is a very low value for 1976. What is your interpretation of these values?

G. THOMA: The Pilat dune shifts with a velocity of 6 m/a, covering more and more of the forest at the eastern slope. This organic material produces CO_2 by bacterial decay. The plant material whose CO_2 was sucked up may have been covered between 1950 and 1960.

J. JOUZEL: I would like to ask you two questions about the use of the molecular sieve. Have you studied the memory effect occurring between two successive uses? And, secondly, do you not have problems in the case of stable isotopes, in particular with ^{18}O?

G. THOMA: The memory effect with respect to tritium amounts to 3% and can be corrected for. No memory effect has been found with stable isotopes. With respect to your second question, you are right: the ^{18}O values are completely off, probably because of exchange of oxygen between the water vapour and the soil-air CO_2. The CO_2 concentration in the soil-air of the Pilat dune was found to be more than 1.5% by volume, because of the decay of wood from trees recently overrun by the dune.

I. WENDT: For stable isotope measurements it is very important to have 100% collection of water in the molecular sieve. We have applied a very similar method for soil-CO_2 isotope measurements and have observed isotope fractionation. Have you checked whether you have 100% sample recovery in order to be sure that you are measuring the true isotope values?

G. THOMA: The efficiency of a molecular sieve bed is a function of flow rate and sample size. At 30 litres/min the amount of molecular sieve required for quantitative adsorption is approximately the weight of the water multiplied by a factor between 15 and 20. Thus, for a 5-g sample, 75 or 100 g of molecular sieve are needed.

Y. YURTSEVER: The simulation model used for evaluation of tritium data from lysimeters is certainly interesting, and agreement between the actual tritium output and simulated output is quite good. The use of interconnected boxes in fact involves a simulation of dispersion phenomena, and the number of boxes used is directly related to dispersion. How do you decide on the number of boxes to be used for the model? Is it selected arbitrarily, or do you use a trial and error procedure whereby you start with a small number of boxes and increase the number until a good fit is achieved between the observed and computed tritium output curves?

N. ESSER: Selection of the number or width of boxes is in fact performed by a trial and error procedure. The box width is adjusted in such a way that the observed peak width is reproduced correctly. In the case of sandy soil, the plate height H is around 1 cm, so more than 50 boxes should have been used for the 1-m sand lysimeter. We noticed, however, that when we used more than ten boxes, the response function became independent of the number of boxes. This is due to the fact that the data from the rain tritium input histogram (monthly average values) are already too approximate to justify a higher resolution.

T. DINÇER: I think it would be advisable to use a reliable evapotranspiration formula instead of the moisture deficit of the atmosphere for estimating evapotranspiration. This could improve the results of the study and one would have more confidence in the results. The factors used for the Penman formula, for example, are measured at most meteorological and lysimeter stations.

G. THOMA: You are certainly right in saying that evapotranspiration rates calculated from Penman's or any other formula would give more reliable results. However, the meteorological data needed for this study have not been monitored near the lysimeter stations, and the available data certainly do not go back more than 15 years. For this reason we felt that our crude approach, which is based only on local values for air temperature and relative air humidity, was likely to be more useful than Penman's formula applied to meteorological data measured far from the lysimeter stations.

K.O. MÜNNICH: The method provides no information on the soil-moisture content. Since we have a resistance thermometer in the probe tip anyway, we thought of adding a small electric heater to it to warm up the region around the tip artificially. If we then sucked air we would get Rayleigh distillation of water from the warm region. The slope of the changing isotopic composition would then yield information on the initial amount of water present. When discussing this possibility we became aware of another, namely that of estimating water content simply by watching the changing temperature at the tip. Since heat conduction depends strongly on moisture content, particularly with very low moisture contents (the liquid from heat conduction makes a bridge between the grains), this method should be feasible. I would be interested to know whether it has been tried before. Secondly, for this study we had great difficulty getting the probe down to large depths in normal soils. This caused us to try sand dunes where this is easy. Does anybody know of better equipment for lowering such a probe under less favourable conditions?

T. TORGERSEN: I believe the US Army Cold Regions Research and Engineering Laboratory (CRREL), at Hanover, New Hampshire, USA, has developed such a heated-tip probe.

TRITIUM TRACING IN HYDROGEOCHEMICAL STUDIES USING MODEL LYSIMETERS

G. MATTHESS, A. PEKDEGER, H.D. SCHULZ
Geologisch-Paläontologisches Institut
   der Universität, Kiel

H. RAST, W. RAUERT
Gesellschaft für Strahlen- und Umweltforschung mbH,
Institut für Radiohydrometrie,
Neuherberg,
Federal Republic of Germany

Abstract

TRITIUM TRACING IN HYDROGEOCHEMICAL STUDIES USING MODEL LYSIMETERS.
   Tritium was used as a reference tracer for hydrogeochemical studies in the unsaturated zone, since it does not affect the geochemical processes and can be used to identify water of an individual rain event infiltrating and seeping down to the groundwater table. The investigators made use of different lysimeter types (25, 50, 100 cm), with and without suction plates filled with undisturbed soil monoliths of sandy podsol and loamy lessivé. The tritium loss was greater than the evaporation amount determined. Water-logging takes place in lysimeter bottoms increasing the evaporation in lysimeters up to 100 cm filled with loamy lessivé and 25 cm with sandy podsol. After a 20-mm rain event seepage characteristics indicate "by-passing" water besides intergranular seepage. Dispersion coefficients (8.5×10^{-5} cm^2 s^{-1}) are higher than molecular diffusion coefficient. Dispersion takes place mainly in top soil with wide-ranging pore-size distribution. Distribution coefficients of tritium in soil is rather low. Concentrations of anions and dissolved organic substance are different depending on residence time of seepage water in soil. Even a short residence time of seepage water in unsaturated soil is enough for cation exchange reactions to take place.

INTRODUCTION

The geochemical and biochemical reactions during percolation of infiltrated rain-water through the unsaturated zone down to the groundwater table control the quality of recharge water, groundwater and the movement of pollutants. The unsaturated zone can be considered as a three-phase system consisting of solid ground material, liquid soil moisture and ground air. The solute content of soil moisture is so different from groundwater that various physico-chemical changes are to be expected [1].

Weathering processes, such as transformation and translocation, lead to a stratified structure of the uppermost underground layer. The characteristics and type of soil profile depend on the geological substrate, the local climate and the vegetation. In humid climates, the soil genesis includes — as present in the investigated localities — physical, chemical and biochemical weathering with dissolution, precipitation and co-precipitation, adsorption, ion-exchange, gas exchange and biodegradation. These processes mobilize the solid constituents of the unsaturated zone, which move down into the groundwater. The percolating water and groundwater, which are the transport media for the dissolved substances, move according to the physical laws of gravitation, adhesion, dispersion and diffusion. The residence time of soil water in the unsaturated zone depends on rain intensity, soil-water content, permeability of the soil and underlying rock. Recent studies using tritium as a tracer have verified in rather homogeneous media a stratification of percolate water layers of varying ages [2—7]. Precipitation infiltrating the unsaturated zone is stored there and moves downwards very slowly and regularly. It may take several years before the water of a rain event reaches the groundwater table [e.g. 2—8].

Water movement in the unsaturated zone is controlled by gravity and capillary forces. Seepage takes place when a downward gradient of the hydraulic potential occurs, owing to a change in moisture content caused by precipitation.

The seepage velocity of the percolates in the unsaturated zone of porous media is slow enough for lateral molecular diffusion, which is theoretically a fast-mixing process over short distances to equalize the different quality of seepage and adhesive water even in larger pores [8, 9]. The chemical properties, especially the cation content, are controlled by adsorption processes, depending on the nature of soil substance and on soil-moisture content, as given by Freundlich's isothermal equation:

$$C_s = k \cdot C_w^n$$

C_s = concentration of cation adsorbed
C_w = concentration of cation in solution
k, n = specific constants

A stratification of different chemical properties occurs in soil water. After a rain event the soil-moisture content of soil changes and a downward movement occurs, disturbing the chemical equilibrium between soil-water solution and solids. Owing to repeated adsorption and desorption processes and lateral diffusion, a chemical equilibrium will be reached after some time. The considerable changes of the chemical composition of rain are so complete that, for a quantitative evaluation of the migration rates of chemical substances, other criteria are needed. Tritium proves to be a suitable tool for such hydrochemical studies in the unsaturated zone, since it does not affect the geochemical processes and can be used to identify water of an individual rain event seeping down to the groundwater table.

The seepage velocity depends mainly on the gradient of soil-water tension and gravity, and on the permeability of unsaturated soil. Permeability of unsaturated porous media are controlled by texture, structure and moisture content. In saturated soil nearly all pores are water-filled; thus, conductivity is at a maximum. As the soil becomes depleted, the air-filled portion of the pores increases raising the adhesive forces over-proportionally and reducing the conductive cross-sectional area of the medium. Only after heavy precipitations may the hydraulic conductivity of the unsaturated zone allow a more rapid movement of the infiltrated water in large pores so that a chemical equilibrium with adhesive water, present in small capillary pores, cannot be achieved. This is expected mainly in the top soil (0-horizon) with a loose structure and a high moisture content, which rather resembles, hydrologically, a fractured hard rock. Infiltration ranges of heavy precipitations (25 mm/d) reaching 30–40 cm depth have been reported [4].

The purpose of this study was to investigate the relation between seepage water movement and chemistry. Supplementary information was expected about the kinetics of chemical equilibria between rainfall and seepage water, oxidation of organic substances, ion-exchange and impeded oxidation due to water logging.

Tritium was used as a tracer, since the field studies quoted above indicate that tritium migration takes place at nearly the same velocity as soil moisture movement. The following observations concerning the possibility of differences between tritium and soil moisture movement have been made:

(i) Tritium ultimately exists in soil as a tritiated water molecule and is theoretically capable of exchanging for hydrogen and other ions on soil particles.

(ii) Kline and Jordan [10] observed a tritium loss in the free soil water of a clay soil in Puerto Rico. They suggested that the tritium loss is due to isolated immobile interstitial water that has free interchange with the more rapidly moving water in larger pores. Leaching experiments indicate immobile interstitial water with higher ion concentrations than in seepage water [11, 12]. The pH and hydrogen carbonate concentrations of the leachates were the same as in the seepage water; nitrate and sulphate were found only in traces. Other, even poorly adsorbed ions, such as sodium and chloride, showed ten times higher concentrations in the leachates than in the respective seepage water.

(iii) It has been reported [13–16] that tritium is retarded with respect to the movement of chloride ions in soils containing clay and silt. This behaviour was reported as caused by the tritiated water molecule entering adhesive water and replacing non-tritiated water adsorbed on clay surface and interlayers [13, 14] Other authors [15, 16] have suggested that chloride ions were repelled owing to anion exclusion from negatively charged soil particles into the central regions of the soil pores where the flow velocity is greatest, thereby causing higher chloride migration rates than mean flow velocity of water.

EXPERIMENTAL PROCEDURES

To observe the dynamics of chemical changes and transport of ions and colloids in the soil, soil water from different depths has to be sampled. Different methods can be used though none of these methods are perfect [17] because the retention of the soil water has to be interrupted to collect water samples. The use of tension lysimeters involves subpressure with suction plates which disturbs the equilibria of the dissolved gases and causes adsorption of the ions at the suction plates. The lysimeters without subpressure tend to water-log at the bottom of the lysimeters [18]. Drain-water is already groundwater by definition and shows other chemical characteristics. To eliminate mistakes, different samplings methods were used and the results were compared:

1.   Sampling of percolates
1.1. Use of suction plates or tubes
1.2. Lysimeters of different lengths which are set up in situ
1.3. Lysimeters (with and without suction plates) set up in the laboratory with definite precipitation and constant temperature
1.4. Experimental leaching of soil material with deionized water
1.5. Lysimeters with water saturation (groundwater lysimeters)
2.   Groundwater samples from the same localities

Seepage water from two different soil types (sandy podsol and loamy lessivé), which are very common in northern Europe, was investigated. The construction technique of the monoliths is described by Schulz [19]. The soil profiles are dissected at lengths of 25, 50 and 100 cm (dia. 30 cm) according to the soil horizons (A, A+B, A+BC), and the respective leach water seeping out is analysed (Fig.1). Rain events of 19 mm within 20 minutes were given weekly. All lysimeters were tagged in one certain "rain event" using tritium activities between 3.4 and 12.3 μCi. Water samples were taken after 2 h, 24 h and 1 week, following each rain event. After eight weeks only one weekly cumulative sample was collected. A total of 14 lysimeters with and without suction plates was investigated. The same lysimeters were also run a second time, partly covered, to decrease evaporation. In this case samples were taken weekly.

Batch tests were carried through to investigate the interaction between tritiated water, soil water and solid soil components. Various quantities of tritium were used (0.9–4.7 μCi) with 500 g of soil substance in 1000 ml water. After shaking the water-soil mixture for 6 h, 10 cm^3 tritiated water were injected. One set of samples was shaken manually only at the beginning, and a parallel set by a machine with a frequency of 150 rpm. Twenty-millilitre samples were taken after 5 min, 2 h, 24 h, 3 d and 9 d for tritium measurements. Chemical analyses

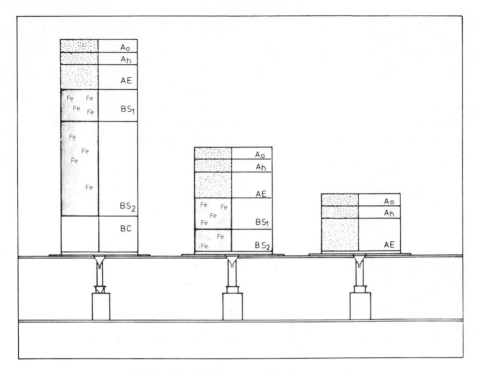

FIG.1.   Monolith lysimeters filled with podsol.

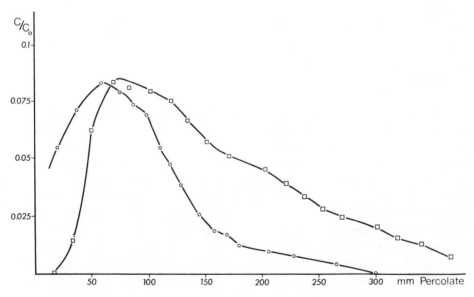

FIG.2.   Seepage characteristics of two podsol monolith lysimeters (50 cm) with (○) and
without (□) suction plates; c_0 = tritium concentration of tagged rain, c = tritium concentration
of the percolate.

of the leachates were made, and the Eh and pH of the solutions were registered
after every sampling.

Tritium was measured by liquid scintillation counting of water samples mixed
with instagel scintillator. In untreated effluent water from the lysimeters, quenching
effects occurred (owing to organic substances), which were small and could be
corrected by internal standardization. Water samples from batch tests contained
higher amounts of organic substances and had to be distilled prior to measurement.

The tritium distribution coefficients K_d were calculated for all soil horizons
according to

$$K_d \left[\frac{ml}{g} \right] = \frac{\bar{n}}{n} = \frac{V}{m} \left(\frac{c_0}{c} - 1 \right)$$

$\bar{n}$ = activity of adsorbed tritium in 1 g soil sample
n = activity of tritium in 1 cm^3 solution
m = mass of dry soil (g)
V = volume of solution (cm^3)
c_0 = initial tritium concentration in the solution ($\mu Ci/cm^3$)
c = tritium concentration after reaching equilibrium ($\mu Ci/cm^3$)

RESULTS

By-passes

Break-through curves of tritium obtained on two lysimeters of different
construction are shown in Fig.2, representing a uniform intergranular seepage flow.
The seepage characteristics in other lysimeters containing loamy lessivé and short
podsol monoliths ($\leqslant 50$ cm) display a flow additional to the uniform intergranular
seepage (first peak in Fig.3). The high precipitation of 19 mm, given as one rain
event within about twenty minutes, causes for a certain time a high water saturation in the lysimeters. Owing to increased hydraulic conductivity the water
moves faster in greater pores and in inter-aggregate pore systems than the
interstitial water; thus, the hydraulic and chemical equilibrium is only partly
reached. These by-passes are only present in soil horizons with wide-ranging pore-
size distribution such as organic A-horizons, or loamy soils such as the investigated
lessivé. The percentage of by-pass flow calculated from the tritium recovery in
effluent one day after tagged rain event is given in Table I.

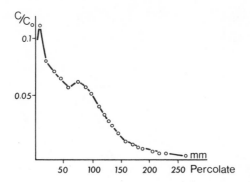

FIG.3.   Seepage characteristics of a 50-cm podsol monolith lysimeter with "by-pass" (first peak).

TABLE I.   EVAPORATION, TRITIUM LOSS (CALCULATED EXCLUDING
THE BY-PASS FLOW WHICH IS PRACTICALLY NOT SUBJECT TO EVAPO-
RATION) AND PERCENTAGE OF BY-PASS FLOW.

Lysimeter	Evapora-tion %	Tritium loss %	By-passes %
Podsol 25 cm	15	45	33
Podsol 25 cm	19	38	8
Podsol 50 cm	17	27	6
Podsol 50 cm	15	30	2
Podsol 100 cm	21	24	0
Podsol 100 cm	17	27	0
Lessivé 25 cm	26	59	69
Lessivé 25 cm	27	61	69
Lessivé 50 cm	30	54	56
Lessivé 100 cm	31	52	21

Equilibrium time

After a rain event it takes some time, depending on the length of the soil
column and the soil type, until the moisture distribution is completed and
percolation ceases. The water which is sampled in the time interval between
two rain events is a mixture of water from by-passes and intergranular seepage.
In the water samples collected after two hours (series I), a higher portion of by-
pass water is to be expected than in the samples which are collected after 24 hours
(series II) and 7 days (series III). In a soil with tritiated soil water, different

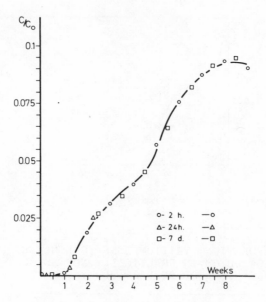

FIG.4. Seepage characteristics of a 100-cm podsol monolith lysimeter.

tritium concentrations may be present in the percolates sampled at different time intervals after each rain event. From samples taken at different time intervals the following results were obtained:

(i) There was no difference in the tritium concentrations in the percolates of series II and III (data of Series III is therefore not given in Table II). In all soil monoliths two hours seem to be enough for equilibrium to be reached.

(ii) In sandy podsols there is also no difference concerning tritium concentrations between series I and II (Fig.4), the same lysimeters also showing no by-pass effects (Table I).

(iii) In soil monoliths with a high percentage of by-pass flow (Table I) the tritium concentrations are lower in the percolates of series I than of series II (Table II).

(iv) In lessivé monoliths percolates of series I are already mixed partly with the tritiated adhesive soil water, as seen in the parallel development of tritium concentrations in Fig.5 (for more detailed discussion see Ref.[12]).

Evaporation and tritium loss

The tritium loss was greater than the evaporation determined hydrologically by the difference between precipitation and seepage (Table I). This is probably due to the evaporation of tritiated water originating from the single tritium tagged

TABLE II. AVERAGE TRITIUM CONCENTRATIONS IN SAMPLES COLLECTED 2 HOURS (I) AND 24 HOURS (II) AFTER EACH RAIN EVENT.

Lysimeter	Tritium concentrations $(\cdot 10^{-6} \mu Ci/cm^3)$	
	Series I	Series II
Lessivé 25 cm	112	218
Lessivé 50 cm	148	249
Lessivé 100 cm	150	180
Podsol 25 cm	194	227

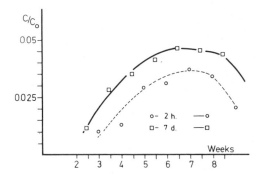

FIG.5. Seepage characteristics of a 100-cm lessivé monolith lysimeter.

precipitation, which ascends with the capillary rise to the surface and is subjected to subsequent evaporation. This agrees with the results reported by Zimmermann et al. [2], who state that the loss of concentrated deuterium tracer due to evaporation is greater than evaporation of moisture in the same soil level, since the tracer concentration in the atmosphere is negligible, so that the evaporation of tracer is not counteracted by the tracer content of the humidity of the air as it is in the case of normal water. In partly covered podsol lysimeters with limited evaporation, only minor differences between hydrologically measured evaporation and tritium loss were observed. This result shows that tritium loss due to tritium fixation at the soil solid can only be minimal.

Capillary rise

In addition, the evaporation of tritium is increased in lysimeters because of water logging in the lysimeter bottoms, causing additional evaporation due to ascending water. This affects lysimeters of sandy podsol down to 25 cm depth.

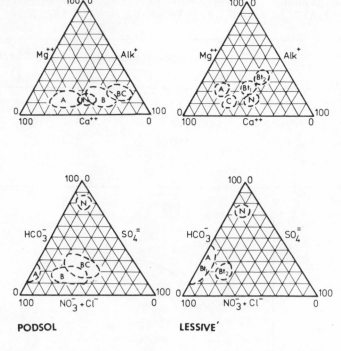

FIG.6.　PIPER diagrams of the investigated percolates A, B, BC, C = soil horizons; N = rain.

The loss of tritium calculated without the by-pass water, which is not subject to continuous evaporation, is higher in 25-cm lysimeters (42±5%) than in 50- and 100-cm lysimeters (27±2%). In loamy lessivé lysimeters, owing to higher hydraulic conductivity at low soil-moisture contents and greater capillary forces, there is increased tritium loss down to 100 cm depth (Table I), the total tritium loss in lessivé being higher than in sandy podsol.

Water logging in lysimeter bottoms also causes a retardation of effluents. The use of suction lysimeters prevents such water logging in monoliths. The average residence time of tritium, defined as time of 50% tritium recovery, d_{50}, in the suction lysimeters, is less than in the other monoliths without suction plates (Fig.2).

Distribution coefficients

The investigation of the distribution coefficients K_d of tritium in various soil horizons showed that tritium adsorption was rather low, being approximately 0.08 to 0.55 ml/g and decreasing with increasing initial tritium concentration.

The adsorption of tritium varied largely with time. The assumption that tritium adsorption could be affected by natural variations of pH and Eh in the system was not verified. The pH and Eh values of the samples showed no particular time dependence. Only a slight decrease of redox potentials in A-horizon samples was measured, although no correlation exists between redox potentials and tritium adsorption. Experimental difficulties make further investigations necessary.

Dispersion coefficients

The average dispersion coefficient is about $(8\pm5) \times 10^{-5}$ cm^2/s. This value is higher than the molecular diffusion coefficient of water (1.5×10^{-5} cm^2/s), probably owing to factors already mentioned. Dispersion coefficients lower than 1.5×10^{-5} cm^2/s were found in field tests by injecting tritium at a depth of 20 cm [20]. In this case the distribution of tritium was only due to vertical diffusion in deeper soil horizons. The A-horizon and the top soil (O-horizon) are the most inhomogeneous parts of the soil, increasing the dispersion coefficients due to by-passes and strongly varying moisture content of humus cover. For chemical studies, however, the behaviour of water movement is of considerable interest.

Residence time and cation exchange

Changes in the chemical composition of percolates within very short distances show that even a low residence time of seepage water in unsaturated soil is enough for cation exchange reactions to take place. This can be seen in Piper diagrams of both soils (Fig.6). The relative sodium and potassium concentrations increase proportionally as the selectively adsorbed calcium and magnesium decrease. Even in A-horizons with a high percentage of by-passing percolate, the elapsing time seems to be adequate for the ion-exchange processes to take place. The cation concentrations of two percolates from two different A-horizon lysimeters of 25 cm length, do not vary remarkably despite different residence time (Table III).

Oxidation of dissolved organic substance

Processes of oxidation, dissolution and precipitation of organic substance and their decomposition products depend on residence time, length of soil column and redox potentials. In the soil columns of A-horizons remarkable differences exist in the organo-chemical composition of the percolates (Table III) owing to different states of decomposition, depending on the average residence time and percentages of by-passing water.

In groundwater and in the seepage of deeper soil horizons, nitrate is the main compound of the present nitrogen compounds, whereas in rain-water other nitrogen

TABLE III. MEAN RESIDENCE TIME (d_{50}), PERCENTAGE OF BY-PASS FLOW AND CHEMICAL COMPOSITION OF THE PERCOLATES FROM PODSOL, A-HORIZON (25 cm) OF TWO LYSIMETERS.

d_{50}	By-passes %	$KMnO_4$-dem	NO_3	NO_2	NH_4	SO_4	PO_4	SiO_2	Ca	Mg	Na	K
					Concentration of solutes (mg/l)							
21 d	33	256	91	0.03	8.46	35	0.23	21.6	3.5	0.6	3.3	2.2
28 d	8	145	108	0.02	4.54	23	0.02	14.7	3.0	0.8	4.2	3.6

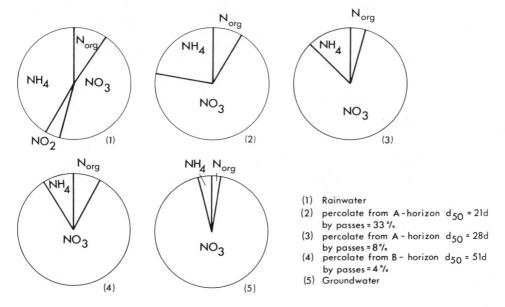

FIG. 7.   *Nitrogen distribution in percolates of podsol and groundwater.*

(1) Rainwater
(2) percolate from A-horizon d_{50} = 21d
    by passes = 33 %
(3) percolate from A-horizon d_{50} = 28d
    by passes = 8%
(4) percolate from B-horizon d_{50} = 51d
    by passes = 4%
(5) Groundwater

TABLE IV.   MEAN RESIDENCE TIME (d_{50}), MOISTURE CONTENT AND
CHEMICAL COMPOSITION OF THE PERCOLATES OF TWO PODSOL LYSI-
METERS (100 cm), WITH DIFFERENT SOIL MOISTURE CONTENTS

d_{50}	Soil moisture content %	Concentration of solutes (mg/l)					
		KMnO$_4$-dem	NO$_3$	SO$_4$	NH$_4$	CO$_2$	Eh (mV)
58 d	12	65	61	31	0.2	61	506
77 d	17	99	105	53	0.4	128	405

compounds are also present in comparable quantities. The nitrate percentage of
the total nitrogen compounds increase, under comparable physico-chemical
conditions, with increasing residence time because of oxidation of organic nitrogen
compounds. In Fig.7 the nitrogen compounds of seepage and groundwater are
presented together with the average residence time (d_{50}).

Transport of silica in the A-horizon takes place in colloidal form and in
organic complexes. The higher silica concentrations only reach deeper horizons
when the protecting organic colloids are not oxidized (Table III). During decomposition of organic substances, phosphorus and sulphur will be set free and will be

oxidized and selectively adsorbed in organic soil horizons [12]. Higher concentrations of these compounds in lysimeter seepages with by-passing water are probably due to their transport in higher permeable pore-systems where the adsorption is not as effective.

Water logging and soil-water chemistry

In water-logged soils with a high moisture content, the water movement is retarded. Furthermore, a higher moisture content decreases the diffusion of gases in soil [21, 22]. Owing to an impeded oxygen supply the redox potential in the water is lower and oxidation of organic substance is retarded. Impeded water migration is correlated with higher concentrations of dissolved organic substances and of their degradation products (carbon-dioxide substances and of their degradation products (carbon-dioxide, ammonia, nitrate — Table IV).

REFERENCES

[1]   MATTHESS, G., Die Beschaffenheit des Grundwassers, Gebrüder Born-
      traeger, Berlin, Stuttgart (1973)

[2]   ZIMMERMANN, U. et al., Science 152 3720 (1965) 346

[3]   ZIMMERMANN, U. et al., "Downward movement of soil moisture traced by
      means of hydrogen isotopes", Geophysical Monograph No.11, Isotope
      Techniques in the Hydrologic Cycle (STOUT, G.E., Ed.) A.G.U.
      Washington (1967) 28

[4]   BLUME, H.P. et al., Z. Pflanzenernaehr. Bodenkd. 112 2 (1966) 156

[5]   BLUME, H.P. et al., "Tritium tagging of soil moisture: the water
      balance of forest soils", Isotope and Radiation Techniques in Soil
      Physics and Irrigation Studies (Proc. Symp. Istanbul, 1967), IAEA,
      Vienna (1967) 315

[6]   BLUME, H.P. et al., Z. Pflanzenernaehr. Bodenkd. 121 3 (1968) 231

[7]   SMITH, D.B. et al., "Water movement in the unsaturated zone of high
      and low permeability strata by measuring natural tritium", Isotope
      Hydrology (Proc. Symp. Vienna, 1970), IAEA, Vienna (1970) 73

[8]   MÜNNICH, K.O., "Isotope studies on groundwater movement in the un-
      saturated and saturated soil zone", Symp. on Trace Elements in
      Drinking Water, Agriculture and Human Life (Cairo, 1977), MERRC/Dokki
      and Goethe Institut (1977)

[9]   FOSTER, S.S.D., J. of Hydrology 25 (1975) 159

[10] KLINE, J.R. & C.F. JORDAN, Science 160 (1968) 550

[11] MATTHESS, G. et al., "Geochemical-biogeochemical processes in seepage water during groundwater recharge", The second International Symposium on Water-Rock Interaction (Proc. Symp. Strasbourg, 1977), Ed. H. PAQUET, Y. TARDY, (1977) I 146

[12] PEKDEGER, A., Labor- und Felduntersuchungen zur Genese der Sicker- und Grundwasserbeschaffenheit, Diss. Kiel (1977)

[13] KAUFMANN, W.J. & G.T. ORLOB, Trans. Am. Geophys. Union 37 3 (1956) 297

[14] KAUFMANN, W.J. & D.K. TODD, "Application of tritium tracer to canal seepage measurements", Tritium in the biological and physical sciences (Proc. Symp. Vienna, 1962), IAEA, Vienna (1962) 84

[15] BEESE, F., R.R. v.d.PLOEG, Mitt. Dtsch. Bodenkdl. Ges., 23 (1976) 65

[16] FRITZ, P. et al., "Deuterium and Oxygen 18 as indicators of leach-water movement from a sanitary fill", Interpretation of Environmental Isotope and Hydrochemical Data in Groundwater Hydrology (Proc. of an advisory group meeting, Vienna 1975), IAEA, Vienna (1976) 131

[17] FRIESEL, P. et al., "Natürliche organische Substanzen im Sickerwasser", Proc. Symp. Organische Verunreinigungen in der Umwelt – Erkennen, Bewerten, Vermindern (Proc. Symp. Berlin 1976) Inst. Wasser-, Boden-, Lufthygiene, Berlin (in print)

[18] CZERATZKI, W., Landbauforschung Völkenrode 23 1 (1973) 1

[19] SCHULZ, H.D., Geol. Mitt. 10 (1970) 151

[20] JAKUBICK, A., Untersuchung der Wasserbewegung in teilweise gesättig-ten Böden – Tritium Tracermethode zur Bestimmung der Grundwasser-spende, Diss. Heidelberg (1972)

[21] ALBERTSEN, M. & G. MATTHESS, DGMK-BErichte, Forschungsbericht 146 Hamburg (1977)

[22] ALBERTSEN, M., Labor- und Felduntersuchungen zum Gasaustausch zwi-schen Grundwasser und Atmosphäre über natürlichen und verunreinig-ten Grundwaessern, Diss. Kiel (1977)

DISCUSSION

C.T. RIGHTMIRE: To what do you attribute the tritium retardation relative to chloride in waters percolating through the unsaturated zone, as mentioned at the end of the Introduction to your paper? Is it not possible for soil mineral composition, particle size, water film thickness or density to influence the rate of ion or molecular movement in unsaturated flow?

G. MATTHESS: The retardation of tritium relative to chloride was observed in clayey and silty soils by several scientists mentioned in the Introduction.

A possible explanation is that chloride ions are repelled because of anion exclusion from negatively charged soil particles into the bigger soil pores, where the flow velocity is higher than mean flow velocity through the whole soil section. We observed only slight indications of tritium fixation on clay surfaces and interlayers in our study.

P. MOUTONNET: Let me just say, with reference to the difference between the dispersive movement of tritiated water and of Cl^- in a clayey soil, that it is not so much the tritiated water which is slowed down but rather the Cl^- which is speeded up as a consequence of the exclusion phenomenon connected with the electrical charges on the clays.

G. MATTHESS: Yes, I agree.

J.P. MOLINARI: I am also inclined to share the opinion of Mr. Moutonnet, since we have ourselves observed, in a *saturated* and *reconstituted* porous medium, that the gap in propagation between the "faster" chlorides and the "slower" tritiated water disappeared when the amounts of chloride injected were increased. This result suggests that the tritium in fact gives a better representation of the movement of percolation water. However, I believe one should pay careful attention to all theories about what happens to tritiated water and the other ions; for example, the theory mentioned by Mr. Rightmire and all those discussed by Mr. Matthess. I do not think these questions have been properly considered so far, and I believe this should be undertaken on a less complex medium than a natural unsaturated soil — namely, on a saturated and reconstituted medium. I am sure that this research will produce results if it is carried out as meticulously and methodically as the studies described by Mr. Matthess.

G. MATTHESS: I agree with you that a comprehensive study of tritium and chloride should be undertaken. In our investigation, tritium proved a suitable tool for comparative studies in the unsaturated zone.

P. MOUTONNET: In the section of your paper entitled "Results (capillary rise)", you mention lysimeters which work by suction at the bottom of the column. How did you determine the underpressure and at what level did you maintain it?

G. MATTHESS: The suction lysimeters changed the composition of seepage where certain readily absorbed components, e.g. phosphate, calcium and dissolved organic substances, were concerned. We measured the underpressure with a manometer and maintained it between -0.2 and -0.3 atm to counter water logging. The underpressure is present in the soil at field capacity.

J.B. DAHL: In connection with studies on the fallout of acidic rain in the southern part of Norway, we have carried out a series of experiments using sulphate ($^{35}SO_4$) and tritiated water tracers mixed together in artificial rain-water with known concentrations of chemical compounds. This solution was evenly poured on to a 5-m^2 field area of soil covered with different kinds of vegetation. The seepage water was collected just below the field area. We observed that the sulphate ions moved at a faster rate than the water itself during the first period of water addition; later on, they moved at the same speed as the water. The same result was obtained when pouring the water on to the surface of a mountain. The explanation of these results is given in the paper that has just been presented. When using calcium (^{45}Ca) instead of sulphate, we observed that nearly 100% of this tracer was taken up in the field. At the same time, however, a significant amount of calcium was leached out from the field by the seepage water. I believe that investigations of the kind described in the paper will contribute to a better understanding of the acidic rainfall problem.

G. MATTHESS: Sulphate is an anion which should behave like chloride. However, it is a biological nutrient and can be absorbed by vegetation and humic substances (see Refs [11, 12] in our paper). It can be seen from the Piper diagrams in Fig.6 that calcium will be set free in the A-horizon owing to decay of soil organic substances and will be adsorbed in the B-horizons. We found the mass balances of each water component investigated for each soil horizon. In this way we discovered that, in acid soils with a calcium depletion in the upper layers, the calcium content of rain is relatively enriched in the B-horizon.

OBSERVATION IN SITU DES MOUVEMENTS VERTICAUX DE L'EAU DU SOL PAR MARQUAGE A L'EAU DEUTEREE ET MESURE GAMMA/NEUTRONS ASSOCIEE

Mise au point en laboratoire
et application sur le terrain

P. MOUTONNET, Ph. COUCHAT, F. BRISSAUD,
M. PUARD, A. PAPPALARDO
CEA, Centre d'études nucléaires de Cadarache,
Département de biologie,
Saint-Paul-lez-Durance, France

Abstract–Résumé

IN SITU OBSERVATION OF VERTICAL GROUNDWATER MOVEMENTS BY LABELLING WITH HEAVY WATER AND BY ASSOCIATED MEASUREMENT OF GAMMA RAYS AND NEUTRONS: DEVELOPMENT IN THE LABORATORY AND APPLICATION IN THE FIELD.

    The gamma/neutron technique for measuring concentrations of heavy water used to label groundwater has been developed successfully both for liquid samples and for in situ applications in the laboratory and in the field. Two measuring devices have been constructed, equipped with high-pressure helium-3 counting tubes and a $^{228}Th/^{208}Tl$ source of 5 mCi. The first device is designed for measuring D_2O concentrations in 5-ml samples with an accuracy of $\pm1\%$. The second is a single-channel counter probe designed for in situ measurements with the help of aluminium access tubes. These devices have been used in the laboratory and in the field. The results given by the first can be used quantitatively to the extent of defining the dispersion coefficient of the porous medium, by applying a model simulating dispersive transport. The second provides valuable qualitative information but, where quantitative data are concerned, there is still a problem connected with material balance conservation which it should be possible to solve by less localized surface additions.

OBSERVATION IN SITU DES MOUVEMENTS VERTICAUX DE L'EAU DU SOL PAR MARQUAGE A L'EAU DEUTEREE ET MESURE GAMMA/NEUTRONS ASSOCIEE: MISE AU POINT EN LABORATOIRE ET APPLICATION SUR LE TERRAIN.

    La méthode gamma/neutrons pour la mesure des concentrations d'eau lourde utilisée comme marqueur de l'eau du sol a été développée avec succès, aussi bien sur échantillons liquides qu'in situ en laboratoire ou sur le terrain. Deux dispositifs de mesure ont été réalisés, qui sont équipés de tubes compteurs à haute pression d'hélium-3 et d'une source de 5 mCi $^{228}Th/^{208}Tl$. Le premier ensemble est destiné aux mesures de concentrations en D_2O sur échantillons de 5 ml avec une précision de $\pm1\%$. Le deuxième est une sonde monocompteur destinée aux mesures in situ grâce à des tubes d'accès en aluminium. Ces appareils ont été mis en œuvre au laboratoire et sur le terrain: dans le premier cas l'exploitation quantitative des résultats peut être poursuivie jusqu'à la définition, par le biais d'un modèle de simulation des transferts dispersifs, du coefficient de dispersion du milieu poreux; dans le deuxième cas l'information qualitative est de qualité mais il subsiste au plan quantitatif un problème de conservation du bilan-matière que l'on devrait pouvoir régler par des apports superficiels moins localisés.

L'humidimètre à neutrons classique permet aux agronomes et hydrologues d'établir, à partir de tubes d'accès préalablement installés, le bilan hydrique d'une couche de sol plus ou moins épaisse selon un pas de temps défini [1, 2, 3]. En fait cette approche néglige totalement l'aspect hydrodynamique du problème lié à l'existence de gradients de potentiel hydraulique qui résultent de l'extraction racinaire ou des apports d'eau. La mise en oeuvre de tensiomètres [4, 5] implantés dans la couche de sol arable permet la mesure du potentiel de l'eau à différentes profondeurs (rarement au-delà de 2m) ; on déduit des gradients de potentiel et de la conductivité hydraulique du sol les déplacements d'eau entre couches élémentaires de sol. Une telle méthode n'est pas applicable aux grandes profondeurs (en particulier au-delà de 10m) ; elle nécessite par ailleurs l'implantation de sondes multiples indépendamment du tube d'accès de la sonde à neutrons. Dès 1964, HASKELL [6] proposa la détection in situ d'un pulse d'eau lourde (D_2O) grâce à une sonde constituée d'une source émettrice de rayonnements gamma (^{24}Na) centrée sur un détecteur de neutrons lents. La période du ^{24}Na n'étant que de 15h, COREY [7] puis HAWKINS [8] lui substituèrent -pour un travail de laboratoire- le ^{228}Th dont la demi-vie est de 1,9 an. En ce qui nous concerne nous avons voulu promouvoir cette technique non destructive de contrôle des déplacements de la solution de sol : la méthode gamma/neutrons de mesure des concentrations en D_2O in situ à partir d'un tube d'accès de sonde à neutrons classique a été utilisée au laboratoire et sur le terrain en vue de la mesure quantitative des transferts dispersifs.

I - MATERIELS ET METHODES

On met à profit la réaction γ/n pour créer un flux de neutrons à partir d'une source gamma (d'énergie supérieure à 2,23 MeV) plongée, par l'intermédiaire d'un tube d'accès, dans un sol dont la solution est enrichie localement en D_2O

$$h\nu + {}_1D^2 \longrightarrow {}_1H^1 + {}_0n^1 - 2,23 \text{ MeV}$$

Les gamma-photons sont émis avec une énergie de 2,61 MeV par le ^{208}Tl en équilibre avec le ^{228}Th, lequel détermine la période de la source. Les photo-neutrons produits ont une énergie de 0,2 MeV et sont thermalisés par l'eau du sol puis détectés par un tube compteur ^3He haute pression (4 ou 10 atmosphères suivant les besoins). Les études préliminaires ont montré que la meilleure sensibilité est obtenue avec une sonde compacte regroupant, comme pour la mesure neutronique de l'humidité du sol, la source et le détecteur. Sur ce principe nous avons réalisé deux dispositifs : une sonde monocompteur de terrain (fig. n°1) dont le diamètre extérieur est de 39 mm ; un dispositif de laboratoire à 4 compteurs destiné aux déterminations de teneurs en D_2O sur 5 ml de solution de sol (fig.n°2). Dans les deux cas nous utilisons une source scellée de 5 mCi de $^{228}Th/^{208}Tl$. Pour l'appareil de terrain, la protection biologique est assurée par un étui en plomb portable qui limite les risques d'irradiation à 70 mRad/h au contact et 1 mRad/h à 1 m. Pour le dispositif de laboratoire la protection biologique limite la dose à 35 mRad/h au contact et 2 mRad/h à 1 m.

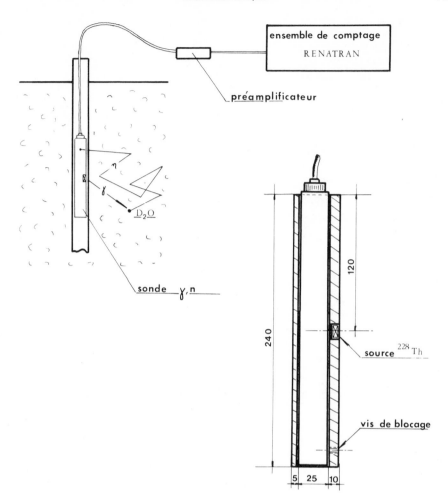

*FIG.1. Sonde monocompteur de terrain pour la mesure in situ des concentrations en D₂O
(dimensions en mm).*

II - ETALONNAGE ET RESOLUTION DES MESURES

1. Bruit de fond :
　　　　　Les dispositifs réalisés présentent un bruit de fond non
négligeable résultant de deux causes essentielles : la sensibilité des tubes
compteurs aux rayonnements gamma et l'émission neutronique propre à la source
(indépendamment des photo-neutrons) liée aux réactions (α, n) sur les éléments légers en particulier l'oxygène. Des deux composantes, la dernière est
prépondérante (7 contre 1 dans l'eau deutérée) et varie avec l'humidité
volumique du milieu : nous avons dû établir une courbe de correction du
comptage en fonction de l'humidité du sol (fig. n°3).

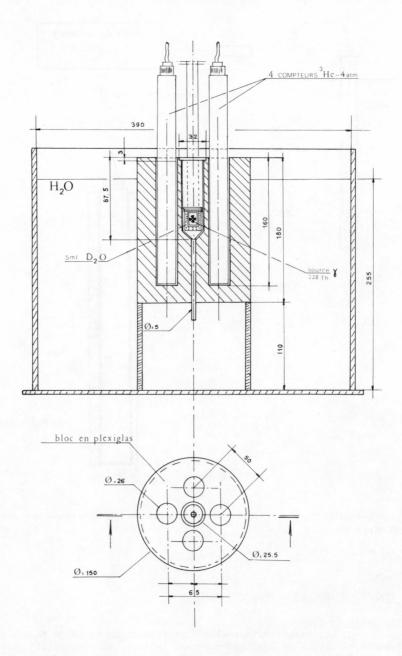

FIG.2. *Dispositif de laboratoire pour la mesure des concentrations en D₂O des solutions
(dimensions en mm).*

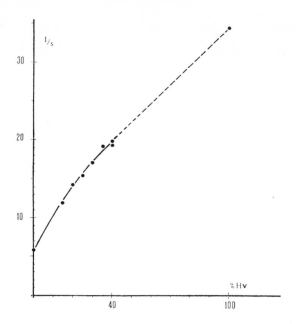

FIG.3.   Courbe de bruit de fond de la sonde monocompteur en fonction de l'humidité du sol.

2. Etalonnage :
        a. Le dispositif de laboratoire a été étalonné à partir de solutions
enrichies en D_2O de 0 à 100% par pas de 10% (fig. n°4). Pour un temps de
comptage de 350 s, la précision estimée est de 1%. L'équation de régression
de la droite est :

$$Y = 1,234 \ X + 0,907 \qquad r \sim 1$$

avec Y : vitesse de comptage en c/s, et X : concentration en D_2O.

        b. La sonde monocompteur de terrain a été étalonnée sur colonne de
sol pour des concentrations en D_2O variant de 5 à 30% et des humidités
volumiques comprises entre 10 et 30%. Les dimensions du container sont
telles (30 cm de diamètre sur 30 cm de haut) que les mesures sont représen-
tatives du milieu infini. La méthode gamma-neutrons est sensible à la
concentration en D_2O d'une unité de volume de sol centrée sur la source de
rayonnements. Nous avons donc rapporté (fig.n°5) nos vitesses de comptages
nettes, aux humidités volumiques en D_2O des échantillons de sols correspon-
dants :

$$HvD \ = \ Hv \ x \ [D_2O]$$

        Théoriquement la vitesse de comptage nette en neutrons lents
résulte à la fois de HvD et de Hv, puisque à l'émission de photoneutrons de
200 KeV succède une thermalisation et une diffusion des neutrons qui sont
liées à la teneur en eau du sol. En fait, on note sur la figure 5 que la

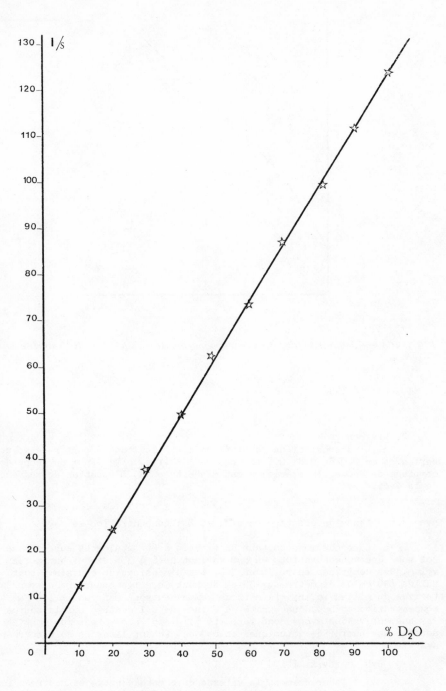

FIG.4.   *Courbe d'étalonnage du dispositif de laboratoire.*

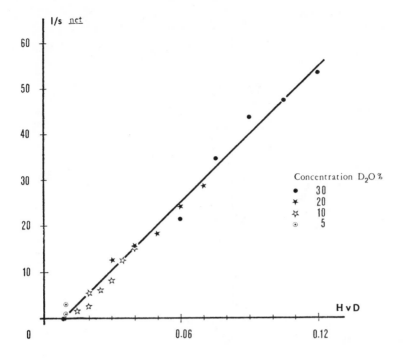

FIG.5. *Courbe d'étalonnage de la sonde monocompteur.*

vitesse de comptage est peu dépendante de Hv ; la régression linéaire en
HvD donne un coefficient de corrélation de 0,985 pour une équation :

$$Y = 520 \times X - 5,47$$

Y : vitesse de comptage en c/s mesurée sur 350 s,
X : humidité volumique en D_2O.

3. Résolution des mesures in situ :

Du fait du libre parcours moyen du rayonnement gamma entre
2,61 et 2,23 MeV (de l'ordre de 10 cm dans un sol à 20% Hv) et de la longueur de ralentissement et de diffusion des photoneutrons produits, la mesure
γ/n est peu focalisée. Nous avons pu expérimentalement déterminer la zone
d'influence des mesures grâce au dispositif représenté en figure 6A : il
s'agit d'éléments de colonne superposés de 30 cm de diamètre ; le module
intermédiaire (h = 10,20 ou 30 cm) contient de l'eau à 10% D_2O, les autres de
l'eau pure. La sonde γ/n est mise en place dans le tube d'accès vertical et
les comptages sont effectués avec un pas de 5 cm. Les profils attachés aux
différentes valeurs de h sont donnés en figure 6B. Si les conditons du
milieu infini sont atteintes pour h = 30 cm, on note une diminution du
comptage maximum de 5,4 et 24,2% pour 20 puis 10 cm. Par ailleurs, dans les
3 cas, le profil expérimental est sensiblement déformé par rapport aux cré-
neaux des concentrations théoriques. Il faut donc déconvoluer les profils
expérimentaux pour obtenir les concentrations réelles in situ. Nous avons
traité ce problème en deux étapes :

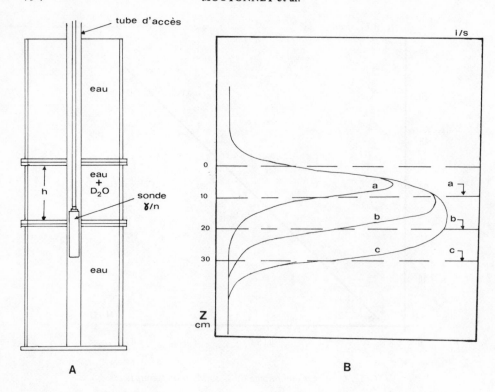

FIG.6. A — Dispositif pour l'étude du pouvoir de résolution de la sonde gamma/neutrons.
B — Profils des comptages obtenus pour différentes épaisseurs de la lame d'eau deutérée.

a. Définition du poids de chaque couche élementaire d'eau deutérée
dans la mesure globale. Soit un profil de concentrations quelcon-
que. Faisons l'hypothèse simplificatrice que l'humidité est constante le
long de ce profil. On peut admettre que la réponse de la sonde à la profondeur z_i est une combinaison linéaire des concentrations en D_2O de ce profil.
On pose en outre que l'influence de la concentration à la cote z ($C(z)$) sur
le comptage de la sonde ($C_S(z_i)$) n'est fonction que de la distance $[z - z_i]$:
soit $\alpha([z - z_i])$ cette fonction. On a alors

$$C_S(z_i) = \int_0^L \alpha \left([z - z_i]\right) C(z)\, dz$$

avec L, la longueur du profil de mesure. Discrétisons ce profil en n mailles
de hauteur Δz. La réponse de la sonde au droit de la maille i est

$$C_{s_i} = \sum_{j=1}^{n} A([j - i]) C_j$$

l'écriture de l'équation précédente pour les n mailles constitue un système
d'équations linéaires de forme matricielle

$$[C_S] = [A]\,[C]$$

ou encore,

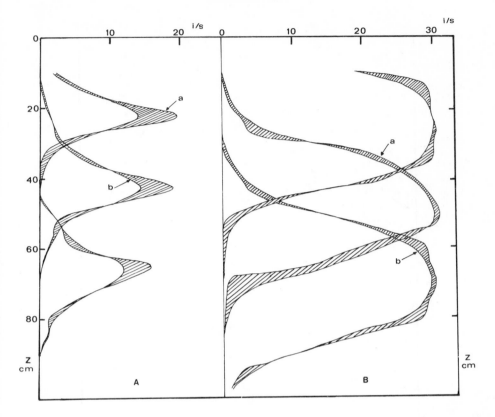

FIG.7. *Profils expérimentaux (notés b) et profils théoriques en milieu infini (notés a) obtenus
par déconvolution. A – Cas d'un pulse de D₂O de 10 cm à mi-hauteur. B – Cas d'un pulse
de D₂O de 30 cm à mi-hauteur.*

$$[C] = [A]^{-1} [C_s]$$

la matrice $[A]^{-1}$ a été déduite des essais présentés en figure 6B, pour
lesquels les valeurs de C et de C_s sont connues. Δz a été pris égal à
à 2,5 cm. On observe, dans la pratique, que les coefficients A peuvent être
considérés comme nuls pour $[j-i] > 4$.

    b. Application à la déconvolution des profils expérimentaux.
       Pour une expérience quelconque, on peut retrouver le profil des
concentrations réelles à partir des réponses de la sonde ; il suffit d'effectuer le calcul suivant :

$$[C] = [A]^{-1} [Cs]$$

MOUTONNET et al.

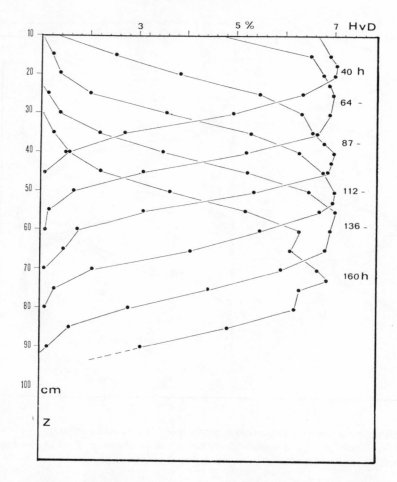

FIG.8. Evolution au cours du temps d'un pulse de D_2O entraîné dans le sol d'une colonne conduite en régime permanent.

Ce procédé a été appliqué au traitement des données relatives à deux essais d'infiltration d'impulsions d'eau deutérée dans une colonne de sol. La première expérience consiste à déplacer 2 l d'une solution d'eau deutérée à 20% dans un écoulement d'infiltration stationnaire. L'humidité est constante le long de la colonne (35%) ; le débit est de 0,127 l/h. L'expérience n°2 est semblable à la précédente, mais la dimension du pulse d'eau deutérée dans l'écoulement est plus importante (8,31). Trois profils pour chacune des expériences 1 et 2 sont reportés sur la figure 7 (A et B). A l'aide de la méthode précédente, nous avons reconstitué les profils de comptage "en milieu infini". On passe de ceux-ci aux concentrations par utilisation de la courbe d'étalonnage de la sonde. La figure 7 montre que, pour une impulsion de dimension réduite, les corrections qu'il est nécessaire d'apporter

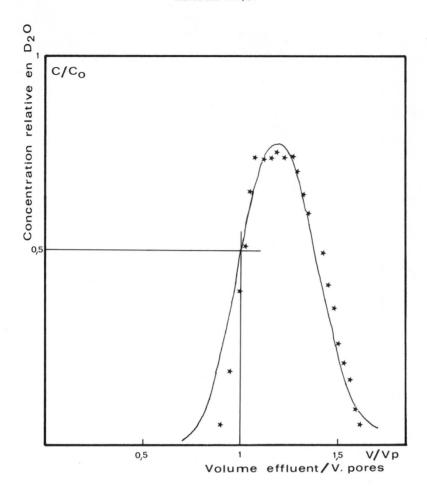

FIG.9.   Concentration relative en D_2O de l'effluent collecté à la sortie d'une colonne de sol conduite en régime permanent (*: valeurs expérimentales; —: courbe théorique).

aux comptages sont relativement importantes, particulièrement au voisinage du pic de concentration ; plus le pic est étroit, plus la correction est grande. En revanche, pour une impulsion de taille plus importante, il n'est pas indispensable de corriger le profil de mesure : c'est ce que nous nous efforcerons de réaliser pratiquement.

III - <u>MISE EN OEUVRE AU LABORATOIRE</u>

Nous avons étudié sur colonnes de sol le transport dispersif par de l'eau normale d'un pulse d'eau deutérée : l'expérience a été conduite sur le sol limoneux-sableux de Cadarache compacté à 1,6 g/cm³ dans des

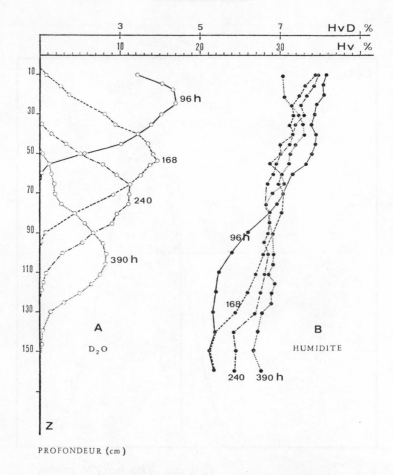

FIG.10. Evolution des concentrations en D_2O et de l'humidité volumique d'un sol irrigué, dans un premier temps par de l'eau deutérée et ultérieurement par de l'eau normale.

colonnes (30 cm de diamètre, 120 cm de hauteur) équipées d'un tube d'accès axial (aluminium, 41/45 mm). Après avoir humidifié le sol à 35% Hv sur toute sa hauteur, grâce à un débit d'eau permanent de 1,83 mm/h, on procède au pompage sous le même débit des 120 mm d'eau deutérée (hauteur du pulse d'eau marquée : 34 cm) puis on continue l'expérience avec de l'eau normale, en régime permanent. Les profils de concentration en D_2O, mesurés in situ au cours du temps grâce à la sonde mono-compteur γ/n, sont donnés en figure 8. La progression du traceur est aisément mise en évidence : elle est en moyenne de 4,65 mm/h. A l'extrémité de la colonne, les effluents ont été recueillis sur collecteur de fractions et nous avons procédé aux mesures de $[D_2O]$ sur le dispositif 4 compteurs déjà décrit : nous donnons en figure 9 la courbe expérimentale d'élution du D_2O, comparée à la courbe théorique obtenue par

TABLEAU I. BILAN MATIERE EN D₂O POUR UN DEPLACEMENT DU PULSE
SUR LE TERRAIN

Temps (heures)	0	77	160	260	390
Pourcentage d'eau deutérée détectée in situ %	100	68	58	52	46

simulation avec une dispersivité du milieu K = 1,44 x 10⁻⁴cm²/s (LE CARDINAL
[9]). A noter que la détermination de K peut se faire aussi bien sur la courbe
d'élution que sur les courbes de concentration mesurées in situ par la méthode
γ/n.

IV - MISE EN OEUVRE SUR LE TERRAIN

            L'expérience consiste à suivre l'évolution d'un pulse de D₂O,
poussé par une irrigation. Deux anneaux concentriques de surfaces respectives
691 cm2 et 1 m2 sont enfoncés de 10 cm dans le sol, le centre du dispositif
étant occupé par un tube d'accès vertical. On apporte dans l'anneau central
120 mm d'eau deutérée à 20% à raison de 2 mm/h, puis on poursuit l'irrigation
pendant 236 h : le pulse D₂O est alors à 90 cm de profondeur et on cesse les
apports aussi bien dans l'anneau central, qui a reçu le D₂O; que dans l'anneau de garde qui a été soumis à une irrigation continue équivalente depuis
le début de l'expérience. Les résultats (fig. n°10) sont paramétrés en temps,
suivant des courbes Hv D (Z) et Hv(Z) occupant respectivement les zones A
et B du graphique. On observe en A l'évolution des [D₂O] avec le temps : pour
des conditions d'expérience assez voisines de celles retenues en laboratoire
on note une vitesse d'avancement du traceur inférieure (3,6 mm/h contre
4,65 précédemment) et une dispersivité du milieu apparemment plus forte que
celle du sol remanié utilisé en laboratoire (on rend compte par simulation
de la forme des pulses de [D₂O] avec une valeur de K double de celle utilisée précédemment). Enfin lorsqu'on s'attache au bilan matière, il apparaît
que les pertes en D₂O sont relativement importantes et ce dès les premiers
profils selon les indications du tableau n°1. On peut en conclure qu'à la
dispersion verticale s'ajoute -malgré le flux d'eau équivalent entretenu
sur l'anneau de garde- une dispersion horizontale qui réduit sensiblement
la quantité de D₂O présente dans le cylindre de sol défini par l'anneau
central. Les expériences en cours sont conduites avec des anneaux de plus
grand diamètre (50 cm pour la partie centrale ; 2,5 m pour la zone de garde) :
on peut espérer dans ces conditions une incidence moindre des transports
dispersifs horizontaux.

REFERENCES

[1]   RAKOTOFIRINGA, H., MOUTONNET, P., Essai de détermination neutronique de
      l'évapotranspiration réelle d'un terrain gazonné des hauts plateaux malgaches, Terre
      Malgache 12 (1971) 147.
[2]   MOUTONNET, P., COUCHAT, Ph., MARCESSE, J., Etude expérimentale et simulation
      numérique de l'économie de l'eau sous culture, Agric. Meteorology 16 (1976) 143.

[3]   MARCESSE, J., COUCHAT, Ph., «Etude hydrodynamique des sols à l'aide d'un
      humidimètre à neutrons automatique», Isotope and Radiation Techniques in Soil
      Physics and Irrigation Studies 1973 (C.R. Coll. Vienne, 1973), AIEA, Vienne (1974) 277.
[4]   MOUTONNET, P., Contribution à l'étude des remontées capillaires sous culture
      cotonnière de décrue du nord-ouest de Madagascar, Terre Malgache 12 (1971) 161.
[5]   MOUTONNET, P., COUCHAT, Ph., «Etude expérimentale et théorique de l'absorption
      racinaire de l'eau du sol par la plante», Isotope and Radiation Techniques in Soil Physics
      and Irrigation Studies 1973 (C.R. Coll. Vienne, 1973), AIEA, Vienne (1974) 373.
[6]   HASKELL, C.C., HAWKINS, R.H., D_2O − Na^{24} method for tracing soil moisture move-
      ment in the field, Soil Sci. Soc. Am., Proc. 28 (1964) 725.
[7]   COREY, J.C., HAWKINS, R.H., OVERMAN, R.F., A gamma photoneutron method for
      laboratory studies of soil water, Soil Sci. Soc. Am., Proc. 34 (1970) 557.
[8]   HAWKINS, R.H., OVERMAN, R.P., COREY, J.C., ^{208}Tl and ^{24}Na gamma sources for
      identifying soil water tagged with deuterium, Soil Sci. Soc. Am., Proc. 35 (1971) 199.
[9]   LE CARDINAL, M.G., MODIS: un modèle de dispersion en milieu poreux saturé,
      Rapport CISI, CEN Cadarche n°109 (1975).

DISCUSSION

T. FLORKOWSKI: The method you have described seems very promising
for field use. The response of the ^{3}He counter is a function of two parameters:
(1) the concentration of deuterium; and (2) the moisture content of the medium.
I notice that you take into account the background correction (background due
to neutrons from the source), which is related to the moisture content. I wonder
whether a correction for the moisture content or total hydrogen content should
not also be taken into account for measurements of D_2O concentrations.

P. MOUTONNET: The gamma/neutron field counter is in fact calibrated
for moisture per unit volume of the soil in D_2O. Also, the background is a
function of moisture per unit volume of soil. So when doing an experiment, one
must perform gamma/neutron counting and also measure the water content of
the soil. From the latter value one calculates the gamma/neutron background,
which is then subtracted from the total count in order to find the net count and
hence the moisture per unit volume of soil in D_2O (or the D_2O concentration,
since the moisture per unit volume of soil is known).

T. FLORKOWSKI: Can you tell me how much your equipment costs, i.e.
the price of the source and deuterated water needed for a single injection
experiment as compared with that of using a standard neutron moisture probe?

P. MOUTONNET: The probe used is equipped with a 5-mCi ^{228}Th/^{208}Tl
source costing 5000 French francs and a 10-atm ^{3}He counter costing
4000−5000 francs. The scaler is portable and of standard design. The equipment
might cost about 6000 francs more than a neutron moisture meter. Between one
and two litres of D_2O, at 1400 francs per litre, are used.

L. KOSTOV: Have you used your method on soils of organic origin,
especially turf, which has a porosity around 70−80%?

P. MOUTONNET: We have no special experience with the method on soils rich in organic matter. However, I can say that at the flow-rates used we have never had a significant amount of isotopic exchange between the tracer dissolved in the soil and the (OH) of the organic matter. In fact, at the levels of porosity you mentioned there must be macroporous and microporous layers with preferential paths in the former. The gamma/neutron technique could give you some information about this, in the form of a double peak.

L. KOSTOV: What is the minimum layer thickness you require in order to follow soil-moisture variations with sufficient confidence?

P. MOUTONNET: For a quantitative determination of D_2O the soil layer labelled should be about 30 cm thick. However, for qualitative information about the migration of water in the soil, a labelled pulse 10 cm high is sufficient; in this case a smaller quantity of D_2O is needed.

GROUNDWATER RECHARGE STUDIES IN MAHARASHTRA

Development of isotope techniques and field experience

A. RADHAKRISHNAN NAIR, A.S. PENDHARKAR,
S.V. NAVADA, S.M. RAO
Isotope Division,
Bhabha Atomic Research Centre,
Bombay, India

Abstract

GROUNDWATER RECHARGE STUDIES IN MAHARASHTRA — DEVELOPMENT OF ISOTOPE TECHNIQUES AND FIELD EXPERIENCE.

Some regional groundwater recharge investigations both in hard rock and in alluvial areas were carried out in the State of Maharashtra, India, by using isotope techniques. A simple stable isotope approach was evolved and applied in an investigation on the extent of artificial recharge in the command area of Shindawane percolation tank in the Pune district. A similar investigation in the Bhangarwadi percolation tank in the Osmanabad district was made using radioactive tracers. A method for the in-situ determination of soil-moisture transport rates using $K_3{}^{60}Co(CN)_6$ as tracer has been proposed, which compares well with the tritiated water method in laboratory investigations, and the results obtained in the limited field studies carried out in the Tapti-Purna alluvial tract are promising.

INTRODUCTION

Maharashtra, situated in western India (Fig.1) covers an area of over 300 000 km^2. Physiographically the State can be divided into three distinct zones — the Konkan strip along the Arabian sea coast, the Deccan plateau and the Tapti-Purna alluvial trough.

Geologically, over 80% of the State is occupied by the Deccan trap in the form of basaltic lava flows with intertrappean beds, dykes and sills. Each lava flow constitutes a top vesicular zone followed by a massive zone towards the bottom. The weathered, vesicular jointed and fractured trap horizons are found to be productive for groundwater development. Groundwater occurs in these fractured rocks under both water-table and confined conditions. The erratic distribution of these water-bearing formations, coupled with the difficulties of hydrological investigations in hard rocks, make it difficult to plan for a full

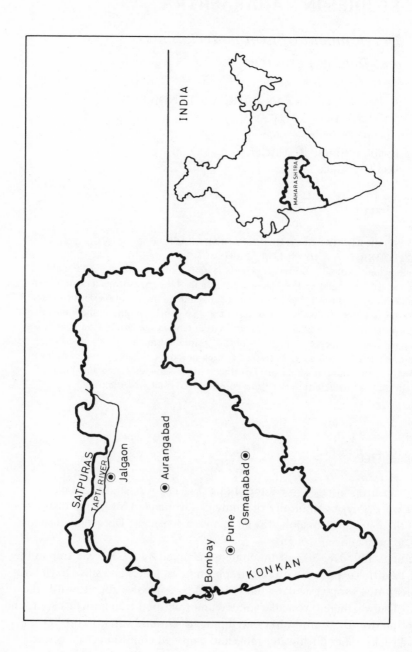

FIG.1. Location of investigation sites in Maharashtra.

exploitation of the groundwater potential. We propose to indicate in this paper
how isotope techniques could be of value in regional groundwater supply programmes in hard rock areas. These pertain to studies on the beneficial effects
of percolation tanks. The Quarternary deposits constituting the alluvia of the
Tapti-Purna trough lie between the Satpura and Satmala ranges in the north
and northeastern Maharashtra. This region, occupying only about 5% of the
total area, has probably the most important formation in the State for groundwater exploitation. The alluvial deposits, consisting of clays, silt, sand, gravels,
pebbles and boulders, vary in thickness from 0 to 300 m and have a number of
water-bearing sand formations with interspersed clay lenses. The main problem
in the region, where heavy lift irrigation is in progress, is the rapid depletion
of the water table, particularly in the case of the dug wells nearer to the Satpuras.
There is some doubt concerning whether the groundwater reservoir is already
overexploited. It appears essential at this stage to identify the recharge zones
(Satpura foothills?) and to estimate the direct recharge to the aquifers from
precipitation. In this paper, we also discuss the development and application
of a gamma tracer complex method for recharge studies and compare the
technique with the tritium tracer method.

1.   PERCOLATION TANKS

        The construction of percolation tanks is a common practice in several
parts of India for the artificial replenishment of groundwater for lift irrigation
in small agricultural tracts. A percolation tank is created by constructing a
small earthen dam across a natural stream in a suitable cross-section. It is
located upstream of an existing cluster of dug wells usually with poor yields.
The surface run-off during the short monsoon period of June-September is
collected in the tank and consists of around a few million cubic metres. Under
favourable soil and rock conditions, the water is expected to percolate and
recharge the groundwater system and to improve the well yields, particularly
in summer. The command area, consisting of 20 to 50 dug wells, covers an
area of a few square kilometres. Many such percolation tanks have been
constructed in the semi-arid regions of Maharashtra and it has been reported
that some of them have brought increased benefits to the farmers while some
others made no significant contribution to the wells. Most of these wells are
in the hard rock areas of the Deccan Trap whose hydrological characteristics
are not as easily determined as, for example, those in the alluvial aquifers. The
Groundwater Survey and Development Agency and the Directorate of Irrigation
Research and Development, Government of Maharashtra, have undertaken a
detailed study of the behaviour of some percolation tanks in order to make
suitable recommendations on the locations of future tanks.

The authors carried out isotope studies in some representative tanks using both stable isotope and artificial radiotracer methods. The methodology evolved for the application of the techniques and the results obtained are indicated in the following paragraphs.

2.   THE STABLE ISOTOPE METHOD

Since the hydrological system of a percolation tank consists of an evaporating body (the tank) contributing to a non-evaporating groundwater system, a stable isotope method could offer an elegant approach for studying the effectiveness of the tank and for delineating the extent of its beneficial effect. One way would probably be to have an analysis programme of the well-water samples for both deuterium and ^{18}O and to compare the data with the known D-^{18}O relationships for the evaporating and precipitation (or ground-) waters in the region. In principle, the D-^{18}O line for the well-water samples should lie between the D-^{18}O lines for the evaporating and precipitation waters. It is felt that this procedure would pose obvious difficulties regarding analytical facilities and lack of adequate precision in determining the D-^{18}O relationships which are not readily available.

Alternatively, the enrichment of either deuterium or ^{18}O in the tank can be considered to be a natural labelling process and an investigation on the distribution of either of the enriched tracers in the groundwater domain can yield the necessary data for identifying the wells recharged by the tank and for delineating the command area. It is also possible, under favourable conditions, to estimate at least semi-quantitatively the surface water contribution to the well yield. From considerations of isotopic balance Eq.(1) has been derived [1] for the contribution, m_s, of surface water to the well water.

$$m_s = \frac{R_{PG} - R_{G(d,t)}}{R_{PG} - R_{L(t-t_d)}}$$ (1)

$$= 1 - m_g$$

where

m_g = original groundwater fraction in well water

R_{PG} = isotopic composition of pure groundwater samples beyond the recharge zone of the tank

$R_{G(d,t)}$ = isotopic composition of the admixture at a distance d from the tank and at time t

$R_{L(t-t_d)}$ = isotopic composition of the tank water that is found at distance d
and at time t

t_d = average travel time of seeping water to reach distance d.

Equation (1) can be used to estimate the tank contribution to each
irrigation well.

The above treatment assumes a well-mixed situation in the tank and in
the exploited wells, and that equilibrium seepage conditions are attained. This
situation is usually true as most of the percolation tanks are not deep and the
wells are exploited more or less continuously. The fast-mixing in shallow tanks
is experimentally confirmed in a radiotracer study [2].

The applicability of the method to field conditions is derived from the
following considerations:

Highly localized nature of the problem;
Short duration of the study which ensures that the method is not affected
by large seasonal variations of stable isotope concentrations in groundwater;
No precipitation during the investigation period;
No other known surface body recharging the groundwater zone.

3.   APPLICATION OF THE METHOD TO THE
      SHINDAWANE PERCOLATION TANK

The enriched stable isotope method was applied to the case of the
Shindawane percolation tank in the Pune district of Maharashtra.

Figure 2 shows the plan of the tank and the command area under irrigation. The tank, which is shallow with an average depth of 3–4 m, impounds
approximately 0.5 to 1 × 10^6 m³ of water during the June-September monsoon
season. The command area under the tank was envisaged to be about 2.5 to
3 km², extending up to 2 km from the dam.

The dug wells (about 40) in the command area penetrate soft rock over-
burden up to the water table and then fissured hard rock which is exposed in the
tank bed. The seepage medium thus consists of fissured hard rock with jointed
planes and fractures.

It is learnt that the acreage under lift irrigation doubled after the tank was
constructed.

A sampling programme for the deuterium analysis of the tank and well
waters was initiated in March 1972 and was continued at regular intervals up
to March 1974.

For a preliminary analysis of the deuterium data, two sets of results, one
for March 1972 and another for May 1972, were chosen since the period satisfied

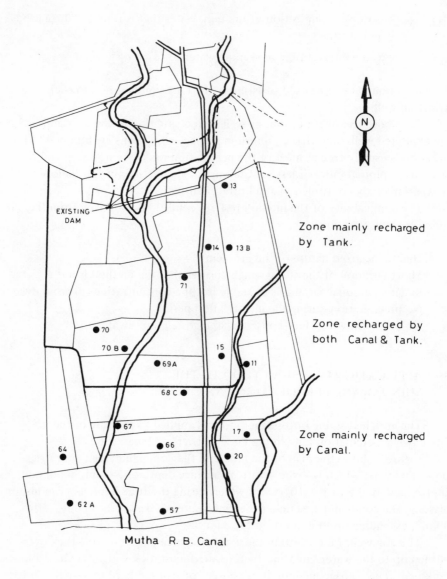

EXISTING
DAM

● 13

Zone mainly recharged
by Tank.

● 14 ● 13 B

●
71

Zone recharged by
both Canal & Tank.

● 70

70 B ●

15
●

● 69A

● 11

68 C ●

● 67

17 ●

64
●

● 66

● 20

Zone mainly recharged
by Canal.

● 62 A

● 57

Mutha R. B. Canal

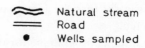

~~ Natural stream
═══ Road
● Wells sampled

FIG.2. *Plan of Shindawane percolation tank and its command area.*

the requirements of the technique. The analysis of the data indicated [1] that
the wells nearer to the dam were being recharged by the tank. The effect of the
Mutha Right Bank Canal, flowing at the southern edge of the command area,
on the recharge characteristics of the wells nearer to it could also be clearly
noticed. The latter situation was not anticipated in the earlier part of the
sampling programme and hence a fresh analysis of the data for the period
December 1973/March 1974 has been carried out, also taking into consideration the presence of the Mutha R.B. Canal.

The canal, which is unlined, has an average width of about 16 m and carries
a discharge of about 14 m^3/s.

4.   QUALITATIVE ANALYSIS OF THE DATA FOR
DECEMBER 1973–MARCH 1974

Table I gives the deuterium data of the samples for December 1973 and
March 1974. The percentage enrichment of deuterium ($\Delta\delta D\%$) in the tank
and in the canal, as well as in the sampled wells, is also indicated. Though the
two surface water bodies show the same per cent enrichment, it may be noted
that the tank water is enriched from $-41\%_0$ to $-19\%_0$ during the period whereas
the canal water is enriched from $-19\%_0$ to $-9\%_0$. This considerable difference
in deuterium labelling of the two surface water bodies helped in identifying
three distinct groups of wells:

Group I — mainly recharged by the tank;
Group II — recharged to varying degrees by the tank and the canal; and
Group III — mainly recharged by the canal.

It may, however, be noted that the surface water contribution to wells
13B, 69 and 68C is not significant. In hard rock areas, such a situation is possible
since water can only flow in fractures and jointed planes. Also the abnormally
high per cent enrichment in well No.15 cannot be explained, and its D values
are a poor fit.

Figure 3 shows the enrichment variation along a traverse from wells 14 to
57. Here again, at least two groups of wells can be identified.

Equation (1) cannot be easily applied to the present situation owing to
the complications caused by the presence of the canal and its contribution to
the wells.

5. RADIOTRACER TECHNIQUE

This is a more direct approach in which an artificial radioactive tracer is
used to study the groundwater conditions in the presence of a recharging surface

TABLE I. DEUTERIUM VALUES OF SAMPLES COLLECTED IN DECEMBER 1973 AND MARCH 1974

Serial No.	Distance from dam (m)	Distance from canal (m)	Source of sample (Well No.)	δD‰ (Dec.'73)	δD‰ (March '74)	$\Delta\delta D$‰ (enrichment)	Remarks
1	0	2100	Tank	−41	−19	54	
2	0	2100	13	−52	−33	37)	Group I — Mainly recharged by the tank
3	360	1740	14	−54	−28	48)	
4	360	1740	13B	−44	−39	11)	
5	560	1540	71	−44	−32	27)	
6	800	1300	70	−44	−28	36)	
7	960	1140	70B	−41	−18	56)	Group II — Appears to be recharged to varying degrees by both the tank and the canal
8	1060	1040	69	−31	−28	10)	
9	1060	1040	69A	−25	−17	32)	
10	1120	980	15	−57	−11	81)	
11	1240	860	11	−42	−28	33)	
12	1280	820	68C	−30	−25	17)	
13	1480	620	67	−39	−19	51)	Group III — Mainly recharged by the canal (when the water-table level is low to permit seepage)
14	1600	500	66	−43	−19	56)	
15	1640	460	17	−35	−25	29)	
16	1660	440	64	−38	−11	71)	
17	1740	360	20	−50	−18	64)	
18	1960	140	62A	−42	−16	62)	
19	2040	60	57	−44	−9	80)	
20	2100	0	Canal	−19	−9	53	

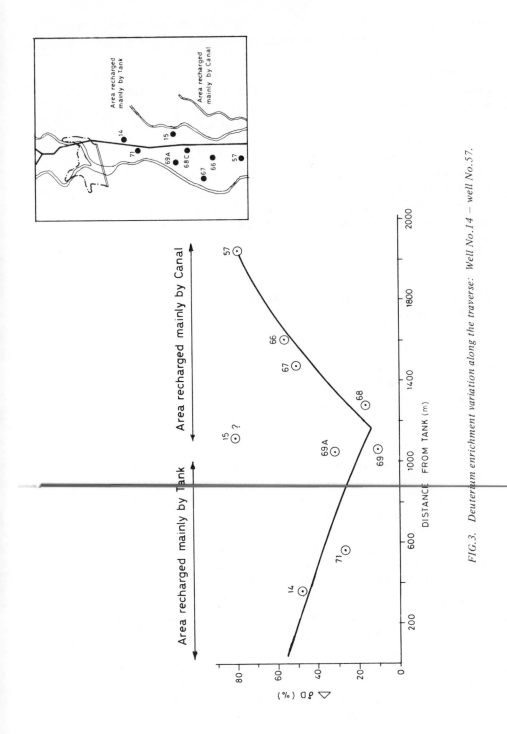

FIG.3. Deuterium enrichment variation along the traverse: Well No.14 — well No.57.

water body like the percolation tank. The necessity for a radiotracer approach arises from the erratic nature of groundwater distribution even in limited zones of hard rock formations.

One way is to determine the quantum of groundwater flow in each well using the point dilution method [3] and compare the results with well yields obtained by pumping to evolve a general flow pattern for the command area. This is not possible on a large scale since most of the wells are pumped every day to meet the irrigation requirements of the crops.

Alternatively, the tracer can be introduced into the tank followed by an investigation of the tracer distribution in the wells in the command area and its variations with time. The pumping pattern in the command area can be expected to have an effect on the tracer distribution. The advantage of the method over the stable isotope technique is that the duration of the investigation can be much longer.

6.   APPLICATION OF THE METHOD TO THE BANGARWADI PERCOLATION TANK

The radiotracer method was applied to the Bangarwadi percolation tank in the Osmanabad district of Maharashtra.

The tank (Fig.4) is situated in a micro-basin of which the NNW and NE parts are surrounded by 50 to 65 metre high hillocks. The tank has a capacity of 0.4×10^6 m^3 and the percolating medium mainly comprises highly weathered vesicular zeolitic basalt followed by moderately weathered and jointed massive basalt.

There are twenty dug wells in the command area, which extends to a distance of about 1.6 km from the dam. It is reported that the yields of the wells improved considerably after the percolation tank was constructed.

As a first step to study the groundwater flow under non-pumping conditions, the radiotracer point-dilution method was applied, in a rather unorthodox way, to three wells (Nos 7, 12 and 13) using bromine-82. Mixing is ensured by recirculating the well water with an irrigation pump.

Table II gives the results of the experiments and compares them with the well yields.

Wells 7 and 12 penetrate the weathered fractured and jointed trap, whereas well No.3 draws water mainly from a lower formation consisting of zeolitic trap.

It can be seen that each well yields more quantity of water than the quantity flowing through the fractures encountered in the well. This indicates, as could be expected in hard rock areas, that the groundwater body and the tank act like a cistern and yield water only when pumped. Thus, the percolation tank which

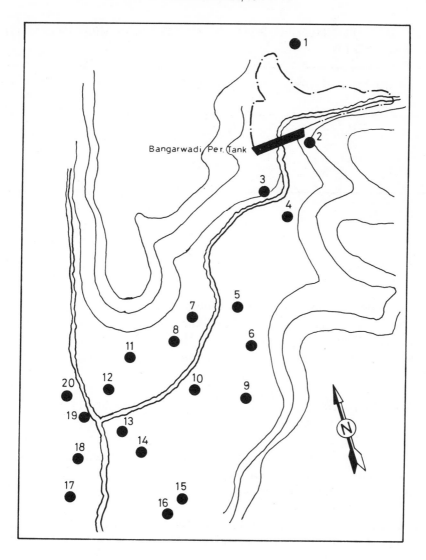

FIG.4.   Plan of Bangarwadi percolation tank and its command area.

contributes to the groundwater body does so only when the wells in the command area are exploited. This is an almost ideal situation since there is a minimum loss of percolating water during non-pumping hours.

The above is a general understanding arrived at by the point-dilution measurements. However, to comprehend the distribution of the tank water

TABLE II. EXPERIMENTAL RESULTS AND THEIR COMPARISON WITH
THE WELL YIELDS

S.No.	Well No.	Dia. of the well (m)	Volume of the well (m^3)	Flow-rate m^3/d	Well yield m^3/d
1.	7	7	270	79	200
2.	12	6	90	38	200
3.	13	8	290	122	250

within the command area it was felt necessary to label the tank water with a
radiotracer and to follow its appearance in the different wells. Accordingly,
tritiated water was injected into the tank in March 1978. The tank and the
well waters are being continuously analysed for tritium concentrations.

7. SOIL-MOISTURE MOVEMENT STUDIES – DEVELOPMENT OF A GAMMA TRACER TECHNIQUE

This part of the paper reports on the investigations carried out for the
development of a gamma complex tracer method for soil-moisture transport
investigations and the at present limited experience in the application of the
method under field conditions.

The application of tritiated water for tracing soil-moisture movement in
the unsaturated zone is now well established [4] and extensively used by
many investigators.

The main disadvantage of the tritium method is that it is not amenable
to in-situ and non-disturbing detection. The procedure involves collection of
soil samples by augering, extraction of moisture in the laboratory and counting
for tritium activity. The procedure does not take advantage of the properties
of a radioactive tracer. The main advantage – that the tritiated water is an
ideal tracer for water movement – could sometimes paradoxically be a disadvantage. For example, in low-moisture areas such as arid zones, the tracer
could get into vapour phase during the day and may condense at points above
the injection point during the night and thus cause complications in tracer
profile interpretation.

COBALT-60   METHOD

TRACER :- $K_3{}^{60}Co(CN)_6$ COMPLEX OR
           ^{60}Co-EDTA   COMPLEX

● Labelling of water layer around a bore hole with tracer at desired depth

● 'INSITU' detection of tracer position with scintillation detector

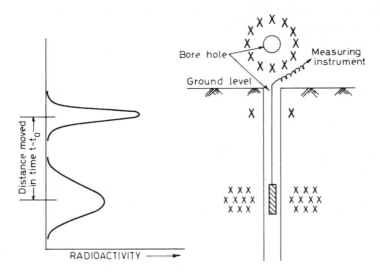

Bore hole

Measuring instrument

Ground level

RADIOACTIVITY ⟶

Distance moved in time $t-t_0$

FIG.5.   Technique of soil-moisture movement study with $K_3(Co(CN)_6)$.

In view of the above, we carried out laboratory and limited field investigations on the suitability of radiolabelled cobalt complexes, particularly $K_3Co(CN)_6$ for soil-moisture transport investigations. The basic technique (Fig.5), adopted both in laboratory models and in the field study, is as follows:

The selected radiotracer is injected around an access hole at several points on a circle at a suitable depth from the surface;

The tracer is allowed to diffuse and form a ring-like layer around the access hole;

The position of the tracer peak is determined at regular intervals by simply logging the hole with a scintillation detector.

The advantages of the gamma tracer method over the tritium method can be:

No repeated augering to extract soil samples and hence no disturbance to soil conditions;
No elaborate laboratory support as is required for vacuum distillation of soil samples for soil-moisture extraction and liquid scintillation counting of water samples;
The same borehole can be used for soil-moisture movement measurement using a neutron moisture gauge.

8.   LABORATORY STUDIES

Initial column studies carried out with $K_3Co(CN)_6$ labelled with shorter-lived [58]Co, and reported elsewhere [5], showed that the tracer maintained its position integrity in a dry soil column for over four months and it was possible to follow its movement with intermittent spraying at the surface of the column.

To compare the transport characteristics of the $K_3Co(CN)_6$ with those of tritiated water, a special laboratory model was constructed. Figure 6 gives details of the experimental set-up. It consists of a large drum filled with soil and with a bottom outlet for water. Rigid plastic tubes with nearly 30–40% perforations are placed on a circle around a central access hole. The cobalt-cyanide tracer labelled with ^{60}Co and the tritiated water are injected together at points between the perforated tubes at a depth of 15 cm from the soil surface. Measured quantities of water are sprayed on to the soil surface. Besides moving vertically down, under the action of the sprayed water, the tracers diffuse laterally, also creating a labelled ring around the access hole. The position of the gamma-emitting tracer is located by logging the access hole with a NaI scintillation probe. At specific intervals, one of the perforated plastic tubes is taken out, cut into 10-cm sections, the soil moisture is extracted by vacuum distillation and the tritium concentrations are measured by liquid scintillation counting. The tritium profile thus determined is compared with the gamma profile. Figure 7 compares the transport of the two tracers in sandy soil. It can be seen that their transport rates are comparable. Similar studies carried out on clayey soils yielded similar results. Figure 8 gives an example of such a comparison.

Laboratory studies were also carried out on the behaviour of ^{60}Co-EDTA in different types of soil. It was generally found that the EDTA complex is retarded with respect to the cyanide tracer. Figure 9 compares the movement of ^{60}Co-EDTA and $K_3^{60}Co(CN)_6$ in clayey soil columns.

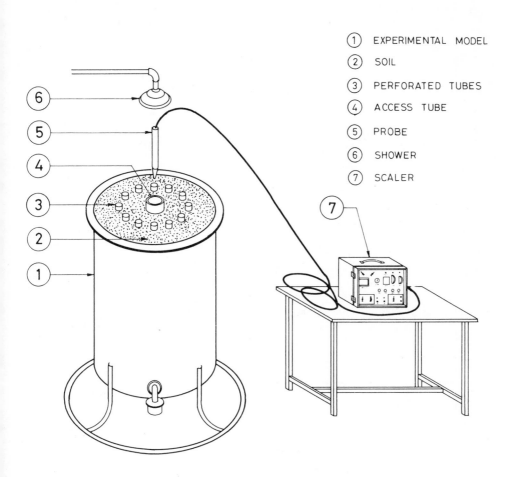

FIG.6. Laboratory model for soil-moisture movement study with ^{60}Co and HTO.

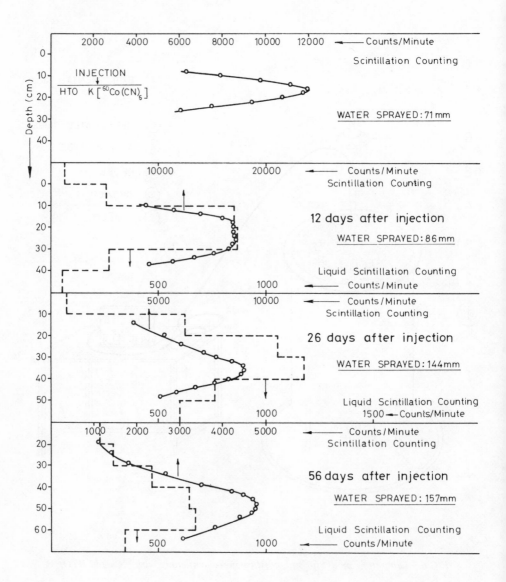

FIG.7.   Comparison of ^{60}Co tracer with HTO in sandy soil.

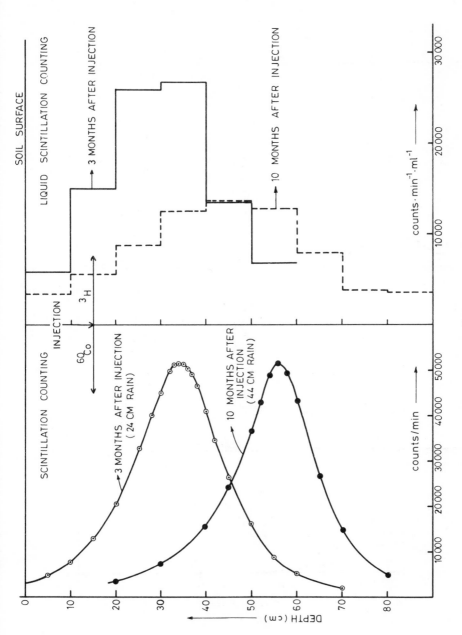

FIG.8. Comparison of ^{60}Co tracer with HTO in clayey soil.

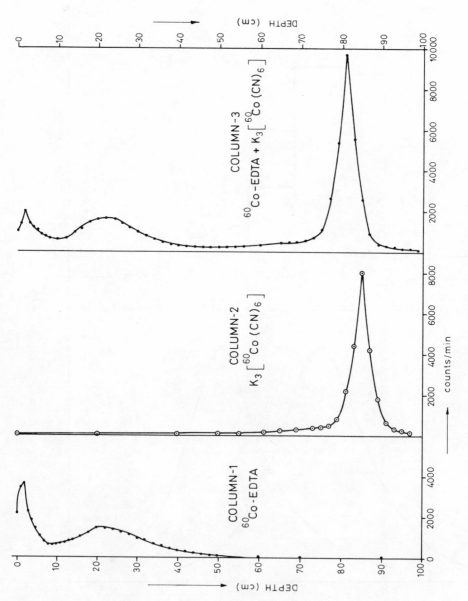

FIG.9.   Comparison of ^{60}Co-EDTA and $^{60}Co(CN)_6$ in column studies.

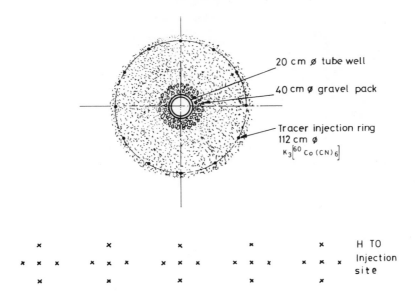

FIG.10. Plan of $K_3[^{60}Co(CN)_6]$ and HTO injections at the Chikli site.

9.   APPLICATION OF THE TRITIUM AND ^{60}Co METHODS
      IN TAPTI ALLUVIUM

        A tracer study to estimate rainwater infiltration in the Tapti alluvial tract
was undertaken for the Groundwater Survey and Development Agency,
Government of Maharashtra. Tritiated water was injected at two sites — Chikli
and Sathod villages in the Jalgaon district in September 1976. At the Chikli
site near the area of tritium injection, an unused 20-cm-dia. bore-well is being
used for field application of the gamma tracer method. The cobalt-cyanide
tracer was injected around the bore-well on a circle of 56 cm radius (Fig.10).
        The soil at both the sites consists of low-permeability silt and clay with
calcareous nodules.
        Figure 11 shows the tracer injection system used for both tritiated water
and $K_3^{60}Co(CN)_6$. The tracer solution is contained in a 100-ml serum capped
vial. With the help of hypodermic needles at the inlet and outlet ends of the
system the required volume of the tracer is injected into the soil through a thin
stainless-steel tube by operating the peristaltic pump for a calibrated time.
This method is very simple and reduces the risk of contamination.
        The position of the gamma tracer is determined by logging the bore-well
with a gamma-scintillation detector. The tritium profile is obtained by the
usual core sampling and tritium analysis procedures. Figure 12 shows the

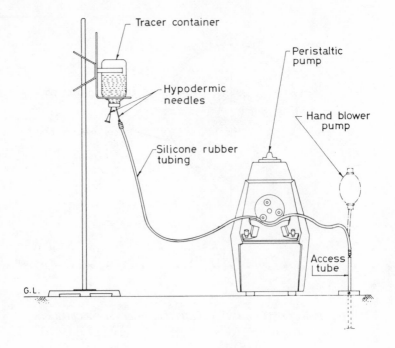

FIG.11. Radiotracer injection system for soil-moisture movement studies.

results obtained at Chikli with the cobalt and the tritium tracers for the period
December 1976 to April 1978. Figure 13 shows the results for the Sathod site
for the same period. Table III compiles the recharge data obtained for the
two sites.

10. DISCUSSION

The difference between the transport rates of the cobalt complex and the
tritiated water cannot be readily explained in view of their similar behaviour
in laboratory experiments carried out on the same Chikli soil. It should, however, be noted that the accuracy in peak positioning in the HTO method is
already ± 5 cm. Such differences in soil-moisture movement between two sites,
even though separated by a few metres, are not entirely unexpected. In view of
this, the investigation is being continued on a larger scale. Cobalt-60 injections
are being made around 10-cm-dia. cased bore-holes in five different sites in the
Jalgaon district. Tritium injections are also being made for comparison purposes.

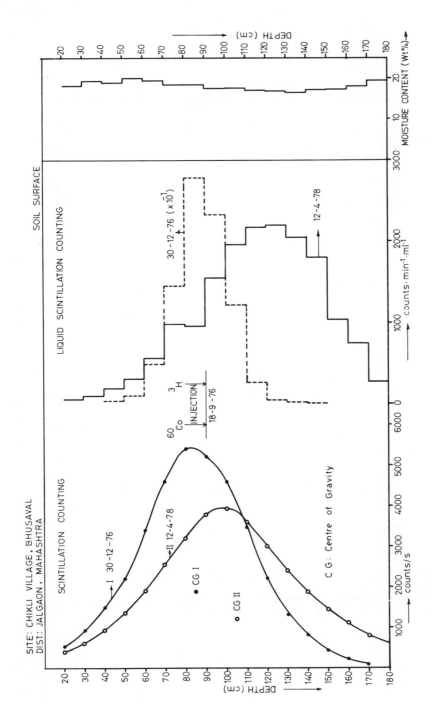

FIG.12.   Results of soil-moisture movement studies conducted at the Chikli site with HTO and $K_3[Co(CN)_6]$.

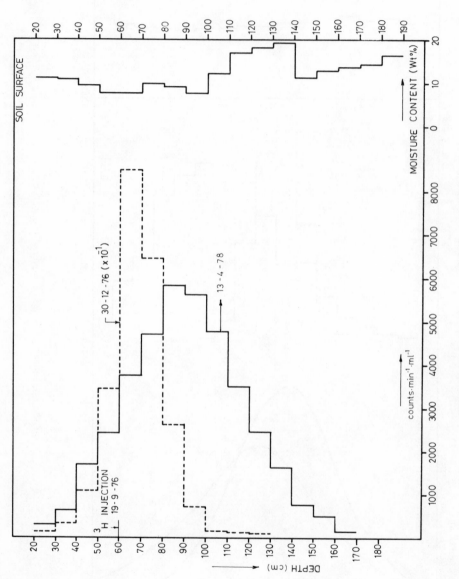

FIG.13.   Results of studies conducted at the Sathod site.

TABLE III. RECHARGE EVALUATION FOR DECEMBER 1976
TO APRIL 1978

Site	Tracer	Tracer patch displacement (cm)	Moisture content (cm)	Rainfall as recorded at Sathod (cm)	Recharge % of rainfall
Chikli	^{60}Co	20 (C.G.)[a]	6.1	74	8
	HTO	35 (peak)	10.6	74	14
Sathod	HTO	20 (peak)	3.0	74	4

[a] C.G. = Centre of gravity.

ACKNOWLEDGEMENTS

We thank Dr. V.K. Iya, Director, Isotope Group, BARC, for his guidance in
the studies reported here. Our thanks are due to the authorities of the Groundwater Survey and Development Agency and the Directorate of Irrigation Research
and Development both of the Government of Maharashtra for their help in field
investigations. We thank our colleagues Shri M.S. Krishnan, Shri S.K. Jain and
Shri U.P. Kulkarni for their help in many aspects of these studies.

REFERENCES

[1] KRISHNAMURTHY, K., RAO, S.M., "A simple stable isotope approach to study the
effectiveness of artificial recharge from percolation tanks", Proc. First National Symp.
Applications of Isotope Techniques in Hydrology and Hydraulics, Central Water and
Power Research Station, Poona (India), Bhabha Atomic Research Centre (1974) 123–33.

[2] NAIR, A.R., PENDHARKAR, A.S., RAO, S.M., JOSHI, N.R., "Investigations of the
beneficial effect of the Badji percolation tank using tritium tracer", National Symp.
Isotope Applications in Industry, BARC, Bombay (1977).

[3] KRISHNAMURTHY, K., RAO, S.M., Theory and experiment in canal seepage estimation
using radioisotopes, J. Hydrol. 9 (1969) 277–93.

[4] ZIMMERMANN, U., MUNNICH, K.O., ROETHER, W., "Downward movement of soil
moisture traced by means of hydrogen isotopes", Isotope Techniques in Hydrological
Cycle (Proc. Symp. University of Illinois) (1965) 28–36.

[5] NAIR, A.R., NAVADA, S.V., RAO, S.M., "Evaluation of K_3Co(CN)$_6$ and HTO as tracers
for soil moisture investigations", National Symp. Isotope Applications in Industry, Bhabha
Atomic Research Centre, Bombay (1977).

DISCUSSION

K.E. WHITE: Could you please explain why you chose to use ^{60}Co for your pilot investigations lasting only a few weeks? ^{60}Co has a half-life of 5.26 years, whereas ^{58}Co has a half-life of only 77 days, which is appropriate for studies lasting between a few weeks and a few months. Furthermore, in-situ detection of ^{58}Co is no more difficult than that of ^{60}Co.

S.M. RAO: The half-life of ^{58}Co is too short for our pilot or other field investigations as they are to be continued for several years. As mentioned in the section of the paper entitled "Laboratory Studies", we did use ^{58}Co in our early column studies; later we changed over to ^{60}Co for reasons of convenience.

Session IX, Part 1

SURFACE WATERS

Chairman

E. LINARES
Argentina

MODELS FOR THE RUNOFF FROM A GLACIATED CATCHMENT AREA USING MEASUREMENTS OF ENVIRONMENTAL ISOTOPE CONTENTS

H. BEHRENS, H. MOSER, H. OERTER,
W. RAUERT, W. STICHLER
Gesellschaft für Strahlen- und Umweltforschung mbH,
Neuherberg, Federal Republic of Germany

W. AMBACH, P. KIRCHLECHNER
Physikalisches Institut der Universität Innsbruck,
Innsbruck, Austria

Abstract

MODELS FOR THE RUNOFF FROM A GLACIATED CATCHMENT AREA USING MEASUREMENTS OF ENVIRONMENTAL ISOTOPE CONTENTS.

For several years, in the glaciated catchment area of the Rofenache (Oetztal Alps, Austria), measurements have been made of the environmental isotopes ^{2}H, ^{18}O and ^{3}H in precipitation, snow and ice samples and in the runoff. Furthermore, the electrolytic conductivity of runoff samples was measured and tracing experiments were made with fluorescent dyes. From core samples drilled in the accumulation area of the Vernagtferner, the gross beta activity was investigated and compared with the data from ^{2}H, ^{3}H and ^{18}O analyses and the data from mass balance studies. It is shown that the annual net balance from previous years can be recovered on temperate glaciers using environmental isotope techniques. From the diurnal variations of the ^{2}H and ^{3}H contents and the elctrolytic conductivity, the following proportions in the runoff of the Vernagtferner catchment area were obtained during a 24-hour interval at a time of strong ablation (August 1976): about 50% ice meltwater, 25% direct runoff of firn and snow meltwater, and 7% of mineralized groundwater. The rest of the runoff consists of non-mineralized meltwater seeping from the glacier body. The annual variations of the ^{2}H and ^{3}H contents in the runoff of the glaciated catchment area permit conclusions on the time sequence of the individual ablation periods, and on the residence time, on the basis of model concepts. The residence times of approximately 100 days or four years, respectively, are obtained from the decrease in the ^{2}H content at the end of the ablation period and from the variation of the ^{3}H content in the winter discharge.

1. INTRODUCTION

1.1. Investigation area and studies to date

The Rofenache catchment area in Oetztal Alps (Austria) is 96.2 km^2, 44% of which is covered by glaciers. Figure 1 shows the catchment area with its

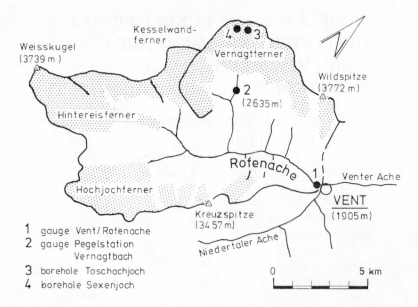

*FIG.1.   Catchment area of Rofenache creek in the Oetztal Alps (Austria). The glaciated
area (44%) is dotted, with 1 being the Vent/Rofenache gauge station; 2 the Pegelstation
Vernagtbach gauge station; 3 the Taschachjoch borehole; and 4 the Sexenjoch borehole.*

glaciers and creeks. The discharge of the catchment area is measured at the
runoff gauge station of Vent/Rofenache (built in 1963 and operated by the
Hydrogeographischer Dienst Tirol[1]). Vent has also a precipitation gauging
station (mean precipitation 706 mm/a) providing precipitation samples for
isotope measurements. Figure 1 also shows the runoff gauge station, Pegelstation
Vernagtbach, run by the Kommission für Glaziologie der Bayrischen Akademie
der Wissenschaften, together with the Institut für Radiohydrometrie [1]. In this
part of the catchment area, covering 11.44 km², is the 9.03-km² Vernagtferner
glacier. The gauge station thus comprises a catchment area, 82% of which is
glaciated.

In this catchment area, isotope hydrological studies were made from 1968
to 1977 by the Institut für Radiohydrometrie together with the Physikalisches
Institut der Universität Innsbruck. The content of environmental isotopes
^{2}H, ^{18}O and ^{3}H in the precipitation, in the runoff and in snow and ice samples,
was determined; furthermore, the gross beta activity was determined in drill
cores. The results of these studies have been published in, among others,
Refs [3–5]. By comparison with the isotope hydrological studies, tracer

---

[1]   The measured data have been published in the various yearbooks of Ref.[2].

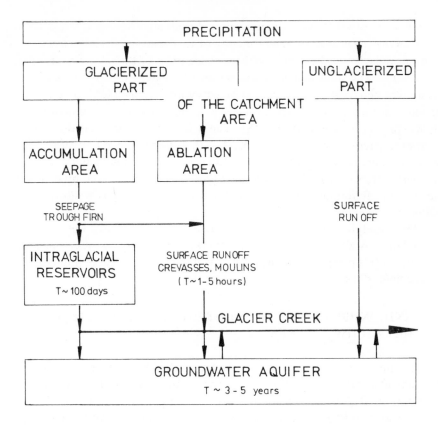

FIG.2. Schema of the runoff within the catchment area of Rofenache. The mean residence time T is explained in Section 4.

experiments with fluorescent dyes have been made [6, 7] and the electrolytic conductivity of the runoff was determined [3, 8]. The results of all these studies up to 1975 have been reviewed [9].

1.2. Runoff from a glaciated catchment area

The Rofenache catchment area consists of glaciated and non-glaciated parts (Fig.2). In the non-glaciated part, groundwater discharge is accompanied by a surface water runoff after snow melt and after rainfall. The non-glaciated part, however, shall not be described here. The glaciers located in the catchment area decisively influence the discharge characteristics by the change of the precipitation stored and by snow and ice melt.

The runoff from the glaciated part is mainly governed by the melting conditions on the glacier. The ice meltwater runs off on the ice surface of the

ablation area until it flows through crevasses or moulins into an intraglacial or
subglacial discharge system. The ice body of the ablation area has a low
retention capacity so that only short residence times of the ice meltwater may
be expected. The snow meltwater, however, percolates through the firn body.
The path of percolation is strongly affected by intermediate ice horizons. At
the depth of the firn-to-ice transition the existence of a water-saturated layer
was also proved on Vernagtferner [8]. The firn body certainly has the highest
retention capacity of all the glacier parts so that the residence time of the
meltwater there must be expected to be longer. The moraine material beneath
the glacier offers only limited space for a groundwater system fed by meltwater.
The amount of groundwater in the total runoff is only small as is shown in
Section 4.

When studying the hydrology of glaciated catchment areas, the snow
accumulation as well as the runoff and storage of meltwater must be taken into
account. In this paper, the application of isotope hydrological methods to
this problem is described.

2.   ENVIRONMENTAL ISOTOPE CONTENT OF PRECIPITATION IN THE CATCHMENT AREA

Precipitations in the catchment area are labelled by their content of
environmental isotopes. Figure 3 shows the monthly means of ^2H content in the
precipitations at Vent (for location see Fig.1) in 1972–1977. The typical
seasonal variation is evident. The amplitudes of the summer/winter periodicity
were roughly equal each year throughout the whole measurement period. The
isotope contents of the individual precipitations vary considerably owing to
specific weather situations (cf. Ref.[10]) but are largely smoothed out by
averaging over one month.

In Fig.4, the time fluctuation of the monthly mean values of the ^3H content
in the precipitations at Vent is shown for the years 1970–1977. Here, too, the
seasonal fluctuations are evident. The decrease in the ^3H contents, which has
been generally observed since the nuclear weapon test ban, has not been evident
since about 1972. Of note is the occurrence of some individual ^3H content
peaks in the winter, which we also found at other measuring stations for the
same periods [11].

3.   RETENTION SYSTEMS IN THE CATCHMENT AREA

3.1. Retention in the glacier

When snow is being deposited on the glacier surface, the snow layers that
correspond to the individual precipitations are largely preserved and labelled

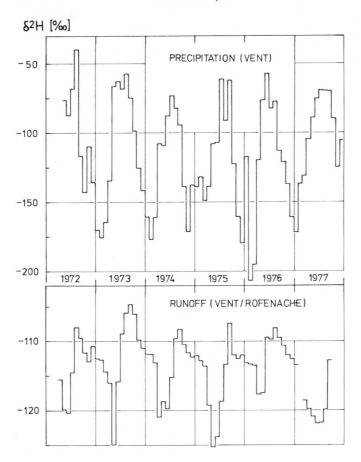

FIG.3.   Variation of the ^2H content (monthly means) in the precipitation at Vent and in the Rofenache runoff.

by their original environmental isotope content, at least until the onset of ablation, sometimes also longer. Possibly the stratified structure of the varying amounts of net accumulation of former years may be traced back in drill cores by means of their isotope contents.

For this purpose, core samples were taken on Taschachjoch, 17 May 1976, and on Sexenjoch, 9 June 1976 (for location see Fig.1) by means of a core drill described in Ref.[12]. The drilling sites were positioned at an altitude of about 3160 m above sea level, with drilling depths at 14.10 and 15.14 m, respectively. The cores were cut into 10-cm pieces, on which the density, gross beta activity and content of environmental isotopes ^2H, ^{18}O and ^3H, were determined.

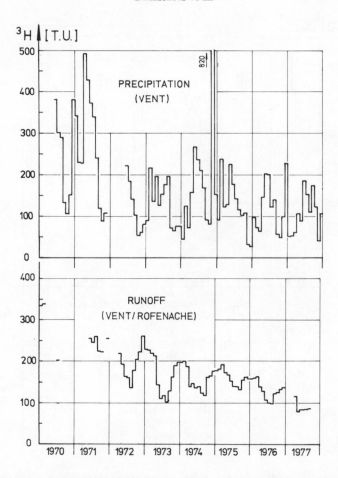

FIG.4.   Variation of the ^3H content (monthly means) in the precipitation at Vent and in the
Rofenache runoff.

The results measured on the Sexenjoch samples are given in Fig.5. For a
time assignment of the individual layers the amount of net accumulation in the
neighbourhood of the drilling sites is also plotted. From the mass budget studies
on Vernagtferner [13, 14][2], these net accumulation values have been traced as
far back as the summer horizon of 1966. From 1966 back to 1953, the net
accumulations were extrapolated from mass budget studies on the adjacent
Hintereisferner glacier [15], the horizons of reference being the peaks of the
gross beta activity (cf. below). These summer horizons of Fig.5 are plotted in

───────────
    [2] Dipl.-Met. O. Reinwarth, Kommission für Glaziologie der Bayerischen Akademie
der Wissenschaften, is thanked for making unpublished data available.

dashed lines. From the position of the summer horizons it may be seen that hardly any net accumulation exists from 1958–1959, 1962–1964 and 1969–1971, which makes an assignment of these layers difficult on account of their gross beta activity or their ^3H contents.

The gross beta activity reaches its maximum at a depth of slightly more than 10 m. The profiles of gross beta activity described in Ref.[12] on drill cores from Kesselwandferner glacier show that, above the summer horizon of 1964, no high values of gross beta activity occur. Hence, it was assumed that also the peak in the Sexenjoch profile, where the summer horizons of 1964 and of 1963 almost coincide, indicates the radioactive fallout of 1963. In comparison with precipitation fallout values, the observed value of 385 pCi/kg is high, being, however, of the same order of magnitude as the maximum values of the gross beta activity measured on core samples from Kesselwandferner. The fallout of several summer horizons must be assumed to have accumulated in the layer of maximum gross beta activity; indeed, the dry residue of the sample showing the peak yields a mass about ten times that of samples collected from the layer above it, although their activity values are also high. Similarly, the peak of gross beta activity found at a depth of 12 m (134 pCi/kg) suggested that the summer horizons of 1957 and 1958 also coincide.

The main peak of the ^3H profile differs only slightly from that of the gross beta activity by its greater depth and width. If one assigns the ^3H peak to the period from 1962 to 1964, an original concentration of about 1700 TU is obtained which agrees well with the mean ^3H contents of precipitation during the winter halves of 1962/63 and 1964/65 in the Davos area 80 km west of the Rofenache catchment area [18]. As a result the ^3H peak would cover an accumulation period about two years less than that which corresponds to the time scale shown in Fig.5. However, if one keeps to this time scale then it must be assumed that the great quantity of tritium deposited in 1962 and 1963 was removed by the meltwater percolating into the firn, and was then fixed in the form of ice layers. The ^3H peak in the precipitations of the years 1958 and 1959 cannot be found in the core, probably owing to the low accumulation and high ablation during this period. On the other hand, comparatively high accumulations of winter and even summer snow are reflected in the ^3H contents of the 1966 and 1967 layers.

The annual variations of the ^2H content in the precipitation (cf. Fig.3) cannot be traced so clearly in the ^2H profile of the core as it can in ice cores from Greenland (e.g. Ref.[16]) or in a firn core from Mont Blanc [17], which covers a depth of 16 m, but comprises only three years of accumulation. This is because the stratification sequence of winter and summer precipitations is sometimes largely disturbed by ablation processes. Thus, either it may happen that the entire summer precipitation melts away, which would cause an absence of the δ^2H maximum of that year's accumulation, or that the winter precipitation

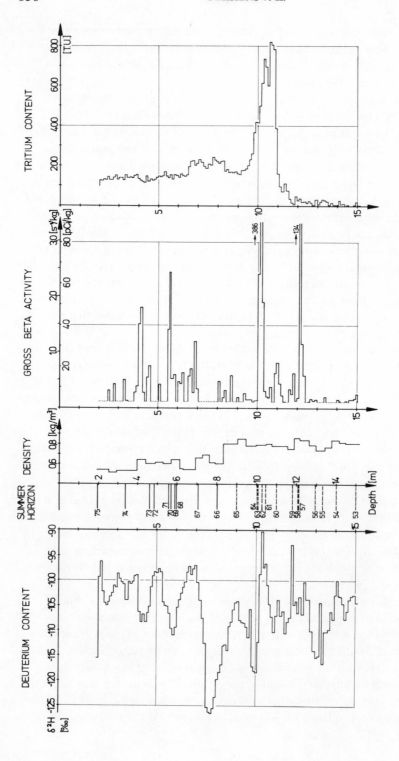

FIG.5.　Drill core Vernagtferner (Sexenjoch) June 1976 (see Fig.1): Profile of the ^2H content, the summer horizons, the density, the gross beta activity and the ^3H content. The drilling depth is given for the glacier surface on 9 June 1976. The ^3H content corresponds with the time of drilling. The gross beta activity values below 3 pCi/kg are marked by a dotted line.

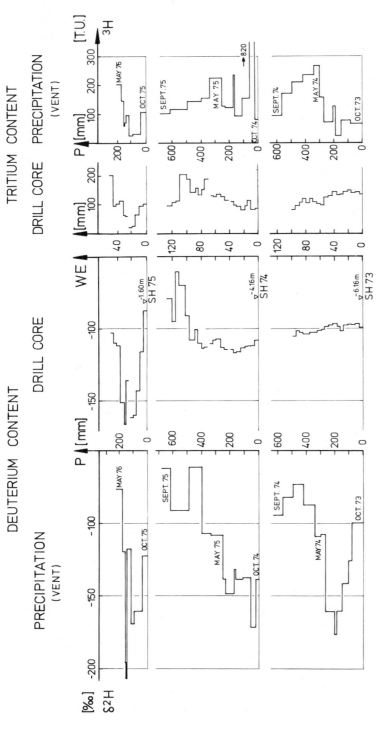

FIG.6. Drill core Vernagtferner (Taschachjoch) May 1976 (cf. Fig.1): Comparison of ^2H and ^3H contents in the precipitation of Vent (October 1973 — May 1976) and in the core (summer horizon 1973 — summer horizon 1974) with respect to the precipitation depth (P) and the water equivalent (WE) of firn layers, given separately for three hydrological years. Regarding precipitation, one stage always corresponds to the precipitation of one month. The depths given for the drill cores refer to the glacier surface on 17 May 1976. The ^3H contents given for the boreholes refer to the time of drilling.

may be almost completely melted by early summer ablation so that only summer
precipitation is accumulated, thus causing an absence of the winter minimum
in ^{2}H contents. This rare phenomenon was observed on Vernagtferner in
summer 1977. It is remarkable that the absolute minimum in the δ^2H value
of this drill core which was $-126.5\%_0$ in the winter of 1966/67, was found to
be repeated in a drilling sample of Taschachjoch with the same order of magnitude.
This minimum might be due to a special meteorological situation, e.g., precipitations of largely polar origin.

The drilling site on Taschachjoch shows higher accumulation rates,
approximating the summer horizon of 1966, which was determined only from
accumulation values obtained from mass balance studies. The ^{3}H contents in
the 1973—1976 layers show the same slightly increasing tendency (about 150 TU)
as those obtained on Sexenjoch. In contrast to the latter, however, the ^{2}H
profile of the Taschachjoch drill samples at 0—7 m depth, showed the annual
variation of the ^{2}H content in the precipitations to be traceable also in the snow
layers (Fig.6). The seasonal amplitude of the ^{2}H contents in the precipitations
here is maintained so long as there was no ablation. This seasonal variation is
also revealed in the ^{3}H profile.

What has been said about the ^{2}H content is also valid for the ^{18}O content
in the two drill cores. The slope of the δ^{18}O/δ^2H relation is approximately 7.5,
which might be due to an enrichment caused by melting and evaporation.

3.2. Retention in the groundwater reservoir

Apart from the glacier, the groundwater aquifer also acts as a reservoir.
In the winter months, when there is no surface water runoff and when the runoff
from the glaciated area is fed only by the groundwater and by the meltwater
retained in the ice and firn body of the glacier, the groundwater reservoir has
a decisive effect on the discharge.

3.3. Labelling of the discharge components

Discharge from a glaciated catchment area consists of various components,
depending on the state of ablation: snow meltwater, ice meltwater, immediately
running-off rainwater and emerging groundwater. These components may be
characterized by their content of environmental isotopes and their electrolytic
conductivity; their quantitative proportion in the total runoff can be thus
determined (cf. Section 4). In this way, the ^{3}H content helps to distinguish
between ice meltwater from old glacier ice formed long before the nuclear
weapon tests, and therefore practically free of tritium, and other discharge
components. The ^{2}H content and ^{18}O content differentiates immediately running-
off snow meltwater from meltwater with longer retention, because these two

FIG.7.   Photograph of Vernagtferner glacier on 26 August 1976, taken with an automatic camera set up at the edge of the glacier by O. Reinwarth.

originate from precipitations of different seasons and have a different exposure time on the glacier surface in the form of snow. Finally, groundwater in contact with mineral rock is characterized by a higher electrolytic conductivity than meltwater running off within the glacier or on its surface.

4.   RUNOFF FROM THE GLACIATED CATCHMENT AREA

4.1. Runoff from an entirely glaciated part

The total runoff of Vernagtferner is measured at the Vernagtbach gauge station (location, see Fig.1). As 82% of the catchment area of this gauge is glaciated, the influence of the non-glaciated part is negligible, especially at times with no precipitation runoff and no snow melt in that area; the runoff there may thus be considered as a specific glacier runoff. Figure 7 shows a photograph of Vernagtferner taken on 26 August 1976. The dark ice surface contrasts well with the lighter firn and the white snow surface. The picture, taken after a week of fine weather, shows that ice ablation, firn and snow ablation took place simultaneously on that day.

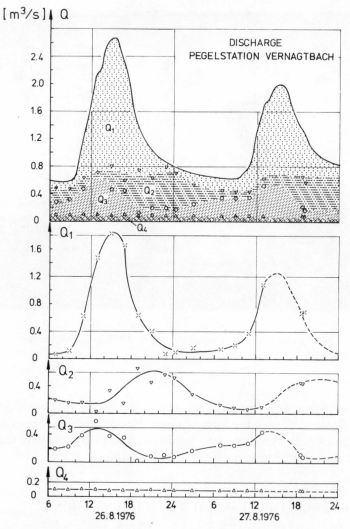

FIG.8.   Variation of the total runoff Q at the Vernagtbach gauge station (cf. Fig.1) on 26/27 August 1976 and variations of the individual runoff components such as ice melt-water (Q_1), directly running-off snow meltwater (Q_2), longer retained meltwater (Q_3), and groundwater parts (Q_4), marked by different patterns.

Figure 8 gives the runoff Q of the Vernagtbach gauge station on 26/27 August 1976, which actually shows a considerable daily variation. To determine the daily variations of ice meltwater, (Q_1), directly running-off firn and snow meltwater, (Q_2), meltwater of longer retention in the glacier, (Q_3), and ground-water, (Q_4), the fluctuations were measured of the isotope contents and the electrolytic conductivity marking these components (Fig.9). It can be seen

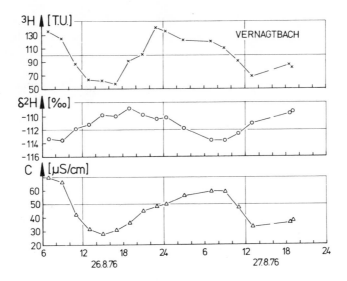

FIG.9.   Variation of the ^2H and ^3H contents and of the electrolytic conductivity in the runoff at the Vernagtbach gauge station (cf. Fig.1) on 26/27 August 1976.

that all three measuring quantities show diurnal fluctuations. Expecially significant are the variations in ^3H content owing to the changing admixture of ice meltwater in the total runoff.

The mixing proportion of the runoff components at a certain time was calculated from the individual isotope contents and electrolytic conductivity, respectively, by using the following balance equation (cf. [3]):

$$\Sigma \; Q_i \cdot c_{iK} = Q \cdot c$$

with Q_i = runoff portion of the component i,

    c_{iK} = isotope content or value of electrolytic conductivity of the component i for the tracer K (^3H, ^2H, mineralization) and

    c   = isotope content or value of the electrolytic conductivity of the total runoff Q.

The following values were used for the quantities c_i:

(a) ^3H: $c_1 = 0$; $c_2 = c_3 = c_4 = 150$ TU corresponding to the ^3H content in the discharge of Vernagtferner in winter and the mean value of the ^3H contents in the precipitations between April and August 1976 (cf. Figs 10 and 3).

(b) ^{2}H: $c_1 = c_2 = -108‰$ corresponding to the δ^2H value of a surface runoff on the glacier snout; $c_3 = c_4 = -118‰$ corresponding to the mean δ^2H value in the discharge of Vernagtbach in winter [5].

(c) Electrolytic conductivity: $c_1 = c_2 = c_3 = 20 \ \mu S$ corresponding to values measured on meltwater from Vernagtferner $c_4 = 350 \ \mu S$, corresponding to the average value measured on groundwater from springs in the catchment area.

Figure 8 reviews the variations of the individual runoff components determined from balance equations by using the above values and the variations of the total runoff and its ^{3}H and ^{2}H contents as well as its electrolytic conductivity.

The direct runoff of firn and snow meltwater (Q_2) has a time lag of about six hours as compared with the ice meltwater (Q_1). The peak of this meltwater wave from glacier areas of higher altitudes passed through the Vernagtbach gauge station at about 21.00 hours (9 p.m.). The fluctuations of the residual amount, Q_3, cannot be clearly interpreted. The increase in Q_3 in the morning hours may be due to onsetting ablation, which certainly activates the entire hydraulic system in and below the glacier so that the meltwater with longer residence times runs off more intensively.

The total quantitative balance for the 24-hour interval from 26 August 1976 (6 a.m.) to 27 August 1976 (6 a.m.) is given in Table I, clearly showing the specific influence of the ice meltwater on the amount of the total runoff and on the shape of diurnal variations.

TABLE I. BALANCE OF THE RUNOFF COMPONENTS IN VERNAGTBACH CREEK FROM 6 a.m., 26 AUGUST 1976 TO 6 a.m., 27 AUGUST 1976

	Q	Q_1	Q_2	Q_3	Q_4
Mean runoff (m^3/s)	1.23	0.60	0.33	0.21	0.09
(%)	100	49	27	17	7

4.2. Total runoff of the catchment area

The runoffs of the individual claciers (location: Fig.1) in the catchment area are characterized by different seasonal fluctuations of the ^{2}H values. In this respect, detailed studies were made between 1968 and 1974, with the results published in Ref.[5]. The elements are the ^{2}H and ^{3}H values (Figs 3 and 4) measured in the Rofenache runoff at the Vent/Rofenache gauge, and in the precipitation of the sampling site of Vent, during a period of several years.

The variations of the isotope contents in the runoff, together with those in the precipitations, demonstrate what runoff components prevail at what times:

(1)   The onset of snow melt in spring, beginning at lower altitudes and gradually growing upwards, is clearly marked by a sudden decrease of the δ^2H values in the runoff. In the period thereafter, the runoff mainly consists of melting winter snow. The decrease in ^{3}H content in the runoff of spring is also due to the melting winter snow.

(2)   The period of ice ablation on the glaciers is marked by minimum ^{3}H contents. Such ^{3}H minima, however, occur only during fairly long periods of fine weather since, in a period of precipitation, the surface runoff from the non-glaciated part of the catchment area, causes an increased ^{3}H content. Thus, two mechanisms are activated during the summer. Furthermore, the period of heavy ablation is characterized by maximum ^{2}H contents.

By means of fluorescent dye tracers the residence time of ice meltwater in the glacier was found to be several hours, and the percolation time in firn was found to take up to several weeks [6, 7].

(3)   After the end of the ablation period, which is usually set by new snow falls down to lower altitudes of the catchment area, runoff no longer consists of directly running-off ice meltwater, firn and snow meltwater, or of rain running off directly. Normally, the catchment area is covered with snow from November to March. The runoff is then fed only by water retained in the catchment area. During this period, the ^{2}H content decreases until the beginning of snow melt in the following spring. The ^{3}H content behaves conversely: it increases up to a winter maximum. Figure 10 shows the trend of these maximum ^{3}H contents in the Rofenache catchment area from 1968 to 1977/78 and the annual mean values of the ^{3}H contents in the precipitation of various stations. All precipitation values are lower than the corresponding winter runoff values, which shows that the winter runoff is fed from precipitations of former years.

If it is assumed that the runoff after the ablation period is largely fed by two different reservoirs, which are characterized by their respective ^{2}H and ^{3}H contents, the residence time of meltwater can be calculated [19] from the decrease of ^{2}H contents in late autumn (Fig.3) and the decrease in the winter values of the ^{3}H content over a period of several years. In the former case, a mean residence time of 100 days can be calculated from the ^{2}H values. It seems reasonable that this residence time should be attributed to the meltwater, which is still retained in the glacier at the end of the ablation period. From the ^{3}H values of the years 1971/72 to 1974/75 residence times of 3.6 years and 4.8 years, respectively, result, depending on whether the winter runoff is assumed to be

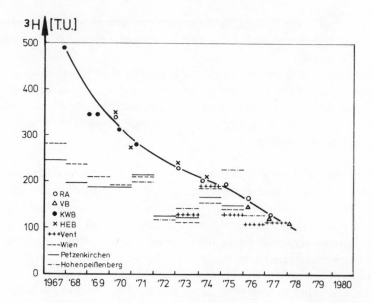

FIG.10.　Decrease in ^3H content in the winter runoff of creeks in the Rofenache catchment area (cf. Fig.1) and the annual mean values of the ^3H contents in the precipitation of Vent, Vienna, Petzenkirchen (Austria) and Hohenpeißenberg (Bavaria, F.R. Germany). (The ^3H concentrations for Vienna and Petzenkirchen were provided by the IAEA.)
RA = Rofenache creek, VB = Vernagtbach creek, KWB = Kesselwandbach creek, HEB = Hintereisbach creek.

due to the summer or to the winter precipitation. This residence time of about four years is suitably attributed to the outflowing groundwater. From the two residence times of 100 days and 4 years it becomes evident that the winter runoff originates from two different reservoirs. The runoff of meltwater from the glaciated part decreases gradually up to midwinter so that the groundwater portion in the runoff prevails at that time. This can be proved by conductivity measurements.

ACKNOWLEDGEMENT

Gratitude must be expressed to all assistants helping with the field and laboratory work. Special gratitude must be expressed to Dipl.-Met. O. Reinwarth, Kommission für Glaziologie der Bayerischen Akademie der Wissenschaften, for advice and discussions in evaluation. The studies were supported by the Deutsche Forschungsgemeinschaft (Sonderforschungsbereich 81 der Technischen Universität München, Teilprojekt A1) and by the Österreichische Akademie der Wissenschaften and the Österreichischer Bundesminister für Inneres.

REFERENCES

[1] BERGMANN, H., REINWARTH, O.Z., Gletscherkd. Glazialgeol. **12** (1977) 157.
[2] Hydrographisches Jahrbuch für Österreich 1964, Vienna (1968).
[3] BEHRENS, H., BERGMANN, H., MOSER, H., RAUERT, W., STICHLER, W.,
     AMBACH, W., EISNER, H., PESSL, K., Z. Gletscherkd. Glazialgeol. **7** (1971) 79.
[4] MOSER, H., RAUERT, W., STICHLER, W., AMBACH, W., EISNER, H., Z. Gletscherkd.
     Glazialgeol. **8** (1972) 275.
[5] AMBACH, W., EISNER, H., ELSÄSSER, M., LÖSCHHORN, U., MOSER, H., RAUERT, W.,
     STICHLER, W., J. Glaciol. **17** (1976) 383.
[6] AMBACH, W., ELSÄSSER, M., BEHRENS, H., MOSER, H., Z. Gletscherkd. Glazialgeol.
     10 (1974) 181.
[7] BEHRENS, H., LÖSCHHORN, U., AMBACH, W., MOSER, H., Z. Gletscherkd. Glazial-
     geol. **12** (1976) 69.
[8] OERTER, H., "Wasserbewegung in einem Gletscher, dargestellt an den Feldarbeiten auf
     dem Vernagtferner", Sonderforschungsbereich 81 der Techn. Universität München,
     Vortragsveranstaltung 9.2.1977, Munich (1977) 67.
[9] MOSER, H., AMBACH, W., Z. Gletscherkd. Glazialgeol. (in press).
[10] AMBACH, W., ELSÄSSER, M., MOSER, H., RAUERT, W., STICHLER, W., TRIMBORN, P.,
     Wetter Leben **27** (1975) 186.
[11] BAUER, F., RAJNER, V., RANK, D., Naturwiss. **62** (1975) 526.
[12] EISNER, H., Z. Gletscherkd. Glazialgeol. **7** (1971) 65.
[13] REINWARTH, O., Z. Gletscherkd. Glazialgeol. **8** (1972) 43.
[14] REINWARTH, O., "Hydrologische Untersuchungen am Vernagtferner (Ötztaler Alpen)",
     Sonderforschungsbereich 81, Techn. Universität München, Vortragsveranstaltung
     12.5.1976, Munich (1976) 7.
[15] HOINKES, H., Z. Gletscherkd. Glazialgeol. **6** (1970) 37.
[16] DANSGAARD, W., JOHNSON, C.J., CLAUSEN, H.B., GUNDESTRUP, V., Stable
     isotope glaciology, Medd. Groenland **197** (1973) 1.
[17] JOUZEL, J., MERLIVAT, L., POURCET, M., J. Glaciol. **18** (1977) 465.
[18] MARTINEC, J., SIEGENTHALER, U., OESCHGER, H., TONGIORGI, E., "New
     insights into the runoff mechanism by environmental isotopes", Isotope Techniques in
     Groundwater Hydrology (Proc. Symp. Vienna, 1974) **1**, IAEA, Vienna (1974) 129.
[19] LÖSCHHORN, U., AMBACH, W., MOSER, H., STICHLER, W., Z. Gletscherkd.
     Glazialgeol. **12** (1977) 181.

DISCUSSION

J. MARTINEC: Were the tritium concentrations in the winter baseflow as illustrated in Fig.10 not affected by the presence of meltwater from glaciers containing practically no tritium? It seems from Fig.2 that part of the ground-water recharge would come from the glacier.

W. AMBACH: Yes, you are right. During the ablation period, a small part of the ice meltwater contributes to groundwater recharge.

P. THEODORSSON: With reference to Fig.5, which shows the isotopic content of the glacier ice core, you ascribe the single pronounced tritium peak to the summer precipitation of 1963. Judging from my experience with many similar tritium profiles for glaciers in Iceland, something seems to be missing. The sharp peak you show indicates very little vertical mixing from percolating meltwater, especially where there is a sharp decrease in tritium at the lower edge of the peak. Under such circumstances I would certainly also expect the summer peak of the years after 1963 to show up clearly in the ice.

W. AMBACH: The ^{3}H peak covers several years of ^{3}H fallout. In this Alpine region it is mainly winter precipitation that is accumulated, whereas spring and summer precipitation (particularly that of May and June) melts and is not accumulated.

J.Ch. FONTES: How do you explain the fact that the ^{3}H peak attributed to the high level of ^{3}H in precipitation in 1963–4 remained so remarkably sharp in a profile in which processes of melting and recrystallization occur?

W. AMBACH: We assume that the firn layers beneath can store only 1% per volume of meltwater. In such a situation, contamination moving downwards is small.

J. MARTINEC: What was the model used to determine the residence time of groundwater?

W. AMBACH: We used the exponential model described in Ref.[18].

B.R. PAYNE: Could you please indicate the errors in your estimates of the relative amounts of the different runoff components?

W. AMBACH: The values in Table I are integrals of the discharge curves in Fig.8. The scatter in the measurements can also be seen in Fig.8. The isotope concentration values for the discharge components were found from a group of samples (see Ref.[5]).

B.R. PAYNE: I would say that in the case of deuterium the range in delta values is only about 10‰, which would give rise to quite large errors in the estimates.

P. FRITZ: I find it very surprising that the discharge of the fourth component, namely the groundwater, in the runoff you describe is constant in time. If I remember correctly, similar studies also involving measurements of Alpine environments show very great seasonal variations, whereby the groundwater discharge increases with river discharge.

W. AMBACH: The period shown in Fig.8 is only 42 hours. We also observed seasonal variations in groundwater discharge.

J.Ch. FONTES: Did you measure conductivity for the snow profile and, if so, in that case did you observe a relationship between ^{2}H and total dissolved solids (TDS)?

W. AMBACH: We did not measure conductivity.

ESTUDIO DEL COMPORTAMIENTO DINAMICO DEL ESTUARIO DEL RIO GUADALQUIVIR

E. BAONZA, A. PLATA BEDMAR, J.A. JIMENEZ,
A. RUIZ MATEOS
Gabinete de Aplicaciones Nucleares
    a las Obras Públicas,
Madrid, España

Abstract—Resumen

STUDY OF THE DYNAMIC BEHAVIOUR OF THE ESTUARY OF THE GUADALQUIVIR RIVER.

One of the largest rivers in Spain, the Guadalquivir, is showing signs of two types of contamination. The first is caused by effluents from industry and settlements on the river, which include Seville. The second is due to penetration by salt water from the Atlantic, which is carried more than 100 km by tides. Both types of contamination, and especially the second, seriously affect the water collected for irrigating rice fields, which can tolerate a maximum salinity of 0.8 g/l. The flow of fresh water carried by the river is regulated in its final stretch by the Alcalá del Río dam upstream from Seville. To determine the influence of variations in river discharge on the penetration of sea-water and on the dilution of salt water and contaminated water, twelve ^{82}Br point injections of 4 Ci each were carried out at different spots; the passage of the radioactive cloud and its dilution were then monitored along the river. A single tritium injection of 30 Ci was also made at the foot of the Alcalá del Río dam and the passage of the tracer was studied from three stations, the last of which was situated at the mouth of the river where it enters the Atlantic. The field data were used to prepare a mathematical model which yielded very interesting information, suggesting ways in which the use of the river could be improved.

ESTUDIO DEL COMPORTAMIENTO DINAMICO DEL ESTUARIO DEL RIO GUADALQUIVIR.

Uno de los ríos más importantes de España, el Guadalquivir, presenta en la actualidad dos clases de contaminación. La primera ocasionada por los vertidos de las industrias y de las poblaciones ribereñas, entre las cuales se encuentra Sevilla. La segunda es debida a la penetración, más de 100 km, de las aguas saladas del Atlántico, movidas por las mareas. Ambos tipos de contaminación, y en especial el segundo, afectan gravemente a las captaciones de agua para riego de plantaciones de arroz, que soportan una salinidad máxima de 0,8 g/l. El caudal de agua dulce que lleva el río está regulado en su tramo final por la presa de Alcalá del Río, situada aguas arriba de Sevilla. Con objeto de conocer la influencia de este caudal variable del río en la penetración del agua del mar y en los procesos de dilución de agua salada y de agua contaminada, se realizaron 12 inyecciones puntuales, de 4 Ci de ^{82}Br cada una, en diversos puntos, controlándose el paso de la nube de trazador y su dilución a lo largo del río. Tambien se realizó una inyección única de 30 Ci de tritio, al pie de la presa de Alcalá del Río, estudiando su paso por tres estaciones, la última de las cuales situada en la desembocadura en el Atlántico. Con los datos de campo fue preparado un modelo matemático que ha permitido obtener datos muy interesantes para mejorar la utilización del río.

1.   INTRODUCCION

El Guadalquivir, tercer río de España en longitud, recoge las aguas de una
cuenca sujeta a grandes desigualdades climatológicas y pluviométricas, estando
regulado a lo largo de los 650 km de su curso por un total de 37 presas. Hasta
la última (presa de Alcalá del Río) llega la influencia de las mareas del Atlántico.
Por consiguiente, el estuario del Guadalquivir, con una longitud de 110 km, queda
limitado por dicha presa y por la desembocadura del río al Atlántico, en San-
lúcar de Barrameda (fig. 1).

La contaminación de las aguas del estuario alcanza frecuentemente niveles
muy elevados, debido a los vertidos industriales y domésticos de las poblaciones
ribereñas, entre las cuales se encuentra Sevilla con más de 500 000 habitantes.
Esta contaminación constituye un serio inconveniente para la utilización del
estuario con fines recreativos y de pesca comercial, como se venía haciendo
hasta ahora.

Por otra parte, el agua dulce del estuario se utiliza para regar unas 30 000 ha,
dedicadas en su mayor parte al cultivo del arroz, zona de cultivo que se proyecta
duplicar. Este uso del estuario exige la regulación del caudal del río por medio
de su sistema de presas, dejando correr un caudal continuo de agua dulce
suficiente para mantener el agua salada lejos de la zona donde se encuentran las
estaciones de bombeo para regadío. El problema principal se plantea principalmente durante los periodos de estiaje, debido al reducido caudal de agua dulce
disponible en esta época del año.

El estudio de ambos problemas, contaminación del río e intrusión de agua
del mar, exige conocer el comportamiento dinámico del estuario, en cuya
determinación intervienen tanto el caudal de agua dulce del Guadalquivir como
el caudal variable de agua salada aportada por las características dispersivas del
flujo que ocasionan las mareas del Atlántico.

Con este fin, se han realizado una serie de experiencias con trazadores
radiactivos, las cuales se describen a continuación. Para fijar posiciones a lo largo
del estuario, se ha definido una escala de distancias con origen en la presa de
Alcalá del Río (kilómetro cero). La desembocadura se encuentra, por tanto, en
el kilómetro 110.

2.   EXPERIENCIAS REALIZADAS

2.1. Experiencias con bromo-82

Se realizaron 12 experiencias, inyectando en cada caso dosis variables de
2 a 5 curios de bromo-82 en forma de ion bromuro. Las inyecciones se realizaron
de forma instantánea, precisamente en el momento de producirse el cambio de
sentido del flujo correspondiente a una bajamar (BM) o a una pleamar (PM).

Debe indicarse que la inercia inherente al desplazamiento del agua hace que el
cambio de sentido del flujo se produzca con cierto retraso respecto a la PM o
BM correspondiente, excepto naturalmente en el kilómetro 110, donde ambos
fenómenos coinciden.

Los puntos de inyección (fig. 1) se eligieron de forma que los recorridos de
la nube de trazador se superpusieran, con objeto de estudiar la totalidad del
estuario. En dos experiencias, el trazador se inyectó en PM en el kilómetro cero
y en otras dos en BM en el kilómetro 110. Las ocho inyecciones restantes se
realizaron en puntos intermedios.

Una vez inyectado el bromo-82 en uno de los puntos, se determinó la
posición de la nube de trazador durante los sucesivos ciclos de marea, incluyendo
todas las PM y BM siguientes, hasta que la concentración de trazador se reducía
por el fenómeno dispersivo a niveles no detectables. Las medidas se realizaron
con un detector de centelleo sujeto al costado de un barco y sumergido en el
agua a un metro de profundidad. El barco se desplazaba a lo largo del estuario
con velocidad constante. La respuesta del detector alimentaba una escala, de la
cual se tomaba lectura cada vez que el barco pasaba frente a uno de los puntos
P-1, P-2, P-3, etc., localizados en el plano. Si la distancia entre dos puntos
consecutivos era demasiado grande, se tomaban lecturas intermedias, y los tramos
correspondientes a cada lectura se determinaban por interpolación, basándose
en el tiempo transcurrido. El uso de un registrador gráfico no resultó apropiado,
debido a los bajos niveles de recuento que era preciso medir, sobre todo en los
extremos de la nube radiactiva. Se comprobó que el estado de marea (PM o
BM) se transmitía en sentido ascendente por el estuario a una velocidad ligera-
mente superior a la del barco. Por tanto, cuando éste se desplazaba en el mismo
sentido, era posible localizar la posición de la nube de trazador con aguas para-
das en casi todo el recorrido.

2.2. Experiencias con tritio

La concentración de tritio de origen termonuclear se midió en ocho muestras
de agua, tomadas en distintos puntos del estuario durante la BM. La finalidad de
estas determinaciones era conocer la proporción de agua de mar existente a lo
largo del estuario. Pero, además, era necesario conocer la concentración de tritio
natural como valor de fondo para una experiencia única en la cual se inyectaron
28 curios de agua tritiada en el kilómetro cero. En este caso, se determinó la
curva de paso del tritio por tres puntos del estuario situados, respectivamente,
en los kilómetros 33, 67 y 106. Dicha determinación se realizó tomando
muestras en las sucesivas bajamares durante un periodo de dos meses para los
dos primeros puntos y de tres meses para el tercero. Las muestras de los dos
primeros puntos se midieron directamente con un detector de centelleo líquido,
mientras que las del tercer punto tuvieron que someterse a un proceso previo de
enriquecimiento, por vía electrolítica, debido a su baja concentración.

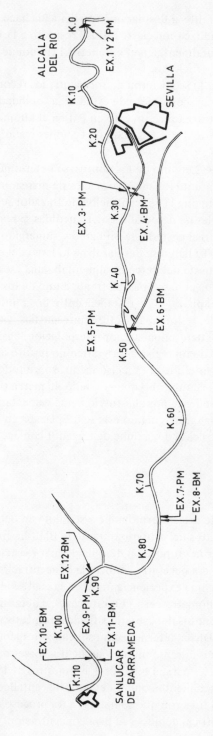

FIG. 1. Situación de los puntos de inyección en el estuario del río Guadalquivir.

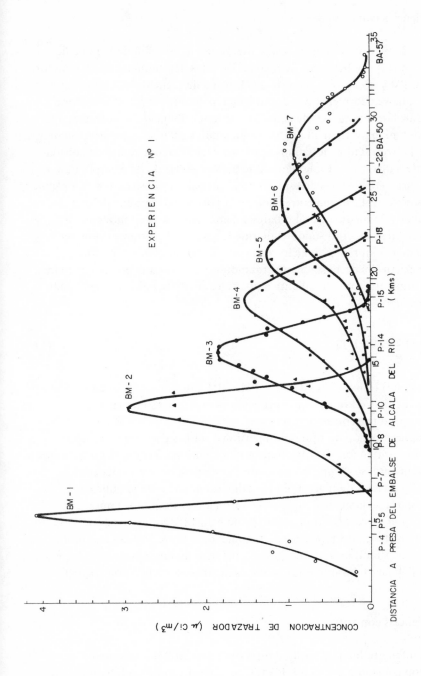

FIG.2. Movimiento de la nube de trazador.

3.   RESULTADOS OBTENIDOS

3.1. Experiencias con bromo-82

En las 12 inyecciones realizadas, se determinó la posición de la nube de
trazador en un total de 96 ciclos de marea. De estas determinaciones, 47 fueron
realizadas en PM y 49 en BM. La velocidad media de recuento obtenida para
cada tramo, una vez corregida por desintegración del bromo-82, se transformó
en unidades de concentración, de acuerdo con la sensibilidad del detector, que
había sido obtenida previamente (705 cuentas/min para una concentración de
1 $\mu Ci/m^3$). Los resultados obtenidos se representaron en diagramas de concentración frente a distancia. En la figura 2 se presenta a título de ejemplo uno de
estos diagramas. Por razones de espacio no se presentan los restantes. En algunos
casos, fue preciso realizar una corrección en la posición de la nube de trazador,
debido a haber sido medida, total o parcialmente, con aguas en movimiento.
Dicha corrección se efectuó suponiendo que la velocidad de transmisión del
estado de la marea, durante un ciclo, varía en función del tiempo de forma
senoidal. A partir del desplazamiento total observado durante el ciclo de marea,
$\bar{x}$, el desplazamiento correspondiente a una fracción de tiempo t/T viene dado
por la expresión

$$x_t = \frac{\bar{x}}{2}.(1 - \cos \pi \, t/T) \tag{1}$$

siendo t el tiempo transcurrido desde la PM o BM anterior y T la duración del
ciclo de marea (unas 6 horas). Las correcciones realizadas son escasas y de
pequeña magnitud (inferiores a 500 m).

Esta representación de tipo unidimensional solo es correcta si la concentración de trazador en los distintos puntos de una misma sección es constante.
Experimentalmente, se comprobó que esto se cumple de forma satisfactoria
después de 2 o 3 ciclos de marea. En las primeras medidas de la nube de
trazador, después de la inyección, se observaron a veces cambios importantes
de concentración en sentido transversal, principalmente en tramos de sección
elevada. Para restar importancia a este fenómeno, la inyección del trazador se
realizó de forma simultánea en tres puntos distintos de una misma sección, y
el recorrido del barco durante estas primeras medidas se realizó con trayectorias
en zig-zag.

3.2. Experiencias con tritio

El perfil longitudinal de tritio natural obtenido en BM reveló una
concentración constante de 35,0 ± 1,3 UT en el tramo comprendido entre los

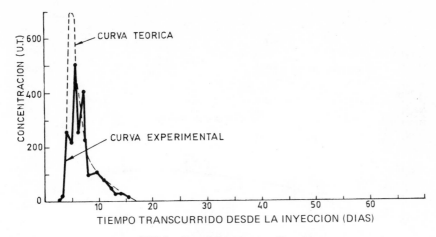

FIG.3. Paso del tritio por el km 33.

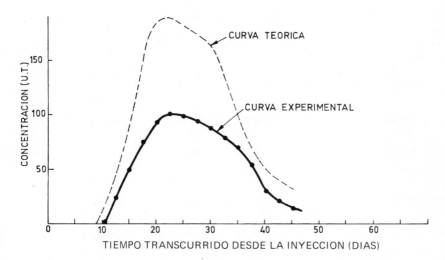

FIG.4. Paso del tritio por el km 67.

kilómetros cero y 86. En el kilómetro 91 la concentración descendió a 32,4 UT
y en el 106 a 28,0 UT. Teniendo en cuenta que la concentración en el mar es
del orden de 2 UT, las concentraciones anteriores corresponden a una proporción
de agua de mar del 7,9 y 21%, respectivamente. Estos valores están de acuerdo
con los obtenidos con el modelo de cálculo que se describirá más adelante.

Las curvas de paso del tritio inyectado en el kilómetro cero por los kilómetros
33, 67 y 106 se muestran en las figuras 3, 4 y 5, junto con las curvas teóricas
deducidas mediante el modelo antes mencionado. Estos resultados se discuten
más adelante.

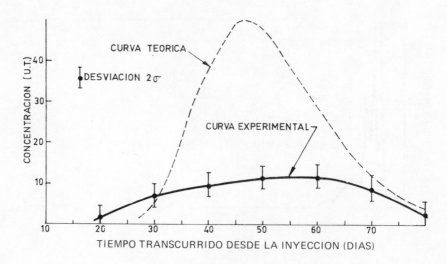

FIG.5. Paso del tritio por el km 106.

4. PARAMETROS DEL ESTUARIO

4.1. Influencia de la carrera de marea en el desplazamiento del trazador

El desplazamiento medio de la nube de trazador en un ciclo de marea depende de los siguientes parámetros:

a) Sección media del estuario en el tramo recorrido por el trazador, $S_{x_1 - x_2}$;
b) Sección media $S'_{0 - x_1}$ entre el kilómetro cero y el comienzo del tramo anterior, referida solo a la capa superficial evacuada o llenada, según se trate de un ciclo PM-BM o BM-PM;
c) Carrera de marea en el mar, Δh;
d) Caudal de agua dulce aportado por el río durante el ciclo de marea.

Como es lógico, el caudal de agua dulce hace que el desplazamiento aumente en el ciclo PM-BM y disminuya en el ciclo BM-PM. Si este caudal es cero, o despreciable, y la carrera de marea permanece constante, el desplazamiento depende de la relación $S'_{0 - x_1}/S_{x_1 - x_2}$. Si tanto el caudal como Δh y la sección $S_{x_1 - x_2}$ permanecen constantes, el desplazamiento medio para los ciclos PM-BM y BM-PM viene dado por el producto de x_1 por la relación anterior de secciones, como puede demostrarse fácilmente. A la inversa, a partir del desplazamiento de la nube de trazador obtenido experimentalmente, puede obtenerse información cuantitativa de la mencionada relación de secciones.

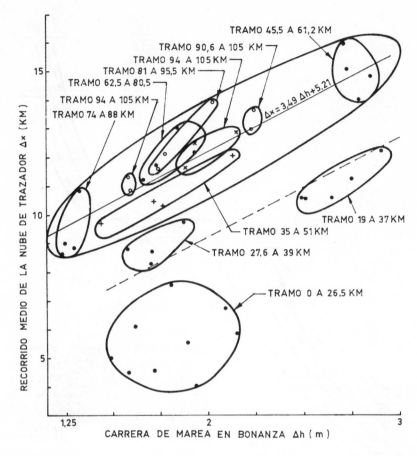

FIG.6. Relación entre el recorrido de la nube de trazador y la carrera de marea en bonanza.

Para reducir el efecto del caudal de agua dulce aportado por el río, se calcularon los valores medios de los desplazamientos observados para cada dos ciclos PM-BM consecutivos. Estos valores medios se han representado en función de Δh (fig. 6). Se observa que, en el tramo km 0–km 30 (experiencias 1 a 4) el valor de Δx, recorrido por la nube de trazador, varía entre 2 km y unos 6 km. En cambio, a partir del kilómetro 35 (experiencias 5 a 12) los puntos forman aproximadamente una línea recta que obedece a la ecuación $\Delta x = 3,49\,\Delta h + 5,21$. Por tanto, en este último tramo el recorrido de la nube de trazador permanece aproximadamente constante para un valor fijo de la carrera de marea, lo que indica que, para Δh constante, la relación $S_{x_1-x_2}/S'_{0-x_1}$ ha de ser proporcional a la distancia x_1. Según se comprobó, la constante de

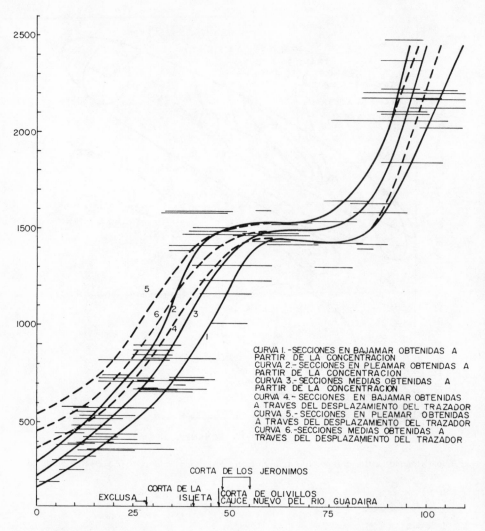

FIG.7. Secciones medias del estuario obtenidas a partir de la concentración de trazador y de los desplazamientos del mismo.

proporcionalidad K varía con Δh (en metros) según la recta $K = 0{,}00213\,\Delta h + 0{,}078$, y por tanto la relación anterior viene dada por la expresión

$$\frac{S_{x_1-x_2}}{S'_{0-x_1}} = (0{,}00213\,\Delta h + 0{,}078)\,x_1 \tag{2}$$

4.2. Secciones medias y volúmenes embalsados

Como no se disponía de datos batimétricos, las secciones medias y los volúmenes embalsados en el estuario han sido obtenidos de forma aproximada, basándose en las curvas que reflejan la posición de la nube de trazador. Estas curvas han sido integradas por un procedimiento gráfico, obteniéndose una concentración media referida al tramo del estuario ocupado por la nube de trazador. Suponiendo que la sección transversal en el tramo es constante, su valor se obtiene dividiendo la actividad de bromo-82 inyectada por el área de la curva correspondiente. Las secciones medias obtenidas por este procedimiento se muestran en la figura 7, representadas en función de la distancia. Tales secciones son válidas solo para tramos del orden de unos 10 km, por ser esta la amplitud media de las nubes de trazador. Debe indicarse que la forma de las curvas se justifica satisfactoriamente de forma cualitativa teniendo en cuenta los anchos medios del estuario. La curva 1 da la sección media en BM y la 2 en PM. La curva 3 representa el valor medio de las anteriores.

Como es obvio, las secciones medias del estuario varían con la altura de marea. Por tanto, la distribución real de estas secciones tendría que realizarse en forma de una familia de curvas, cada una de las cuales correspondería a una altura de marea determinada. Debido a la dispersión de los puntos de la figura 7, no fue posible determinar de forma precisa la influencia de la altura de marea. Ello introduce una limitación en el grado de precisión de las curvas. No obstante, la altura de marea llega a tener una importancia secundaria cuando se consideran largos periodos de tiempo, porque entonces los errores se compensan entre sí. Las curvas 1 y 2 están obtenidas con alturas de marea variables, y representan por tanto una situación media aproximada.

Por otra parte, las curvas anteriores obtenidas en base a las concentraciones medias de trazador han sido ajustadas de forma que justifiquen aproximadamente los tiempos de tránsito del tritio inyectado en la presa de Alcalá del Río, así como los desplazamientos de las nubes de trazador en los sucesivos ciclos de marea. En el caso del tritio, se observó que el volumen de agua almacenado en BM en el tramo km 0 – km 33, calculado a partir de la curva 1, era alrededor del 30% inferior al aportado por el río entre el momento de la inyección y el tiempo medio de paso del tritio por el kilómetro 33. Esto indicaba que la curva 1 debía ser corregida en este tramo. La comprobación de las curvas a partir de los valores de Δx – desplazamientos de la nube de trazador durante los ciclos de marea – se ha realizado por dos procedimientos distintos:

1) La diferencia ΔV entre el volumen de agua desplazado durante el ciclo PM-BM en el recorrido Δx y el volumen análogo correspondiente al ciclo BM-PM, una vez corregido por diferencias de carrera de marea Δh, equivale al volumen aportado por el río durante el ciclo PM-BM-PM, el cual es

conocido. El valor de ΔV puede obtenerse de forma aproximada a partir de las secciones indicadas por la curva 3. La relación entre ΔV y el volumen aportado por el río durante el ciclo PM-BM-PM proporciona un factor de corrección que permite trazar la curva 6, curva que representa los valores medios de las secciones medias corregidas para PM y BM.

2)   De acuerdo con lo expuesto en el apartado 4.1, el valor medio de Δx para dos ciclos consecutivos PM-BM y BM-PM, suponiendo Δh =constante, viene dado por la ecuación siguiente

$$(\Delta x)_{medio} = x_1 \; \frac{S'_{0-x_1}}{S_{x_1-x_2}} \tag{3}$$

Puesto que $(\Delta x)_{medio}$ y x_1 son conocidos, el valor de S'_{0-x_1} puede calcularse en función de $S_{x_1-x_2}$ obtenido a partir de la curva 6 ajustada anteriormente. Para realizar este cálculo, el valor de $(\Delta x)_{medio}$ se ha corregido por cambios de carrera de marea mediante la expresión $\Delta x = 3,49\Delta h + 5,21$. Todos los valores de Δx se han referido a un valor $\Delta h = 1,9$ m, que representa una situación de tipo medio. Los valores de S'_{0-x_1} se han representado en función de la distancia, y la curva obtenida ha sido diferenciada gráficamente, obteniéndose así las secciones $S'_{x_1-x_2}$ válidas para la capa de agua influenciada por las mareas en el tramo x_1-x_2. Esta última sección equivale a la diferencia entre las secciones indicadas por las curvas 4 y 5, lo cual permite ajustar la separación entre ambas curvas que, por otra parte, han de quedar equidistantes con la curva 6.

Por último, los volúmenes medios embalsados en PM y BM, desde el origen hasta una distancia x, se han obtenido a partir de las secciones indicadas por las curvas 4, 5 y 6. Los resultados se reflejan en la figura 8.

5.   PARAMETROS DE DISPERSION

5.1. Coeficiente de asimetría de Pearson

Como se sabe, el coeficiente de asimetría de Pearson viene dado por la expresión K_P = (mediana − moda)/desviación típica (σ). El valor de K_P se determinó solo para las nubes de trazador medidas en BM, de acuerdo con las necesidades del programa de cálculo elaborado. En las experiencias 1 a 3 todos los valores de K_P fueron negativos, a excepción de uno de ellos. En cambio, en las experiencias 4 a 12 se dieron 5 valores negativos y 24 positivos. Un análisis del signo y valor absoluto de K_P reflejó una marcada influencia del estado de marea existente en el momento de la inyección del trazador. Cuando la inyección se realiza en PM, los coeficientes de asimetría son más negativos que si se realiza en BM. Por otra parte, se observó que en líneas generales el valor

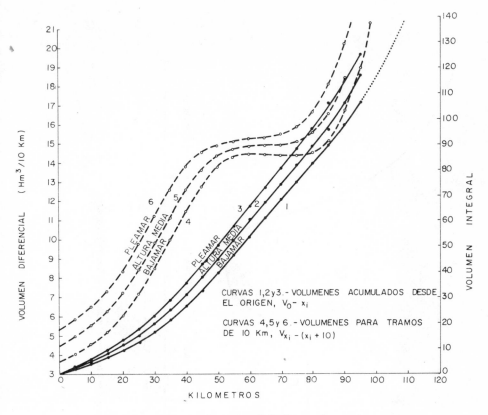

FIG.8.   Volúmenes de agua embalsados en el estuario.

de K_p tendía a aproximarse a cero a medida que aumentaba el número de ciclos de marea, es decir que las curvas se hacen cada vez más simétricas. La asimetría de las curvas se debe al fenómeno dispersivo y a los cambios de sección del estuario. El hecho de que el fenómeno dispersivo actúe en los dos sentidos puede explicar la disminución de la asimetría con el número de ciclos de marea.

5.2. Desviación típica

La desviación típica de todas las curvas correspondientes a BM se determinó por medio de la expresión

$$\sigma_x = \sqrt{\frac{C_1(x_1 - x_m)^2 + C_2(x_2 - x_m)^2 + \ldots + C_n(x_n - x_m)^2}{C_1 + C_2 + \ldots + C_n}} \tag{4}$$

donde C_1, C_2, etc., representan las concentraciones existentes en los puntos x_1, x_2, etc., y x_m es la mediana o esperanza matemática de la curva.

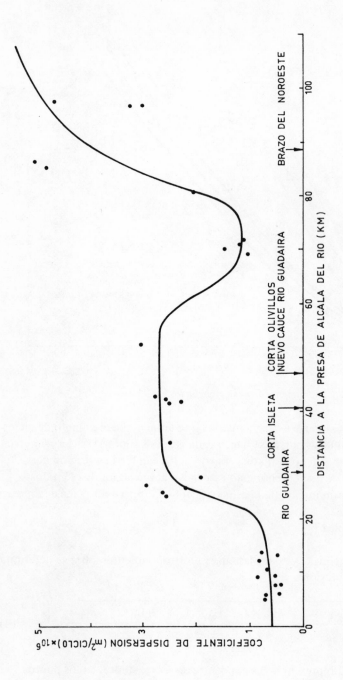

FIG.9.   *Variación del coeficiente de dispersión a lo largo del estuario.*

5.3. Modelo y coeficiente de dispersión

El modelo matemático utilizado para estudiar la renovación del agua en el estuario es el correspondiente a una distribución normal definida por la ecuación

$$C(x, t) = \frac{A}{S\sqrt{4\pi Dt}} \; \exp\left[-\frac{(x - \bar{v}t)^2}{4Dt}\right] \tag{5}$$

donde

C = concentración de trazador
A = actividad inyectada
S = sección media del estuario en el tramo ocupado por el trazador
t = tiempo (en nuestro caso expresado en ciclos de marea)
D = coeficiente de dispersión
$x - \bar{v}t$ = distancia desde el centro de la nube de trazador a otro punto
     cualquiera de esta.

Como se sabe, en este modelo la desviación típica se relaciona con el coeficiente de dispersión por medio de la relación $\sigma_x = \sqrt{2Dt}$. El valor de este coeficiente, en función de la distancia x, se representa en la figura 9. En algunas experiencias, los valores de D se han incrementado en los sucesivos ciclos de marea, debido a un mal mezclado inicial del trazador. En tales casos, se han tomado los valores mayores. Como era de esperar, se observa una estrecha relación entre el valor de D y la sección media del estuario.

6.   COMPROBACION DEL MODELO DISPERSIVO

La comprobación del modelo dispersivo definido por la ecuación (5) se ha realizado tratando de justificar con el mismo las curvas experimentales de paso del tritio, inyectado en el kilómetro cero, por los puntos kilométricos 33, 67 y 106. Las curvas teóricas de paso por estos puntos se han calculado por medio de la ecuación (5). Para ello, se han utilizado las secciones medias en BM indicadas por la curva 6 de la figura 7 y los volúmenes medios en BM de la figura 8. Estos volúmenes se han utilizado para obtener los desplazamientos sucesivos de la nube de trazador durante un cierto incremento de tiempo, en función del caudal de agua dulce aportado por el río.

En los dos primeros casos (33 y 67 km) se ha utilizado un coeficiente de dispersión medio Dm dado por la media aritmética de los valores de la figura 9 entre el punto de inyección y la posición de la mediana de la nube de trazador. Dicha posición de la mediana equivale a una distancia x tal que el volumen medio de agua embalsado en BM sea igual al volumen total aportado por el río hasta el momento que se considera. Con estos valores de S y Dm se han obtenido las concentraciones teóricas de tritio en los puntos anteriores durante los sucesivos ciclos de marea. Los resultados se muestran en las figuras 3 y 4, que se discuten más adelante.

Este método de cálculo no era utilizable en el caso del kilómetro 106, porque debido a su proximidad al mar la curva de paso por este punto se produce de forma casi simultánea con la salida al mar del trazador. Se ha seguido el criterio de suponer que el trazador que llega a mar abierto, es decir más allá del kilómetro 110 durante un ciclo PM-BM, no retorna al estuario durante el ciclo siguiente BM-PM. Esto no es del todo exacto, pero introduce un error poco importante. La curva de paso se ha calculado siguiendo un método de incrementos finitos basado en dividir el estuario en 220 compartimentos, cada uno de ellos de 500 m de longitud. Mediante un programa de cálculo para ordenador, desarrollado al efecto, se van obteniendo las distribuciones del trazador en los sucesivos ciclos de marea. Se supone que la masa de trazador existente en un momento dado en uno cualquiera de estos compartimentos evoluciona como si se tratase de una inyección puntual realizada en el centro del compartimento, y por tanto da lugar en el siguiente ciclo de marea a una nueva distribución parcial definida por la ecuación (5). Esta distribución parcial se obtiene considerando un valor de (5) igual a la sección media de todos los parámetros afectados dentro del intervalo x ± 2σ, situando la mediana a una distancia Δx tal que el volumen embalsado en BM sea igual al aportado por el río durante el ciclo BM-PM-BM. Como valor de D se toma el valor medio correspondiente al tramo recorrido por el trazador durante dicho ciclo. La nueva distribución del trazador se obtiene por suma de todas las distribuciones para todas las BM siguientes a la inyección, las cuales se corrigen por un factor próximo a 1 a fin de que se conserve la masa total de trazador. Cuando éste llega al mar, el programa de cálculo elimina en el siguiente ciclo la parte de la curva de distribución que ha sobrepasado el kilómetro 110.

La curva teórica de paso del tritio por el kilómetro 106 se muestra en la figura 5. Debe indicarse que las tres curvas teóricas obtenidas pueden estar afectadas por un cierto error, debido a que los caudales extraídos para regadío, así como los excedentes vertidos a lo largo del estuario, solo eran conocidos de forma aproximada. En líneas generales, se observa que las curvas teóricas se ajustan satisfactoriamente a las experimentales, en lo que se refiere a la escala de tiempos. En cambio, en lo que respecta a las concentraciones, las curvas reflejan pérdidas importantes de trazador. Estas pérdidas, obtenidas por integración de las curvas, ascienden al 20,5, al 42,9 y al 62,0%, respectivamente, para los kilómetros 33, 67 y 106. Las captaciones de agua existentes entre los kilómetros 30 y 65 son responsables de una fracción de pérdidas que puede oscilar entre el 30 y el 40%. Tomando el valor medio de este intervalo resultaría

un excedente de pérdidas igual al 27%, referido a la totalidad del estuario.
Como el caudal medio aportado por el río fue de 35,2 m³/s, dicho excedente de
pérdidas asciende a unos 9,5 m³/s. Tales pérdidas tienen que justificarse por
conceptos tales como infiltración, evaporación, fijación temporal del tritio en
la capa de lodos del estuario, consumo por la biota, etc. Puesto que el trabajo
se realizó en época de estiaje puede pensarse que la infiltración sea la causa más
importante de estas pérdidas. El valor estimado de la superficie media cubierta
por las aguas en el estuario es del orden de 4×10^7 m². Por tanto, el caudal
anterior representaría una velocidad media de infiltración igual a unos 2 cm/d,
que puede ser aceptable.

En definitiva, opinamos que el modelo utilizado se ajusta satisfactoriamente
a la dinámica del estuario, pudiendo ser utilizado para estudiar la renovación del
agua en el mismo.

7. APLICACION DEL MODELO A CASOS HIPOTETICOS

7.1. Renovación del agua en el estuario

El modelo descrito, junto con los datos de coeficientes de dispersión y
secciones del estuario determinados experimentalmente, permite estudiar el
comportamiento de cualquier tipo de trazador o agente contaminante vertido
en cualquier punto del estuario de una forma instantánea. El modelo proporciona
la curva de paso del agente contaminante por cualquier punto del estuario, la
distribución de dicho agente a lo largo del estuario en cualquier momento y su
salida al mar. A título de ejemplo, se han calculado con el modelo las curvas
que reflejan la salida al mar de una masa conocida de agente contaminante
inyectado en cada uno de los siguientes puntos:

a) Presa del embalse de Alcalá del Río (km 0).
b) Desembocadura actual del río Guadaira (km 29), realizando inyección en
    PM y BM.
c) Desembocadura del nuevo cauce del río Guadaira (km 47), realizando la
    inyección en PM y BM.

Los cálculos se han realizado empleando un programa para ordenador,
siguiendo el método de incrementos finitos al cual se ha hecho referencia en el
apartado anterior. Los resultados obtenidos se reflejan en la figura 10. Las
curvas indican el tanto por ciento de trazador que ha salido al mar, en función
del tiempo transcurrido desde el momento de la inyección, y para diferentes
caudales de agua dulce. Como era de esperar, se observa que las curvas disminuyen
de pendiente a medida que disminuye el caudal de agua dulce. Con caudales muy

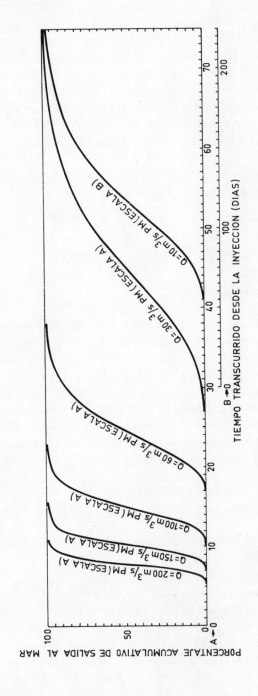

FIG. 10a). Curvas de salida al mar de un trazador inyectado en el km 0 para diferentes caudales del río.

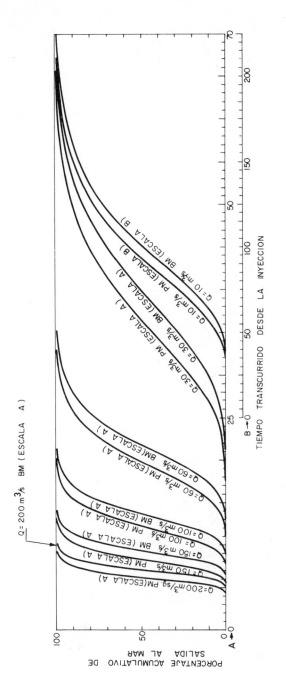

FIG.10b). Curvas de salida al mar de un trazador inyectado en el km 29 para diferentes caudales del río.

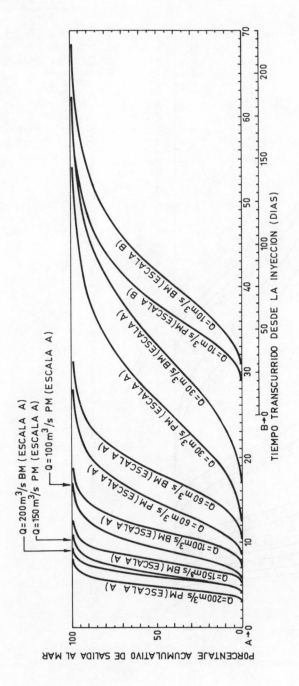

FIG.10c). Curvas de salida al mar de un trazador inyectado en el km 47 para diferentes caudales del río.

elevados el trazador llega rápidamente a la desembocadura, y el tiempo disponible para su dispersión es corto. Con caudales pequeños sucede lo contrario.

El proceso de salida al mar del trazador, en función del caudal, ocurre de forma aproximadamente asintótica por ambos extremos. Al aumentar el caudal, el tiempo de salida tiende asintóticamente a cero. Cuando disminuye dicho caudal, el tiempo tiende hacia un valor muy elevado, que será el correspondiente a un caudal cero, para el cual la salida del trazador tiene lugar solamente por efecto del fenómeno dispersivo. De acuerdo con esto, puede esperarse que la variación del tiempo de salida en función del caudal corresponde a una hipérbola ($Q \cdot t =$ constante). Para comprobar este extremo se han representado los tiempos de salida del 95 y del 50% del trazador en función del caudal, en escala logarítmica, con el fin de obtener una línea recta. Los resultados obtenidos se representan en la figura 11. Los puntos delimitan líneas rectas, muy bien ajustadas, que obedecen a ecuaciones del tipo $Q^{K_1} \cdot t = K_2$.

La constante K_1 es muy próxima a 1 en el caso del 95% y alrededor de 0,9 para el 50%. Por tanto, en ambos casos los puntos delimitan aproximadamente hipérbolas equiláteras. El valor de la constante K_2 en el primer caso disminuye a medida que el punto de inyección se aproxima al mar, como era de esperar. Dicha constante debe ser cero cuando la inyección se efectúa en el punto de la desembocadura (km 110) en el momento de la pleamar. Tal inyección correspondería a una inyección en bajamar efectuada 12 km aguas abajo (recorrido medio de marea), es decir en el kilómetro 122. Teniendo en cuenta este dato, y los restantes valores de K_2, se ha representado la curva de la figura 12, que permite obtener de forma aproximada el valor de dicha constante en función del punto donde se realiza la inyección. Con el valor de K_2 obtenido de esta gráfica puede calcularse el tiempo de salida del 95% del trazador, haciendo uso de la expresión aproximada $Q \cdot t = K_2$.

7.2. Intrusión de agua del mar

7.2.1. Proceso de cálculo

Para la aplicación del modelo al caso de la intrusión de agua del mar, es preciso definir primero las condiciones iniciales del proceso. En nuestro caso, hemos supuesto que en el tiempo inicial $t = 0$ nos encontramos en una BM y en todo el estuario ne hay más que agua dulce, es decir en el kilómetro 110 existe una interfase nítida entre agua dulce y agua salada. Esta última corresponde a una concentración de ion cloruro igual a 20 g/l. Durante el próximo ciclo BM-PM-BM, la interfase sube por el estuario hasta una distancia de unos 12 km (recorrido medio de marea) y luego baja, de forma que la interfase nítida inicial se transforma en una zona de transición. Durante el recorrido de dicha interfase, la probabilidad de que el agua dulce penetre por dispersión en la zona

BAONZA et al.

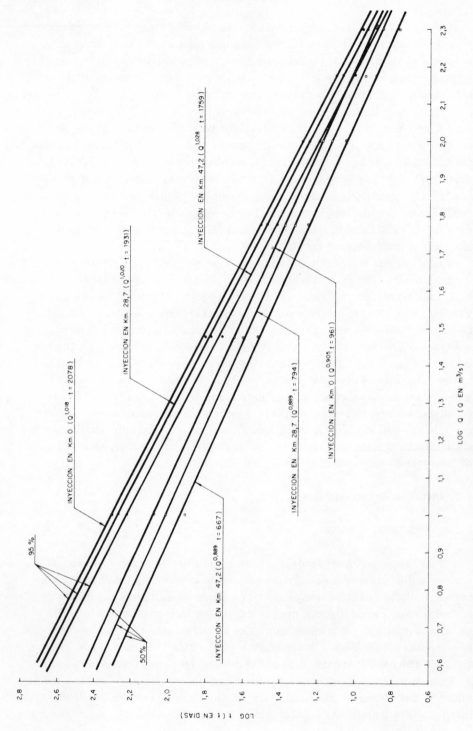

FIG.11. Tiempo de salida del 95 y del 50% del trazador inyectado en diferentes puntos del estuario en función del caudal de agua dulce.

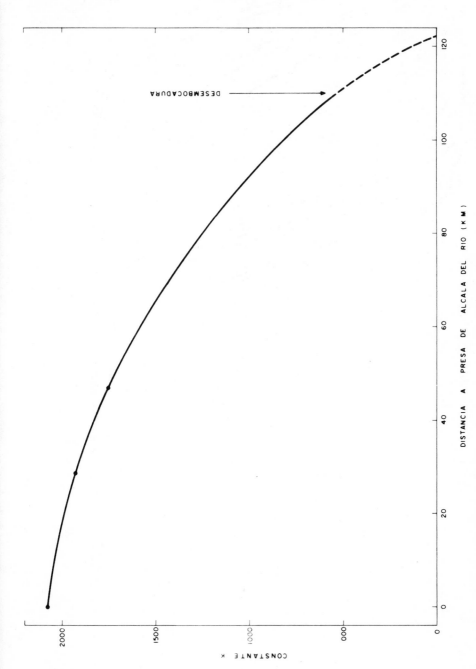

FIG.12. Variación aproximada de la constante K_2 con la situación del punto de inyección (inyección en BM).

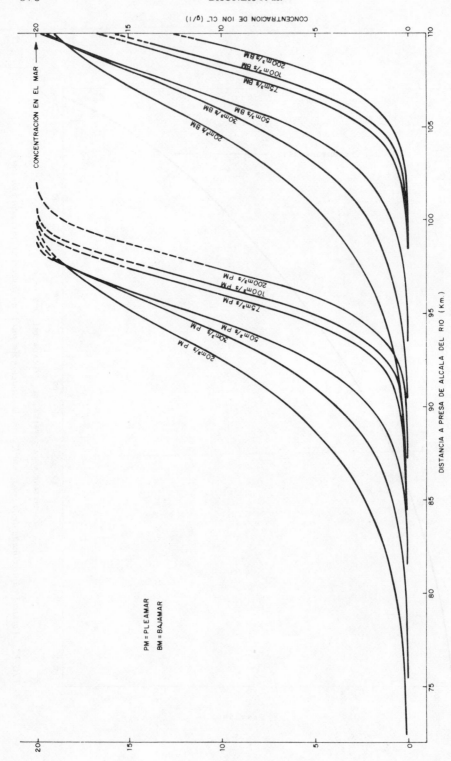

FIG.13. *Intrusión del ion Cl⁻ en el estuario para una marea de tipo medio (Ah = 1,94) para diferentes caudales de agua dulce.*

de agua salada es la misma que la correspondiente al caso inverso. Por tanto,
al finalizar el ciclo anterior, la concentración de ion cloruro en el kilómetro 110
será igual a 10 g/l, suponiendo que el caudal de agua dulce aportado por el río
es igual a cero y que la carrera de marea se mantiene constante. Si el caudal de
agua dulce es distinto de cero, la concentración en el kilómetro 110 será menor,
porque la interfase inicial equivalente ahora al centro de la zona de transición
no se estabiliza en este punto kilométrico, sino que penetra en el mar debido al
desplazamiento motivado por el caudal aportado. Basándose en estas considera-
ciones, el cálculo de las concentraciones medias en los distintos compartimentos
se realiza como sigue:

a)   A partir del caudal constante de agua dulce se calcula la penetración de
     agua salada en el estuario, que será como se ha indicado al principio de
     esta sección de 12 kilómetros, menos el desplazamiento debido al volumen
     aportado. Se obtiene así el punto kilométrico alcanzado por el centro de
     la zona de transición que se encuentra en proceso de formación en el ciclo
     BM-PM.
b)   Se determina el coeficiente de dispersión correspondiente al centro del
     tramo comprendido entre el punto kilométrico anterior y el kilómetro 110.
     Este valor se utiliza para calcular la desviación típica σ_x.
c)   Se sustituye en la ecuación (5) el valor de $C(x, t)$ por 10 g/l y el de la
     expresión $A/S \sqrt{4\pi Dt}$ por 20 g/l, obteniéndose:

$$10 = 20 \cdot \exp\left[-\frac{x_0^2}{2\sigma_x^2}\right]$$

     Con el valor σ_x obtenido anteriormente, se determina el valor de x_0 que
     cumple esta ecuación. Esta es una distancia ficticia, medida desde el
     kilómetro 110, a la que debería encontrarse la concentración máxima de
     20 g/l, para el caso de un caudal de agua dulce igual a cero. Si dicho caudal
     es distinto a cero, hay que sumar al valor de x_0 el desplazamiento debido
     al volumen de agua dulce aportado durante el ciclo de marea. Este puede
     obtenerse de forma aproximada dividiendo el volumen aportado por la
     sección media del compartimento 109–110, lo cual equivale a suponer
     que el estuario se prolonga con una sección constante igual a esta última.
d)   Las concentraciones medias de ion cloruro en los compartimentos
     anteriores al kilómetro 110 equivalen al segundo miembro de la expresión
     anterior incrementando x_0 con la distancia entre tales compartimentos y
     el kilómetro 110. Las masas de ion Cl^- existentes en cada compartimento
     m_{11}, m_{12}, etc., se obtienen multiplicando las concentraciones medias por
     los correspondientes volúmenes de agua almacenados en bajamar. A
     continuación se calcula la distribución de estas masas en los diferentes

compartimentos al finalizar el segundo ciclo de marea, mediante la ecuación
(5). Lógicamente, una parte de estas masas retorna al mar durante este
ciclo. El cálculo se realiza considerando que toda la masa de cada compartimento está concentrada en el centro del mismo, como si se tratase de una
inyección puntual. Para cada masa tienen que obtenerse los parámetros
posición de la mediana, S y σ_x, lo cual se realiza del mismo modo que se
ha descrito anteriormente. Sumando la contribución en un mismo
compartimento de todas estas masas, se obtienen las nuevas masas m_{21}, m_{22},
etc. El cálculo se continúa hasta que los valores de m_i sean despreciables,
lo cual indica que se ha llegado a una situación de equilibrio. Las expresiones
$M_1 = m_{11} + m_{21} + m_{31} + \ldots$, $M_2 = m_{12} + m_{22} + m_{32} + \ldots$, etc., propor-
cionan las masas totales de ion cloruro en cada compartimento, una vez
alcanzado el equilibrio. El cociente entre M_i y el volumen de agua correspondiente para la bajamar da la distribución de concentraciones de ion
cloruro en el equilibrio.

7.2.2. Resultados obtenidos

Siguiendo el procedimiento descrito se ha calculado la distribución de
concentraciones de ion cloruro en el estuario, una vez alcanzado el equilibrio,
para diferentes caudales de agua dulce. Lógicamente, cuando el caudal es cero
todo el estuario se saliniza, con la misma concentración de agua de mar, al cabo
de un número suficientemente elevado de ciclos de marea. Para caudales
distintos de cero se alcanza siempre un equilibrio de concentración, en el cual
la tendencia del ion cloruro a ascender por el estuario, motivada por el fenómeno
dispersivo, se compensa con el desplazamiento hacia el mar de las masas de agua
debido al caudal inyectado.

Una muestra de los resultados obtenidos se presenta en el figura 13. El
número de ciclos de marea necesarios para alcanzar el equilibrio aumenta cuando
disminuye el caudal de agua dulce, siendo alrededor de 10 para un caudal de
200 m^3/s y de 100 para 20 m^3/s.

La concentración máxima de ion cloruro en el agua para riego que admite
el cultivo del arroz es 0,8 g/l. Por tanto, era interesante conocer el desplazamiento de esta concentración en función del caudal de agua dulce una vez
obtenido el equilibrio. Lógicamente, cuando el caudal aumenta este valor de
concentración tiende a aproximarse al kilómetro 110, y cuando disminuye
se aproxima al kilómetro cero; en ambos casos de forma casi asintótica. El
punto de estabilización de dicha concentración en función del caudal varía
según la ecuación de una hipérbola, dada por la función

$$(110 - x) \cdot Q^{0,599} = 106,6$$

siendo $(110 - x)$ la distancia a la desembocadura.

Esta ecuación solo es válida para caudal constante y marea de tipo medio. Pero, con los datos disponibles, es posible elaborar un programa de cálculo que permita resolver el problema para caudal e índice de marea variables.

DISCUSSION

K.E. WHITE: We conduct similar dispersion studies and agree with your findings that the dispersion coefficient settles down from an apparently high initial value to a lower and more constant one with distance from the source of release of the tracer; this effect may take place over a distance of many tens of kilometres, possibly 50 km in some cases. Could you give more information on how you concluded that you had lost 30–40% of the tritium?

A. PLATA BEDMAR: In our case, the dispersion coefficient increases with time until a constant value is reached after two or three tidal cycles. The tritium losses are shown by the existence of a delayed flow in the mud layer of the estuary bed.

T. FLORKOWSKI: Concerning the loss of tritium during the experiment, it is important that water should not be removed from the river after injection (e.g. for irrigation) until complete mixing of the tracer has occurred. After the "mixing length" is achieved, the water can be taken from the river. I wonder whether this could account for the tritium loss as seen in your case.

A. PLATA BEDMAR: In view of the location of extraction stations for irrigation water (35–70 km), mixing must be complete by the time the water reaches the stations.

T. FLORKOWSKI: A considerable tritium loss can be observed as a result of molecular exchange with atmospheric moisture, especially if the tritium concentration in the river is much higher than that in the atmosphere and if the time of the experiment is long (one to three months).

A. PLATA BEDMAR: This could only have occurred during the first five or ten days after injection. Later, the tritium concentrations in the estuary water were similar to those in atmospheric water vapour or only slightly higher.

T. DINÇER: How reliable were the discharge measurements in the Guadalquivir estuary?

A. PLATA BEDMAR: The amount of fresh water discharged by the river and the quantities extracted for irrigation are known, as mean daily values, with an error below 5%.

K.E. WHITE: Could you please confirm that, using the same methods of calculation as for the tritium study, you did not lose ^{82}Br, which was studied over a shorter period?

A. PLATA BEDMAR: Logically, losses of ^{82}Br must have occurred, but in relative terms these losses would be small. ^{82}Br was measured only for two or three days whereas tritium was measured over three months.

D.B. SMITH: Have you observed whether vertical mixing in the estuary is complete or incomplete, particularly where the saline wedge enters the estuary on a rising tide?

A. PLATA BEDMAR: Vertical mixing was observed in many cases and was always complete after several tidal cycles, though not at the beginning of the experiment. We did not observe saline wedges in the estuary; the small cross-section of the estuary may be one reason why this phenomenon is absent.

CHANGES IN THE SEDIMENTATION HISTORIES OF LAKES USING LEAD-210 AS A TRACER OF SINKING PARTICULATE MATTER

R.S. BARNES, P.B. BIRCH*, D.E. SPYRIDAKIS
Department of Civil Engineering,
University of Washington

W.R. SCHELL
Laboratory of Radiation Ecology,
College of Fisheries,
University of Washington,
Seattle, Washington,
United States of America

Abstract

CHANGES IN THE SEDIMENTATION HISTORIES OF LAKES USING LEAD-210 AS A
TRACER OF SINKING PARTICULATE MATTER.

A detailed study of man's impact over the last 150 years on six lakes in western
Washington State has been made using ^{210}Pb dating methods and historical records. These
lakes represent a gradient in watershed usage from pristine natural environments to heavily
urbanized areas. Fine structure in the sediment profiles of ^{210}Pb measurements were found
to correlate with changing watershed land use. Contemporary sedimentation rates varied
from 50−679 g/m^2·a (0.37−2.9 mm/a) and were generally higher than pre-cultural rates.
The highest average sedimentation rates (1230−1800 g/m^2·a or 5.6−8.3 mm/a) were
simultaneous with suburbanization. Construction of roads and houses appeared to be the
major cause of increased erosion in the watersheds. The present-day sediments of all lakes
were enriched in lead compared with older background material. The stable lead profiles
from all lakes except Lake Union were consistent with the local history of lead pollution
based on the ^{210}Pb geochronologies. Water column residence times for ^{210}Pb and stable lead
were almost identical and were consistent with algal settling rates and the sinking rates of
fine silts and clays.

1. INTRODUCTION

Lake sediments preserve a record of processes which have occurred in the overlying water column as well as events which have occurred in the watershed. To reconstruct the history of these processes the sedimentation rate is necessary. This has been accomplished by counting varves, by palynological (pollen) methods or by independently dating some sediment

---

* Present address: 16 Woodlands St., Woodlands 6018, Western Australia.

anomaly (volcanic ash, landslides, glaciers). Radioactive dating of deposited carbonaceous materials by the C-14 method has been used to give sedimentation rates over the last one to fifty millenia but does not extend to the present time. However, more recently the Pb-210 dating method has emerged as the most promising means of determining sedimentation rates over the last 100 to 150 years [1,2,3,4,5].

The Pb-210 technique has been applied primarily to the sediments of large lakes with low sedimentation rates such as the Laurentian Great Lakes. A few smaller lake systems have been measured but the frequency of measurements within the core profiles has precluded the determination of more than a single sedimentation rate averaged over the last one hundred or more years. The Pb-210 method has been used in this study to determine the detailed sedimentation histories of six lakes in western Washington State; correlations have been made with known anthropogenic or natural events. These lakes represent a gradient in cultural usage ranging from pristine alpine environments to heavily urbanized watersheds in large metropolitan areas. Thus, tests of the method can be made since all cultural impacts have occurred within the last 125 years. By using the Pb-210 sedimentation rates in conjunction with the profile of trace metal concentrations in the sediment it has been possible to calculate removal rates of trace metals from the overlying water column. If trace metal concentrations in the overlying water column are known, average trace metal residence times may also be calculated.

2. BACKGROUND OF PB-210 METHOD

The use of decay products of naturally occurring Ra-226 for sediment geochronologies was first outlined by Goldberg [1]. The decay chain of Ra-226 and half lives of the radionuclides are:

$$\overset{226}{Ra} \xrightarrow[1620 \text{ a}]{} \overset{222}{Rn} \xrightarrow[3.8\text{d}]{} \overset{218}{Po} \xrightarrow[3.05 \text{ m}]{} \overset{214}{Pb} \xrightarrow[26.8 \text{ m}]{} \overset{214}{Bi} \xrightarrow[19.7 \text{ m}]{} \overset{214}{Po}$$

$$\overset{214}{Po} \xrightarrow[1.6 \ 10^{-4}\text{s}]{} \overset{210}{Pb} \xrightarrow[22.26 \text{ a}]{} \overset{210}{Po} \xrightarrow[138 \text{ d}]{} \overset{206}{Pb} \text{ (stable)}$$

The basic principle for Pb-210 dating is that radon gas, Rn-222, is emitted to the atmosphere from the lithosphere, surface waters and airborne dust. This change in phase isolates radon from its radium precursor and destroys the initial secular equilibrium. A new secular equilibrium is then reached by the radon and its decay products with the longest lived product becoming the predominant radionuclide. Radon remains in the atmosphere until it decays. Pb-210 and other Rn-222 daughter products rapidly become attached to atmospheric aerosols and are removed by dry fallout and precipitation. It has been estimated that the residence time of Pb-210 in the lower atmosphere may be less than a week [6]. This continuous source provides a widespread flux of Pb-210 to land and water surfaces. Pb-210 is efficiently scavenged by organic-rich soil horizons and can only be lost from the soil by erosion [7]. Pb-210 entering riverine systems is rapidly removed from the water [7,8]. Thus it is generally concluded that most of the Pb-210 entering lakes comes from direct atmospheric input across the lake surface.

For dating purposes, it must be assumed that Pb-210 is scavenged by suspended particulate matter and accumulated as lake sediments. From a

knowledge of the present Pb-210 input to a region the age of a sediment
horizon which has been isolated from the contemporary input can be determined. By measuring the Pb-210 concentrations in several different sediment
horizons the rates of sediment accumulation can be calculated. Because
Pb-210 also is formed in situ from the small but finite amounts of terri-
genous Ra-226 present in the sediment, this supported amount must be subtracted from the total concentration to obtain the unsupported Pb-210; the
unsupported Pb-210 represents the amount added to the water column from
atmospheric sources.

The unsupported Pb-210 radioactivity in the sediment decreases by a
factor of two every 22.26 years (the half life of Pb-210). When the loga-
rithm of the unsupported Pb-210 activity (disintegrations per minute per
gram, dpm/g) is plotted as a function of the total mass of dry sediment
accumulated above a sediment core section, a log-linear relationship is
found. This log-linear relationship is true only if the independent variable is the mass of overlying sediment and is not valid if the linear depth
in the sediment core is used. This arises because of the compaction of the
sediment due to the mass of overlying material which produces an exponential
decrease in porosity thus the bulk sediment density increases with increasing
depth in the sediment until a constant compaction is reached.

Often the profiles unsupported Pb-210 measurements obtained from the
logarithmic plot of the unsupported activity as a function of the overlying
mass of dry sediment accumulated show variations which are outside the analytical errors expected from the measurement of radioactive decay. Possible
reasons for these discrepancies are:

1) Disturbance or incomplete recovery of the upper layers of sediment
by the coring device.
2) In situ mixing of the upper sediment layers due to burrowing
organisms, gas bubbles, etc.
3) Post-depositional mobility of Pb-210.
4) Presence of foreign material such as wood chips, pebbles, shells,
etc.
5) Actual changes in sedimentation rate.

Low surface values of Pb-210 in profiles of marine sediments from Baja,
California and the Santa Barbara Basin were attributed to mobilization of
Pb-210; the low Pb-210 values found in Lake Tahoe and Lake Titicaca sediments
were attributed to incomplete recovery of the uppermost sediment horizons
[3,4]. Later it was determined that the surface Pb-210 anomaly in the
Santa Barbara Basin sediments was probably due to a higher sedimentation
rate occurring at the present time and the possibility of post-depositional
migration of Pb-210 was unlikely [9]. It has been suggested that Pb-210
mobility in Lake Michigan sediments was probably insignificant and that
post-depositional redistribution by physical mixing or bioturbation could
account for the observed anomalies [5].

In the Pb-210 lake sediment geochronologies reported by the aforemen-
tioned papers, it was possible to derive only a single average sedimentation rate over the last 80-100 years; no direct indication from the Pb-210
measurements that anthropogenic influences may have affected the sedimentation rates over this time were found. This failure to observe fine structure which is possible in sedimentation rate analysis may be due to several
factors including lake size and sampling frequency within the core. Large
lake systems are inherently less sensitive to local watershed disturbances

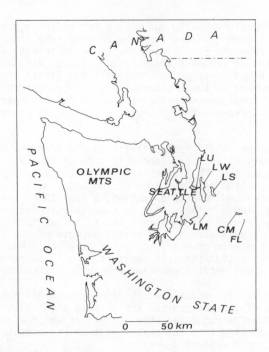

FIG.1.   Location of the study area.
(LU = Lake Union, LW = Lake Washington, LS = Lake Sammamish, CM = Chester Morse
Reservoir, FL = Findley Lake, and LM = Lake Meridian)

that might tend to alter profundal sedimentation rates. This inherent
insensitivity in observing changes in sedimentation rates is due to distance
between the sediment source and the profundal basin as well as to the physical dispersion of a given sediment source over a relatively large sediment
accumulating area. The sampling frequency involves the vertical thickness
of each sediment horizon being measured relative to the rate of sediment
deposition at that location. In the case of Lake Mendota and Trout Lake,
two of the smaller lakes examined, the thickness of the sediment sections
was 5 cm and only every other 5 cm section was measured [4]. If a minimum
of three points is necessary to define an average sedimentation rate over a
period of time, then it is evident that for these lakes the minimum time
period for which a sedimentation rate could be obtained corresponds to the
deposition of 25 cm of material, or forty years at the reported sedimentation rate of 0.6 cm/a [4]. In both cores from a pond in Belgium, and cores
from Lake Biwa and Lake Shinji, Japan, measured sections were used whose
thicknesses corresponded to as much as 40 years deposition, which at best,
represented a 36 year average sedimentation rate [10,11,12].

     In larger lakes with generally lower sedimentation rates an analogous
problem is encountered even if the sediment core is dated at consecutive
two centimeter intervals. In Lakes Titicaca and Tahoe, for example, the
2 cm incremental samples used represented about 60 years [4].   Lake

Michigan cores were measured at one centimeter intervals but with the observed linear sedimentation rates of 0.03 to 0.05 cm/a, three consecutive
one centimeter sampling intervals still amounted to 60 to 100 years deposition [5]. Thus, no evidence of fine structure in the Pb-210 sedimentation
rates has been reported, nor have any correlations been made on anthropogenic
increases in sedimentation rates of lakes based on Pb-210.

3. METHODS

3.1. The study area

The lakes studied are located in the west central part of Washington
State about 160 km east of the Pacific Ocean and to the east of the Olympic
Mountains, Figure 1. Topographically, the area is divided into two regions;
the lowland region near Puget Sound and the mountainous regions of the
Cascade Mountains to the east. The landforms and lakes of the lowland region
have resulted from continental glaciation that moved southward along the
Puget Sound trough and retreated during the late Pleistocene. At the higher
elevations many lakes were formed as a result of alpine glaciation. The
physical characteristics of the study lakes are summarized in Table I.

The area was first permanently settled in the mid-1800's. Fisheries
and timber have always been important elements of the local economy. The
late nineteenth and early twentieth century saw the pioneer economy expand
to include light industry; at present the area has a diversified economy
including both light and heavy industry, aircraft manufacture and extensive
port facilities of a major shipping terminus. Most of the population of
about 1,000,000 reside in the lowlands.

3.2. Sample collection and measurement

Sediment cores were collected using a Shapiro-type piston corer [13]
from the profundal basins of all lakes except for Findley; a 7 kg gravity
corer of a modified Phleger design was used to collect sediments from the
profundal basin of Findley Lake. The coring chambers for both corers were
made of clear tenite butyrate tubing, 35 mm inner diameter. The corers
sampled the upper layers of soft sediments with minimum mixing or disturbance. After the cores were examined for gross physical characteristics
they were extruded and sectioned at 0.5, 1.0 or 2.0 cm intervals.

For cores from Lakes Washington and Union the outer 3 mm was peeled
from each of the sections and combined for later Ra-226 measurement by
gamma ray spectrometry. The center portions of each section were weighed
wet and then dried to constant weight at 70°C. Dried samples were placed
in beakers with a known amount of Po-208 tracer and wet ashed using the
$HClO_4$-HNO_3 procedure [14]. This solution was then evaporated to dryness,
dissolved in 100 ml of 0.3N HCl and filtered through a pre-rinsed filter.
The polonium isotopes were spontaneously electro-deposited onto 2.2 cm
diameter silver discs at 50°C after the addition of about 100 mg of ascorbic
acid to complex the iron. After plating, the discs were removed, rinsed
with distilled water and air dried. The plating onto the silver discs was
not quantitative since recoveries were found to vary from 50-90% thus a
chemical yield tracer, Po-208, was required for each sample. The Pb-210
activity was assumed to be identical with that of its Po-210 daughter product. The plating solutions were subsequently analyzed for stable lead
using a Perkin-Elmer Model 303 atomic absorption spectrophotometer.

TABLE I. RELEVANT CHARACTERISTICS OF THE STUDY LAKES AND THEIR SEDIMENTS

Lake	Eleva-tion,[a] m	Mean Depth, $\bar{z}$, m	Maximum Depth, m	AREA Lake Surface, km^2	AREA Catch-ment, km^2	Volume, m^3	Mean Annual Precipita-tion, cm	Hydraulic Residence Time, a
Findley	1131	8.2	28	0.158	2.6	$1.29\ 10^6$	234	0.2-0.6
Chester Morse Reservoir	474	18.8	35.4	6.81	217	$1.28\ 10^8$	265	0.25
Meridian	113	12.3	27.4	0.61	2.2	$7.5\ 10^6$	92	4.
Sammamish	8.5	16.6	30.5	19.8	253	$3.28\ 10^8$	115	1.7
Washington	4.3	32.9	65.	87.6	1560[b]	$2.88\ 10^9$	88	2.35
Union	4.3	10.1	16.5	2.42	1572[c]	$2.45\ 10^7$	88	0.02

[a] Mean sea level datum.
[b] 1073 km^2 prior to 1916; consists of 451 km^2 direct drainage, 622 km^2 of the Sammamish River Drainage including 253 km^2 tributary to Lake Sammamish.
[c] 12 km^2 prior to 1916.

The determination of the sample Po-210 and the Po-208 tracer was made using Si-surface barrier detectors and alpha pulse-height analysis spectro-scopy. Counting times ranged from 400 to 1000 minutes with typical counter backgrounds of 1 to 10 counts per 1000 minutes. The outer 3 mm portions of the sections were combined, dried and packed into standard counting holders 5.08 cm in diameter and 1.27 cm in height. After sealing the counting holders were stored for at least a month to allow the gamma-emitting daughter products of Ra-226 to reach secular equilibrium. The samples were then counted using a calibrated Ge(Li) detector and pulse-height analyzer system to determine the supported Pb-210 which was in equilibrium with the Ra-226.

Cores from Lakes Sammamish, Findley, Meridian and the Chester Morse Reservoir were treated slightly different from the Lakes Washington and Union cores. After being dried to constant weight the 1 cm thick core sec-tions from these lakes were comminuated and homogenized. Subsamples were then removed for Po-210 and stable elemental analysis. A 100-200 mg subsam-ple was digested in acid using the procedure of Bortleson and Lee [15] ex-cept that all the acid digestions were carried out in the same teflon cru-cible. After filtration and solution was diluted to a known volume with distilled deionized water. This solution was analyzed for lead using an Instrumentation Laboratory Model 353 atomic absorption spectrophotometer. The supported Pb-210 activity in these cores was taken to be the average Pb-210 content of the deeper sections where a constant value was found.

4. RESULTS AND DISCUSSION

4.1. Lake Washington

The Pb-210 results from Lake Washington are based on two replicate cores taken near Madison Park at a depth of 62 meters by Dr. W. T. Edmondson. Both cores contained a density anomaly in the 18-20 cm horizons

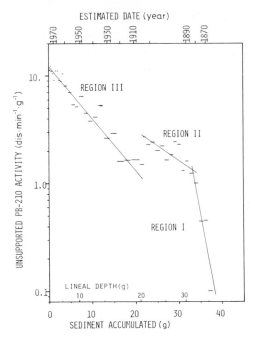

ESTIMATED DATE (year)

FIG.2. Unsupported ^{210}Pb activity in Lake Washington sediment cores. Estimated sedimentation rates: Region I, 150 g/m² · a or 0.63 mm/a (pre-1889); Region II, 1800 g/m² · a or 8.3 mm/a (1890-ca 1902) and Region III 644 g/m² · a (ca 1916–1973).

where a layer of silt was present. The origin of this layer has been attributed to the erosion of littoral areas which followed the lowering of the level of the lake by about 2.4 m in 1916 when the Cedar River was diverted into the lake and a ship canal and locks were constructed which connected the lake to Puget Sound [16]. The data plotted in Figures 2 and 3 have been obtained by averaging the Pb-210 and stable lead values from both cores after the silt band had been juxtaposed. One core was sectioned at 0.5 cm intervals over the top 7 cm and then at 1 cm intervals for the remaining 40 cm; the second core was sectioned at 1 cm intervals. The upper 5 cm of the second core were not analyzed, so the values for these horizons represent only a single core.

Since the atmospheric input of Pb-210 must have remained constant over the past 150 years, the deviations from a single log-linear relationship of the unsupported Pb-210 activity versus the dry mass of sediment accumulated, must be due to some property of the lake or watershed, as mentioned earlier. Scanning electron microscopy has showed that the seasonal stratigraphy was preserved in recent Lake Washington sediments from this sampling location [17]. It was noted that one of the alternating bands consisted primarily of diatom frustules and the other appeared to contain rough clay and silt particles indicative of allochthonous material. The lack of bioturbation in Lake Washington sediments was thought to be due to the scarcity of burrowing organisms of the Pisidium genus [17]. None of the sediment samples from the lakes studied contained any living or dead members of this genus.

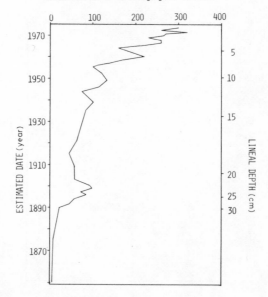

CONCENTRATION OF LEAD mg/kg DRY SEDIMENT

FIG.3. Stable lead profile in Lake Washington sediment cores.

The silt band, present at 18-20 cm, is the only known addition of for-
eigh material to these lake sediments outside of normal sedimentation processes. The Pb-210 profile is diluted by this material over the 18-20 cm
horizons. The almost constant value for unsupported Pb-210 activity in this
material suggest that either the dilution occurred over a period of about
fifteen years based on the total unsupported activity present in these horizons or that the heavier silt penetrated the flocculent surface sediment
layers for a distance of a few centimeters. In either case the nature of
the in-lake sedimentation for a period of about fifteen years from 1903 to
1918 is indeterminable. The remaining changes in the Pb-210 profile are
presumably attributable to real changes in the rates of sediment accumulation. It is evident from Figure 2 that rather than a single log-linear
relationship fitting the data, there are, in fact, three relationships with
each corresponding to a distinctly different sedimentation rate. The oldest
of these sedimentation rates Region I (150 g/m^2·a or 0.63 mm/a), represents a Pb-210 time interval of about 1840-1889, we believe that this is the
natural pre-cultural sedimentation rate. Near this vicinity of the lake,
a volcanic ash layer overlain by 480 cm of sediment has been found [18].
Radiocarbon dating of adjacent carbonaceous material determined that the
ash layer was deposited about 6700 years ago which would attribute it to
the Mt. Mazama (Crater Lake) eruption [18,19]. Thus, the sedimentation rate
during the 6600 years prior to 1890 was 0.68 mm/a [(480-30)cm/6600 a] if
deposition from 1890-1958 is not included; this value compares well with the
0.63 mm/a estimated from Pb-210.

Between 1890 and at least 1902 the sedimentation rate Region II,Figure 2
averaged 1800 g/m^2·a or 8.3 mm/a.   By 1895 most of the land which drains

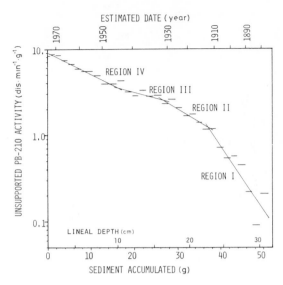

FIG.4. Unsupported ^{210}Pb activity in Lake Sammamish sediment core. Estimated
sedimentation rates: Region I, 229 g/$m^2 \cdot a$ or 1.1 mm/a (pre-1909), Region II, 568 g/$m^2 \cdot a$
or 2.7 mm/a (1909–1932), Region III, 1230 g/$m^2 \cdot a$ or 5.6 mm/a (1932–1944), and
Region IV, 689 g/$m^2 \cdot a$ or 3.3 mm/a (1944–1973).

into the lake had been logged and the suburbs of Seattle had reached the
lakeshore. Very rapid land development in the watershed occurred about the
turn of the century. Since 1916 the sedimentation rate Region III, Figure 2,
has averaged 644 g/$m^2 \cdot a$ or 3.0 mm/a which is substantially above the pre-
cultural rate. The diversion of the Cedar River (average flow 20 m^3/s) into
the lake in 1916 provided the water necessary to operate the ship canal
locks and contributes an estimated 4-5·10^7 kg/a of allochthonous material
[20]. This would correspond to a bulk sedimentation rate of about 1000
g/$m^2 \cdot a$ if it were averaged over the entire sediment accumulating area of
Lake Washington. It is probable that much of this sediment load settles at
the southern end of the lake at the delta near the river mouth and does not
reach the deep profundal basin; sedimentation rates are believed to be about
30-40% higher 7 km south of the coring location toward the Cedar River
[21,22].

Stable lead also serves as a cultural geochronometer. Atmospheric lead
pollutant particles have entered the lake sediments as fallout from local
sources which include fossil fuel combustion, industrial operations, and
combustion of leaded gasoline. Coal fly ash may be an important source of
lead [23]. Coal has been burned locally since the fields were developed in
the study area in the late 1800s. A major industrial source of lead has
been a smelter located about 35 km upwind of the lake. This installation
produced lead bullion from its inception in 1890 until 1913 when it became
a copper smelter; current emissions are 40% PbO [24]. Since the advent of
leaded fuel additives in 1923 increases in lead concentrations in Lake
Washington sediments closely parallel increases in local gasoline consumption [25].

CONCENTRATION OF LEAD, mg/kg DRY SEDIMENT

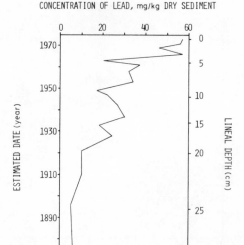

FIG.5. *Stable lead profile in Lake Sammamish sediment core.*

The stable lead profile shown in Figure 3 appears to consist of a pre-cultural background of less than 10 mg/kg until about 1885 followed by a rapid increase during the 1890s. The decline in lead concentrations from about 1902-1916 appears to be a result of dilution by the silt band rather than due to a reduction in external sources. By about 1930 lead levels in the sediment returned to pre-silt band concentrations. However, since the sedimentation rate at this later time averages about one third of pre-silt band material a genuine and substantial decline in lead input to the lake after 1916 is indicated. From 1930 to 1973, when the cores were collected, lead levels increased by about threefold.

The lead profiles in the sediment, when interpreted using Pb-210 dating and sedimentation rates, suggest that a source of lead pollution was present in the area from about 1890-1916 and that a growing source or sources was present since about 1930. This interpretation is consistent with the known history of anthropogenic lead emissions in the area and shows that the sediments preserve the pollutional record; it also shows the sensitivity and accuracy of geochronologies developed using Pb-210 measurements.

4.2. Lake Sammamish

The Pb-210 results from Lake Sammamish are shown in Figure 4 where a similar sedimentation history to that of Lake Washington is found. Four different sedimentation rates are present. A sedimentation rate, Region I, of 229 $g/m^2 \cdot a$ or 1.1 mm/a existed prior to 1909. The sedimentation rate increased in Region II to an average of 568 $g/m^2 \cdot a$ or 2.7 mm/a during the

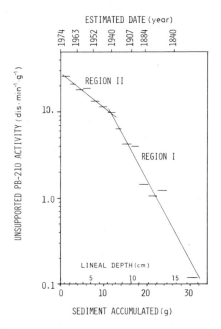

FIG.6.   Unsupported ^{210}Pb activity in Chester Morse Reservoir sediment core. Estimated sedimentation rates: Region I, 150 g/m^2· a or 0.73 mm/a (pre-1940) and Region II, 360 g/m^2· a or 2.3 mm/a (1940–1973).

1909-1932 time interval. From 1932 to 1944 a much higher sedimentation rate, Region III, of 1230 g/m^2·a or 5.6 mm/a is present. Since 1944 the average sedimentation rate, Region IV, has been 679 g/m^2·a or 3.3 mm/a.

The cultural history of the Lake Sammamish watershed is similar to that of Lake Washington except that land development was somewhat slower. Much of the watershed was still in virgin timber in 1895. By 1933 about 85% of the watershed had been logged. During the late 1930s it was observed that all the big timber had been removed and that the barren hillsides were ready for use by a growing suburban population [26]. The fact that the highest sedimentation occurred well after the immediate lakeshore and adjacent up-lands had been logged suggests that suburbanization may have been respon-sible for the highest sedimentation rates (Region III).

The stable lead profile in the Lake Sammamish core shown in Figure 5 suggests that about a 20% increase in lead occurred by 1910, and that four-fold increase in lead is present by the mid-1930s. Between the mid-1940s and 1973 lead levels in the sediment have tripled. In the absence of local sources the simultaneous increases in lead and in the sedimentation rate (1932-1944) suggest that the source for much of the lead entering the lake was contained in easily erodable surface soils which had previously collec-ted lead from fallout.

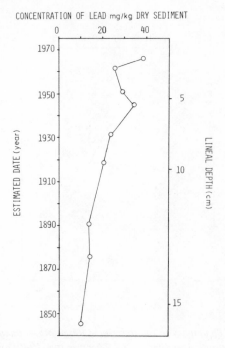

FIG. 7.   Stable lead profile in Chester Morse Reservoir sediment core.

4.3.   Chester Morse Reservoir

The Pb-210 measurements of sediments collected from Chester Morse Reservoir indicates two different sedimentation rates, Figure 6. A background rate of 150 $g/m^2 \cdot a$ or 0.73 mm/a was present until 1940; since 1940 a higher rate of 360 $g/m^2 \cdot a$ or 2.3 mm/a is indicated. As part of the public water supply for the City of Seattle the watershed has been closed to public entry since the early 1900s. With the exception of some mineral exploration prior to the turn of the Century, logging is the only activity which has been allowed.

The water level of Chester Morse Reservoir has been raised several times; first in 1903-4 by a log crib dam and later in 1917 by a masonry structure. Most of the timber was removed from the inundated land prior to raising the water level. Logging was begun along the new shoreline in 1917, by working to progressively higher elevations 15% of the watershed had been cut by 1947. The logging method used has been clear cutting by sky boom and donkey engine which transported the logs from where they fell to a railroad. From 1940 to the present spur roads have been constructed and trucks have been utilized to transport the logs; such practices have required the construction and maintenance of an extensive road system. It is important to note that the Chester Morse watershed which is adjacent to the long axis of the lake is exceptionally steep with hillside slopes of 1:2. Virgin timber covering these slopes was logged shortly after the railroad bed was

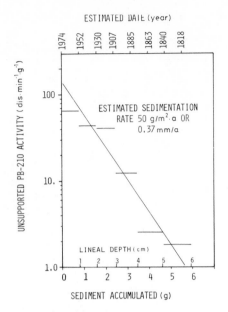

FIG.8.   Unsupported ^{210}Pb activity in Findley Lake sediment core.   Estimated sedimentation rate is 50 g/m$^2 \cdot$ a or 0.37 mm/a.

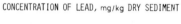
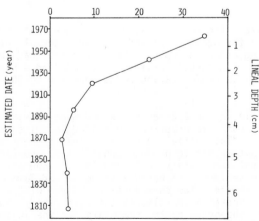

FIG.9.   Stable lead profile in Findley Lake sediment core.

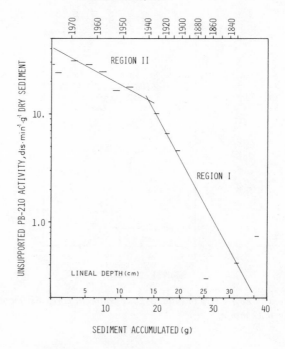

FIG.10.   Unsupported ^{210}Pb activity in Lake Meridian sediment core. Estimated
sedimentation rates are: Region I, 77 g/m² · a or 1.6 mm/a (pre-1941) and Region II,
262 g/m² · a or 3.9 mm/a (1941–1977).

built near the lake.   The Pb-210 measurements show that there were no pronounced changes in the quantity of sedimenting material until about 1940.
Thus, road building and maintenance rather than railroad construction appear
to be the most probable causes for the increased land erosion and consequent increase in lake sedimentation rates from 150 to 360 g/m²·a prior to
and after 1940.

The lead profile in the Chester Morse sediment core, Figure 7, indicates
that increases in lead have occurred since 1940 in spite of the increased
sedimentation rate.   The initial increase in lead concentrations may predate
1940 but it does not appear to predate about 1912.   The rapid increase in
lead in the interval immediately following 1940 is counter to the general
trend found in the sediments of Lakes Washington and Sammamish and thus must
represent a local source.   The construction and use of the logging road network by internal combustion vehicles is the most likely source.   This contention is further supported by the temporary decline in the lead levels of
the sediments after 1950, which corresponds to the completion of the road
network.

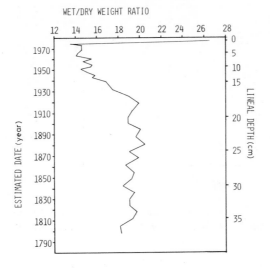

FIG.11. Wet- to dry-weight ratios in Lake Meridian sediment core.

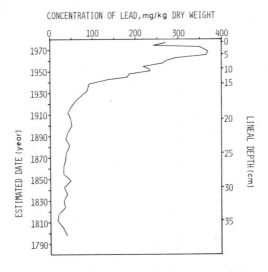

FIG.12. Stable lead profile in Lake Meridian sediment core.

BARNES et al.

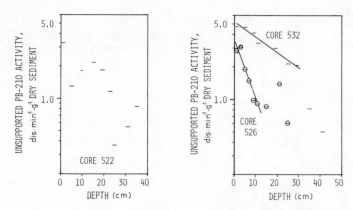

FIG.13.   *Unsupported* ^{210} *Pb activities in three Lake Union cores. Estimated sedimentation rates — core 532, 8.0 mm/a (ca 1935–1975); core 526, 1.6 mm/a (ca 1900–1975).*

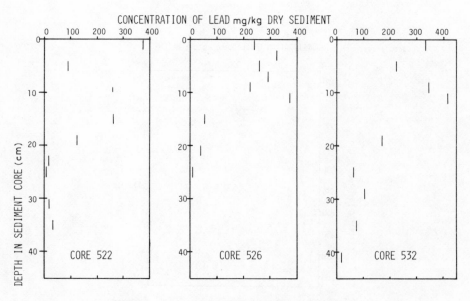

FIG.14.   *Stable lead profiles in Lake Union cores.*

4.4. Findley Lake

The Pb-210 profile of the Findley Lake core is shown in Figure 8 where
one apparent sedimentation rate of 50 g/m^2·a or 0.37 mm/a is found. This
is in agreement with the value determined by the C-14 method of 0.27 mm/a
[27]. Since the C-14 dated core was not taken at the deepest part of the
lake, we would expect a lower average sediment accumulation rate than from
our core; sediments are being redistributed continually from the shallower
to the deeper regions. A single sedimentation rate is consistent with the
undisturbed nature of the watershed. The stable lead profile of the Findley
Lake core shown in Figure 9 indicates that lead enrichment is present at
least as early as 1920. Since the lake is subjected to no local lead pollution, the increased lead levels must be due to general aeolian sources;
this increase can be identified easily over the background levels because
of the low sedimentation rate.

4.5. Lake Meridian

Pb-210 measurements of the Lake Meridian core indicated that an increase
in the sedimentation rate occurred about 1941, Figure 10. Wet to dry weight
ratios of the sediment sections, Figure 11, exhibit a marked decrease simultaneously with the change in sedimentation rate. The more recent material
is apparently more compact or more dense than the sediment deposited prior
to 1941. The sedimentation rate prior to 1941 averages 77 g/m^2·a or
1.6 mm/a ; since 1941 the sedimentation rate has averaged 262 g/m^2·a or
3.9 mm/a .

Land use changes in the Lake Meridian watershed were begun by settlers
arriving in the 1880s. By 1920 the entire drainage basin had been clear
cut and about 90% of the watershed was either under cultivation or in pas-
tureland. Land use changed from agricultural to suburban after 1943. The
Pb-210 measurements again indicate that it was roads and house construction
rather than either logging or agriculture which resulted in the increase in
sedimentation rates.

The stable lead profile in the Lake Meridian core indicates a rapid
increase in lead concentrations after 1920, Figure 12. The 100% increase
in the lead values from 1937 to 1945 is atypical. Regional gasoline consumption, the presumed driving force for lead pollution, increased by only
16% over this interval. Because this anomaly occurred simultaneously with
the changes in the nature of the sedimenting material, it probably reflects
the erosion of surfical soils which were enriched from past lead deposition.

4.6. Lake Union

The Pb-210 concentration profiles of three cores from Lake Union were
measured. The results, shown in Figure 13, require the use of both Pb-210
as well as stable lead to arrive at any meaningful interpretation. The
three cores were taken at water depths ranging from 13.5 to 16.5 m in the
relatively flat profundal area. The upper 30 cm of Core 532, collected
where water enters Lake Union through the ship canal from Lake Washington,
has been deposited in less than forty years based on the Pb-210 sedimentation rate of 8.0 mm/a . There is insufficient data to determine the sedimentation rate prior to about 1935. If it is assumed that the initial
stable lead enrichment occurred simultaneously in Lake Union and Lake
Washington at about 1890 and that pre-1890 lead levels in Lake Union were on
the order of 5-10 mg/kg then it is apparent that the 42 cm core did not
reach pre-1890 material as can be seen from Figure 14.

TABLE II. PARAMETERS USED IN ESTIMATING THE SEDIMENT FLUX
AND MEAN WATER COLUMN RESIDENCE TIMES OF ^{210}Pb AND STABLE LEAD

Parameter	Lake Washington Pb-210	Lake Washington Stable Lead	Lake Sammamish Stable Lead	Chester Morse Reservoir Stable Lead
Water Column Height, z (cm)	6200	6200	3050	3540
Concentration in Water C_w(dpm/cm^3)Pb-210 (μg/cm3) Pb[a]	2.5-4.0 10^{-5}	0.7-1.0 10^{-3}	0.65 10^{-3}	0.27 10^{-3}
Concentration in Surface Sediment C_s	12.3 dpm/g	320 μg/g	56 μg/g	39 μg/g
Sedimentation Rate, q, g/cm^2/a	0.0644	0.0644	0.0610	0.0360
Flux, r	0.79 dpm/cm$^2\cdot$a	20.6 μg/cm$^2\cdot$a	3.42 μg/cm$^2\cdot$a	1.4 μg/cm$^2\cdot$a
Integrated Water Column Concentration per Unit Area	0.155-0.248 dpm/cm^2	4.34-6.2 μg/cm^2	1.98 μg/cm^2	0.96 μg/cm^2
Mean Water Column Residence, τ (days)	72-114	77-110	211	250

[a] Data from Barnes [28].

Core 526, taken in the middle of the lake, apparently penetrated the
1890 horizon. It appears that no systematic change in the Pb-210 occurred
over about a 13 cm increment in the core from 25 to 12 cm, though based on
the stable lead concentrations all this material is post-1890. In the upper
12 cm of this core the sedimentation rate appears to have stabilized at
1.6 mm/a . This would give an estimated date of 1900 for the 12 cm horizon,
and suggests that the pertubation in the underlying material may date from
about the same time.

Core 522, taken nearer to the western lakeshore, produced a confusing
profile of both Pb-210 and stable lead. The simultaneous decreases in both
stable lead and Pb-210 in the 4-6 and 24-26 cm horizons suggest dilution by
material low in both lead and Pb-210. The remainder of the profile is almost impossible to decipher. Lead and Pb-210 levels in the 34-36 cm section
suggest that the overlying material has been deposited since 1890.

The effects of the canal construction in 1913-1916 and a slight increase
in the lake level at that time are not as readily apparent as in Lake
Washington. Lake Union has been the site of industrial and commercial activity for over 90 years--repeated dredging and dumping are known to have
occurred in the lake since at least 1916. We believe that these activities
probably contributed to the very anomalous data for the Lake Union cores.
Unfortunately, the Pb-210 and stable lead determinations in these cores were
not carried out at small enough sampling intervals to give the measure of
detail available in some of the other study lakes.

If it is assumed that the precultural background sedimentation rates in Lake Union were on the order of those from the other study lakes there is little question but what the present data represent a substantial increase in these rates. All the Lake Union cores exhibit Pb-210 activities in the surface sediment of 3 to 5 dpm/g. This is a factor of 2 to 3 lower than the Pb-210 activities in surface material from the other study lakes with comparable sedimentation rates. The most probable explanation of this is that Lake Union has a very short hydraulic residence time of about a week and that this results in high advective losses of Pb-210 entering the water column.

4.7. Mean residence times

It has been assumed that both Pb-210 and stable lead are rapidly sca-venged from the water column and serve as tracers of sinking particulate material. This assumption implies that lead, Pb-210 and suspended parti-culate matter have similar residence times in the water column. The assumption may be tested if the sedimentation rate, water column concentra-tions and surface sediment concentrations of the tracers are known. The mean residence time τ of a tracer in the water column is

$$\tau = \frac{C}{r} \tag{1}$$

where C is the concentration of the tracer in the water column per unit area, dpm/cm^2 or $\mu g/cm^2$ defined by

$$C_w = z \cdot b \tag{2}$$

where z is the water column height in cm and b is the respective tracer con-centration in dpm/cm^3 or $\mu g/cm^3$. The rate of deposition or flux of the tracer to the sediment is given in dpm/$cm^2 \cdot a$ or $\mu g/cm^2 \cdot a$ by

$$r = C_s \cdot q \tag{3}$$

where C_s is the concentration of the tracer in the surface sediment in dpm/g or $\mu g/g$ and q is the sedimentation rate in g/$cm^2 \cdot a$. If the density and wet to dry weight ratio of the sediment is known then q can be calculated from linear sedimentation rates.

The residence time of Pb-210 and stable lead may be calculated using equation (1) and the data given in Table II. Because some unknown fraction of the stable lead and Pb-210 may be present in the soluble form in the water column, and because the water column values are given for the total metal, the resulting residence times must be taken as a maximum estimate. By using the maximum depth of each lake for the height of the water column, z, the estimate of the residence times also tends to be maximized.

The calculated mean residence time of Pb-210 in the Lake Washington water column is approximate since the average water concentration of Pb-210 is not accurately known. Pb-210 measurements of water samples collected in the water column over a two year period indicate a mean concentration of 25 dpm/m^3 with a maximum of 40 dpm/m^3; thus, the residence time of 72 days is probably a more typical value. The almost identical residence time estimates for Pb-210 and stable lead in Lake Washington suggest that the two tracers behave similarly in the water column and that their reactions with particulate matter are similar.

Independent estimates of particulates matter residence times can be made by considering that the Lake Washington sediment is a gyttja consisting of algal remains, clays and fine silts. Using Stoke's Law and assuming that 1) the clay and fine silt particles 2-5 μm in diameter, have a density of 2.65 g/cm^3, 2) the average water temperature is 10°C, 3) the particles are not affected by mixing, then the residence times would be 40-260 days in the 6200 cm water column. Algal settling rates of 0.3 to 3.0 m/d have been measured in Lake Sammamish using sediment traps; the rates depended on season and algal species [29]. On this basis algal residence times of 20-200 days are reasonable for Lake Washington. Thus, since the residence times of the clay and silt and the algae particles are comparable to those estimated for Pb-210 and stable lead, the assumption that these nuclides serve as tracers of suspended material is valid.

With water column residence times in the smaller lakes of 100-200 days significant amounts of Pb-210, stable lead and suspended particulate matter may be lost due to flushing of lakes with hydraulic residence times less than one year. The flux of Pb-210 to the sediments of a given lake is also dependent on the ratio of the lake surface area over which the unsupported Pb-210 is entering to the actual area over which sediment is being permanently accumulated. Thus, for Lake Washington where the hydraulic residence time ($\tau_W = 2.35$ a) is sufficiently long so that almost all the entering lead will reach the sediments then the ratio would be:

$$\frac{\text{Unsupported Pb-210 flux to sediments}}{\text{Unsupported Pb-210 flux to lake surface}} = \frac{\text{Lake surface area}}{\text{Permanent sediment area}}$$

Since the numerators of both terms are readily determined it is possible to determine either denominator if the other is known. Using the estimate of 50 km^2 [20] for the permanent sediment accumulating area of Lake Washington and the data in Tables I and II, the unsupported Pb-210 flux to the lake surface is

$$\frac{50 \text{ km}^2}{88 \text{ km}^2} \cdot (0.79 \text{ dpm/cm}^2 \cdot a) = 0.45 \text{ dpm/cm}^2 \cdot a$$

This estimate is in agreement with the estimate obtained from using the concentration of Pb-210 in marine precipitation at the latitude multiplied by the average precipitation

$$(4.7 \cdot 10^{-3} \text{ dpm/cm}^3)(88 \text{ cm/}) = 0.41 \text{ dpm Pb-210/cm}^2 \cdot a$$

and that measured value of 0.3 dpm Pb-210/cm$^2 \cdot$a by integrating monthly bulk precipitation samples collected near the shore of Lake Washington [30,31].

5. CONCLUSIONS

Information on the uses and limitations of the Pb-210 dating technique has been derived from sediment studies of six lakes in western Washington State. Fine structure of the Pb-210 concentration profiles which signify changes in the sedimentation rates was found in four of the lakes. These changes in sedimentation rates correlated with cultural activities in the watersheds; thus, the changes are believed to indicate real differences in sedimentation rates. The detection of Pb-210 fine structure in the sediment cores was found to require both frequent sampling intervals and individual sample sections which represented at most a few years of sediment deposition.

Natural background precultural sedimentation rates were found to range from 50 to 229 $g/m^2 \cdot a$ or 0.37 to 1.6 mm/a depending on the lake and watershed. Current sedimentation rates which were different from background rates, ranged from 262 to 679 $g/m^2 \cdot a$ or 2.4 to 3.9 mm/a. The highest sedimentation rates observed were 1230 $g/m^2 \cdot a$ (6.7 mm/a) and 1800 $g/m^2 \cdot a$ (8.3 mm/a) and occurred in Lake Sammamish (1932-1944) and Lake Washington (1890-ca 1902) respectively. A sedimentation rate of 8.0 mm/a was also noted in Lake Union since 1935 but appeared to be highly localized. In both Lakes Washington and Sammamish the highest sedimentation rates were coincident with the onset of suburbanization immediately after or during the last stages of timber harvesting. In both cases after a decade or so at the very high sedimentation rates, the rate declined substantially but still remained higher than the precultural background rates. We suspect that this reduction in sedimentation resulted from revegetation, landscaping, construction of impervious surfaces and drainage alterations. The relatively constant sedimentation rates in Lake Washington since about 1916, Lake Sammamish since 1944, Chester Morse Reservoir since 1940 and Lake Meridian since 1941 suggest that new equilibria between sedimentation and environmental conditions in a watershed may be rapidly attained.

Limitations on the Pb-210 method were found in Lake Union where intense cultural activities including repeated dredging and filling apparently have disrupted normal sedimentation processes. Stable lead analysis could only confirm the strategraphic alterations suggested by Pb-210 anomalies by identifying more recent lead enriched material from older low lead material. We believe that analysis for other elements in the core profiles such as carbon and aluminum could greatly assist in the interpretation of sedimentary geochronologies. Stable lead profiles found in the sediments of the other lakes were consistent with the local history of lead pollution based on the Pb-210 geochronologies. The sediments of all lakes were found to be enriched in lead compared to precultural material.

The water column residence times for Pb-210 and stable lead in Lake Washington were almost identical, and were consistent with independent estimates of algal settling times and the expected residence times for fine silt and clay particles. This observation confirms the suitability of Pb-210 and stable lead as tracers of suspended particulate material in lakes. The input of Pb-210 across the surface of Lake Washington calculated from the Pb-210 flux to the sediments and the ratio of permanent sediment accumulating area to lake surface area was consistent with the expected Pb-210 input due to marine precipitation at this latitude.

ACKNOWLEDGEMENTS

We would like to acknowledge the assistance of Mr. Charles Vick for sample counting and Dr. W.T. Edmondson for sample collection and technical discussion.

Financial support for the program was provided by the Municipality of Metropolitan Seattle (METRO), and the State of Washington Water Research Center (OWRT-A-083-WASH).

REFERENCES

[1] GOLDBERG, E.D., Geochronology with lead-210, In Radioactive Dating, IAEA STI/PUB/68 (1963) 121.

[2]   KRISHNASWAMI, S., et al., Geochronology of lake sediments, Earth
      Planet. Sci. Lett. 11 (1971) 407.
[3]   KOIDE, M., SOUTAR, A., GOLDBERG, E.D., Marine geochronology with Pb-
      210, Earth Planet. Sci. Lett. 14 (1972) 442.
[4]   KOIDE, M., BRULAND, K.W., GOLDBERG, E.D., Th-228/Th-232 and Pb-210
      geochronologies in marine and lake sediments, Geochem. Cosmochim.
      Acta 37(1973) 1171.
[5]   ROBBINS, J.A., EDGINGTON, D.N., Determination of recent sedimentation
      rates in Lake Michigan using Pb-210 and Cs-137, Geochim. Cosmochim.
      Acta 39(1975) 285.
[6]   POET, S.E., MOORE, H.E., MARTELL, E.A., Lead 210, bismuth 210, and
      ^{210}Po in the atmosphere: accurate ratio measurement and application
      to aerosol residence time determination, J. Geophys. Res. 77(1972)
      6515.
[7]   LEWIS, D.M., The use of ^{210}Pb as a heavy metal tracer in the Susque-
      hanna River system, Geochim. Cosmochim. Acta 41(1977) 1557.
[8]   KOIDE, M., GOLDBERG, E.D., Lead-210 in natural waters, Science 134
      (1961) 98.
[9]   KRISHNASWAMI,S., et al., Geochronological studies in Santa Barbara
      Basin: ^{55}Fe as a unique tracer for particulate settling, Limnol.
      Oceanog. 18(1973) 763.
[10] PETIT,D., ^{210}Pb et isotopes stable du plomb dan des sediments lacus-
      trus, Earth Planet. Sci. Lett. 23 (1974) 199.
[11] MATSUMOTO, E., Accumulation rate of Lake Biwa-ko sediments by the
      ^{210}Pb method, J. Geol. Soc. Japan 81 (1975) 301.
[12] MATSUMOTO,E., ^{210}Pb geochronology of sediments from Lake Shinji,
      Geochemical J. Japan 9 (1975) 167.
[13] KAJAK,Z. "Benthos of standing water", in A Manual on Methods for the
      Assessment of Secondary Productivity in Fresh Waters, (EDMONDSON, W.T.,
      WINBERG, G. G., eds) International Biological Program Handbook No.17
      (1971) 25.
[14] SMITH,G.F., The wet ashing of organic matter employing hot concentra-
      ted perchloric acid--The liquid fire reaction, Anal. Chemica Acta
      8 (1953) 397.
[15] BORTLESON, G.C., LEE,G.F., Recent sedimentary history of Lake Men-
      dota, Wisconsin, Environ, Sci. Technol. 6 (1972) 799.
[16] EDMONDSON, W.T., ALLISON, D.E., Recording densitometry of X-radio-
      graphs for the study of cryptic laminations in the sediments of Lake
      Washington, Limnol. Oceanog. 15 (1970) 138.
[17] EDMONDSON, W.T., Microstratification of Lake Washington sediments,
      Verh. Internat. Verein. Limnol. 19 (1975) 770.
[18] GOULD, H.R., BUDINGER, T.F., Control of sedimentation and bottom con-
      figuration by convection currents, Lake Washington, Washington, J.
      Mar. Res. 17 (1958) 183.
[19] POWERS,H.A., WILCOX,R.E., Volcanic ash from Mount Mazama (Crater Lake)
      and from Glacier Peak, Science 144 (1964) 1334.
[20] CRECILIUS, E.A., Arsenic geochemical cycle in Lake Washington, Limnol.
      Oceanog. 20 (1975) 441.
[21] DAVIS, M.B., Pollen evidence of changing land use around the shores
      of Lake Washington, North West Sci. 47 (1973) 133.
[22] BARNES, R.S., SCHELL,W.R., Physical transport of trace metals in the
      Lake Washington watershed, in Cycling and Control of Metals, U.S.
      Environmental Protection Agency, National Resource Center, Cincin-
      natti, Ohio, (1973) 45.
[23] BLOCK, C., DAMS,R., Lead contents of coal, coal ash and fly ash, Water
      Air, Soil Pollution 5 (1975) 207.

[24] CRECILIUS, E.A., PIPER. D.Z., Particulate lead contaminations recorded in sedimentary cores from Lake Washington, Seattle, Environ. Sci. Technol. 7 (1973) 1053.

[25] SCHELL, W.R., BARNES, R.S., "Lead and mercury in the aquatic environment of Western Washington State, in Aqueous Enviornmental Chemistry of Metals, (RUBIN, A.J., ed), Ann Arbor Science, Ann Arbor, Michigan, (1974) 43.

[26] FISH, E.R., The Past at Present in Issaquah, Washington, Kingsport, Press, Kingsport, Tennessee (1967).

[27] ADAMS, D., The Paleoecology of Two Lakes in Western Washington, M.S. thesis, University of Washington, Seattle, Washington, (1973).

[28] BARNES, R.S., A Trace Elemental Survey of Selected Waters, Sediments and Biota of the Lake Washington Drainage, M.S. Thesis, University of Washington, Seattle, Washington, (1976).

[29] BIRCH, P.B., The Relationship of Sedimentation and Nutrient Cycling to the Trophic Status of Four Lakes in the Lake Washington Drainage, Ph.D. Dissertation, University of Washington, Seattle, Washington, (1976).

[30] SCHELL, W.R., Concentrations, physico-chemical states and mean residence times of ^{210}Pb and ^{210}Po in marine and estuarine waters, Geochim. Cosmochim. Acta 41 (1977) 1019.

[31] NEVISSI, A., SCHELL, W.R., The use of ^{210}Pb and ^{210}Po as tracers of atmospheric processes. (In press, Proceedings, Natural Radiation Environment III, Houston, Texas April 23-28, 1978).

DISCUSSION

D.B. SMITH: Have you undertaken any measurements of radioactive fall-out components such as ^{137}Cs as a cross-check on the sedimentation rates over the last twenty years? Such measurements might provide more information on the near-surface mixing which you describe, e.g. in Fig.10, since the ^{137}Cs input has varied in a known way.

R.S. BARNES: Yes, we have made preliminary ^{137}Cs measurements on some of these cores. In general, they agree very well with the ^{210}Pb results. I might add that, although the upper two centimetres of the Lake Meridian core shown in Fig.10 might be interpreted as lower near-surface ^{210}Pb activity due to bioturbation, there is also a 20% increase in aluminium between the 1−2 cm horizon as compared with the 2−3 cm horizon. This increase is believed to coincide with a once-in-fifty-years flood which occurred in December 1975 and which unquestionably could have produced a similar dilution of sedimentary ^{210}Pb activities.

T. TORGERSEN: I am concerned about the discontinuity that occurs in the Lake Washington data. I agree that a change in sedimentation rate occurred in about 1880 and again about 1903, but does the discontinuity not also indicate a change in ^{210}Pb fallout rate? I would be interested to know how you explain this discontinuity.

R.S. BARNES: The discontinuity in unsupported ^{210}Pb occurring in the Lake Washington sediment cores (Fig.2) is the result of dilution by silt. This

dilution affects four consecutive centimetres of material in the 17 to 21 cm
horizons in this particular case. The unsupported ^{210}Pb activity in the 16 and
22 cm horizons then becomes continuous. Thus, the discontinuity is more
apparent than real.

A.W. ELZERMAN: As the previous question indicates, your results depend
on a constant flux of ^{210}Pb to the lake surface. In Section 4.1 of the paper you
state that the flux has been constant for the last 150 years. Could you explain
the basis of this assumption? Both the sources of ^{210}Pb (and therefore of ^{222}Rn)
and the atmospheric deposition processes must be considered.

R.S. BARNES: The assumption is based on an unchanged ^{222}Rn emanation
rate from lithospheric material. In view of the total amount of ^{222}Rn emitted
to the atmosphere it appears unlikely that anthropogenic effects would be noted.

A.W. ELZERMAN: Have you looked at the possibility of using ^{210}Pb and
stable Pb ratios in order to investigate the sources of the Pb in lakes, along the
lines of the approach used by Robbins and Edgington (Ref. [5]), and would you
care to comment on the validity of this type of approach?

R.S. BARNES: We are currently examining our data with such an approach
in mind.

Session IX, Part 2

PROGRESS AND INNOVATION IN ISOTOPE HYDROLOGY

Chairman

K.O. MÜNNICH
Federal Republic of Germany

ETUDE PAR TRACEURS RADIOACTIFS
DE L'ACTION DE LA HOULE
SUR LES FONDS MARINS

G. COURTOIS, C. MIGNIOT*, G. SAUZAY
CEA, Centre d'études nucléaires de Saclay,
Section d'application de la radioactivité,
Gif-sur-Yvette, France

Présenté par J. Guizerix

Abstract–Résumé

RADIOACTIVE TRACER STUDY OF THE EFFECT OF SWELLS ON THE SEA BED.
Little is known at present about the limits of the effect of swells on the sea bed and the subject is a controversial one in scientific circles connected with maritime hydraulics. A programme of general studies using radioisotope-labelled sediments has therefore been put in hand to determine the extent of this action and to find a definitive semi-empirical formulation of the relations $Q_s = f(H, S, p)$, i.e. the solid flow-rate Q_s as a function of hydraulic conditions H (especially the swell characteristics), sediment conditions S and depth p. This programme has consisted in carrying out several series of immersions of labelled sediments perpendicular to a carefully selected coastline, at a number of depths down to 25 m. A study of this nature was done one year in a region subject to tides (in the Atlantic near Bayonne) and one year in a non-tidal region (the Mediterranean near Sète). A number of transport characteristics were found, for example the limits of the effect in particle size terms, cumulative transport as a function of energy balances in the swell and the existence of a critical depth below which solid transport becomes negligible (in hydraulic, not geological, terms). These parameters are discussed quantitatively and the repercussions they may have on problems of coastal civil engineering projects — especially the creation of artificial beaches by the use of solid matter deposited off shore — are evaluated.

ETUDE PAR TRACEURS RADIOACTIFS DE L'ACTION DE LA HOULE SUR LES FONDS MARINS.
La limite de l'action de la houle sur les fonds marins est actuellement mal connue et sujette à controverse dans les milieux scientifiques de l'hydraulique maritime. Un programme d'études générales a donc été développé qui utilise les sédiments marqués radioactifs pour déterminer cette action et obtenir en définitive une formulation semi-empirique des relations $Q_s = f(H, S, p)$, c'est-à-dire une formulation du débit solide Q_s, enregistré en fonction des conditions hydrauliques H et plus particulièrement des caractéristiques des houles, des conditions sédimentologiques S, et de la profondeur p. Ce programme a consisté à effectuer des séries

---

* Laboratoire central d'hydraulique de France, Maisons Alfort, France.

d'immersions de sédiments marqués sur une perpendiculaire à un littoral judicieusement choisi à différentes profondeurs allant jusqu'à 25 m. Une telle étude a été effectuée une année sur site soumis à marée (Atlantique, région de Bayonne), et une année sur site non soumis à marée (Méditerranée, région de Sète). Un certain nombre de caractéristiques du transport ont été mises en évidence, entre autres les limites d'action sur le plan granulométrique, les transports cumulés en fonction des bilans d'énergie de la houle, et l'existence d'une profondeur critique en dessous de laquelle le transport solide devient négligeable (à l'échelon des hydrauliciens et non des géologues). L'ensemble de ces paramètres sont discutés en termes quantitatifs; on évalue les répercussions que peuvent avoir ces conclusions sur les problèmes de génie civil d'aménagements côtiers et notamment sur la création de plages artificielles par apports solides de masses rejetées au large.

1. INTRODUCTION

Les études par traceurs radioactifs des déplacements sédimentaires sont très connues et nous ne reprendrons ici aucun point de la théorie ni de la méthodologie d'utilisation. Leur emploi a été le plus souvent décrit pour l'approche d'un problème précis tel que l'envasement d'un port ou la dérive d'un littoral.

Mais ce remarquable moyen permet d'aborder également des problèmes de recherche de base qui fassent progresser la connaissance générale d'un phénomène hydraulique donné.

Telles sont les études menées en commun au Commissariat à l'énergie atomique français et au Laboratoire central d'hydraulique de France, portant sur:
— les modalités de transport sédimentaire en estuaire,
— le devenir des métaux lourds en tant que polluants,
— l'action de la houle sur les fonds marins.

C'est ce dernier sujet qui va être sommairement abordé ici.

2. GENERALITES SUR L'ACTION DES HOULES SUR LES FONDS MARINS

Sous l'action des houles, les sédiments peuvent être soumis à des mouvements très divers allant de la simple oscillation sur les fonds sous l'effet des mouvements arbitraires à des déplacements dans le profil perpendiculairement à la côte, ou à des entraînements massifs le long du littoral sous l'action de houles obliques.

Les caractéristiques de la houle (notamment période et amplitude), la nature des sédiments (cohésifs, non cohésifs, diamètre moyen et étendue de la courbe granulométrique), la morphologie des fonds et du littoral vont intervenir dans les phénomènes de transport, l'équilibre sédimentaire final n'étant qu'une résultante d'actions hydrodynamiques où fluides et solides réagissent les uns sur les autres.

Il est par contre possible, en utilisant un certain nombre de lois théoriques et empiriques, de mieux comprendre les phénomènes physiques qui vont contribuer aux mouvements sédimentaires et de quantifier l'importance de ces mouvements dans des cas relativement simples.

Les expériences de terrain et plus particulièrement les expériences de traceurs radioactifs permettent de jauger les hypothèses faites et surtout de préciser les valeurs des paramètres à introduire dans les formules de charriage.

3. LISTE DES EXPERIENCES DE TRACEURS RADIOACTIFS AYANT DONNE LIEU A UNE INTERPRETATION D'ACTION DE LA HOULE

Depuis 1965, environ une dizaine d'expériences de traceurs radioactifs ont pu donner lieu à une interprétation de l'action de la houle. Les principales sont les suivantes (fig. 1):

— 1 (1965): Campagne de galets sur les plages des Bas-Champs (en France, Manche au nord de Dieppe), permettant d'étudier l'action cumulée d'une houle oblique et d'un courant de marée important.

— 2 (1967): Etude de l'action de la houle sur le Cap Ferret, cap nord d'entrée du Bassin d'Arcachon; cinq points d'immersion au verre au ^{192}Ir perpendiculairement au rivage de − 5 m à − 22 m.

— 3 (1973): Etude des transports sur le littoral entre St-Palais et Bonne-Anse (au nord de Royan); travail en sédiments fins permettant de préciser les épaisseurs sur lesquelles a lieu le transport.

— 4 (1974): Expérience dans un pays étranger de Méditerranée orientale; six points d'immersion au verre au ^{192}Ir de granulométrie variable permettant d'étudier le cas particulier de la présence d'une barre sableuse.

— 5 (1975): Expérience volontairement mise en œuvre pour obtenir des résultats généraux sur la limite d'action de la houle en fonction de la profondeur; six points d'immersion, pour deux granulométries, de − 18 m à − 22 m; expérience située juste au nord de Bayonne, donc dans une mer soumise à marée.

— 5 bis (1973): Recherche du comportement de la madrague (matériaux grossiers de 2 à 4 mm de diamètre) sous l'action de la houle; site identique au site n° 5.

— 6 (1975): Programme complémentaire du programme n° 5; mêmes conditions opératoires mais dans une mer non soumise à marée, en Méditerranée, au voisinage de Sète.

Les études n° 5 et n° 6, spécialement conçues pour étudier la limite de l'action de la houle en fonction de la profondeur, ont bénéficié d'une aide partielle du Ministère de l'équipement, Service central technique.

4. LIMITE VERS LE LARGE DE L'ACTION DE LA HOULE

Les campagnes n° 2, n° 5 et n° 6 ont permis d'étudier ce point et de s'en faire une idée relativement précise.

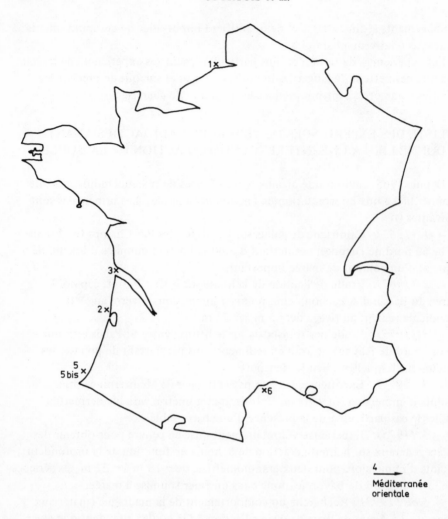

FIG.1.   Sites expérimentaux des études d'action de la houle sur les fonds marins.

4.1. Résultats généraux obtenus par les traceurs radioactifs — Déplacements dans le profil

Au Cap Ferret, des sables de 0,25 mm déposés par des profondeurs supérieures à − 10 m ne sont pratiquement pas remaniés sur des périodes de six mois bien que les houles aient agi pendant ces six mois, avec des amplitudes pouvant atteindre 5 m.

Pendant la campagne de Bayonne (n° 5) on a observé une tempête avec des houles de 7 m d'amplitude suivies ou précédées de plusieurs tempêtes plus petites avec des amplitudes de 4 à 5 et 6 m.

Le bilan d'énergie de la houle $H^2 \cdot T \cdot t$, dans lequel H est l'amplitude de la houle, T sa période et t le temps pendant lequel elle a agi, a été, sur une période de 3 mois: 420×10^6 m$^2 \cdot$s^2, soit environ 20% du bilan annuel moyen $(2,1 \times 10^9$ m$^2 \cdot$s^2) des houles du Golfe de Gascogne.

Il ressort de cette étude que des sables fins que l'on dépose artificiellement sur des fonds marins de 15 à 20 m peuvent être remaniés temporairement et transportés lors de tempêtes sur quelques centaines de mètres tant qu'ils ne sont pas incorporés aux sédiments naturels.

Les expériences de traceurs radioactifs montrent que, une fois ces sables intégrés aux matériaux du fond et par voie de conséquence représentatifs de ces derniers, la limite granulométrique de début de déplacement des grains se situe à 0,4 mm pour les profondeurs de -15 m CM (CM = cartes marines, c'est-à-dire -17 à -18 m au niveau moyen) et à 0,8 mm pour les profondeurs de -6 à -8 m.

Mis en oscillation par la houle les sédiments plus grossiers sont remontés vers la côte, mais les déplacements restent négligeables au-delà des fonds de -12 à -15 m   (environ 1 m^3/m de largeur en 3 mois pour le bilan d'énergie précédent de 420×10^6 m$^2 \cdot$s^2). Ces déplacements augmentent très rapidement lorsqu'on se rapproche de la côte.

Ainsi la figure 2 montre l'allure des courbes de transport dans le profil en m^3/m pour plusieurs détections successives correspondant à des bilans énergétiques de houles différents: on observe très nettement un coude vers les profondeurs de 11 à 12 m (800 à 1000 m de la côte dans le cas particulier étudié) en dessous duquel le transport peut être considéré comme négligeable.

En première approximation on peut dire que

$$\frac{Q_{p \text{ à } 8m}}{Q_{p \text{ à } 15m}} \sim 30$$

Q étant le transport cumulé en m^3/m.

A une même profondeur les transports peuvent être plus importants pour des particules de 0,3 à 0,5 mm que pour des particules de 0,1 à 0,15 mm (réduction du transport par suite d'un pavage ou d'une certaine cohésion entre les grains).

4.2. Tarage de la formule de bilan d'énergie pour un transport dans le profil

Le transport cumulé Q_p à une profondeur p est donné par

$$Q_p = K \sum (H^2 T t) \, g(\alpha)$$

où $\sum (H^2 T t)$ est le bilan d'énergie des houles précédemment défini,

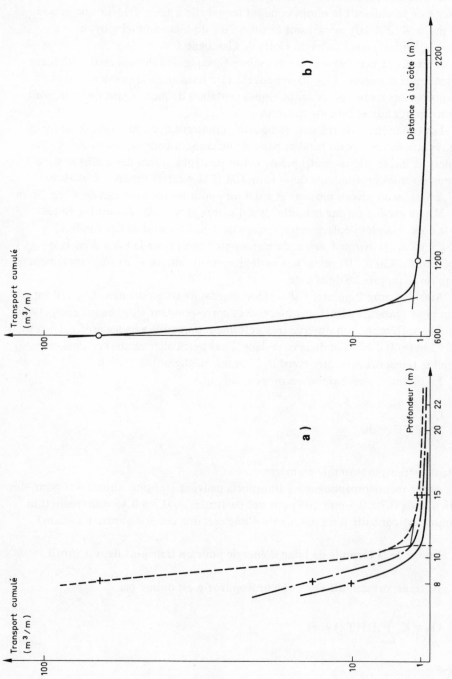

FIG.2. a) Variation du transport cumulé en fonction de la profondeur. Bilan énergétique: ——— 50 × 10⁶ m² · s² (deuxième détection); — · — 100 × 10⁶ m² · s² (troisième détection); – – – 300 × 10⁶ m² · s² (quatrième détection). b) Variation du transport cumulé en fonction de la distance à la côte.

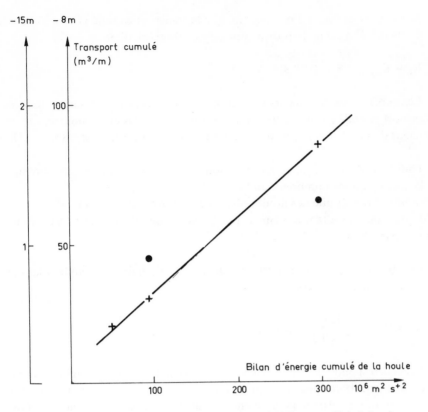

FIG.3.   Variation du transport cumulé avec le bilan cumulé de la houle.

$g(\alpha)$ est une fonction qui représente l'influence de l'obliquité α des houles par
  rapport au rivage;

K est normalement une constante, qui varie en pratique avec la profondeur.

Dans la figure 3, on a porté pour la profondeur $p = -8$ m le transport cumulé
en m^3/m en fonction du bilan d'énergie des houles. On observe effectivement une
droite qui passe approximativement par l'origine et donne

$$K_p = K_{8m} = 3 \times 10^{-7} \, s^{-2}$$

Sur la même figure, on a reporté (points noirs) les deux valeurs correspondant
à la même profondeur de 15 m. Certes l'imprécision est plus grande mais en
prenant $Q_p = 1 \; m^3/m$ pour un bilan d'énergie de $100 \times 10^6 \, m^2 \cdot s^2$ on a

$$K_p = K_{15m} = 10^{-8} \, s^{-2}$$

En utilisant la figure 2 on constate que le transport cumulé pour un bilan de 300×10^6 m²·s² à 10 m de profondeur est de 10 m³/m, d'où

$$K_p = K_{10m} = 3 \times 10^{-8} \text{ s}^{-2}$$

Les expériences de traceurs radioactifs ont ainsi permis d'étalonner la formule de transport cumulé par dissipation énergétique des houles et de montrer que le coefficient de profondeur varie d'un facteur 30 quand on passe de -8 m à -15 m.

Pour l'expérience n° 6, en mer non soumise à marée, les résultats obtenus sont beaucoup moins nombreux du fait
— d'un bilan énergétique des houles observées faible $\sim 40 \times 10^6$ m²·s²,
— d'un transport résultant également très faible et notable seulement sur un des points par -8 m.

On observe alors un transport cumulé de 1,3 m³/m linéaire pendant la période d'observation de 5 mois, correspondant à

$$K_p = K_{8m} = \frac{1,3}{4 \times 10^7} = 3 \times 10^{-8} \text{ s}^{-2}$$

Ceci correspond à la valeur trouvée pour 10 m dans le cas de milieu soumis à marée. Il apparaîtrait donc, sans qu'on puisse en avoir la certitude, que l'effet de marée viendrait amplifier le phénomène de transport et augmenter en conséquence les valeurs du coefficient K_p.

4.3. Utilisation des lois de charriage

En régime établi (fleuve par exemple), on a l'habitude de calculer le débit de charriage g_s en kg/s par mètre de largeur à partir de formules semi-empiriques de charriage:
— soit du type déterministe: on admet qu'à chaque débit instantané correspond un débit de charriage; l'exemple type en est la formulation de Meyer-Peter

$$g_s = A(\tau_0 - \tau_c)^{3/2}$$

(A = constante, τ_0 = force tractrice, τ_c = force tractrice de début d'entraînement);
— soit du type probabiliste, qui associe à tout débit liquide, si faible soit-il, une probabilité de transport; l'exemple type en est la formulation d'Einstein.
On peut appliquer ces formules instantanément sur le fond de la mer et cumuler l'effet, sur l'ensemble des oscillations au voisinage du fond, dû à la houle.

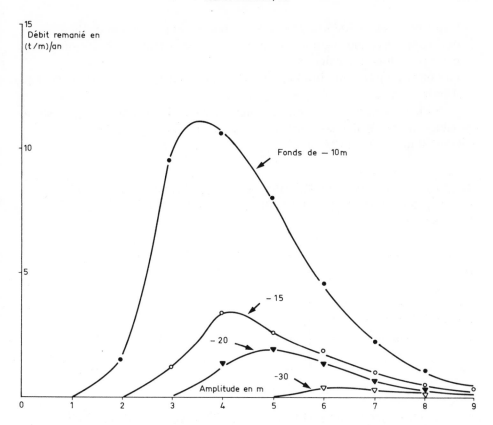

FIG.4. Remaniements théoriques à différentes profondeurs en fonction de l'amplitude de la houle (côte des Pyrénées atlantiques).

En tenant compte de la fréquence d'une amplitude donnée des houles dans la zone intéressée on peut ainsi obtenir le débit remanié en t/m par an en fonction de la profondeur. C'est ce que donne la figure 4 pour différentes amplitudes de houle.

Ce graphique permet un calcul simple des volumes totaux remaniés (tableau I). On ne saurait attacher une valeur absolue à ces résultats, les conditions d'emploi des formules des types Meyer-Peter et Einstein étant très éloignées des conditions hydrauliques réellement rencontrées. On constate cependant que:
— les ordres de grandeur des transports solides sont respectés, comme le montre la figure 5 qui donne le volume des sédiments sableux déplacés mesuré à l'aide de traceurs d'une part, théorique d'autre part, obtenu à l'aide des calculs précédents développés par le Laboratoire central d'hydraulique de France (LCHF);

— les quantités de sables susceptibles d'être remaniées par l'action des houles
   diminuent très rapidement avec la profondeur; elles sont encore importantes
   par − 10 m et pratiquement nulles par − 30 m;
— plus on se rapproche du littoral, plus les houles de faible amplitude interviennent
   dans le remaniement global des fonds.

   Sur le plan pratique, on ne peut donc que conseiller de limiter les extractions
de sables aux profondeurs supérieures à − 20 m. En prenant des limites situées
entre − 20 m et − 30 m CM on se trouve dans un secteur de remaniement très faible.

TABLEAU I. COMPARAISON DES VOLUMES THEORIQUES REMANIES
PAR LA HOULE D'APRES LES FORMULES DE MEYER-PETER ET
D'EINSTEIN

Profondeur (m)	− 7	− 10	− 15	− 20	− 30
Volumes remaniés ($(m^3/m)/a$) Formule de Meyer-Peter	385	38	13	5,5	0,5
Volumes remaniés ($(m^3/m)/a$) Formule d'Einstein	180	72	19	10	2

5. LE TRANSPORT PARALLELE AU LITTORAL − TRANSPORT AU VOISINAGE DU DEFERLEMENT

   La zone de déferlement correspond à un véritable pôle d'attraction sédimentaire
vers lequel les sédiments du large peuvent remonter et ceux du littoral redescendre
pour certaines caractéristiques des vagues. Par suite de la concentration de l'énergie
des houles, les sédiments sont remis en suspension, brassés par les vagues, la plus
grande partie du transport s'effectuant entre cette zone et le premier tiers de la
distance entre les brisants et la côte.

   En 1958 Zenkovitch, utilisant des traceurs fluorescents [1], donnait la répartition du transport littoral (fig. 6). Zenkovitch a montré que environ 50% du
transport s'effectuait en suspension au voisinage du premier rouleau de déferlement
avec des turbidités moyennes de 3 à 4 g/l, que 20% du transport avait lieu au
voisinage de la deuxième barre et 6% sur l'estran bien que l'on puisse y mesurer des
concentrations de 15 à 30 g de sable par litre d'eau prélevé. Le reste du transport
était supposé se faire par charriage sur les fonds. Dans ces mesures en nature, les
houles avaient des valeurs ne dépassant pas 3 m et les courants de houle des
valeurs de 0,60 m/s.

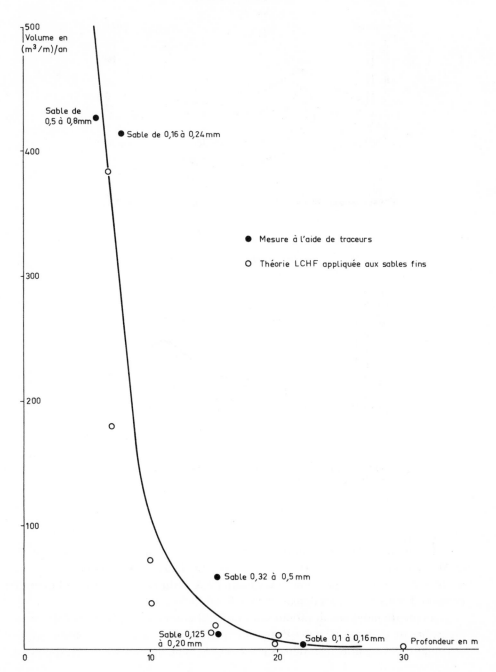

FIG.5.   Volume des sédiments sableux déplacés dans le profil en fonction de la profondeur (côte des Pyrénées atlantiques). (Résultats des expériences à l'aide de traceurs radioactifs de 1975 ramenés à une année moyenne.)

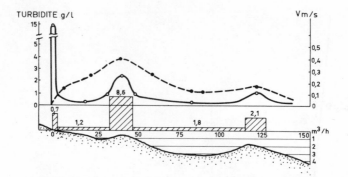

Vitesse du courant en surface et débit solide par houle de 0,5 à 1m

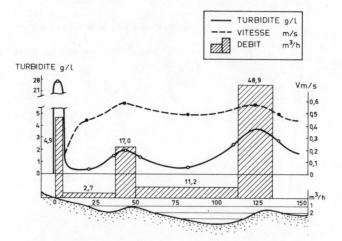

Vitesse du courant en surface et débit solide par houle de 2 à 3 m
d'après ZENKOVITCH [1]

FIG.6. Débit solide le long d'un profil.

L'expérience la plus quantitative menée en ce domaine est l'expérience n° 4
en Méditerranée orientale. Avec l'aide contractuelle de l'AIEA, un pays
méditerranéen et les organismes français du CEA et du LCHF ont suivi et quantifié
pendant 3 mois d'hiver les déplacements de sables locaux.

La pente moyenne des fonds était de 1/80 et les sédiments constitués de
sables fins de 0,25 à 0,10 mm. Pendant la période d'observation, les houles avaient
des amplitudes significatives dépassant 3 m et le bilan de transport correspondant

à la somme $\sum (H^2 Tt) g(\alpha)$ atteignait $12,5 \times 10^6$ m$^2 \cdot$ s^2. La figure 7 et le tableau II

illustrent les résultats obtenus. Ils font apparaître les faits suivants:

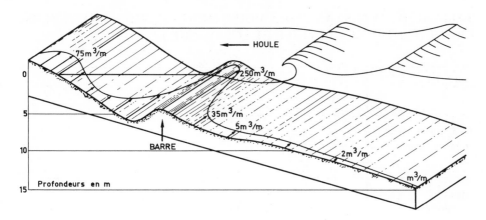

SCHEMA _ Répartition des transports le long d'un profil
(entre le 4 février et le 23 avril 1974)

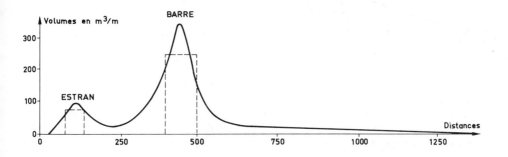

Graphique de transit vers le nord à différentes distances
de la côte

FIG.7. *Etude effectuée en Méditerranée orientale.*

— le transport en dessous de 8 m de profondeur est faible; il commence à progresser
par fond de -6 m à -7 m où il atteint 35 m³/m parallèlement au rivage;
— dans la barre et sur la face interne de cette barre le débit solide par charriage
atteint 250 m³/m (en 80 jours); c'est de loin la zone où on observe le transport
maximal;
— après être redescendu à 25 m³/m dans la zone de lévigation, située à 250 m de la
barre vers la terre, le débit remonte ensuite à 75 m³/m sur l'estran.

Donc environ 75% du transport parallèle au rivage s'est effectué au voisinage
du déferlement et notamment sur sa face interne.

TABLEAU II.   IMPORTANCE DES MOUVEMENTS SEDIMENTAIRES VERS
LE NORD ENTRE LE 4 FEVRIER ET LE 23 AVRIL 1974

Profondeur (m)	− 15	− 12	− 8	− 7 à −6 (face extérieure)	− 5 (barre)	Estran
Epaisseur maximale en mouvement (cm)	1	1 à 2	5 à 10	15	25	12
Quantité transportée (m³/m)	1	2	5 (évaluée)	35	250	75 (évaluée)
Largeur de transport (m)	250	300		250	100	30
Volume de transport (m³)	375	1050		2 à 5000	25 000	2250
Total			30 000 à 33 000 m³ (en 3 mois)			

Le coefficient K de proportionnalité avec l'énergie de la houle pour des sables de granulométrie comprise entre 0,25 mm à l'estran et 0,15 mm au large atteint des valeurs comprises entre 3,3 et 5 × 10⁻⁶. Les estimations théoriques du LCHF donneraient une valeur moyenne de 4 × 10⁻⁶ en partant d'une influence du diamètre des grains proportionnelle à $D^{-1/2}$.

$$K = 0,18 \times 10^{-5} \, D^{-1/2} \qquad \text{(D en mm)}$$

Cette expérience est particulièrement intéressante, car l'association d'une expérience de traceurs radioactifs qui intègre l'ensemble des phénomènes de transport solide en fonction des conditions hydrodynamiques et d'un régime de houles traduit par des périodes successives de tempêtes et de calme a permis d'aller, pendant les périodes de calme, examiner les phénomènes résultants en zone déferlante, ce qui est particulièrement rare. On a pu ainsi montrer que le débit de transport dans la barre parallèlement au rivage était de 3 à 4 fois supérieur au transport d'estran, et le transport total 10 fois supérieur au transport d'estran.

Cette conjoncture d'observations est assez rare; notamment en régime atlantique les observations des transports en zone déferlante sont extrêmement difficiles, voire impossibles. Nous cherchons actuellement à mettre en place des moyens auxiliaires qui permettent de telles observations.

6. QUELQUES NOTIONS SUR LE DEPLACEMENT SUR L'ESTRAN

L'estran est caractérisé par des modalités de transport tout à fait particulières qui se traduisent non seulement par des vitesses de transport mais surtout par des épaisseurs de transport particulièrement importantes, et apparemment d'autant plus grandes que le matériau est plus grossier.

Nous n'en voulons pour preuve que les trois exemples suivants.

Dans la campagne des Bas-Champs (campagne n° 1) le matériau en transit était constitué de galets de 5 à 15 cm de diamètre. Les galets marqués (fil de ^{182}Ta, 100 μCi par galet) ont été suivis pendant six mois d'hiver et l'épaisseur moyenne de transport a été trouvée égale à 55 cm ± 9 cm, les houles observées pendant cette époque ayant atteint 3 m au maximum, 50% étant comprises entre 0,3 et 0,6 m. Le transport des galets pendant cette période a eu lieu sur un kilomètre et le transit résultant annuel était de 23 000 tonnes, valeur en accord avec une estimation de cubage des épis et qui correspondrait à un calcul par le bilan énergétique des houles avec un facteur K $\sim$ 1,5 $\times$ 10^{-4}.

Dans la campagne de l'Adour (campagne n° 5 bis) les matériaux tracés étaient de la madrague, matériau grossier de 2,5 à 4 mm de diamètre. Ces matériaux avaient été immergés au large par des profondeurs de − 5 m environ. Ils ont été retrouvés à la côte sur des épaisseurs de 40 cm (houle de 2 à 3 m).

Enfin, dans une expérience réalisée sur le massif de la Coubre (campagne n° 3), où le matériau en déplacement sur l'estran était un sable de 0,2 à 0,25 mm de diamètre, tant par traceurs fluorescents que par traceurs radioactifs, on a été conduit à admettre une couche mobile de 8 cm.

Ces phénomènes ont également été retrouvés dans d'autres expériences, non rapportées ici. Bien qu'on ne puisse leur associer une loi, ils fournissent quand même une indication à caractère général qu'il nous semblait utile de préciser.

7. CONCLUSIONS

Les expériences de traceurs radioactifs en sédimentologie rapportées ici ont permis de mettre en évidence des phénomènes d'ordre général, utiles à la compréhension de l'action de la houle sur les fonds:

a) La limite d'action de la houle se situe vers des profondeurs de 12 à 15 m. Non point qu'au-delà de cette profondeur les mouvements résultants soient nuls: ils existent mais ont des valeurs très faibles (de l'ordre de 1% du charriage nominal au au niveau de la barre).

b) Le transport parallèle à la côte a lieu sur la ou les barres et sur l'estran. La barre assure environ 75% du débit de transport (c'est-à-dire du débit par mètre de largeur) et la région de l'estran environ 25%.

c) Les matériaux grossiers peuvent être amenés à être transportés sur l'estran avec des épaisseurs relativement importantes pouvant dépasser 50 cm.

d) Enfin les résultats généraux obtenus permettent de tarer les formules de transport sédimentaire par dissipation de l'énergie des houles.

Ainsi, pour le transport dans le profil on trouve des valeurs de K, coefficient de proportionnalité entre le débit de charriage et l'énergie des houles, de l'ordre de 10^{-7} à 10^{-8} entre -8 m et -15 m en milieu soumis à marée, et des valeurs environ 10 fois plus faibles à même profondeur pour des mers non soumises à marée.

Pour le transport parallèle au rivage et sur l'estran, les valeurs de K trouvées satisfont bien à une relation de la forme

$$K = 0,18 \times 10^{-5} \; D^{-1/2} \qquad (\text{D en mm})$$

Ces résultats, pour importants qu'ils puissent apparaître, ne font que renforcer les connaissances déjà acquises et apporter une pierre à l'édifice des données considérables sur l'action de la houle sur les fonds marins. Une très importante étude bibliographique en a été récemment faite [2], à laquelle le lecteur pourra se reporter.

Les traceurs radioactifs sont maintenant entrés dans la pratique de telles études. On aurait tort de les considérer comme une panacée. Ce n'est en rien diminuer leur mérite extraordinaire que de reconnaître qu'ils ne sont ni d'application universelle, ni toujours le mieux adaptés à la solution d'un problème donné. Dans le cas particulier de l'action de la houle sur les fonds marins, ils ont permis et continuent de permettre d'accroître les connaissances dans un domaine difficile.

REMERCIEMENTS

Ont également participé à des titres divers à ces expériences: le Service central technique du Ministère de l'équipement (M. Graillot), le Centre de recherches et de sédimentologie marine (Mme Duboul-Razavet, M. Barusseau), M. Bellessort, du LCHF, et MM. Caillot, Quesney, Jeanneau, Tanguy et Massias, du CEA (Section d'application de la radioactivité), ainsi que les Services du Port de Sète et de Bayonne. Qu'ils en soient ici remerciés.

REFERENCES

[1]   ZENKOVICH, V.P., Emploi des luminosphères pour l'étude du mouvement des alluvions sableuses, Bulletin du COFC X 5, Service hydrographique de la Marine, Paris (1958).

[2]   LABORATOIRE CENTRAL D'HYDRAULIQUE DE FRANCE, Action de la houle sur les sédiments, Publication du Centre national d'exploitation des océans (CNEXO).

THE TRITIUM/HELIUM-3 METHOD
IN HYDROLOGY

T. TORGERSEN
Woods Hole Oceanographic Institute,
Woods Hole, Massachusetts,
United States of America

W.B. CLARKE
McMaster University, Hamilton,
Ontario, Canada

W.J. JENKINS
Woods Hole Oceanographic Institute,
Woods Hole, Massachusetts,
United States of America

Abstract

THE TRITIUM/HELIUM-3 METHOD IN HYDROLOGY.
The ^{3}H-^{3}He parent-daughter radiotracer pair is discussed in relation to various hydrological systems. The criteria for evaluating mass-spectrometrically obtained ^{3}H-^{3}He measurement and the procedures for calculating a ^{3}H-^{3}He age are described. This age constitutes the mean gas residence time for open systems and the water residence time for closed systems. These measurements can be used to calculate a wide variety of dynamic parameters, and examples of net supply rates, gas exchange rates and vertical diffusivity calculations are given. Data for two meromitic lakes are presented and discussed with respect to groundwater influence. In one lake, groundwater can be ruled out; in the other it contributes a major source. Possible other uses of this ^{3}H-^{3}He tracer pair in hydrological systems are discussed.

INTRODUCTION

One of the major tools of modern geochemistry has been the use of stable and radioactive tracers to elucidate the time parameters of earth processes. Not only are naturally occurring isotopes being extensively used, but also artificial ones, especially those created by atmospheric nuclear weapons testing in the 1950's and 1960's.

One of the tracers made available by these weapons tests is the radioactive isotope of hydrogen, ^{3}H (tritium). Although present naturally at low levels, the increase caused by weapons testing allowed a much broader and more precise set of studies to be made, among these are groundwater storage and flow [1-3] surface ocean mixing [4-7] and the dynamics of lake water, balances [8,9].

917

More recently, the technology to accurately measure ^{3}He, the stable daughter product of ^{3}H, has been developed. While the previously discussed ^{3}H experiments required either large-scale or slowly acting processes, the combined use of ^{3}H-^{3}He is not so restricted. Furthermore, the ^{3}H-^{3}He method usually allows an assessment of process dynamics to be made without an extended time-series study.

THE ^{3}H-^{3}He METHOD

The method involves the collection of about 10 grams of water. The dissolved gases are extracted and purified. ^{3}He, ^{4}He, and Ne are then measured mass spectrometrically to determine the concentrations in cc STP of gas per gram of water. The remaining water having been totally degassed is sealed in vacuo in an alumino-silicate breakseal vessel for ^{3}H analysis by the ^{3}H grow-in method of [10]. Oftentimes it is convenient to report the helium data as a helium saturation anomaly:

$$\Delta^4He \ \% \ = \ \left\{ \frac{[^4He]}{[^4He]_{sat.}} - 1 \right\} \ x \ 100 \tag{1}$$

the per cent deviation from saturation at that temperature and pressure, and the isotope ratio anomaly;

$$\delta^3He \ \% \ = \ \left| \frac{(^3He/^4He)}{(^3He/^4He)_{air}} -1 \right| \ x \ 100 \tag{2}$$

the per cent deviation of the measured isotopic ratio from the air standard $(^3He/^4He)_{air}$ = 1.384 x 10^{-6}, [10]). Water at equilibration with air has a $\delta^3He \ \%$ = -1.4 [11].

The following criteria are then used to separate the various helium components, (the subscripts tot, sat and air refer to the total concentration, the saturation concentration [12] and the air injection concentration):

(1) Neon is assumed to contain only a saturation component and an air injection component

$$Ne_{tot} = Ne_{sat} + Ne_{air} \tag{3}$$

(2) Helium-4 is assumed to represent a mixture between that present by saturation, air injection, ^4He_c, a crustal helium component ($^3He/^4He \sim 10^{-7}$), and ^4He_m, a mantle helium component ($^3He/^4He \sim 10^{-5}$). These last two components are assumed not to occur together. Hence,

$$^4He_{tot} = {}^4He_{sat} + {}^4He_{air} + {}^4He_c + {}^4He_m \tag{4}$$

In the well-defined case where either ^4He_c or ^4He_m may be present (but not both)

$$^4He_{air} = (He/Ne)_{air} (Ne_{tot} - Ne_{sat})$$

where the volume ratio $(He/Ne)_{air}$ is known to be 0.288 and $^4He_{sat}$ is known [12]. The third component is then calculable.

(3) ^3He is assumed to have one additional helium component, $^3He_{tri}$, that from the β^- decay of tritium.

$$^3He_{tot} = {}^3He_{sat} + {}^3He_{air} + {}^3He_c + {}^3He_m + {}^3He_{tri} \qquad (5)$$

In this case $^3He_{air} = ({}^3He/{}^4He)_{air} ({}^4He_{air})$ and likewise for ^3He_c or ^3He_m.

In many cases the Ne measurement can be eliminated and the $^3He/^4He$ ratio can be used to define a binary ^4He mixture

$$^4He_{tot} = {}^4He_{sat} + {}^4He_x$$

where $\qquad ^4He_x = {}^4He_{air}$ if $^3He/^4He \sim 10^{-6}$

$\qquad\qquad\qquad = {}^4He_c$ if $^3He/^4He \le 10^{-7}$

$\qquad\qquad\qquad = {}^4He_m$ if $^3He/^4He \sim 10^{-5}$

Provided a range of samples can be obtained (as is the usual case for a lake), plots of ^3He vs. ^4He can be used to determine the helium isotopic ratio and hence ^4He_x. $^3He/^4He$ vs. ^4He and $^3He/^4He$ vs. $[^4He]^{-1}$ plots are used to test this binary assumption (a parabola and a straight line respectively confirm a binary mixture assumption, [13]).

Using this analysis scheme, $^3He_{tri}$ can be determined and a water mass age calculated according to the equation

$$t = 17.69 \ln \left| \frac{4.01 \, [^3He_{tri}]}{[^3H]} \times 10^{14} + 1 \right| \qquad (6)$$

where t is the age in years, $[^3He_{tri}]$ is the tritiugenic helium-3 concentration in cm^3 at standard temperature and pressure and $[^3H]$ is the tritium concentration in tritium units (1 T.U. = 1 $^3H/10^{18}$ ^1H). Because the $^3H-^3He$ age represents the effective time since water parcel equilibration with the atmosphere, this method supplies the time parameter necessary to compute a wide variety of physical parameters in hydrologic and limmologic systems. The method is, however, subject to constraints involving the assumptions about the number and ratio of the He components and also to interpretational problems due to the mixing of different waters. The "ages" obtained by this method should therefore not be taken at face value but reviewed in the context of the system being studied.

NET SUPPLY RATES

Any situation in which chemical species are removed or supplied to a water parcel under conditions similar to helium for either a closed system or an open system, the net supply rate of that species can be calculated. For example, if we assume that conditions that bring oxygen in equilibrium with air also bring helium to equilibrium, the net oxygen consumption rate for a closed system hypolimnion can be calculated. [14] report ^{3}H-^{3}He ages for Lake Erie during September 4-6, 1973, and a mean hypolimnion age of 102 days was found for Station "A." During that cruise, Dr. N. M. Burns found the average oxygen concentration to be 0.19 mg/l and the oxygen depletion rate over a 26-hour period to be 0.098 mg O_2 l^{-1} day^{-1} (Burns, pers. comm.). If both He and oxygen equilibrated to the same degree during spring circulation, and if the O_2 depletion rate was constant, the net summer season O_2 depletion rate is

$$\frac{[O_2]_{sat} - [O_2]}{age} = \frac{(10-0.19) \text{ mg } l^{-1}}{102 \text{ days}} = 0.096 \text{ mg } l^{-1} \text{ day}^{-1}$$

in agreement with that measured directly by Burns.

The problem posed by [15] in Teggau Lake is a similar problem. In this case, crustal helium was found in the lake such that the closed system hypolimnion (the deepwater below the thermocline, volume 1.4 x 10^8 m^3) contained 2 x 10^7 cm^3 STP ^4He from crustal sources. Using the mean hypolimnion age of 302 ± 100 days, the input flux of crustal He was found to be $\sim$ 2 x 10^7 cm^3 STP ^4He/year.

GAS EXCHANGE RATES

For a steady-state epilimnion, the surface boundary condition determines the gas exchange rate. This point is determined by the atmospheric equilibrium of ^3He and also by the gas exchange coefficient. The flux equation is:

$$\text{Flux} = k_{He} \left([^3He]_o - [^3He]_{equil} \right) \qquad (7)$$

where k_{He} is the gas exchange coefficient, m/day

$[^3He]_{equil}$ is the ^3He concentration of water at atmospheric equilibrium,

$[^3He]_o$ is the ^3He concentration at the surface.

The total gas exchange flux for an entire epilimnion (depth = z_{epi}) at steady-state is:

$$\text{Flux} = A_o (z_{epi}) + F$$

where A_o is the production rate per unit volume. In most cases F, the flux from the thermocline is negligible, and the equation reduces to:

$$\text{Flux} = A_o z_{epi} \qquad (8)$$

Balancing the flux of the epilimnion (Equation (8)) and the transport Equation (7)

$$k_{He} = A_o z_{epi} \Big/ \left([^3He]_o - [^3He]_{equil} \right)$$

However,

$$\left([^3He]_o - [^3He]_{equil} \right) \Big/ A_o$$

constitutes the mean residence time of ^3He at $z = 0$ and the effective age. Therefore

$$k_{He} = z_{epi}/age_{z=o} \qquad (9)$$

To increase the precision of the gas exchange calculation by the ^3H-^3He method, the average epilimnion age can be used. This is equivalent to assuming the epilimnion to be a well-mixed layer.

VERTICAL DIFFUSIVITY

The case of the steady-state epilimnion is described where the rate of supply of ^3He by the decay of ^3H is exactly balanced by the loss of ^3He by diffusion. If K_v, the vertical eddy diffusion coefficient, is assumed to be constant, with $\partial[^3He_{tri}]/\partial t = 0$, the descriptive equation is:

$$+K_v \frac{\partial^2 [^3He_{tri}]}{\partial z^2} = A_o \qquad (10)$$

If input to the epilimnion by diffusion across the thermocline is assumed to be zero, Equation (10) can be solved with the conditions

$$K_v \frac{\partial[^3He_{tri}]}{\partial z} = 0 \text{ at } z = z_{epi}$$

$$[^3He_{tri}] = [^3He_{tri}]_o \text{ at } z = 0$$

to give

$$[^3He_{tri}] - [^3He_{tri}]_o = \frac{A_o z_{epi} z}{K_v} - \frac{A_o z^2}{2 K_v} \qquad (11)$$

where $[^3He_{tri}]$ is the surface tritiugenic ^3He value, and z_{epi} is the depth. If it is desired to work in terms of δ^3He % rather than $[^3He_{tri}]$ the approximate production term is:

$$A_o = \frac{\partial \delta}{\partial t} = \frac{T}{^4He} (2.768 \times 10^{-3})$$

Equation (11) allows us to match epilimnion ^3He profiles to corresponding values of vertical eddy diffusivity as is shown in Figure 1 for Lake Huron taken from [14].

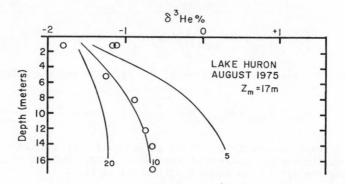

FIG.1.   *Profiles of $\delta^3He\%$ versus depth for the epilimnion of Lake Huron, August 1975,
compared with curves of constant K_v, m/d. A value of $z_{epi} = 17$ m (z_m) and
$[^4He] = 4.65 \times 10^{-8}$ cm^3 STP/g was used to calculate the curves from Eq.(11).*

The discussion will now leave the more general cases (see [14], for a
more complete discussion) to examine specific measurements in two meromictic lakes (lakes which only partially mix at the autumnal and vernal circulation periods).

LAKE 120

Lake 120 is a meromictic lake near Kenora, Ontario, Canada. The
water below 15-16m is continually stratified by high concentrations of
iron (1260 µmoles/1), dissolved inorganic nitrogen (822 µmoles/1) and dissolved inorganic carbon (7353 µmoles/1) roughly half as as ΣCO_2 and half
as CH_4 [16].

Table I shows ^3H-^3He data collected in this lake over a three-year
period. The lake is also obviously stratified with respect to ^3H and ^3He.
Mean ^3H-^3He ages of 3 ± 1.5 years are calculated below 15m while the
waters above this strata reach a maximum of 170 days.

Groundwater is often a cause of meromixis and, in this case, its effect can be assessed. As seen in Figure 2, the excess helium anomaly in
Lake 120 appears along an air ratio line ($^3He/^4He \sim 1.4 \times 10^{-6}$) and not
the crustal line ($^3He/^4He \leq 10^{-7}$) as would be expected if a groundwater
input were dominant. The dynamics of this meromictic lake can be summarized in the following manner: Water supersaturated in atmospheric helium
is injected at approximately 15m (a cooling center also occurs at this
depth during June-July to substantiate this injection, [16], and is incorporated in the 15-19m water. The ^3H thus injected decays to ^3He over the
$\sim$ 3-year residence time and is then removed by mixing with the upper mixo-
limnion water. Because the $^3He/^4He$ ratio in this injected water is atmospheric, a near surface water source is suggested and not a deep groundwater source.

TABLE I. ³H-³He DATA FOR MEROMITIC LAKE 120

Note that in some cases samples were stored in the sampling vessel for up to 50 days. The ages have been corrected for this storage

	Depth (m)	Temp (°C)	^{4}He cm³STP/g x 10^8	^{3}He cm³STP/g x 10^{14}	Δ^4He%	δ^3He%	Tritium[+] T.U.	Age (days)	Storage	Age Corrected
8/6/72	0.5	19.0	4.63	6.94	+ 3.1	+ 7.2	(368)*	39	41	- 2
	3	19.0	3.87	5.91	-13.8‡	+ 9.3	(368)*	-	47	—
	6	11.0	4.84	9.69	+ 4.7	+43.2	(368)*	212	47	165
	13	4.8	8.87	15.34	+86.3	+23.7	(368)*	218	47	171
	17	4.8	6.99	24.07	+46.9	+146.3	(390)*	956	47	909
7/30/73	0	19.0	5.80	7.98	+29.2	-1.55	(260)	- 1.2	-	- 1.2
	1	19.0	5.16	7.11	+14.9	-1.44	260	- 0.3	-	- 0.3
	2	19.0	7.31	10.13	+62.8	-0.85	(260)	5.6	-	5.6
	3	19.0	4.92	6.82	+ 9.6	-0.91	(260)	3.4	-	3.4
	4	19.0	10.57	14.78	+135.4++	+0.03	(260)	—	-	—
	5	15.0	4.38	6.12	- 3.7‡	-0.04	(271)	—	-	—
	6	11.0	5.09	8.05	+10.2	+13.18	280	95.8	-	95.8
	7	8.5	5.84	8.85	+25.1	+8.44	(280)	74.2	-	74.2
	11	5.5	4.98	8.15	+ 4.8	+17.13	(283)	117.8	-	117.8
	14	4.8	4.96	7.66	+ 4.2	+10.45	284	74.8	-	74.8
	15	4.8	8.59	12.53	+80.5	+4.37	267	67.1	-	67.1
	17	4.8	7.33	18.31	+54.0	+78.66	343	618	-	618
9/5/74	2	13.7	4.6	6.44	+ 0.7	+ 0.2	(197)	13.7	5	8.7
	4	13.7	4.4	6.13	- 3.7‡	- 0.3	(197)	-	6	-
	6	11.8	5.1	7.25	+10.6	+ 1.7	197.3	28.9	7	21.9
	8	7.5	5.6	9.14	+19.2	+16.8	(211)	175	6	169
	10	6.2	5.5	8.67	+16.3	+12.7	(215)	131	6	125
	12	5.1	6.3	9.77	+32.6	+10.9	(219)	128	6	122
	13	4.9	6.8	10.30	+42.9	+ 8.4	219	110	7	103
	14	4.8	6.9	10.44	+45.0	+ 8.3	(260)	93	6	87
	15	4.7	8.3	19.03	+74.0	+64.0	(302)	649	6	843
	16	4.7	8.5	23.52	+78.2	+98.0	301.8	1010	7	1003
	17	4.7	7.3	26.64	+53.0	+161.0	(302)	1418	5	1413
	18	4.7	5.3	13.78	+11.1	+86.0	(302)	554	7	547

*   Extrapolated

+   Parenthesis indicate interpolated values.

‡   Incomplete extraction.

++   Probable air contamination on sampling

Using the 3 ± 1.5 year residence time of the monimolimnion water, monimolimnion volume of 19.8 x 10^6 l [16] and the mean iron concentration 1260 µmoles/l, an iron supply rate of 8.30 kmoles Fe/yr is calculated. This value is seven times the net sedimentation of 1.19 kmoles Fe/yr [16]. Since sedimentation is the sink for Fe in this lake, the discrepancy in the supply rate (8.30 kmoles/yr) and the removal rate (1.19 kmoles/yr) for the 15-19m

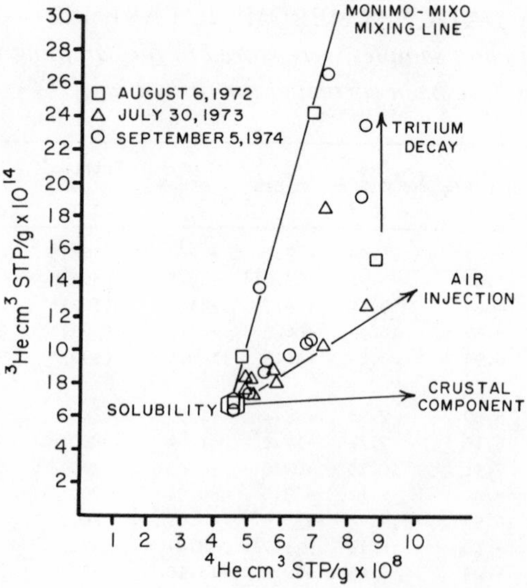

FIG.2. Helium isotope data for Lake 120. Notice the air injection-solubility mixing line which produces the air injection component at 15 m. Tritium decay then raises the ^3He concentration of the monimolimnion while circulation-induced mixing completes the cycle along the monimo-mixo mixing line.

TABLE II. ^3H-^3He DATA FOR MEROMITIC GREEN LAKE, OCTOBER 29, 1974
Helium concentrations have an error of ± 3% and tritium concentrations are ± 1 TU. Excess ^3He is defined as $^3He_{tot}$-$^3He_{sat}$. $\Delta^4He\%$ and $\Delta^3He\%$ are defined as in Eq.(1)

Depth m	Temp. °C	^4He cm³STP/g x 10^8	$\Delta^4He\%$	^3He cm³STP/g x 10^{14}	$\Delta^3He\%$	Excess ^3He cm³STP/g x 10^{14}	T T.U.
1	11.6	8.06	74.8	6.87	46.15	0.52	107
7	11.6	8.14	76.6	6.86	50.77	0.51	107
13	12.0	8.47	84.1	6.85	58.81	0.51	113
15	13.7	81.4	1677	10.23	127.9	3.92	80.2
17	10.6	84.9	1734	9.07	99.10	2.69	69.5
20	8.2	110.6	2261	10.03	114.8	3.58	46.5
23	7.8	118.0	2416	9.45	101.3	2.99	35.5
26	7.7	126.3	2593	9.63	105.6	3.17	25.3
30	(7.7)	146.4	3022	9.53	105.6	3.07	23.3
35	(7.7)	161.8	3350	10.58	129.4	4.12	20.9
40	(7.7)	199.1	4145	10.20	120.8	3.74	16.9
47	(7.7)	224.1	4678	10.62	129.4	4.16	15.0

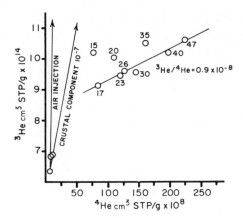

FIG.3. ^3He *vs* ^4He *for Green Lake, October 29, 1974. The helium data plot on a line whose* $^3He/^4He$ *ratio is 0.9 × 10^{-8} indicative of groundwater sources rich in crustal helium. Discrete sources are apparent at 15, 20 and 35 m. (Numbers indicate sample depth.)*

water can only be rationalized by recycling the Fe approximately seven times across the 15m boundary by the reactions involving oxidation of Fe II to Fe III in the oxygenated water above 15m, subsequent precipitation of Fe(OH)$_3$ and reaction of Fe(OH)$_3$ to Fe II upon reentry to the anaerobic deep water.

GREEN LAKE

Green Lake is another meromictic lake located near Syracuse, New York, U.S.A. Previous workers have attributed its meromictic nature to highly saline groundwater inputs rich in Na$^+$, K$^+$, Cl$^-$, Mg^{2+}, Ca^{2+}, HCO$_3^-$ and SO$_4^{-2}$ and estimated the monimolimnion water residence time to be 4 to 30 years, [17].

Table II shows ^3H and helium data collected in this lake. Distinct changes are obvious at the 15m chemocline (stratification boundary). Figure 3 shows that the helium data plot on a line whose $^3He/^4He$ ratio is 0.9 x 10^{-8} indicative of groundwater sources rich in crustal helium. Discrete sources are apparent at 15m, 20m and 35m. Figure 4 shows the data are consistent with a binary helium mixture and Table III gives the calculated ^3H-^3He ages, 4-9 years for the monimolimnion. Figure 5 plots these ages on a ^3H-^3He evolution diagram, and it can be seen that the monimolimnion water lies on mixing lines consistent with an endmember containing 0 T.U. and 2.2 x 10^{-14} cm^3 STP $^3He/g$, a value that would be expected for dead pre-bomb water (7 T.U.). This analysis makes it clear why low ^3H and high helium values are present in the monimolimnion and why the abundances and ratios indicate strong groundwater interactions.

TABLE III. ^{3}H-^{3}He AGE PROFILE FOR GREEN LAKE

Using a $^3He/^4He$ ratio of 0.9×10^{-8} for the groundwater component, the calculated age has an error of ± 5%.

Depth m	Total Excess ^{3}He cm³STP/gx10¹⁴	Groundwater ^{3}He a cm³STP/gx10¹⁴	Net Excess ^{3}He cm³STP/gx10¹⁴	Tritium T.U.	Age years
1	0.52	.07	.45	107	0.30
7	0.51	.07	.44	107	0.29
13	0.51	.07	.44	113	0.27
15	3.92	.73	3.19	80.2	2.62
17	2.69	.76	1.93	69.5	1.87
20	3.58	1.00	2.58	46.5	3.56
23	2.99	1.06	1.93	35.5	3.49
26	3.17	1.14	2.03	25.3	4.94
30	3.07	1.32	1.75	23.3	4.66
35	4.12	1.46	2.66	20.9	7.30
40	3.74	1.79	1.95	16.9	6.73
47	4.16	2.02	2.14	15.0	9.01

a   Excess ^{4}He x $0.9x10^{-8}$

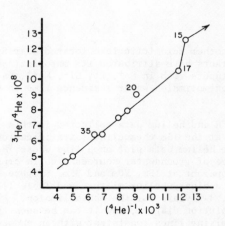

FIG.4. $^3He/^4He$ vs $(^3He)^{-1}$ for Green Lake showing the data are consistent with the binary mixing assumption. The break between 15 m and 17 m marks the chemocline.

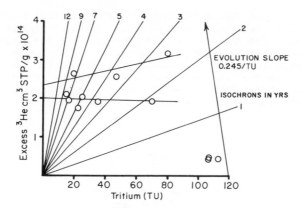

FIG.5. *The evolution diagram for the* ^3H-^3He *data of Green Lake. The monimolimnion water is seen to be consistent with an end-member containing 0 TU and* 2.2×10^{-14} cm^3 *STP excess* $^3He/g$, *a value that would be expected for dead pre-bomb water (7 TU). The evolution slope indicates the slope of closed-system decay. Isochrons are in years.*

OTHER POSSIBLE APPLICATIONS

Anywhere that water resides, the ^{3}H-^{3}He method can be used to calculate a model age which may be interpretable as a residence time, provided
sampling and measurements can be done. In closed systems, this model age
is equivalent to the ^{3}H residence time in the system. In semi-closed or
open systems, the model age is equivalent to the helium residence time
and therefore is a lower limit on the water residence time. It should
therefore be possible to extend this method to direct groundwater dating,
subpermafrost groundwater dating and possibly permafrost dating.

In groundwater and subpermafrost groundwater, high ^{4}He concentrations
and a crustal ratio would be expected as was seen in Green Lake. It may
not be possible under sampling conditions available to obtain a range of
samples sufficient to accurately define the ^{3}He/^{4}He ratio. However, it is
known that the crustal ratio is constrained, and a value of 10^{-8} - 10^{-7} can
be used with some degree of accuracy. Care should be exercised in the analysis and interpretation and a "maximum" ^{3}H-^{3}He age assuming ^{3}He$_{tot}$ =
^{3}He$_{sat}$ + ^{3}He$_{tri}$ can be calculated. It should be noted that very old low
^{3}H water mixed with new water will result in a bias towards the younger
average age [18]. Plots similar to Figure 5 should facilitate this type
of analysis.

Permafrost conditions may or may not be closed with respect to ^{3}H, but
are probably open with respect to ^{3}He. Provided uncontaminated samples can
be obtained, a ^{3}H-^{3}He model age might be obtained. Since this age would
reflect the gas residence time in the permafrost, it may be applicable to
problems of vapor transport.

928   TORGERSEN et al.

We conclude that the ^{3}H-^{3}He method is a useful tool for hydrology and
can be used to calculate gas exchange rates, diffusivities and water mass
residence times. The ^{3}He/^{4}He ratio is also valuable for identifying
sources of excess helium which may be broadly categorized as crustal (10^{-8}
- 10^{-7}), atmospheric ($\sim 10^{-6}$), or mantle ($\sim 10^{-5}$). Caution should be exercised, however, in the interpretation of data and ^{3}H-^{3}He ages, since the
model ages are subject to a number of complicating factors arising from
mixing and variations in the added helium components.

ACKNOWLEDGEMENTS

The work was supported by NSF Grant OCE76-81774 to Woods Hole Oceanographic Institution and by NSF and ERDA Grants GB36348, GA33124, E2185,
and NGR33-008-91 to Lamont-Doherty Geological Observatory. Woods Hole
Oceanographic Institution Contribution Number 4140.

REFERENCES

[1]   ERIKSSON, E., The possible use of tritium for estimating groundwater
        storage, Tellus 10 4 (1958) 472-478.

[2]   BROWN, R.M., Hydrology of tritium in the Ottawa Valley, Geo. Cosmo.
        et Acta 21 3/4 (1961) 199-216.

[3]   BLUME, H.P., ZIMMERMAN, V., MUNNICH, K.O., "Tritium tagging of soil
        moisture: The water balance of forest soils" in Isotope and Radia-
        tion Techniques in Soil Physics and Irrigation Studies.   IAEA,
        Vienna (1967).

[4]   ROOTH, C.G., ÖSTLUND, H.G., Penetration of tritium into the Atlantic
        thermocline, Deep-Sea Res. 19 (1972) 481-492.

[5]   ROETHER, W., The tritium and carbon-14 profiles of the GEOSECS I
        (1969) and GOGO I (1971) North Pacific stations, Earth and Planet.
        Sci. Letters 23 (1974) 108-115.

[6]   ÖSTLUND, H.G., DORSEY, H.A., ROOTH, C.G., GEOSECS North Atlantic
        radiocarbon and tritium results, Earth and Planet. Sci. Letters 23
        (1974) 69-86.

[7]   MICHEL, R.L., SUESS, H.E., Bomb tritium in the Pacific, J. Geophys.
        Res. 80 (1975) 4139-4152.

[8]   SIMPSON, H.J., Closed basin lakes as a tool in geochemistry. Ph.D.
        Thesis, Columbia University (1970) 325 p.

[9]   IMBODEN, D.M., WEISS, R.F., CRAIG, H., MICHEL, R.L., GOLDMAN, C.R.,
        Lake Tahoe geochemical study 1. Lake chemistry and tritium mixing
        study, Limnology and Oceanography 22 6 (1977) 1039-1051.

[10]   CLARKE, W.B., JENKINS, W.J., TOP, Z., Determination of tritium by
        mass spectrometric measurement of ^{3}He, Inter. Jour. of App. Radi-
        ation and Isotopes 27 (1976) 515-522.

[11]   WEISS, R.F., Helium isotope effect in solution in water and seawater,
        Science 168 (1970) 247.

[12]   WEISS, R.F., Solubility of helium and neon in water and seawater,
        J. Chem. Eng. Data 16 (1971) 235.

[13]   BELL, K., POWELL, J.L., Strontium isotopic studies of alkalic rocks:
        the potassium-rich lavas of the Birunga and Toro-Ankole regions,
        east and central equatorial Africa. J. Petrol. 10 (1969) 536-572.

[14]   TORGERSEN, T., TOP, Z., CLARKE, W.B., JENKINS, W.J., BROECKER, W.S.,
        A new method for physical limnology - tritium-helium-3 ages - results
        for Lakes Erie, Huron and Ontario, Limnology and Oceanography 22
        (1977) 181-193.

[15]   TORGERSEN, T., CLARKE, W.B., Excess helium-4 in Teggau Lake: Possi-
        bilities for a uranium ore body, Science 199 17 (1978) 769-771.

[16]   CAMPBELL, P., Descriptive limnology of Lake 120, a meromictic lake
        on the preCambrian shield in northwestern Ontario. M.S. Thesis,
        Dept. of Zoology, University of Manitoba, Winnipeg, Manitoba, Canada
        118 p.

[17]   TAKAHASHI, T., BROECKER, W.S., LI, Y.-H., THURBER, D.L., Chemical
        and isotopic balances for a meromictic lake, Limnology and Oceanog-
        raphy 13 2 (1968) 272-292.

[18]   JENKINS, W.J., CLARKE, W.B., The distribution of ^{3}He in the western
        Atlantic Ocean, Deep-Sea Res., 23 (1976) 481-494.

DISCUSSION

P. FRITZ: Could you please comment on the significance of Fig.4?

T. TORGERSEN: Figure 4 merely justifies the assumption of a binary ^{4}He mixture. I refer you to Ref. [13] for the mathematics.

U. SCHOTTERER: In connection with Fig.2, did you try to identify the surface water admixture using ^{18}O or chemistry?

T. TORGERSEN: The cold runoff injection observed in the ^{3}H and helium data and the ^{3}H-^{3}He ages are supported by δD measurements. Water at 14−16 m is about 6‰ lighter in deuterium than water above or below this level. Furthermore, chemical data (Campbell [16]) indicate that this injected water more closely resembles runoff than groundwater.

U. SCHOTTERER: We think that the ^{3}He/^{4}He method will also prove to be a very useful tool in glaciology.

T. TORGERSEN: I would certainly like to attempt some glacier work with the method, and I agree that it may have potential.

K.E. WHITE: What reaction do you get from water supply and planning authorities when you suggest revision of age estimates by several orders of magnitude?

T. TORGERSEN: I think that the method has distinct potential for water-supply problems. So far, however, I have not had the opportunity to use it in this field.

IAEA-SM-228/50

ARGON-39, CARBON-14 AND KRYPTON-85 MEASUREMENTS IN GROUNDWATER SAMPLES

H.H. LOOSLI, H. OESCHGER
Physikalisches Institut der
 Universität Bern,
Bern, Switzerland

Abstract

ARGON-39, CARBON-14 AND KRYPTON-85 MEASUREMENTS IN GROUNDWATER SAMPLES.

The ^{39}Ar and ^{85}Kr activities were measured in the gas samples extracted out of 12 groundwater wells. In addition, ^{3}H, (^{32}Si), ^{14}C, δ^{13}C and δ^{18}O measurements and chemical analyses were performed. For a confined sandstone aquifer in the Franconian Albvorland, Federal Republic of Germany, decreasing ^{39}Ar activities (from 100 to 17% modern) were found; the ^{14}C activities correspondingly decrease from 80 to 0.3% modern. The calculated ^{39}Ar and ^{14}C ages disagree. The discrepancies, especially for the older samples, cannot be explained by a simple model assuming mixing of waters of different ages. Arguments against and in favour of both dating methods are discussed.

INTRODUCTION

For the cosmic-ray-produced atmospheric ^{39}Ar activity a value of 0.112 ± 0.012 dis/min[1] ^{39}Ar/l argon has been measured [1]. A man-made contribution, mainly due to nuclear weapon tests, cannot be excluded, but it is of minor importance since measurements show that modern argon samples (up to 1975) contain less than 5% man-made ^{39}Ar activity when compared with argon collected in 1940.

^{39}Ar ($T_{1/2} = 269$ a) was first applied to dating of glaciological samples. The ^{39}Ar data of two depth profiles (from Greenland and from Antarctica) show agreement with ages obtained from ^{14}C measurements or from ^{18}O/^{16}O profiles [2, 3]. The upper limit of the ^{39}Ar dating range is about 1200 years, which was reached for some of the ice samples.

---

[1] In this paper disintegrations (dis/min) have been used. In the S.I. system 1 disintegration per second = 1 Bq.

931

Application of the ^{39}Ar dating method to hydrological problems was started in 1973. The first results obtained for one groundwater sample and for water from a thermal spring showed large discrepancies between ^{39}Ar and ^{14}C data [4]. To see whether or not these discrepancies would be substantiated and to find possible explanations, new samples were collected. In particular, a horizontal profile of samples in the are between Hilpoltstein and Beilngries in the Franconian Albvorland in southern Fed. Rep. of Germany was taken. This confined aquifer was selected because it is well studied by hydrological and isotopic methods [5−7]. The aquiclude is up to about 100 m thick and the ^{14}C ages (corrected for 85% modern) increase from about 800 to about 25000 years.

The aim of this paper is to present recently measured data. The set of isotopes, which usually consists of ^3H, ^{14}C, $\delta^{13}C$ and $\delta^{18}O$, could be increased by adding ^{39}Ar, ^{85}Kr and, for some of the samples, ^{32}Si.

1. EXPERIMENTAL PROCEDURE

The water is outgassed by heating it up to about 95°C and the dissolved gases are extracted in a closed system at a small overpressure to avoid contamination by air. The extraction system, which was described earlier [4], could be improved. The outgassing of about 15 m^3 of water for one sample takes now about 8 hours. Typical values for the amount of gases obtained are between 13 and 20 litres gas/m^3 water, showing that the extraction yield for the dissolved air is between 50 and 85%.

The large quantities of CO_2 in the gas are removed by passing it through a trap with a molecular sieve, 5 A at room temperature. The CO_2 can be recovered by heating the molecular sieve and then used for ^{14}C measurements, but the CO_2 samples obtained by this procedure show isotopic fractionation. The separation of argon and krypton from the extracted gases was described earlier [4]. Argon and krypton are separated and purified by distillation, gas-chromatography and chemical procedures. At the beginning of the procedure 10 cm^3 of pre-bomb krypton are added as carrier. The separation yield is about 90% for argon and about 50% for krypton. The present capacity of our laboratory allows for the processing of about 20 samples per year for ^{39}Ar measurements.

The ^{39}Ar and ^{85}Kr activities are measured by proportional counting. Considerable improvement could be reached owing to the availability of an underground laboratory with a depth of about 70 m water equivalent [2]. The obtained reduction of the background allows the ^{39}Ar activity to be measured even in 1-litre samples (about half the amount used before). This means that in future "only" about 7−9 m^3 of water need be outgassed for one ^{39}Ar sample. Some characteristics of the counting systems for ^{39}Ar, ^{85}Kr and ^{14}C are given in Table I. It can be seen that operation of smaller counters, but filled with correspondingly higher pressures, reduces the background values. The diameter of the 100-cm^3 Cu counter is smaller than that of the 100-cm^3 Lucite counter, but its length is correspondingly larger. Therefore, the ratio of surface to volume is higher and it is not surprising that the background, which is partly due to wall effects, is about 20% higher. But since the counting yield of the longer counter could be improved by about 30%, its figure of merit (S^2/B), given for 2-litre argon samples, is higher.

After counting, the krypton concentrations of the samples were measured by mass-spectrometry. From these values and the measured ^{85}Kr activity an estimate can be made for a possible contamination of the samples by air during extraction in the field or during laboratory processing. Another possibility of estimating this contamination is through determining the O_2 content in the gas samples by gas-chromatography. If the ^{85}Kr results are also used to calculate the specific ^{85}Kr activity of the water samples, it is necessary to measure the absolute krypton concentration in the gas samples extracted from the water. These measurements are done together with other mass-spectrometrical noble gas isotope analyses.

2. RESULTS

In Table II some information on the confined aquifer in the Franconian Albvorland is given together with the measured composition of the extracted gases. The high O_2 content of the two first samples is not surprising, because the two locations are near the recharge area. The CO_2 concentrations are in the range of 10 to 30%. Part of the fluctuations observed for the CO_2 concentrations are due to the fact that the extraction conditions during the outgassing of the water could not be kept constant. The He concentrations show a remarkable increase from the samples collected in the recharge area up to those in Greding.

In Table III some results of the isotopic measurements are given. The ^{3}H value observed for Hilpoltstein agrees with an earlier measurement [8] and reflects a component of fresh water. The well in Jahrsdorf showed < 2 TU in 1968 [5], raised to 7.3 ± 0.5 TU in 1974 [4] and now contains 22 TU. One must conclude that the contribution of fresh water for this well is increasing. The water in Laibstadt shows no detectable amount of ^{3}H. The measured O_2 content and ^{85}Kr activity therefore must be attributed to a contamination during extraction or processing of the gas sample. A very small contribution of fresh water may be found in the water of the well, Greding I, supported by the ^{3}H and ^{85}Kr results.

Values for the ^{14}C concentrations decrease from Hilpoltstein and Laibstadt to Beilngries, corresponding to increasing ^{14}C ages from the recharge area in the flow direction. This was already observed earlier by Geyh and Andres [5, 6] and was used to calculate flow velocities of the groundwater. The ^{14}C results given in

TABLE I. PROPORTIONAL COUNTING SYSTEMS FOR ^{85}Kr, ^{39}Ar AND ^{14}C.

| Counter | | Isotope | Filling pressure (bar) | Composition of gas | Background (B) (counts/min) | Net modern counting rate (S) | S^2/B |
Volume (cm^3)	Material						
22	Cu	^{85}Kr	5	P-10	0.03	70% yield	
100	Lucite	^{39}Ar	18	P-5	0.12	(counts/min) 0.13	0.14
100	Cu	^{39}Ar	19	P-5	0.10	0.18	0.32
50	Cu	^{39}Ar	36	P-5	0.08	0.14	0.24
1 lt	Lucite	^{14}C	1.7	CH$_4$	0.6	10.1	170

TABLE II. DESCRIPTION OF THE CONFINED AQUIFER IN THE FRANCONIAN ALBVORLAND IN SOUTHERN FED. REPUBLIC OF GERMANY AND COMPOSITION OF THE EXTRACTED GASES.

(The chemical analyses have been made by Dr. Schmeing, Bayer. Landesamt für Wasserwirtschaft, Munich.
The gas-chromatographic measurements of the N_2, O_2, Ar and CO_2 contents have been made by W. Berner, the mass-spectrometric analysis of the He-content (preliminary results) by Dr. P. Bochsler, Physikalisches Institut der Universität Bern.)

	Hilpoltstein Feb. 2. 74	Laibstadt Sept.17. 77	Jahrsdorf Sept.14. 77	Möning Sept.15. 77	Greding II Sept.9. 77	Greding I Sept.7. 77	Beilngries Sept.9. 77
Thickness of aquiclude (m)	16	45.2	18.9	96	94.5	96.7	78
Hardness of water (°dH)[a]		6.9	13.5	8.3	15.7	21.3	13.4
Free H_2CO_3 (mg/lt)		47.5	47.5	35.2	33.4	49.3	29.9
Temperature of water (°C)	10.0	12.4	12.1	10.8	18.3	17.0	15.5
pH		6.69	6.83	7.01	7.22	7.09	7.28
N_2 in extracted gas (%)	66.2	76.5	66.8	73.8	81.4	74.6	76.6
O_2 in extracted gas (%)	9.3	9.23	≤0.06	6.57	≤0.3	≤0.13	≤0.07
Ar in extracted gas (%)	1.36	1.33	1.32	1.32	1.62	1.49	1.55
CO_2 in extracted gas (%)	≤23.1[b]	≤12.9[b]	≤31.8[b]	≤18.3[b]	≤16.7[b]	≤23.7[b]	≤21.8[b]
He in extracted gas (ppm)	2.5	2.4	4.2			17.4	

[a] dH = Deutsche Härtegrade.
[b] The true CO_2-contents are probably close to the values given. Because the gas-chromatograph was not calibrated for such high CO_2 contents the values given were obtained by calculating the differences between total gas and (N_2+O_2+ Ar) contents, and therefore represent only upper limits.

TABLE III. ISOTOPIC RESULTS OF THE SAMPLES OF THE AQUIFER IN THE FRANCONIAN ALBVORLAND

	Hilpoltstein	Laibstadt	Jahrsdorf	Möning	Greding II	Greding I	Beilngries
^{3}H (TU)	35.4 ± 1.4	<2.6	22.2 ± 3.4			5.7 ± 2.7	
^{14}C (% modern)	79.7 ± 1.0[a]	79.4 ± 1.2	49.3 ± 0.6[a] 60.1 ± 0.8	54.3 ± 0.7[a] 56.1 ± 0.8	12.8 ± 0.27[a]	2.16 ± 0.15[a]	0.3 ± 0.1[a] 2.2 ± 0.2
^{14}C ages (85% modern) (a) [earlier measurements: [6]]	870	820	1810	3470	8120	17400	21200
δ^{13}C (‰ PDB)	-19.02[a]	-19.06[a] -17.86	-16.91[a] -14.43	-19.17[a] -17.91	-15.20[a]	-13.63[a]	-11.02[a]
δ^{18}O (‰ SMOW)	-9.61	-9.86	-9.47	-9.61	-10.23	-11.11	-10.86
^{39}Ar (% modern)	102 ± 4	97.8 ± 5	79.9 ± 4.8	49.6 ± 3.8	35.3 ± 3.6	44.2 ± 3.9	17.2 ± 3.9
Estimated contamination of extracted gases by modern air (%) O$_2$	51.8	44.6	<0.3	(32.9)	1.5	0.6	<0.4
Estimated contamination of extracted gases by modern air (%) ^{85}Kr	22.2	31.2	19.1	1.15	1.0	2.6	0.3
Corrected ^{39}Ar age (a)	<30	<55	85 ± 25	275 ± 35	410 ± 40	325 ± 35	685 ± 100

[a] Using the CO$_2$ which was separated out of the extracted gases.
δ^{13}C values show isotopic effects of the separation procedure.
^{3}H values have been measured by U. Schotterer, δ^{13}C- and δ^{18}O values by K. Hänni.

this paper agree in general with the earlier measured values, although some changes during the last few years can be observed, e.g. [6, 8]:

Greding II:	31%	(1968);	16%	(1975);	12.8%	(1977)
Greding I:	9.7%	(1968);	6.0%	(1975);	2.16%	(1977)
Beilngries:	6.1%	(1970);	2.2%	(1977).		

These changes are probably connected with the mining of the well for water supplies and the disturbance of the natural water flow (for a groundwater mining model see [9]). Surprisingly, our ^{14}C results for Jahrsdorf and Beilngries show some disagreement for the two samples for each location (one using CO_2 obtained by degassing of water, and the other using precipitated carbonates). This disagreement may partly also be a mining effect during the collection of the gas samples, although some contamination by atmospheric air of the samples obtained by chemical precipitation cannot be excluded.

As mentioned previously, the $\delta^{13}C$ results of the CO_2 samples, which were trapped from the extracted gas (marked with a), are on an average lower by about 2.5‰ than those of the precipitated samples due to isotopic fractionation. The adjusted values and those obtained from the precipitated carbonates agree well with the values measured earlier by Geyh [8]; in particular, the trend of increasing $\delta^{13}C$ values for older water samples is again observed. This means that additional carbonate is going into solution, in agreement with the increasing water hardness (see Table II).

The $\delta^{18}O$ values show remarkably lower values for samples with ^{14}C ages older than about 10000 years.

The ^{39}Ar concentrations show a trend similar to the ^{14}C results. This parallel behaviour of both concentrations can also be seen from Fig.1. The slight discontinuity of the ^{39}Ar data for Greding I may be explained by a small contribution of fresh water, which was concluded from the presence of ^{3}H and ^{85}Kr.

With the exception of the samples collected in Hilpoltstein and in Jahrsdorf, a possible falsification of the ^{39}Ar ages owing to a contamination of the samples by air can be corrected, based on ^{85}Kr and O_2 results. For most of the samples a correction smaller than a few per cent had to be applied. These corrected ^{39}Ar ages indicate a clear discrepancy with the ^{14}C ages. Errors due to counting can be excluded because, after a first counting, most of the samples were re-purified by gas-chromatography and were then re-measured in a different counting system. All re-measured results (^{39}Ar and ^{85}Kr) agreed with the first obtained values within statistical limits. From the well in Beilngries two gas samples were extracted and both gave ^{39}Ar ages of 658 ± 100 years. A sample collected in Jahrsdorf in 1974 gave an age of 40 ± 60 years, in agreement with the results of the sample collected in 1977.

In Table IV some results of a few other samples are summarized. Schnaittach and Eckenhaid are wells north-east of Nüremberg. Based on the earlier measured

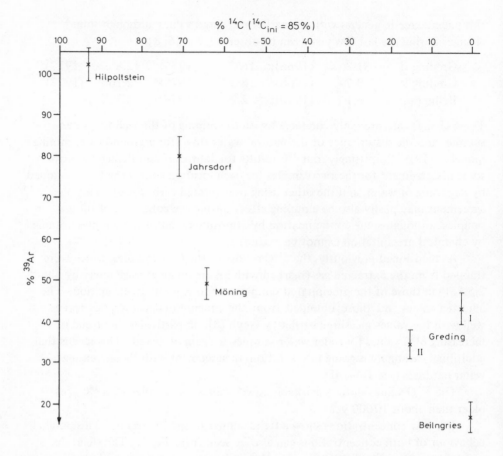

FIG.1.   *Measured ^{14}C and ^{39}Ar concentrations in a confined aquifer in the Franconian Albvorland.*

^{14}C results [8], we hoped to find in Schnaittach an argon sample without any ^{39}Ar activity. The ^{39}Ar results of 26 ± 5% modern is preliminary, but even with the correction for a contamination by modern air (which was noticed already during outgassing and was then confirmed by the O_2 and ^{85}Kr results), an activity of about 20% modern remains. This again is in disagreement with the ^{14}C age. Remarkable for the gas extracted from the water of the well in Schnaittach is its extremely high He content. The ^{14}C and ^{39}Ar data of the Eckenhaid sample are also included in the Table; however, the difference in the δ^{18}O values makes it questionable whether they can be compared with the Schnaittach results (for a possible interpretation with a diffusion-hydraulic model, see Ref.[9]). For these two samples the δ^{13}C, δ^{18}O and ^{14}C values given in this paper agree well with values measured earlier by Geyh and Andres [8].

In Table IV the results of two samples collected in Grafendorf (in the Styrian Basin, Austria) are also included. Together with the Vereinigung für hydrogeologische Forschung in Graz this well-studied confined aquifer [10] was selected for a common project. The positions of the filters are about 53 m below ground for well I and about 84 m for well III. Again, a large He content was found in the gas extracted from well III. Our ^{14}C results are considerably higher than the results measured earlier [10] and recently by the IAEA [11]. Exchange of samples between the two laboratories is planned. As preliminary explanation a mining effect during sample collection can be considered. Again the ^{39}Ar and ^{14}C ages of the two samples are in disagreement, although the concentrations in per cent modern are quite close.

Zurzach is a thermal spring in Switzerland. Its ^{39}Ar result of 3.8 times atmospheric concentration shows that underground produced ^{39}Ar through the ^{39}K(n,p)^{39}Ar process can be found in thermal water. Independently, gas samples were collected from two wells, both were measured in two different counting systems and a laboratory error for the high ^{39}Ar value can therefore be excluded. This value is also surprising because a sample collected in 1973 showed "only" about 100% modern ^{39}Ar activity [4]. It is planned to repeat the sampling to study further the variations of the ^{39}Ar activity in this thermal water.

The last sample mentioned in Table IV was extracted in Monaco from ocean surface water. Based on known atmosphere-ocean exchange models a value for the ^{39}Ar activity in the mixed layer of the ocean of about 100% could be expected [12], which is confirmed by the measurements.

In addition to the data given in the Tables the following information may be useful:

(a)   Three ice samples collected in Devon Island, Canada, allowed the correction procedure of ^{39}Ar ages to be checked for a possible contamination by air, based on the measured ^{85}Kr activities. All ice samples showed ^{39}Ar values around 20% modern although the ^{14}C ages were between 2700 and 5800 years BP. From the ^{85}Kr activities in the same samples, components of 19 to 27% of modern air could be calculated, which explains the high ^{39}Ar results. Therefore, ^{85}Kr measurements can be applied as sensitive indicators of a possible contamination of the samples.

(b)   The results of the He content in the samples given in Tables II and IV are measured for the gases extracted out of water. A high extraction yield for He during the outgassing of water can be assumed and therefore the He values should be reliable. Absolute He concentrations can be expected from special investigations, which are planned.

(c)   For Grafendorf I and III, ^{32}Si values were measured in Copenhagen [13]. The results are 0.037 ± 0.013 and 0.028 ± 0.015 dis/min per t of H_2O, respectively. As in the case of the ^{14}C dating method, for ^{32}Si the determination of an initial activity is also a problem. Depending on the adopted values,

TABLE IV. RESULTS FOR OTHER HYDROLOGICAL SAMPLES (COMMENTS: SEE TABLES II AND III)

	Eckenhaid Sept.9, 77	Schnaittach Sept.19, 77	Grafendorf I Aug.13, 76	Grafendorf III Aug.12, 76	Zurzach II th. spring Aug.23, 76	Monaco Ocean surface water April 4, 74
N_2 in extracted gas (%)	82.2	75.1	73.8	71	~93	66.8
O_2 in extracted gas (%)	<0.15	2.2	<0.5	<0.5	<0.5	26.3
Ar in extracted gas (%)	1.65	1.45	1.41	1.33	1.47	1.41
CO_2 in extracted gas (%)	≤16	≤21	≤25	≤26	~2.4	≤5.4
He in extracted gas (ppm)		1150		450		(1.8)
^3H (TU)				3.9 ± 4	<1	8.6 ± 4.7
^{14}C (% modern)	55.9 ± 0.7	4.7 ± 0.16	<0.5	a) 46.5 ± 0.6 46.3 ± 0.6	4.67 ± 0.16	
^{14}C ages (85% modern) (years) (earlier results: [8,10])	5715 ± 135	>37900	4080 ± 110	6150 ± 150		
$\delta^{13}C$ (‰ PDB)	a)-19.21 -16.66	a) -13.41 -11.12	a) -13.16 -12.10	a) -14.01 -11.13	-9.38	
$\delta^{18}O$ (‰ SMOW)	- 9.07	-10.74	- 8.92	- 8.50	-10.3	
^{39}Ar (% modern)	prel. 88 ±	prel. 26 ± 5	28.4 ± 4.3	36.5 ± 4.2	380 ± 10	101 ± 4.4
Estimated contamination of extracted gases by modern air (%) { O_2 / ^{85}Kr }	<0.7 <0.4	10.8 7.4	<2.5 0.5	<2.6 1.3	<0.6 <0.7	
Corrected ^{39}Ar age (years)	45 ± 35	600 ± 80	490 ± 65	390 ± 50		

the ^{32}Si ages of these two samples range from about 300 to about 1200 years. A contamination of the samples by fresh precipitation during collection is very unlikely. More information may be obtained from the Franconian Albvorland profile where samples were collected in Laibstadt, Jahrsdorf, Greding I and Beilngries.

3. DISCUSSION AND CONCLUSIONS

In particular, two objections to the ^{39}Ar ages of hydrological samples must be considered — underground production of ^{39}Ar and gas exchange with the atmosphere through the aquiclude.

The ^{39}Ar result for the thermal spring in Zurzach of 3.8 times atmospheric concentration shows that underground produced ^{39}Ar can be found in ground-water. The main mode of production would be the ^{39}K(n,p)^{39}Ar reaction with neutrons originating mainly from (α, n) processes [14]. The expected concentration in the groundwater would therefore depend on the U, Th and K content of the rocks, on the diffusion of the produced ^{39}Ar out of the target matrix, and on the porosity of the aquifer. To estimate an upper limit for the underground production rate, the following assumptions are used — amount of neutrons in sandstone: 3000 n/kg · a [15], 2% K content in the rock, 2% pore volume and a ratio of 30 mb to 2000 mb for the cross-section of the ^{39}K(n,p) reaction to the neutron absorption cross-section. For this rough estimate we further assume that all the produced ^{39}Ar diffuses out of the rock into the water in the pore volume. This calculation shows that a concentration of about 0.4 dis/min ^{39}Ar/litre Ar could be possible, i.e. about four times the cosmic-ray produced atmospheric level. This value could be even higher, if ^{39}Ar would also diffuse from the upper or lower aquiclude into the groundwater, or if the ratio of the used cross-sections is too low. On the other hand, the value should be lower since probably most of the produced ^{39}Ar decays on its diffusion way out of the rock, and since the dilution of the ^{39}Ar could be with more water than that in the pore volume.

Assuming that the ^{39}Ar activity found in the oldest samples of about 20% modern (Beilngries and Schnaittach) represents an equilibrium value for the underground production in the aquifer, the discrepancy between ^{39}Ar and ^{14}C ages would still exist for the samples collected in Möning, Greding II and Greding I. One would have to explain the steadily decreasing ^{39}Ar activities of the three remaining locations, e.g. by assuming that the undergound U, Th and/or K contents decrease for this aquifer in the flow direction. The K concentrations found in the water at least would not support this additional assumption (Jahrsdorf 5 mg/l; Möning 8 mg/l; Greding II 27 mg/l and Greding I 38 mg/l; measured by Dr. Schmeing, Munich).

Another possibility of obtaining ^{39}Ar ages that are too young could be the exchange with air of atmospheric ^{39}Ar concentration through the aquiclude. If the

air would be transported in water diffusing through the aquiclude, the calculations
proposed by Klitzsche et al. [16] can be applied. In a model these authors use a
diffusion coefficient of $D = 1.2 \times 10^{-5}$ cm^2/s for water through the aquiclude
between two well-mixed water layers. Using this value a mean penetration depth
of this radioactive isotope with a 269-a half-life is $\sqrt{D \cdot \tau} \approx 4$ m, which is small
compared with the thickness of the aquiclude in this aquifer of up to 100 m.
Therefore, transport in diffusing water probably can be neglected. A much faster
diffusion of ^{39}Ar in gaseous form would have to be adopted, at least in the upper
part of the aquiclude. Since the hydrostatic pressure of the groundwater in the
Franconian Albvorland is around 50 m or more, gaseous diffusion through the
lower part of the aquiclude can probably be excluded. In addition, preliminary
measurements of the He content in the water suggest an increasing trend in the
flow direction. This indicates that gas exchange with the atmosphere might not
be fast enough to allow significant ^{39}Ar activities to reach the aquifer before
decaying.

The problems of the ^{14}C dating method in hydrology must also be considered.
The observed discrepancy between ^{14}C and ^{39}Ar data cannot be explained by the
uncertainty of the initial ^{14}C concentrations alone. Many careful studies using
δ^{13}C, carbonate chemistry and combinations of both allow the conclusion that
this uncertainty could be around at most a few thousand years.

A possible dilution of the ^{14}C activity by additional dissolution of ^{14}C-free
carbonates during the water flow can be corrected by using the ^{14}C to H$_2$O ratio
instead of the ^{14}C/^{12}C ratio. This correction explains only part of the
^{14}C-^{39}Ar discrepancies, because the water hardness increases, for example, from
Möning to Greding by "only" at most a factor of 2, corresponding to about
5000 years. The observed variations in δ^{13}C and bicarbonate concentrations, and
their correlation with ^{14}C ages, indicate that some processes still occur in the
aquifer.

To explain the discrepancies between high ^{14}C ages and the ^{39}Ar results
one might consider the possibility of carbonate exchange between the solid and
the liquid phases in the aquifer. This chromatographic effect has been investigated
by many authors [e.g. 17—20]. Reversible carbonate exchange, which is assumed
to be restricted to part of the surface molecular layer of the solid phase, would
reduce the average velocity of the ^{14}C atoms by a certain factor. The apparent
^{14}C ages would then be too high by the same factor. In laboratory experiments
the order of magnitude of these delay factors for various materials could be
measured [19]. In these experiments losses of ^{14}C were also observed, which
must be attributed to some irreversible exchange processes.

Up to now no experiments allowed the extent of such carbonate exchanges
in the aquifer concerned [16] to be measured with a high degree of reliability.
An increased δ^{13}C content indicates carbonate exchange. But quantitative interpretations based on δ^{13}C values may lead to an underestimation of the exchange

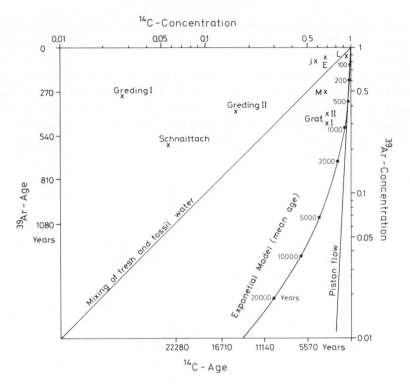

FIG.2.   Theoretical correlation between ^{39}Ar and ^{14}C concentration for three different mixing models. As examples, data are given for the Frankenalb aquifer (corrected for $^{14}C_{ini} = 85\%$) and for the Grafendorf well ($^{14}C_{ini}$ used as in Ref.[10]).

processes [18]. Based on the laboratory experiments by Thilo and Münnich [19], in the case of the Franconian Albvorland aquifer, the delay factor due to the reversible exchange could be about 5–30%. But, from these model estimates, it is difficult to draw conclusions about the natural conditions in the aquifer.

It must be emphasized that there are mainly two arguments against extensive influence of this exchange delay in many aquifers, supporting the given ^{14}C ages — first, there is the long experience of ^{14}C dating in hydrology and the generally observed good agreement between ^{14}C ages and hydrodynamically determined flow velocities. For the aquifer in the Franconian Albvorland the ^{14}C results and their calculated flow velocities and k values agree with the assumption that the limiting factor for the flow velocities is the permeability through the clay materials above the aquifer in the region of the valley of Altmühl [5], which is considered as main drainage channel. On the other hand, the permeability through the sandstone in clefts should be considerably higher [5]. Second, the other reason to believe in the ^{14}C ages is the observation that the $\delta^{18}O$ values often decrease considerably for samples older than 10000 years, as was also measured for the aquifer in the Franconian Albvorland. This hiatus is explained by changes in he palaeoclimatic conditions.

The question arises whether the result of both dating methods could be explained by the mixing of water layers of different ages. In Fig.2 the calculated correlation between ^{14}C and ^{39}Ar concentrations is plotted for three different mixing models. If mixing of waters with different ages but equal carbonate contents should explain a measured pair of concentrations, the point for the ^{14}C and ^{39}Ar concentrations of the sample should be found to the right of the curve calculated for mixing of fossil and fresh water. This is, for example, the case for the two Grafendorf samples. From the hydrogeological aspect, mixing cannot be excluded for this aquifer. The ^{32}Si activity found in the samples would then also be explained.

For the water collected in Laibstadt (L) and Möning (M) this simple explanation also seems to be possible. But simple mixing of different waters with different ages does not explain the results obtained for Eckenhaid (E), Jahrsdorf (J), Greding I and II and Schnaittach.

From the previous discussion it follows that the correction of ^{14}C ages based on calculation models for the initial ^{14}C concentration does not explain all observations. In addition, processes in the aquifer should be studied and included for age corrections. We are planning to measure ^{39}Ar and ^{14}C profiles in aquifers of a type other than sandstone. As in this investigation, all information has to be available — data on radioactive and stable isotopes and chemical analyses. Only the results of such comprehensive studies for different aquifers might lead to a satisfactory understanding of ^{39}Ar, ^{14}C and all other data.

ACKNOWLEDGEMENTS

We thank the following co-workers for sample preparation and processing and for measurements: E. Alt, Dr. P. Bochsler, W. Berner, P. Bucher (Bern); H. Clausen (Copenhagen); A. Etzweiler, K. Hänni, B. Lehmann, M. Moell, E. Moor, Mrs. T. Riesen (Bern); Dr. Schmeing (Munich); U. Schotterer, (Bern); P. Schwarz (IAEA, Vienna) and P. Wittwer (Bern). Valuable discussions were held with Drs. Andres (Munich), Bochsler (Bern), Geyh (Hanover) Payne (IAEA, Vienna), Siegenthaler (Bern) and Zojer (Graz). We thank the Bayer. Landesamt für Wasserwirtschaft, the custodians and owners of the groundwater wells in Beilngries, Eckenhaid, Greding, Hilpoltstein, Jahrsdorf, Laibstadt, Möning and Schnaittach and Mr. Thommeret in Monaco for their valuable support during sample collections.

The work was financially supported by the Swiss National Science Foundation, the IAEA, Vienna, the Vereinigung für hydrogeologische Forschung in Graz, and the Thermalquellen and Thermalbad AG, Zurzach.

REFERENCES

[1] WIEST, W., Messung von Argon-Isotopen in Luftproben und Vergleich mit Modell-
     rechnungen, Dissertation der Universität Bern (1973).
[2] OESCHGER, H., LOOSLI, H.H., "New developments in sampling and low level counting
     of natural radioactivity", Low Radioactivity Measurements and Applications, Proc.
     Int. Conf. High Tatras 6-10 Oct. 1975 (POVINEC, P., USAČEV, S., Eds), Comenius
     University, Bratislava (1975) 13.
[3] OESCHGER, H., STAUFFER, B., BUCHER, P., LOOSLI, H.H., "Extraction of gases
     and dissolved and particulate matter from ice in deep boreholes", Isotopes and Impurities
     in Snow and Ice (Proc. Symp. Grenoble, Aug./Sep. 1975), IAHS 118 (1977) 307.
[4] OESCHGER, H., GUGELMANN, A., LOOSLI, H., SCHOTTERER, U., SIEGENTHALER, U.,
     WIEST, W., "^{39}Ar dating of groundwater", Isotope Techniques in Groundwater Hydrology
     1974 (Proc. Symp. Vienna, 1974), IAEA, Vienna (1974) 179.
[5] ANDRES, G., GEYH, M.A., Untersuchungen über den Grundwasserhaushalt im über-
     deckten Sandsteinkeuper mit Hilfe von ^{14}C- und ^{3}H-Wasseranalysen, Wasserwirtschaft 60
     (1970) 259.
[6] GEYH, M.A., "Basic studies in hydrology and ^{14}C and ^{3}H measurements", 24th Int. Geol.
     Congr. (Montreal) 11 (1972) 227.
[7] GEYH, M.A., Erfahrungen mit der ^{14}C- und ^{3}H-Methode in der angewandten Hydrologie,
     Oester. Wasserwirtsch. 26 (1974) 49.
[8] GEYH, M.A., personal communication.
[9] GEYH, M.A., BACKHAUS, G., "Hydrodynamic aspects of carbon-14 groundwater dating",
     IAEA-SM-228/48, these Proceedings.
[10] PRZEWLOCKI, K., Hydrologic Interpretation of the Environmental Isotope Data in the
     Eastern Styrian Basin, Steirische Beitr. Hydrogeol. 27 (1975) 85.
[11] FLORKOWSKI, T., SCHWARZ, P., IAEA, personal communication.
[12] OESCHGER, H. et al., A box diffusion model to study the carbon dioxide exchange in
     nature, Tellus 27 (1975) 168.
[13] CLAUSEN, H., personal communication.
[14] GEIGE, Y., OLTMAN, B.G., KASTNER, J., Production rates of neutrons in soils due to
     natural radioactivity, J. Geophys. Res. 73 10 (1968) 3135.
[15] BENTLEY, H.W., Some comments on the use of chlorine-36 for dating very old ground
     water, Workshop on Dating Old Groundwater, Tucson, Arizona, 16–18 March 1978,
     unpublished report.
[16] KLITZSCHE, E., SONNTAG, CHR., WEISTROFFER, K., EL SHAZLY, E.M., Grund-
     wasser der Zentralsahara: Fossile Vorräte, Geolog. Rundsch. 65 (1976) 264.
[17] WENDT, I., STAHL, W., GEYH, M., FAUTH, F., "Model experiments for ^{14}C water-age
     determinations", Isotopes in Hydrology (Proc. Symp. Vienna, 1967), IAEA, Vienna
     (1967) 321.
[18] MÜNNICH, K.O., Isotopen-Datierung von Grundwasser, Naturwissenschaften 55 4
     (1968) 158.
[19] THILO, L., MÜNNICH, K.O., "Reliability of carbon-14 dating of groundwater: Effect
     of carbonate exchange", Isotope Hydrology (Proc. Symp. Vienna 1970), IAEA, Vienna
     (1970) 259.
[20] PEARSON, F.J., Jr., HANSHAW, B.B., "Sources of dissolved carbonate species in
     groundwater and their effects on carbon-14-dating", ibid., p.271.

DISCUSSION

S.K. GUPTA: I would like to report some of our recent ^{32}Si measurements on groundwater samples in India. Except in samples that have a distinct component of bomb ^{14}C, we find that, in terms of modern activity, ^{32}Si activity is always less than the ^{14}C activity and that the age calculated by ^{32}Si is always lower than the corresponding ^{14}C age. The age discrepancy increases with increasing ^{32}Si and/or ^{14}C age. This effect can easily be explained as being due to a progressively increasing component, possibly caused by leakage, of young water into the old water in the downflow direction. All the samples lie in the plotting field between the piston and completely mixed water.

R. GONFIANTINI: Perhaps mixing might also explain points above the mixing line in Fig.2, if carbonate concentration is taken into account. In fact, one could imagine that these waters might be the result of the mixing of old highly carbonatic water with modern water having a low carbonate content. Would this be possible in your opinion?

H.H. LOOSLI: Figure 2 is based on the assumption of mixing of waters with different age distributions but with the same carbonate content. For that reason mixing alone would not explain points above the mixing line. A combination of different assumptions, e.g. that suggested by you of different carbonate contents, would be needed to explain points above this line.

R. GONFIANTINI: Some years ago, at a symposium on isotope hydrology, Mr. Oeschger suggested that ^{14}C content should be described in terms of activity per unit volume of water, in order to reduce or eliminate the effects of dead carbon dissolution. Have you tried this approach?

H.H. LOOSLI: Instead of using the ratio of ^{14}C to H_2O, the same information can be obtained by taking into account variations in carbonate content. In the aquifer with which we are concerned here, water hardness increases between the recharge area and the oldest wells by a factor of "just" 2–3, so only part of the ^{14}C-^{39}Ar age differences can be explained in this way.

I. WENDT: Table IV shows a ^{39}Ar value of 380% for Zurzach II. Are you able to estimate the order of magnitude of the ^{39}Ar contribution produced by the ^{39}K(n,p)^{39}Ar reaction?

H.H. LOOSLI: The estimated underground production of ^{39}Ar in the sandstone aquifer of the Frankenalb is about 0.4 disintegrations per minute (dpm) per litre of argon, which corresponds to four times the atmospheric activity. But since this rough estimate does not take into account the fact that the ^{39}Ar, which diffuses out of the crystals into the water, decays during diffusion, it can only be considered an upper limit.

P. THEODORSSON: I would like to ask you to have your work on modern ^{39}Ar values published in an easily accessible form.

H.H. LOOSLI: We are planning to publish a paper on the ^{39}Ar dating technique and its applications in glaciology, hydrology and oceanography within about a year.

P. THEODORSSON: If ^{39}Ar is produced underground by neutrons, might not other isotopes with higher cross-sections and/or higher mother nuclide concentrations be produced at the same time and be measurable?

H.H. LOOSLI: True, other isotopes might also be produced underground by neutron reactions, and we will certainly be looking for detectable activities.

P. THEODORSSON: If ^{39}Ar is produced underground, would not hot rock (above 350°C) or molten magma more readily give off their argon, so that the ^{39}Ar content of groundwater receiving argon from such a source would be high? Could the high value found in Zurzach II, which is a thermal spring, be explained by this mechanism?

H.H. LOOSLI: You are right; it should be easier to find a ^{39}Ar contribution produced underground in water from a thermal spring than in normal groundwater because diffusion of argon out of the rock is higher at higher temperatures. This could well explain the high ^{39}Ar values in the Zurzach water.

P. THEODORSSON: An abnormal stable isotope ratio in argon might reveal its magmatic origin. Have you looked for this?

H.H. LOOSLI: Anomalies in the ^{36}Ar/^{40}Ar ratio are not easy to detect in gases extracted from normal groundwater. However, Dr. Bochsler is at present doing such measurements at our institute; unfortunately, the results are not yet available.

P. FRITZ: Although I have no data to support my suggestion, it seems to me that the low ^{18}O values found by Mr. Loosli in waters with low ^{14}C activity might well confirm his argon results, as they may well reflect the little ice age that occurred during the Middle Ages.

H. OESCHGER: I would like to suggest that unexpectedly low ^{39}Ar ages might not be due exclusively to undergound production. It is possible that a chromatography effect from ^{14}C has been underestimated; such an effect might be indicated by the trend in δ^{13}C values in the profile which Mr. Loosli showed us.

Secondly, although the effort required for one ^{39}Ar measurement is about five times greater than that expended on a ^{14}C measurement, in view of the fundamental nature of the information that can be obtained by means of this new technique, I hope it will become a standard tool in isotope hydrology.

KRYPTON-85 DATING OF GROUNDWATER

K. RÓZAŃSKI, T. FLORKOWSKI*
Institute of Physics and Nuclear
 Techniques, Crakow,
Poland

Abstract

KRYPTON-85 DATING OF GROUNDWATER.

The possibility of dating groundwater by ^{85}Kr is being investigated. The method of gas extraction from a water sample of 200 to 300 litres has been developed. The argon and krypton mixture, separated from the gas extracted from water, was counted in a 1.5-ml volume proportional counter. The amount of krypton gas in the counter was determined by mass spectrometry. A number of surface and groundwater samples were analysed, indicating a ^{85}Kr concentration ranging from present atmospheric content (river water) to zero values. ^{85}Kr "blank value" was determined to be about 5% of present ^{85}Kr atmospheric content. For groundwater samples, the mean residence time in the system was calculated assuming the exponential model and known ^{85}Kr input function. Further improvement of the method should bring higher yield of krypton separation and lower volume of water necessary for analysis.

INTRODUCTION

The possibility of using ^{85}Kr for dating groundwater has been investigated. The method of extracting krypton from water samples has been elaborated and the ^{85}Kr concentration has been determined in a number of water samples to illustrate its dating feasibility. ^{85}Kr (half-life 10.76 years) presents the following advantages as an environmental radioactive tracer of water:

(a) The only and comparatively well-mixed reservoir of krypton is the atmosphere;
(b) The concentration of ^{85}Kr in the atmosphere is steadily increasing and is known with sufficient accuracy;
(c) Since krypton is a noble gas, it does not interact with the matrix of the aquifer.

The specific activity of ^{85}Kr in groundwater, together with the known "input function", is a good base for dating groundwater.

The method of ^{85}Kr determination in water was reported for the first time by Schröder [1] for application in oceanographic studies.

---

* At present with the International Atomic Energy Agency.

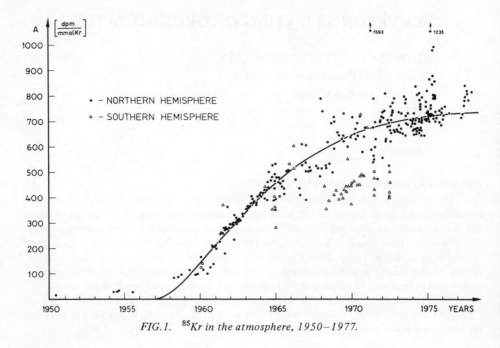

FIG.1. ^{85}Kr in the atmosphere, 1950–1977.

^{85}Kr IN THE ATMOSPHERE AND PRECIPITATION

An increase of the ^{85}Kr concentration in the atmosphere was already noticed many years ago. The natural level of ^{85}Kr before the nuclear age resulted from the production of ^{85}Kr by the (n,γ) reaction of cosmic neutrons with the stable ^{84}Kr in the atmosphere and from the release of ^{85}Kr from the earth owing to spontaneous fission of uranium and thorium.

The increasing nuclear industry (nuclear reactors and fuel reprocessing) contributes considerably to the increase of the ^{85}Kr concentration. Figure 1 summarizes the measurement data published by various authors [2–10]. Rather good mixing of ^{85}Kr in the northern hemisphere and a certain gradient in the southern hemisphere are observed [11, 12] owing to the fact that all nuclear fuel reprocessing plants are situated in the northern hemisphere and the meridional mixing of air masses is slow.

The mean ^{85}Kr concentration in the southern hemisphere is at present equal to 85 to 90% of the ^{85}Kr concentration in the northern hemisphere [5, 13], the latter being 780 dpm/mmol Kr.[1] Table I shows the estimated contribution of

---

[1] Throughout this paper disintegrations per minute (dpm) have been used. In the S.I. system 1 disintegration per second (dps) = 1 Bq.

TABLE I. ^{85}Kr SOURCES AS FOR THE END OF 1976 [14]

Source	Estimated ^{85}Kr activity in atmosphere (MCi)
Natural production	1.4×10^{-5}
Nuclear explosions	2.0
Nuclear power	28.3
Plutonium production	30.3

various ^{85}Kr sources in the atmosphere at the end of 1976. The release of ^{85}Kr into the atmosphere has decreased during recent years from 85 ± 5 dpm/mmol Kr per year in the 'sixties, to about 40 ± 5 dpm/mmol Kr per year in 1976 [8]. The atmosphere can be considered as the only large reservoir of krypton, since only 2.8% of the total mass of krypton is contained in the hydrosphere. The transport of krypton from the atmosphere to the hydrosphere is by precipitation and by molecular exchange processes on the atmosphere-water phase boundary. The annual transport of ^{85}Kr from the atmosphere to the hydrosphere is estimated to be 0.1% of the total amount of ^{85}Kr in the atmosphere [14].

Precipitation (rain) and surface waters, being in equilibrium with the atmosphere, indicate a ^{85}Kr specific activity equal to that of the atmosphere. During infiltration and flow of water in the aquifer, the radioactivity of ^{85}Kr only decreases owing to radioactive decay since krypton as an inert gas practically does not react with the system. Thus, dating of water is possible by measuring the ^{85}Kr specific activity in water samples, providing the sample is not contaminated with atmospheric air. The krypton concentration in the atmosphere is 1.11×10^{-4} vol.%, and its solubility in water at 15°C is 6.9×10^{-6} vol.%. This results in a concentration of 0.07 cm^3 krypton in 1 m^3 water, corresponding to an actual activity of 2.6 dpm.

ANALYTICAL PROCEDURE

The very low level of ^{85}Kr radioactivity in water requires a special procedure for extraction of krypton from the water sample and measurement of the ^{85}Kr specific activity in a low-level counting system. The block diagram of the whole procedure is shown in Fig.2. A more detailed description of each step is given

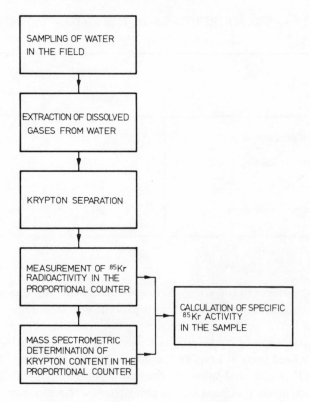

FIG.2.   Block diagram of the procedure.

below. Our aim was to make the method as simple as possible and the field
equipment as light as possible. The task was to check the method and then to
improve the technique.

Sampling of water

Sampling of water in the field should be performed in such a way that
atmospheric contamination is avoided. Large plastic containers (about 120 litres)
are provided with tight metal taps and filled with nitrogen in the laboratory before
sampling. A manually operated water pump is used in the field for filling the
containers with water. Sample size ranged from 120 to 360 litres. Containers
are transported to the laboratory where the next step (extraction of gases) is made
with the minimum delay in order to avoid loss of gases from the water.

Gas extraction from water sample

Figure 3 shows the gas extraction system. Water is pumped from the container
through the furnace where it is heated to +90°C. Then water is sprayed in the

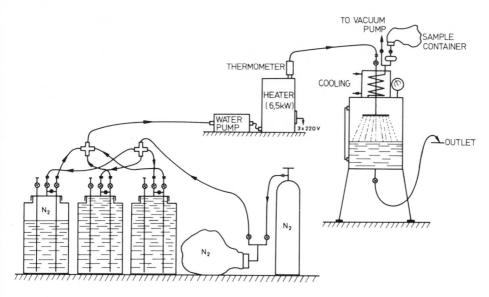

FIG.3. Extraction system for degassing water.

extraction container and allowed to flow outside the system. The evolved gases are accumulated in the upper part of the extraction container equipped with a plastic bulb. After pumping the water from the water containers the gases are collected in the plastic bulb with practically no loss. If necessary, the water can be re-circulated in the system; however, a single run gives a yield of gases of about 80 to 85%. The speed of degassing is about 70 litres/h and depends on the heating power of the furnace (at present 6.5 kW). The degassing of 360 litres of water yields about 7 litres of gases. In principle, the system can also operate in the field if power for the heater can be provided (oil heating is also possible).

Krypton separation

Figure 4 shows the vacuum system used for krypton separation. The plastic sampler with the gas is connected to the system and is pumped by the vacuum oil rotary pump through the traps cooled with liquid nitrogen (separation of water and CO_2), then through the charcoal column cooled with dry ice. Krypton and argon are selectively adsorbed on the charcoal in the column. Pumping is continued until a pressure of about 20 torr above the charcoal is achieved.

In the next step, the charcoal column is heated to +350°C and the desorbed gas is passed to the furnace containing metallic barium heated to +500°C. In the reaction with barium, oxygen and nitrogen are removed from the gas sample. Before and after the furnace metal U-tubes remove radon gas. After the completion of the reaction, about 0.7 cm^3 (NPT) argon and krypton mixture is left in the

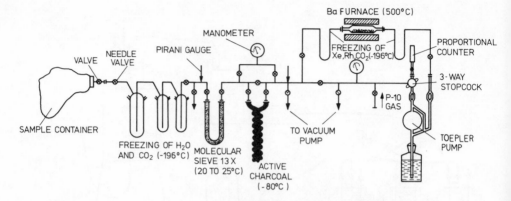

FIG.4.   Vacuum line for separation of argon-krypton mixture.

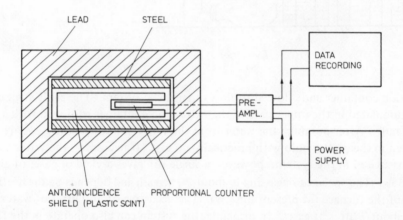

FIG.5.   Low-level counting system for measurement of ^{85}Kr radioactivity.

furnace. The krypton volume is only about 1.3% of the total volume. The argon-krypton mixture is pumped with a Toepler pump to the miniature proportional counter. The krypton recovery in the whole procedure at present amounts to about 50%.

Measurement of ^{85}Kr activity

The low-level counting system as shown in Fig.5 is used for the ^{85}Kr activity measurement. The parameters of the counting system are given in Table II. Figure 5 shows the block diagram of the system.

TABLE II. PARAMETERS OF THE LOW-LEVEL COUNTING SYSTEM USED FOR ^{85}Kr

1. Shield	10 cm lead + 5 cm old steel
2. Anti-coincidence shield	Plastic scintillator
3. Proportional counter:	
Volume	1.5 cm^3
Gas filling	P-10 gas (90% Ar + 10% CH$_4$) + sample
Working pressure	720 torr
4. Background:	0.104 ± 0.007 counts/min (mean value for 1 year operation)
5. Net count-rate for modern ^{85}Kr	ca 0.1 counts/min
6. Measurement time for single sample	5 × 10^3 to 10^4 min

Determination of ^{85}Kr specific activity in the sample

The last step of the analysis is the determination of the amount of krypton in the proportional counter, for which a mass spectrometer is used. The sample gas is compared with standards of known krypton concentration in an argon-krypton gas mixture. The activity measured by the low-level counting system is related to the volume of krypton present in the sample.

Summary of analytical procedure

Using the described procedure, one sample can be prepared and measured within one week. The extraction of gases from water and the separation of the argon-krypton mixture takes 10–12 hours. The mass spectrometric determination of krypton amounts to only a fraction of an hour, while the main time is devoted to counting. The total error of analysis is estimated to be about 10%. As the detection limit of ^{85}Kr, a double standard deviation of the background in the counter is assumed, which corresponds to about 100 dpm/mmol Kr. It is hoped that the detection limit and the analytical error will be decreased by at least a factor of two after improvement of the krypton separation system (higher yield) and by using a better (lower background) proportional counter.

TABLE III. RESULTS OF ANALYSIS

Sample No.	Sample	Sample date	Sample volume (litres)	^{85}Kr (dpm/mmol Kr)	^{3}H (TU)	^{14}C (pmc)
1.	Surface water, Kryspinow reservoir	24.5.77	360	689 ± 84	–	–
2.	„ „ „	02.8.77	240	683 ± 87	139 ± 10	91.2 ± 3.0
3.	„ „ „	09.8.77	120	755 ± 129	–	–
4.	River Raba	06.6.77	360	794 ± 96	95 ± 10	99.2 ± 2.0
5.	Surface water, Bagry reservoir	30.5.77	360	728 ± 83	120 ± 10	100.5 ± 2.0
6.	Karst spring in Tatra (flow 300–600 litres/s)	11.7.77	360	707 ± 83	100 ± 10	89.0 ± 5.0
7.	Groundwater, Lodz well (600 m)	28.6.77	360	30 ± 57	0.8 ± 2	49.2 ± 1.0
8.	„ „ „	04.7.77	360	–13 ± 55	–	–
9.	„ „ „	27.7.77	360	100 ± 61	0 ± 2	49.3 ± 1.0
10.	Mineral spring Slotwinka – Krynica	19.8.77	120	362 ± 143	42 ± 2	2.2 ± 1.0
11.	Pumping well S-4, Crakow	31.1.78	240	209 ± 75	82 ± 3	77.8 ± 2.0
12.	Pumping well S-1, Crakow	10.2.78	240	247 ± 56	105 ± 3	77.4 ± 2.0
13.	Pumping well S-9, Crakow	25.2.78	240	286 ± 70	100 ± 3	78.8 ± 2.0
14.	Nowe spring, Czatkowice	02.3.78	240	151 ± 57	10 ± 2	42.0 ± 1.0

Results of ^{85}Kr dating

Using the procedure described earlier, 14 water samples were measured. The samples were taken from surface and groundwater. Tritium and radiocarbon were also measured. The procedure for measurement of tritium and radiocarbon is described elsewhere [15, 16]. The results are shown in Table III.

The first five samples represent the surface water and have a mean ^{85}Kr concentration of 730 ± 23 dpm/mmol Kr. This value corresponds within the error limit to the actual concentration of ^{85}Kr in the atmosphere. Tritium and radio-carbon data also indicate contemporary water. The same comment applies to the karst spring in the Tatra mountains.

The highest concentration of ^{85}Kr (very similar to that of atmospheric krypton) was found in the sample from the Raba River. This can be explained by the fact that, owing to turbulent flow, the river water is being saturated with the fresh atmospheric krypton during its rapid re-aeration. It might perhaps be possible to measure the re-aeration rate of polluted river by determining the ^{85}Kr in the river at sampling points downstream from the pollution source. In this case the ^{85}Kr activity should be related to the water volume unit. In their experiment Tsivoglou and Wallace [17] used artificial ^{85}Kr injected into water and its escape to study the re-aeration rate coefficient and found that the time for the re-aeration process is a matter of hours.

The lower tritium content in the river as compared with surface reservoirs is probably caused by a certain contribution of low-tritium groundwater base flow.

It should be noted that the surface water reservoirs with older groundwater contribution and low rate of mixing could indicate lower than atmospheric ^{85}Kr concentration. For surface water, tritium and ^{85}Kr yield quantitatively different information. The tritium content depends strongly on the "old" water contribution; on the other hand, ^{85}Kr concentration on the surface is close to the atmospheric value. The ^{85}Kr content in deep profiles, if different, indicates the low vertical mixing.

The next three samples represent deep groundwater from the Cretaceaous aquifer. The aim in sampling this water was to establish the level of contamination with the atmospheric krypton during sampling and processing. The particular well was chosen for two reasons:

(1) Multiple sampling and measurement of tritium and radiocarbon over several years indicated tritium-free water and a radiocarbon concentration of about 49 pmc (per cent modern carbon).
(2) The construction of the well (casing, pumping well) ensures proper sampling.

The mean value obtained for this "krypton-dead" water (39 ± 37 dpm/mmol Kr) is equal to 5% of the actual concentration of ^{85}Kr in the atmosphere and can be

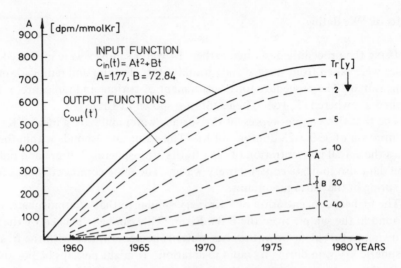

FIG.6.   Curves of mean residence time (exponential model).
        A – Sample No.10
        B – Mean value of samples 11, 12 and 13
        C – Sample No.14.

considered as the present contamination level of the method. As this "blank"
value lies within the assumed detection limit, it was not subtracted from the
results of other samples.

The Slotwinka mineral spring gave an ^{85}Kr concentration of 362 ± 143 dpm/
mmol Kr and 42 ± 2 TU. The low radiocarbon concentration is 247 ± 20 dpm/mmol
content of free CO_2 of volcanic origin (HCO_3^- is 3258 mg/l and free CO_2 is
2700 mg/l).

Samples 11, 12 and 13 originate from a continuously exploited system of
wells (about 20 m deep). Their mean ^{85}Kr concentration is 247 ± 20 dpm/mmol
Kr, which is in the range of "^{85}Kr dating". Also, the last sample from the Nowe
spring can be "dated".

For the last five samples from Table II, the mean residence time in the system
can be calculated, assuming a "good mixing" exponential model and the ^{85}Kr input
function. Figure 6 illustrates the input and output functions for various residence
times. The input function is approximated by the polynomial of second degree [14].
The three points shown in the graph correspond to sample No.10 (A), the mean
value of samples 11, 12 and 13 (B) and sample 14 (C). The krypton age would
correspond to about 10, 20 and 40 years, respectively.

CONCLUSIONS

The results of a limited number of samples prove that the krypton method works satisfactorily and can be used for dating young water bodies. It may provide more unequivocal values than the tritium method where the input function is complicated.

A very serious limitation of the ^{85}Kr method for routine use is the large sample size (120 to 360 litres) and rather complicated and time-consuming analytical procedure. We hope that the low yield of the krypton separation can be improved so as to result in a smaller sample size. On the other hand, a further increase of the ^{85}Kr concentration in the atmosphere can be expected, which will make the procedure easier.

The ^{85}Kr method, although more complicated than the tritium method, can be used as a new tool for the isotope methods in modern hydrology.

REFERENCES

[1] SCHRÖDER, J., Z. Naturforsch. **30a** (1975) 962.
[2] SCHRÖDER, J., MÜNNICH, K.O., EHHALT, D.H., Nature **233** (1971) 614.
[3] CSONGER, E., Acta Phys. Acad. Sci. Hungar. **34** (1973) 249.
[4] FARGES, L., PATTI, F., GROS, R., BOURGEON, P., J. Radioanalyt. Chem. **22** (1974) 147.
[5] SITTKUS, A., STOCKBURGER, H., Naturwissenschaften **63** (1976) 266.
[6] TERTYSNIK, E., SIVERIN, A., BARBANOV, V., At. Ehnerg. **42** (1977) 2.
[7] TELEGADAS, K., FERBER, G.J., Science **190** (1975) 882.
[8] HELLER, D., ROEDEL, W., MÜNNICH, K.O., Naturwissenschaften **64** (1977) 383.
[9] RÓZAŃSKI, K., OSTROWSKI, A., Nukleonika **22** (1977) 343.
[10] STOCKBURGER, H., SARTORIUS, H., SITTKUS, A., Z. Naturforsch. **32a** (1977) 1249.
[11] PANNETIER, R., "Distribution, transport atmosphérique et bilan du Kr-85", CEA-R-3591 (1968).
[12] KORAL, L., BARBANOV, V., ROMANOVSKAYA, A., Fiz. Atom. Okeana **12** (1976) 787.
[13] SCHRÖDER, J., ROETHER, W., in Isotope Ratios as Pollutant Source and Behaviour Indicators (Proc. Symp. Vienna, 1975), IAEA, Vienna (1975) 231.
[14] RÓZAŃSKI, K., Determination of ^{85}Kr in groundwater, Diss. (in Polish), Inst. of Phys. Nucl. Techn., Crakow (1977).
[15] FLORKOWSKI, T., GRABCZAK, J., Nukleonika **20** (1975) 274.
[16] FLORKOWSKI, T., GRABCZAK, J., KUC, T., RÓZAŃSKI, K., Nukleonika **20** (1975) 1053.
[17] TSIVOGLOU, E.C., WALLACE, J.R., in Isotope Hydrology (Proc. Symp. Vienna, 1970), IAEA, Vienna (1970) 509.

DISCUSSION

H.H. LOOSLI: I would like to confirm that improvements can be made which allow higher extraction yields with smaller amounts of water. In Bern we have developed a system for separating krypton from the extracted gases which is almost identical to your system. Our extraction yield for krypton is now about 80%, and, since the backgrounds of our counters are much lower than yours, we can at present measure the ^{85}Kr activities of samples extracted from 50–100 litres of water.

T. FLORKOWSKI: I believe that the sample size of 50 to 60 litres, which is comparable with those used for radiocarbon analysis, would satisfy us in the future.

U. SIEGENTHALER: In Fig.6 you compare measured ^{85}Kr concentrations with values computed by means of the exponential model. By considering ^{3}H and ^{85}Kr together, one could perhaps find out something about the actual mixing pattern (i.e. the age distribution). Have you tried this, for example by computing a tritium age with the help of the exponential model?

T. FLORKOWSKI: Unfortunately we have not been able to calculate the tritium age of the samples owing to difficult hydrological conditions. In fact, our main purpose was to demonstrate the feasibility of krypton dating and to show that there is no contradiction with tritium results. More detailed comparisons between tritium and krypton dating will be made in the near future.

A. PLATA BEDMAR: Table III shows some samples of groundwater with ^{14}C ages of thousands of years but with ^{85}Kr ages of only a few tens of years. What is the reason for this?

T. FLORKOWSKI: In Table III we give the radiocarbon content in per cent modern carbon (pmc) and not ages in years. In general, all our radiocarbon data relate to modern water, except in the cases of the mineral spring, where there is a contribution from volcanic CO_2, and of the Nowe spring, where the radiocarbon content is only 42 pmc, probably because of the limestone aquifer.

A. ZUBER: The tritium value obtained in the case of sample 14 in Table III cannot be interpreted quantitatively, because the input function was not well known. Mr. Florkowski is right in stating that this value does not contradict the ^{85}Kr value. · As far as the ^{14}C of this sampling point is concerned, I would like to add that the water comes from limestones and dolomites. We know from many other measurements in the area that ^{14}C values are much lower than those expected from the hydrodynamic and tritium data. This is because of exchange and dissolution effects.

J. MARTINEC: The residence time of groundwater determined by tritium depends on the distribution of groundwater recharge in winter and summer, which is frequently not known. Parallel ^{85}Kr and tritium measurements could provide this additional information.

T. FLORKOWSKI: I quite agree.

T. DINÇER: In your oral presentation you mentioned the high variability of tritium in precipitation as a drawback of the tritium dating method. However, the tritium content of the recharge to groundwater is not so variable and shows a smooth pulse peaking in about 1963 and 1964, which is easy to use for tritium age calculations.

T. FLORKOWSKI: This is true if only winter precipitation is used for calculation of recharge.

K.O. MÜNNICH *(Chairman)*: I would like to summarize briefly the results of this Symposium. I feel that it has been a good one in many respects. You will have noticed that there are now quite a number of established techniques, available for routine use. It is very gratifying to see them widely applied on this routine basis all over the world and with great success, in countries with highly developed technical facilities as well as in developing regions. You have heard quite a number of very interesting papers on hydrological studies in those regions where isotope methods have played a central role in allowing clear-cut conclusions to be drawn about the hydrological situation — conclusions which then form the basis for planning and for use of the resources available.

This brings me to another observation which I think is worth making. Isotope specialists now seem to have learned to make sufficient use of conventional methods. In other words, close co-operation between people with different scientific backgrounds is now obviously beginning to produce results. I am particularly pleased to see the great success achieved by the combination of isotope methods with refined methods of groundwater chemistry. When used together, they can produce very valuable information on the climatic variations of the past, and they seem to reveal details one could not have dreamed of learning about only a few years ago.

I said in the beginning that isotope methods now seem to have acquired a kind of routine status. I think we may have high hopes for the future in that I cannot see any signs of these methods getting old, or losing their flexibility; on the contrary, we have seen many papers describing not only new and promising methods, but also new ways of thinking and of asking provocative questions about established ways of thinking. This to me is a very positive sign.

CHAIRMEN OF SESSIONS

Session I	F. NEUMAIER	Federal Republic of Germany
Session II	S.M. RAO	India
Session III	R.G. THOMAS	Food and Agriculture Organization of the United Nations (FAO)
Session IV	A. ZUBER	Poland
Session V	E. ERIKSSON	Sweden
Session VI	J.Ch. FONTES	France
Session VII	C.T. RIGHTMIRE	United States of America
Session VIII	D.B. SMITH	United Kingdom
Session IX, Part 1	E. LINARES	Argentina
Session IX, Part 2	K.O. MÜNNICH	Federal Republic of Germany

SECRETARIAT

Scientific Secretary:	W. DROST	Division of Research and Laboratories, IAEA
Administrative Secretary:	Caroline de MOL VAN OTTERLOO	Division of External Relations, IAEA
Editor:	Monica KRIPPNER	Division of Publications, IAEA
Records Officer:	H. ORMEROD	Division of Languages, IAEA

LIST OF PARTICIPANTS

ALGERIA

Demmak, A.   Direction des Etudes de milieu et de la recherche
                              hydraulique,
                              Clairbois, Birmandreij, Algiers

ARGENTINA

Gomez, H.R.   Comisión Nacional de Energía Atómica,
                              Avenida del Libertador 8250, Buenos Aires-1429

Linares, E.   Instituto de Geocronología y Geología Isotópica,
                              Ciudad Universitaria, Pabellón 2, Buenos Aires-1428

AUSTRALIA

Airey, P.L.   Australian Atomic Energy Commission,
                              Lucas Heights, NSW 2232

AUSTRIA

Ambach, W.   Physikalisches Institut der Universität Innsbruck,
                              Schöpfstrasse 41, A-6020 Innsbruck

Boroviczény, F.   Geologische Bundesanstalt,
                              Rasumofskygasse 23, A-1031 Vienna

Kollmann, W.   Geologische Bundesanstalt,
                              Rasumofskygasse 23, A-1031 Vienna

Nussbaumer, W.   Bundesversuchs- und Forschungsanstalt Arsenal,
                              Geotechnisches Institut,
                              Postfach 8, A-1030 Vienna

Rank, D.   Bundesversuchs- und Forschungsanstalt Arsenal,
                              Geotechnisches Institut,
                              Postfach 8, A-1030 Vienna

AUSTRIA (cont'd)

Zoetl, J.

Abteilung für Hydrogeologie, Technische Universität
  Graz,
Rechbauerstrasse 12, A-8010 Graz

Zojer, H.

Institut für Hydrogeologie und Geothermie am
  Forschungszentrum Graz,
Rechbauerstrasse 12, A-8010 Graz

BELGIUM

Bonne, A.

Centre d'étude de l'énergie nucléaire (C.E.N./S.C.K.)
Boeretang 200, B-2400 Mol

BOTSWANA

Dinçer, T.

Department of Water Affairs,
Private Bag 0029, Gaborone

Khupe, B.B.J.

Department of Water Affairs,
Private Bag 0029, Gaborone

BRAZIL

Roldao, J.S.

COPPE/UFRJ, Ilha Universitária, bloco G,
Caixa Postal 1191, 20000 Rio de Janeiro

CANADA

Barker, J.F.

Department of Earth Sciences, University of Waterloo,
Waterloo, Ontario N2L 3G1

Fritz, P.

Department of Earth Sciences, University of Waterloo,
Waterloo, Ontario N2L 3G1

Merritt, W.F.

Atomic Energy of Canada Limited,
Chalk River Nuclear Laboratories,
Chalk River, Ontario K0J 1J0

Weyer, K.U.

Fisheries & Environment Canada,
Hydrology Research Division,
4616 Valiant Drive N.W., Calgary, Alberta T3A 0X9

CHILE

Silva Hennings, C.   Comisión Chilena de Energía Nuclear, Casilla 188-D,
Los Jesuitas 645, Santiago

COLOMBIA

Rodriguez, C.O.   Instituto de Asuntos Nucleares,
Apartado Aéreo 8595, Bogotá, D.E.

CYPRUS

Jacovides, J.S.   Water Development Department, Ministry of
Agriculture and Natural Resources,
Nicosia

CZECHOSLOVAK SOCIALIST REPUBLIC

Holata, M.   Faculty of Constructions, Czech Technical University,
Zikova 4, 160 00 Prague 6

Kumičák, J.   Ustav rádioekológie a využitia hadrovej techniky,
Komenského 9,
040 61 Košice

Šmejkal, V.   Geological Survey,
Malostranské 19, 118 21 Prague 1

DENMARK

Sevel, T.   Danish Isotope Centre,
2 Skelbaekgade, DK-1717 Copenhagen V

FINLAND

Lumiaho, K.   Geological Survey of Finland,
Kivimiehentie 1, SF-02150 Espoo 15

Salmi, M.H.   Geological Survey of Finland,
Kivimiehentie 1, SF-02150 Espoo 15

FRANCE

Artaud, J.	CEA, Centre d'études nucléaires de Grenoble, Département de chimie appliquée, 85 X, F-38041 Grenoble Cedex
Conrad, G.	Laboratoire de géologie, Institut scientifique de Haute-Normandie, Université de Rouen, F-76130 Mont-Saint-Aignan
Delmas, J.	CEA, Centre d'études nucléaires de Cadarache, Section de radioécologie, B.P. 1, F-13115 Saint-Paul-lez-Durance
Dufour, H.	Commissariat à l'énergie atomique, 31–33 rue de la Fédération, F-75752 Paris Cedex 15
Fontes, J.-Ch.	Laboratoire de géologie dynamique, 4 place Jussieu, F-75230 Paris Cedex 05
Gallo, G.M.J.	Centre national de la recherche scientifique (C.N.R.S.), Service de relations internationales, 15 quai Anatole France, F-75007 Paris
Géneau, P.	Etablissements S.P.S., B.P. 17, F-91310 Montlhéry
Guizerix, J.	CEA, Centre d'études nucléaires de Grenoble, Service d'application des radioéléments et des rayonnements, 85 X, F-38041 Grenoble Cedex
Jouzel, J.	CEA, Centre d'études nucléaires de Saclay, Département de recherche et analyse, B.P. 2, F-91190 Gif-sur-Yvette
Madoz-Escande, Chantal J.	CEA, Centre d'études nucléaires de Fontenay-aux-Roses, B.P. 6, F-92260 Fontenay-aux-Roses
Marcé, A.	Bureau de recherches géologiques et minières (B.R.G.M.), Service géologique national, Département minéralogie, geóchimie, analyses, B.P. 6009, F-45018 Orléans Cedex
Marcucci, S.	DE, Service RDE, B III, B.P. 561, F-92542 Montrouge Cedex
Molinari, J.P.	Centre de formation internationale à la gestion des ressources en eau (CEFIGRE), B.P. 13, Sophia Antipolis, F-06560 Valbonne

Moutonnet, P.   CEA, Centre d'études nucléaires de Cadarache, DB/SRA,
                                       B.P. 1, F-13115 Saint-Paul-lez-Durance

GERMAN DEMOCRATIC REPUBLIC

Hübner, H.   Zentralinstitut für Isotopen- und Strahlenforschung
                                          der Akademie der Wissenschaften der DDR,
                                       Permoserstrasse 15, DDR-705 Leipzig

GERMANY, FEDERAL REPUBLIC OF

Aust, H.E.   Bundesanstalt für Geowissenschaften und Rohstoffe,
                                       Postfach 51 01 53, D-3000 Hanover 51

Backhaus, G.   Physikalisch-Chemisches Institut der Universität
                                          München,
                                       Sophienstrasse 11, D-8000 Munich 2

Behrens, H,K.   Institut für Radiohydrometrie der Gesellschaft für
                                          Strahlen- und Umweltforschung mbH,
                                       Ingolstädter Landstrasse 1, D-8042 Neuherberg

Esser, N.   Institut für Umweltphysik der Universität Heidelberg,
                                       Im Neuenheimer Feld 366, D-6900 Heidelberg

Geyh, M.A.   Niedersächsisches Landesamt für Bodenforschung,
                                       Postfach 51 01 53, D-3000 Hanover 51

Giese, K.   Postfach 18 01 45, D-5000 Cologne 1

Grimmelmann, W.F.   c/o United Nations Economic Commission for Africa,
                                       P.O. Box 3005, Addis Ababa, Ethiopia

Gruber, J.   Hahn-Meitner-Institut für Kernforschung Berlin GmbH,
                                       Glienicker Strasse 100, D-1000 Berlin 39

Hauschild, J.   Institut für Geophysikalische Wissenschaften der
                                          Freien Universität Berlin,
                                       Thielallee 50, D-1000 Berlin 33

Herrmann, H.   Bayerische Landesanstalt für Wasserforschung,
                                       Kaulbachstrasse 37, D-8000 Munich 22

Hofius, K.   Bundesanstalt für Gewässerkunde,
                                       Postfach 309, D-5400 Coblenz

Illi, H.   Physikalisch-Technische Bundesanstalt,
                                       Bundesallee 100, D-3300 Braunschweig (Brunswick)

GERMANY, FEDERAL REPUBLIC OF (cont'd)

Kaiser, K.	Varian MAT GmbH, Allacher Strasse 230e, D-8000 Munich 50
Keller, R.	Geographisches Institut I der Universität Freiburg, Werderring 4, D-7800 Freiburg
Klotz, D.	Institut für Radiohydrometrie der Gesellschaft für Strahlen- und Umweltforschung mbH, Ingolstädter Landstrasse 1, D-8042 Neuherberg
Matthess, G.	Geologisch-Palaeontologisches Institut der Universität Kiel, Olshausenstrasse 40/60, D-2300 Kiel
Moser, H.	Institut für Radiohydrometrie der Gesellschaft für Strahlen- und Umweltforschung mbH, Ingolstädter Landstrasse 1, D-8042 Neuherberg
Münnich, K.O.	Institut für Umweltphysik der Universität Heidelberg, Im Neuenheimer Feld 366, D-6900 Heidelberg
Neumaier, F.	Institut für Radiohydrometrie der Gesellschaft für Strahlen- und Umweltforschung mbH, Ingolstädter Landstrasse 1, D-8042 Neuherberg
Nielsen, H.	Geochemisches Institut der Universität Göttingen, Goldschmidtstrasse 1, D-3400 Göttingen
Oerter, H.	Institut für Radiohydrometrie der Gesellschaft für Strahlen- und Umweltforschung mbH, Ingolstädter Landstrasse 1, D-8042 Neuherberg
Oesterle, F.-P.	Physikalisch-Technische Bundesanstalt, Bundesallee 100, D-3300 Braunschweig (Brunswick)
Otto, Karin	Institut für Geophysikalische Wissenschaften der Freien Universität Berlin, Thielallee 50, D-1000 Berlin 33
Rauert, W.	Institut für Radiohydrometrie der Gesellschaft für Strahlen- und Umweltforschung mbH, Ingolstädter Landstrasse 1, D-8042 Neuherberg
Schneider, J.	Hahn-Meitner-Institut für Kernforschung Berlin, Glienicker Strasse 100, D-1000 Berlin 39
Schwarzer, K.	Kernforschungsanlage Jülich GmbH, Postfach 1913, D-5170 Jülich 1

Seiler, K.-P.   Institut für Radiohydrometrie der Gesellschaft für
                             Strahlen- und Umweltforschung mbH,
                             Ingolstädter Landstrasse 1, D-8042 Neuherberg

Sonntag, C.   Institut für Umweltphysik der Universität Heidelberg,
                             Im Neuenheimer Feld 366, D-6900 Heidelberg

Stichler, W.   Institut für Radiohydrometrie der Gesellschaft für
                             Strahlen- und Umweltforschung mbH,
                             Ingolstädter Landstrasse 1, D-8042 Neuherberg

Thoma, G.   Institut für Umweltphysik der Universität Heidelberg,
                             Im Neuenheimer Feld 366, D-6900 Heidelberg

Trimborn, P.   Institut für Radiohydrometrie der Gesellschaft für
                             Strahlen- und Umweltforschung mbH,
                             Ingolstädter Landstrasse 1, D-8042 Neuherberg

Wendt, I.   Bundesanstalt für Geowissenschaften und Rohstoffe,
                             Stilleweg 2, D-3000 Hanover

GREECE

Leontiadis, I.   Nuclear Research Centre "Democritos",
                             Aghia Paraskevi, Athens

HUNGARY

Balla, B.   Institute of Isotopes, Hungarian Academy of Sciences,
                             P.O. Box 77, H-1525 Budapest

Deák, J.   Research Centre for Water Resource Development,
                             Kvassay út 1, H-1095 Budapest

ICELAND

Theodorsson, P.   Science Institute, University of Iceland,
                             Dunhaga 3, Reykjavik

INDIA

Gupta, S.K.   Physical Research Laboratory, Navrangpura,
                             Ahmedabad-380 009

Rao, S.M.   Bhabha Atomic Research Centre, Isotope Division,
                             Trombay, Bombay 400 085

IRAN

Movahhed, M.A.   Ministry of Energy, Water Resources Research Institute,
                                 B.O. Box 64/186, Teheran 16475

ISRAEL

Issar, A.S.   Institute for Desert Research, Ben-Gurion University,
                                 Sede Boqer, Beer-Sheva 89 190

ITALY

Bello, A.   AQUATER S.p.A.,
                                 Via Miralbello, C.P. 20, I-61047 San Lorenzo in Campo

Cassano, G.   C.R.N. TRISAIA del CNEN, C.P. CNEN,
                                 I-75025 Policoro

Cortecci, G.   Laboratorio di Geologia Nucleare,
                                 Via S. Maria 22, I-56100 Pisa

Noto, P.   Istituto Internazionale per le Ricerche Geotermiche,
                                 Via del Buongusto 1, I-56100 Pisa

Nuti, S.   Istituto Internazionale per le Ricerche Geotermiche,
                                 Via del Buongusto 1, I-56100 Pisa

Ricchiuto, T.   AGIP Mineraria, Servizio SGEL,
                                 I-20097 S. Donato Milanese

Tazioli, G.S.   Istituto di Geologia Applicata e Geotecnica,
                                 Via Toma N. 34, I-75125 Bari

JAMAICA

White, M.N.   Water Resources Division,
                                 Hope Gardens, P.O. Box 91, Kingston 6

KENYA

Charania, S.H.   Ministry of Water Development,
                                 P.O. Box 14190, Nairobi

MEXICO

Andreu Ibarra, B.

Centro de Estudios Nucleares, Universidad Nacional
    Autónoma de México,
Circuito Exterior, C.U., Mexico 20, D.F.

Gálvez Cruz, L.

Centro de Estudios Nucleares, Universidad Nacional
    Autónoma de México,
Circuito Exterior, C.U., Mexico 20, D.F.

González Hita, L.

Grupo de Física, Subsecretaría de Planeación,
Secretaría de Agricultura y Recursos Hidráulicos,
Teotihuacan No. 19, Mexico 11, D.F.

Latorre Díez, C.

Grupo de Física, Subsecretaría de Planeación,
Secretaría de Agricultura y Recursos Hidráulicos,
Teotihuacan No. 19, Mexico 11, D.F.

Rivera Carrillo, J.J.

Comisión Federal de Electricidad,
P.O. Box, Tlalnepantla, Edo. de México

NETHERLANDS

Das, H.A.

Netherlands Energy Research Foundation (ECN),
Westerduinweg 3, Petten (N.H.)

Hut, G.

Isotope Physics Laboratory, University of Groningen,
Westersingel 34, 9718 CM Groningen

Mook, W.G.

Isotope Physics Laboratory, University of Groningen,
Westersingel 34, 9718 CM Groningen

Romijn, E.

Provincial Water Board of Gelderland,
Marktstraat 1, Arnhem

Van der Straaten, C.M.

Isotope Physics Laboratory, University of Groningen,
Westersingel 34, 9718 CM Groningen

Van Duijvenbooden, W.

National Institute for Water Supply,
P.O. Box 150, Leidschendam

NIGER

Hassane, A.

Direction Hydraulique, Ministère des Mines et de
    l'Hydraulique,
B.P. 257, Niamey

NORWAY

Berg, E.   Norwegian Water Resources and Electricity Board,
                            P.O. Box 5091, Majorstua, Oslo 3

Dahl, J.B.   Institutt for Atomenergi,
                            P.O. Box 40, N-2007 Kjeller

Steenberg, K.   Isotope Laboratory, Agricultural University of Norway,
                            P.O. Box 26, N-1432 Aas-NLH

Sundøen, K.   Norwegian Water Resources and Electricity Board,
                            P.O. Box 5091, Majorstua, Oslo 3

PAKISTAN

Sajjad, M.I.   Radiation and Isotope Applications Division,
                            Pakistan Institute of Nuclear Science and Technology,
                            (Pinstech), P.O., Nilore
                            (Now with Isotope Hydrology, Division of Research
                                and Laboratories, IAEA)

POLAND

Dowgiallo, J.   Research Centre for Geological Sciences,
                            Polish Academy of Sciences,
                            Zwirki i Wigury 93, 02-089 Warsaw

Makowski, J.   Institute of Hydroengineering, Polish Academy of Sciences,
                            Cystersów 11, 80-953 Gdańsk-Oliwa

Ponikiewska, Hanna M.   Institute of Meteorology and Water Management,
                            ul. Podleśna 61, 01-573 Warsaw

Zuber, A.   Institute of Nuclear Physics,
                            Radzikowskiego 152, Cracow

ROMANIA

Blaga, L.   State Committee for Nuclear Energy,
                            Institute of Isotopic and Molecular Techniques,
                            P.O. Box 243, Cluj-Napoca

SAUDI ARABIA

Shampine, W.J.

Water Resources Development Department,
Ministry of Agriculture and Water,
P.O. Box 5927, Riyadh

SOUTH AFRICA

Smith, P.E.

Nuclear Physics Research Unit,
University of Witwatersrand,
Jan Smuts Avenue, Johannesburg

Verhagen, B.Th.

Nuclear Physics Research Unit,
University of Witwatersrand,
Jan Smuts Avenue, Johannesburg

SPAIN

Plata Bedmar, A.

Gabinete de Aplicaciones Nucleares a las O.P.,
Ministerio de O.P. y Urbanismo,
Paseo Bajo de la Virgen del Puerto 3, Madrid-5

SWEDEN

Burgman, J.O.S.

Uppsala University, Division of Hydrology,
Department of Physical Geography,
Box 554, S-751 22 Uppsala

Eriksson, E.

Uppsala University, Division of Hydrology,
Department of Physical Geography,
Box 554, S-751 22 Uppsala

Kostov, L.

Uppsala University, Division of Hydrology,
Department of Physical Geography,
Box 554, S-751 22 Uppsala

Landstöm, O.K.L.

AB Atomenergi, Studsvik, Fack,
S-611 01 Nyköping

Nordberg, L.O.

Geological Survey of Sweden, Fack,
S-10405 Stockholm

SWITZERLAND

Loosli, H.H.

Physics Institute, University of Bern,
Sidlerstrasse 5, CH-3012 Bern

SWITZERLAND (cont'd)

Martinec, J.   Federal Institute for Snow and Avalanche Research,
                                          CH-7260 Weissfluhjoch/Davos

Oeschger, H.   Physics Institute, University of Bern,
                                          Sidlerstrasse 5, CH-3012 Bern

Schotterer, U.   Physics Institute, University of Bern,
                                          Sidlerstrasse 5, CH-3012 Bern

Siegenthaler, U.   Physics Institute, University of Bern,
                                          Sidlerstrasse 5, CH-3012 Bern

Wildberger, A.   Geological Institute, University of Bern,
                                          Sahlistrasse 5, CH-3012 Bern

THAILAND

Rouysungnern, S.   Watershed Research Section, Watershed Management
                                             Division,
                                          Royal Dorest Department, Bangkok 9

TURKEY

Yurtsever, Y.   General Directorate of Electric Power Resources,
                                          Division of Hydrology,
                                          Demirtepe-Ankara

UGANDA

Kabunduh, E.   Ministry of Land and Water Resources,
                                          Water Development Department,
                                          P.O. Box 19, Entebbe

UNITED KINGDOM

Bath, A.H.   Institute of Geological Sciences, Hydrology Unit,
                                          MacLean Building, Wallingford OX10 8BB

Downing, R.A.   Central Water Planning Unit, Reading Bridge House,
                                          Reading, Berks. RG1 8PS

Elzerman, A.W.   Environmental and Medical Science Division,
                                          Building 364, AERE, Didcot, Oxon OX11 0RA

Evans, G.V.	Hydrology and Coastal Sediment Project, Nuclear Physics Division, Building 7, AERE, Harwell, Oxon OX11 0RA
Ferguson, D.S.	Sir William Halcrow and Partners Consulting Engineers, Newcombe House, 45 Notting Hill Gate, London W11 3JX
Hutton, L.G.	Department of Civil Engineering, University of Technology, Loughborough, Leicestershire LE11 3TU
Lakey, J.R.	Royal Naval College, Greenwich, London SE10 9NN
Lloyd, J.W.	University of Birmingham, Department of Geological Sciences, P.O. Box 363, Birmingham B15 2TT
Nelson, A.	Thames Water Authority, 177 Rosebery Avenue, London EC1R 4TP
Otlet, R.L.	Hydrology and Coastal Sediment Project, Nuclear Physics Division, AERE, Harwell, Didcot, Oxon OX11 0RA
Smith, D.B.	Natural Environment Research Council, Polaris House, North Star Avenue, Swindon
Unsworth, W.D.	VG-Isotopes Limited, Ion Path, Winsford, Cheshire CW7 3BX
White, K.E.	Water Research Centre, Stevenage Laboratory, Elder Way, Stevenage, Herts. SG1 5EB

UNITED STATES OF AMERICA

Barnes, R.S.	Water and Air Resources Division, Civil Engineering Department, FX-10, University of Washington, Seattle, WA 98195
Barr, G.E.	Sandia Laboratories, Division 5311, P.O. Box 5800, Albuquerque, NM 87115
Krueger, H.W.	Krueger Enterprises, Inc., Geochron Labs. Div., 24 Blackstone Street, Cambridge, MA 02139

UNITED STATES OF AMERICA (cont'd)

Rightmire, C.T.

US Geological Survey, Water Resources Division,
Idaho National Engineering Laboratory,
P.O. Box 2230, Idaho Falls, ID 83401

Torgersen, T.

Woods Hole Oceanographic Institution,
Woods Hole, MA 02543

YUGOSLAVIA

Anovski, T.

Centre for Application of Radioisotopes in Science
and Industry,
Karpuš II, YU-91000 Skopje

Boreli, F.

Electrotechnical Faculty, Laboratory of Physics,
University of Belgrade,
Bulevar Revolucije 73, YU-11000 Belgrade

Dolenec, T.

Faculty of Science and Technology, Department of
Geology,
p.p. 311, YU-61001 Ljubljana

Pezdič, J.

Institute Jožef Štefan, University of Ljubljana,
Jamova 39, YU-61000 Ljubljana

Šimunić, Z.

Vodoprivreda Bosne i Hercegovine,
Stup Briješćanska 1, YU-71000 Sarajevo

Žižek, D.

Health Resort "Radenska",
Radenci, Croatia

Organizations

FOOD AND AGRICULTURE ORGANIZATION OF THE UNITED NATIONS (FAO)

Thomas, R.G.

Water Resources, Development of Management Services,
Land and Water Development Division,
Via delle Terme di Caracalla, I-00100 Rome, Italy

INTERNATIONAL ATOMIC ENERGY AGENCY (IAEA)

Dray, M.

Isotope Hydrology, Division of Research and
Laboratories

Florkowski, T.	Isotope Hydrology, Division of Research and Laboratories
Gonfiantini, R.	Isotope Hydrology, Division of Research and Laboratories
Payne, B.R.	Isotope Hydrology, Division of Research and Laboratories
Pozo Contreras, A.	Isotope Hydrology, Division of Research and Laboratories
Quijano, J.L.	Isotope Hydrology, Division of Research and Laboratories
Zuppi, G.M.	Isotope Hydrology, Division of Research and Laboratories

INTERNATIONAL UNION OF PURE AND APPLIED CHEMISTRY (IUPAC)

Lux, F.R.	Institut für Radiochemie der Technischen Universität München, D-8046 Garching Federal Republic of Germany

UNITED NATIONS EDUCATIONAL, SCIENTIFIC AND CULTURAL ORGANIZATION (UNESCO)

Gilbrich, W.H.	Division of Water Sciences, 7 place de Fontenoy, F-75700 Paris, France

AUTHOR INDEX

Roman numerals are volume numbers.
Italic numerals refer to the first page of a paper by an author.
Upright numerals denote the page numbers of participation in discussions.
(Vol.I is paginated from 3—440 and Vol.II from 443—984)

Airey, P.L.: I 158, 199, *205,* 218, 219;
    II 462, 463, 568
Ambach, W.: II *829,* 845, 846
Andreu Ibarra, B.: I *125,* 143, 144
Andrews, J.N.: II *545*
Anovski, T.: I *99*
Aust, H.E.: I 324
Backhaus, G.: II *631*
Baonza, E.: II *847*
Barker, J.F.: I 323; II 629, 642,
    661, 678
Barnes, R.S.: II *875,* 897, 898
Barr, G.E.: II *645,* 660
Bath, A.H.: I 248, 287, 324; II *545,*
    566, 567, 568, 749, 750
Baubron, J.C.: II *585*
Behrens, H.: II *829*
Birch, P.B.: II *875*
Boreli, F.: I *99,* 109, 156, 157
Boreli, M.: I *99*
Borole, D.V.: I *181*
Bortolami, G.C.: I *327, 411*
Bosch, B.: II *585*
Brissaud, F.: II *787*
Brown, R.M.: II *661*
Burgman, J.O.S.: I *27*
Calf, G.E.: I *205*
Campbell, B.L.: I *205*
Carlier, Ph.: II *491*
Carter, J.A.: II *645*
Clarke, W.B.: II *917*

Conrad, G.: I *265,* 285, 286, 287;
    II 581
Couchat, Ph.: II *787*
Courtois, G.: II *901*
Dahl, J.B.: II 784
Datta, P.S.: I *181*
Deák, J.: I *221,* 248, 249
Degranges, P.: II *585*
Del Arenal, R.: I *125*
Desai, B.I.: I *181*
Dinçer, T.: I *3,* 25, 26, 365, 408, 439;
    II *443,* 749, 767, 873, 960
Dowgiallo, J.: II 580
Downing, R.A.: II *679,* 706, 707, 708
Dziembowski, Z.: I 65; II *733*
Edmunds, W.M.: II *545*
El-Shazly, E.M.: II *569*
Elzerman, A,W.: II 898
Eriksson, E.: I *27,* 40, 144, 157, 219,
    248, 323, 348, 365, 407, 409
Esser, N.: II *753*
Evans, G.V.: II *679*
Florkowski, T.: I 324; II 566, 800,
    873, *949,* 960, 961
Fontes, J.Ch.: I 26, 98, 200, 305,
    325, *411,* 436, 437, 439;
    II 524, 580, 846
Fritz, P.: I 81, 217, 218, 248, 263,
    408, 409, 436; II *525,* 543, 544,
    642, *661,* 707, 750, 846, 929,
    947

INDEX OF PAPER NUMBERS

The following conversion table is provided for the convenience of readers and to encourage the use of SI units.

FACTORS FOR CONVERTING SOME OT THE MORE COMMON UNITS TO INTERNATIONAL SYSTEM OF UNITS (SI) EQUIVALENTS

NOTES:

(1) SI base units are the metre (m), kilogram (kg), second (s), ampere (A), kelvin (K), candela (cd) and mole (mol).

(2) ▶ indicates SI derived units and those accepted for use with SI;

   ▷ indicates additional units accepted for use with SI for a limited time.

   [*For further information see The International System of Units (SI), 1977 ed., published in English by HMSO, London, and National Bureau of Standards, Washington, DC, and International Standards ISO-1000 and the several parts of ISO-31 published by ISO, Geneva.*]

(3) The correct abbreviation for the unit in column 1 is given in column 2.

(4) ✻ indicates conversion factors given exactly; other factors are given rounded, mostly to 4 significant figures.

   ≡ indicates a definition of an SI derived unit: [ ] in column 3+4 enclose factors given for the sake of completeness.

Column 1 Multiply data given in:	Column 2	Column 3 by:	Column 4 to obtain data in:
Radiation units			
▶ becquerel	1 Bq	(has dimensions of s^{-1})	
disintegrations per second (= dis/s)	$1 s^{-1}$	$\equiv 1.00 \times 10^0$	Bq ✻
▷ curie	1 Ci	$= 3.70 \times 10^{10}$	Bq ✻
▷ roentgen	1 R	$[= 2.58 \times 10^{-4}$	C/kg] ✻
▶ gray	1 Gy	$[\equiv 1.00 \times 10^0$	J/kg] ✻
▷ rad	1 rad	$= 1.00 \times 10^{-2}$	Gy ✻
sievert *(radiation protection only)*	1 Sv	$[= 1.00 \times 10^0$	J/kg] ✻
rem *(radiation protection only)*	1 rem	$= 1.00 \times 10^{-2}$	J/kg] ✻
Mass			
▶ unified atomic mass unit ($\frac{1}{12}$ of the mass of ^{12}C)	1 u	$[= 1.660\,57 \times 10^{-27}$	kg, approx.]
▶ tonne (= metric ton)	1 t	$[= 1.00 \times 10^3$	kg] ✻
pound mass (avoirdupois)	1 lbm	$= 4.536 \times 10^{-1}$	kg
ounce mass (avoirdupois)	1 ozm	$= 2.835 \times 10^1$	g
ton (long) (= 2240 lbm)	1 ton	$= 1.016 \times 10^3$	kg
ton (short) (= 2000 lbm)	1 short ton	$= 9.072 \times 10^2$	kg
Length			
statute mile	1 mile	$= 1.609 \times 10^0$	km
nautical mile (international)	1 n mile	$= 1.852 \times 10^0$	km ✻
yard	1 yd	$= 9.144 \times 10^{-1}$	m ✻
foot	1 ft	$= 3.048 \times 10^{-1}$	m ✻
inch	1 in	$= 2.54 \times 10^1$	mm ✻
mil (= 10^{-3} in)	1 mil	$= 2.54 \times 10^{-2}$	mm ✻
Area			
▷ hectare	1 ha	$[= 1.00 \times 10^4$	m^2] ✻
▷ barn *(effective cross-section, nuclear physics)*	1 b	$[= 1.00 \times 10^{-28}$	m^2] ✻
square mile, (statute mile)^2	1 mile^2	$= 2.590 \times 10^0$	km^2
acre	1 acre	$= 4.047 \times 10^3$	m^2
square yard	1 yd^2	$= 8.361 \times 10^{-1}$	m^2
square foot	1 ft^2	$= 9.290 \times 10^{-2}$	m^2
square inch	1 in^2	$= 6.452 \times 10^2$	mm^2
Volume			
▶ litre	1 l *or* 1 ltr	$[= 1.00 \times 10^{-3}$	m^3] ✻
cubic yard	1 yd^3	$= 7.646 \times 10^{-1}$	m^3
cubic foot	1 ft^3	$= 2.832 \times 10^{-2}$	m^3
cubic inch	1 in^3	$= 1.639 \times 10^4$	mm^3
gallon (imperial)	1 gal (UK)	$= 4.546 \times 10^{-3}$	m^3
gallon (US liquid)	1 gal (US)	$= 3.785 \times 10^{-3}$	m^3
Velocity, acceleration			
foot per second (= fps)	1 ft/s	$= 3.048 \times 10^{-1}$	m/s ✻
foot per minute	1 ft/min	$= 5.08 \times 10^{-3}$	m/s ✻
mile per hour (= mph)	1 mile/h	$=\begin{cases} 4.470 \times 10^{-1} \\ 1.609 \times 10^0 \end{cases}$	m/s km/h
▷ knot (international)	1 knot	$= 1.852 \times 10^0$	km/h ✻
free fall, standard, g		$= 9.807 \times 10^0$	m/s^2
foot per second squared	1 ft/s^2	$= 3.048 \times 10^{-1}$	m/s^2 ✻

This table has been prepared by E.R.A. Beck for use by the Division of Publications of the IAEA. While every effort has been made to ensure accuracy, the Agency cannot be held responsible for errors arising from the use of this table.

Column 1 Multiply data given in:	Column 2	Column 3 by:	Column 4 to obtain data in:
Density, volumetric rate			
pound mass per cubic inch	1 lbm/in^3	$= 2.768 \times 10^4$	kg/m^3
pound mass per cubic foot	1 lbm/ft^3	$= 1.602 \times 10^1$	kg/m^3
cubic feet per second	1 ft^3/s	$= 2.832 \times 10^{-2}$	m^3/s
cubic feet per minute	1 ft^3/min	$= 4.719 \times 10^{-4}$	m^3/s
Force			
▶ newton	1 N	$[\equiv 1.00 \times 10^0$	m·kg·s^{-2}]✳
dyne	1 dyn	$= 1.00 \times 10^{-5}$	N ✳
kilogram force (= kilopond (kp))	1 kgf	$= 9.807 \times 10^0$	N
poundal	1 pdl	$= 1.383 \times 10^{-1}$	N
pound force (avoirdupois)	1 lbf	$= 4.448 \times 10^0$	N
ounce force (avoirdupois)	1 ozf	$= 2.780 \times 10^{-1}$	N
Pressure, stress			
▶ pascal	1 Pa	$[\equiv 1.00 \times 10^0$	N/m^2] ✳
▷ atmosphere^a, standard	1 atm	$= 1.013\ 25 \times 10^5$	Pa ✳
▷ bar	1 bar	$= 1.00 \times 10^5$	Pa ✳
centimetres of mercury (0°C)	1 cmHg	$= 1.333 \times 10^3$	Pa
dyne per square centimetre	1 dyn/cm^2	$= 1.00 \times 10^{-1}$	Pa ✳
feet of water (4°C)	1 ftH$_2$O	$= 2.989 \times 10^3$	Pa
inches of mercury (0°C)	1 inHg	$= 3.386 \times 10^3$	Pa
inches of water (4°C)	1 inH$_2$O	$= 2.491 \times 10^2$	Pa
kilogram force per square centimetre	1 kgf/cm^2	$= 9.807 \times 10^4$	Pa
pound force per square foot	1 lbf/ft^2	$= 4.788 \times 10^1$	Pa
pound force per square inch (= psi) ^{b}	1 lbf/in^2	$= 6.895 \times 10^3$	Pa
torr (0°C) (= mmHg)	1 torr	$= 1.333 \times 10^2$	Pa
Energy, work, quantity of heat			
▶ joule ($\equiv$ W·s)	1 J	$[\equiv 1.00 \times 10^0$	N·m] ✳
▶ electronvolt	1 eV	$[= 1.602\ 19 \times 10^{-19}$	J, approx.]
British thermal unit (International Table)	1 Btu	$= 1.055 \times 10^3$	J
calorie (thermochemical)	1 cal	$= 4.184 \times 10^0$	J ✳
calorie (International Table)	1 cal$_{IT}$	$= 4.187 \times 10^0$	J
erg	1 erg	$= 1.00 \times 10^{-7}$	J ✳
foot-pound force	1 ft·lbf	$= 1.356 \times 10^0$	J
kilowatt-hour	1 kW·h	$= 3.60 \times 10^6$	J ✳
kiloton explosive yield (PNE) ($\equiv 10^{12}$ g-cal)	1 kt yield	$\simeq 4.2 \times 10^{12}$	J
Power, radiant flux			
▶ watt	1 W	$[\equiv 1.00 \times 10^0$	J/s] ✳
British thermal unit (International Table) per second	1 Btu/s	$= 1.055 \times 10^3$	W
calorie (International Table) per second	1 cal$_{IT}$/s	$= 4.187 \times 10^0$	W
foot-pound force/second	1 ft·lbf/s	$= 1.356 \times 10^0$	W
horsepower (electric)	1 hp	$= 7.46 \times 10^2$	W ✳
horsepower (metric) (= ps)	1 ps	$= 7.355 \times 10^2$	W
horsepower (550 ft·lbf/s)	1 hp	$\simeq 7.457 \times 10^2$	W

- - - - - - - - - - - - - - -

Temperature

▶ temperature in degrees Celsius, t $t = T - T_0$
     where T is the thermodynamic temperature in kelvin
     and T_0 is defined as 273.15 K

degree Fahrenheit	$t_{°F} - 32$	$\Big\}$ $\times \left(\dfrac{5}{9}\right)$ gives $\Big\{$ t *(in degrees Celsius)* ✳
degree Rankine	$T_{°R} (= \Delta t_{°F})$	T *(in kelvin)* ✳
degrees of temperature difference ^{c}	$\Delta T_{°R} (= \Delta t_{°F})$	$\Delta T (= \Delta t)$ ✳

- - - - - - - - - - - - - - -

Thermal conductivity ^{c}

1 Btu·in/(ft^2·s·°F)	*(International Table Btu)*	$= 5.192 \times 10^2$	W·m^{-1}·K^{-1}
1 Btu/(ft·s·°F)	*(International Table Btu)*	$= 6.231 \times 10^3$	W·m^{-1}·K^{-1}
1 cal$_{IT}$/(cm·s·°C)		$= 4.187 \times 10^2$	W·m^{-1}·K^{-1}

---

^{a} atm abs, ata: atmospheres absolute; ^{b} lbf/in^2 (g)   (= psig): gauge pressure;
    atm (g), atü: atmospheres gauge.   lbf/in^2 abs   (= psia): absolute pressure.

^{c} The abbreviation for temperature difference, deg (= degK = degC), is no longer acceptable as an SI unit.

HOW TO ORDER IAEA PUBLICATIONS

An exclusive sales agent for IAEA publications, to whom all orders and inquiries should be addressed, has been appointed in the following country:

UNITED STATES OF AMERICA   UNIPUB, 345 Park Avenue South, New York, N.Y. 10010

In the following countries IAEA publications may be purchased from the sales agents or booksellers listed or through your major local booksellers. Payment can be made in local currency or with UNESCO coupons.

ARGENTINA	Comisión Nacional de Energía Atómica, Avenida del Libertador 8250, Buenos Aires
AUSTRALIA	Hunter Publications, 58 A Gipps Street, Collingwood, Victoria 3066
BELGIUM	Service du Courrier de l'UNESCO, 112, Rue du Trône, B-1050 Brussels
C.S.S.R.	S.N.T.L., Spálená 51, CS-113 02 Prague 1
	Alfa, Publishers, Hurbanovo námestie 6, CS-893 31 Bratislava
FRANCE	Office International de Documentation et Librairie, 48, rue Gay-Lussac, F-75240 Paris Cedex 05
HUNGARY	Kultura, Bookimport, P.O. Box 149, H-1389 Budapest
INDIA	Oxford Book and Stationery Co., 17, Park Street, Calcutta, 700016
	Oxford Book and Stationery Co., Scindia House, New Delhi-110001
ISRAEL	Heiliger and Co., 3, Nathan Strauss Str., Jerusalem
ITALY	Libreria Scientifica, Dott. Lucio de Biasio "aeiou". Via Meravigli 16, I-20123 Milan
JAPAN	Maruzen Company, Ltd., P.O. Box 5050, 100-31 Tokyo International
NETHERLANDS	Martinus Nijhoff B.V., Lange Voorhout 9-11, P.O. Box 269, The Hague
PAKISTAN	Mirza Book Agency, 65, Shahrah Quaid-e-Azam, P.O. Box 729, Lahore-3
POLAND	Ars Polona-Ruch, Centrala Handlu Zagranicznego, Krakowskie Przedmiescie 7, Warsaw
ROMANIA	Ilexim, P.O. Box 136-137, Bucarest
SOUTH AFRICA	Van Schaik's Bookstore (Pty) Ltd., P.O. Box 724, Pretoria 0001
	Universitas Books (Pty) Ltd., P.O. Box 1557, Pretoria 0001
SPAIN	Diaz de Santos, Lagasca 95, Madrid-6
	Diaz de Santos, Balmes 417, Barcelona-6
SWEDEN	AB C.E. Fritzes Kungl. Hovbokhandel, Fredsgatan 2, P.O. Box 16358 S-103 27 Stockholm
UNITED KINGDOM	Her Majesty's Stationery Office, P.O. Box 569, London SE1 9NH
U.S.S.R.	Mezhdunarodnaya Kniga, Smolenskaya-Sennaya 32-34, Moscow G-200
YUGOSLAVIA	Jugoslovenska Knjiga, Terazije 27, POB 36, YU-11001 Belgrade

---

Orders from countries where sales agents have not yet been appointed and requests for information should be addressed directly to:

Division of Publications
International Atomic Energy Agency
Kärntner Ring 11, P.O.Box 590, A-1011 Vienna, Austria